Stem Cell Biology and Regenerative Medicine

Series Editor
Kursad Turksen, Ph.D.
kursadturksen@gmail.com

More information about this series at http://www.springer.com/series/7896

Ivan Bertoncello
Editor

Stem Cells in the Lung

Development, Repair and Regeneration

Humana Press

Editor
Ivan Bertoncello
Lung Health Research Centre,
Department of Pharmacology and Therapeutics
University of Melbourne
Melbourne, Victoria
Australia

ISSN 2196-8985 ISSN 2196-8993 (electronic)
Stem Cell Biology and Regenerative Medicine
ISBN 978-3-319-21081-0 ISBN 978-3-319-21082-7 (eBook)
DOI 10.1007/978-3-319-21082-7

Library of Congress Control Number: 2015950860

Springer Cham Heidelberg New York Dordrecht London

Printed on acid-free paper

Humana Press is a brand of Springer
Springer International Publishing AG Switzerland is part of Springer Science+Business Media (www.springer.com)

Preface

The last decade has witnessed significant progress in understanding the organization and regulation of regenerative cells in the adult lung. The development and refinement of multiparameter flow cytometric cell sorting protocols and *in vitro* clonogenic assays has enabled the identification, prospective isolation, and characterization of candidate adult lung stem and progenitor cells. Powerful gene labelling technologies, cell lineage tracing and cell fate mapping protocols, and genetically engineered mouse models have revealed critical regulatory molecules, mechanisms, pathways, and cellular interactions important in building the lung during fetal development; and in maintaining lung function and rebuilding the lung in adulthood. This in turn has advanced the understanding of the pathophysiology of lung diseases; and how endogenous lung stem and progenitor cells, exogenous stem cells, and bioengineering technologies might be harnessed to attenuate or reverse intractable, life-threatening respiratory diseases.

This is the more remarkable considering that until relatively recently the lung was generally considered a conditionally renewing organ with limited growth potential and regenerative capacity in adult life [1, 2]. Why this paradigm remained so entrenched for so long is puzzling given compelling lung cell kinetic analysis that showed that lung epithelium was continuously replaced, albeit at a very slow rate overall, that could only be explained by continuous renewal [3, 4] of postulated subsets of cells with much higher rates of turnover than the average for the organ [5]. This failure to fully appreciate that the extremely slow turnover of cells in the adult lung was not inconsistent with the existence of continuously renewing endogenous lung stem and progenitor cells able to replace senescent cells in the steady state, and regenerate functional lung cell lineages lifelong following insult or injury "blinded" the field to the potential of stem cells as therapeutic targets for adult lung regeneration and repair. Rather, research predominantly focused on understanding the pathophysiology of intractable lung diseases, and on the development of essentially palliative pharmacologic interventions to resolve inflammation and attenuate fibrotic responses induced by infectious agents, toxicants, or injury. This bias is patently obvious on interrogation of the Pubmed database as of May

7, 2015. Whereas 185,890 publications were retrieved using the search term "lung inflammation OR lung fibrosis," only 6688 publications were retrieved using the search term "lung stem cell OR lung progenitor cell," and 3228 using the search term "lung regeneration." Comparison with an archetypal continuously renewing stem cell hierarchy is also illustrative in this context, showing that 87,492 publications were retrieved in the same period using the search term "hematopoietic stem cell OR hematopoietic progenitor cell."

Interest in the potential of stem cell-based therapies in lung regenerative medicine was initially motivated by studies in the early 2000s purporting to show that various adult and embryonic stem cell populations could be respecified or reprogrammed to express lung epithelial cell lineage markers when cultured in defined media, or engrafted in the lung. Concurrently, various research groups intensified their efforts to develop best-practice tools, assays, and models to identify, isolate, and characterize endogenous adult lung stem and progenitor cells, determine their spatial location(s) in the airway, and assess their regenerative potential. In particular, these studies aimed to determine whether the organization and regulation of these putative stem and progenitor cell compartments conformed with, or deviated from, the organization of classical continuously renewing stem cell hierarchies [6].

The reductionist approach was predominantly concerned with determining and defining the intrinsic properties of adult lung epithelial stem and progenitor cells. But, according to the niche hypothesis first articulated by Schofield in 1978 [7], "stemness" is determined by anchorage of the stem cell to a complex and malleable anatomical niche comprising a matrix scaffold, adhesion molecules, soluble and insoluble factors, and contiguous fixed and circulating cells which engage in crosstalk with the stem cell to specify its fate and constrain its proliferation and differentiation. The importance of these critical niche interactions, and mesenchymal–epithelial crosstalk in particular, have long been appreciated in lung developmental biology, and are likely recapitulated in regeneration and repair of the adult lung [8]. The analysis of mesenchymal–epithelial interactions has also long been a major focus of research in the pathogenesis of fibrosis and airway remodeling in intractable lung diseases such as allergy and asthma [9]. However, the lung is a complex organ comprising as many as 60 cell lineages and less is known about the role of interactions of many of these diverse cell lineages with lung epithelial stem and progenitor cells, and with each other, in regulating lung development and in maintaining, regenerating, and repairing the lung in adulthood. Significant advances have recently been made in understanding the process of vasculogenesis and angiogenesis in the lung, and in the crosstalk with lung epithelial cells. In comparison, the important role of neuronal cells in the regulation of lung development, and in lung regeneration and repair is less well determined and a worthy and fertile subject of further study.

This monograph presents a comprehensive overview of the current understanding of the organization and regulation of endogenous lung stem and progenitor cell compartments during fetal and perinatal lung development, and in the adult and ageing lung. The chapters on fetal lung development aim to elucidate the integrated role of epithelial, stromal, vascular, and neural cell elements in building a functional lung. These chapters provide a basis for identifying critical factors and pathways regulating lifelong lung regeneration and repair, and understanding the fetal and neonatal origin of lung disease. The chapters on adult lung regeneration describe the organization and properties of epithelial stem and progenitor cell compartments distributed along the proximal–distal axis of the airway tree and how they are affected with age. The chapters on lung regeneration and repair describe the role of crosstalk between regenerative cells and cellular and biomatrix elements of their niche in the adult lung in maintaining organ integrity in the steady state, and regulating lung regeneration.

The chapters on cellular therapies for lung disease provide an overview of emerging therapies utilizing exogenous mesenchymal stromal cells, and vascular endothelial cells to repair the diseased lung; and also review studies in animal models exploring the potential of pluripotent stem cells in lung regenerative medicine. The monograph concludes with an overview of recent progress in lung bioengineering describing the challenges in engineering acellular biomatrix scaffolds to replace damaged or diseased airways, or reconstruct a functional lung ex vivo.

It remains for me to thank the many authors at the cutting-edge of the field who enthusiastically contributed the lucid and authoritative chapters in this volume. I also thank Kursad Turksen (Series Editor) for the invitation to edit this volume; and acknowledge Brian Halm and Aleta Kalkstein and their team at Springer for their advice and assistance in developing and preparing this volume for publication.

Melbourne, VIC, Australia Ivan Bertoncello

References

1. Leblond CP, Walker BE (1956) Renewal of cell populations. Physiol Rev 36:255–276.
2. Goss RJ (1966) Hypertrophy versus hyperplasia. Science 153:1615–1620.
3. Bertalanffy FD, Leblond CP (1953) The continuous renewal of the two types of alveolar cells in the lung of the rat. Anat Rec 115:515-541.
4. Evans MJ, Cabral LJ, Stephens RJ, Freeman G (1973) Renewal of alveolar epithelium in the rat following exposure to NO_2. Am J Pathol 70:175–198.
5. Blenkinsopp WK (1967) Proliferation of respiratory tract epithelium in the rat. Exp Cell Res 46:144–154.
6. Bertoncello I, McQualter JL (2013) Lung stem cells: do they exist? Respirology 18:587–595.
7. Schofield R (1978) The relationship between the spleen colony-forming cell and the haemopoietic stem cell. Blood Cells 4:7–25.
8. Warburton D, Tefft D, Mailleux A, Bellusci S, Thiery JP et al (2001) Do lung remodeling, repair, and regeneration recapitulate respiratory ontogeny? Am J Respir Crit Care Med 164: S59–S62.
9. Holgate ST, Davies DE, Lackie PM, Wilson SJ, Puddicombe SM, Lordan JL (2000) Epithelial-mesenchymal interactions in the pathogenesis of asthma. J Allergy Clin Immunol 105:193–204.

Contents

Part I Fetal Lung Development

1 **The Regulation of Branching Morphogenesis in the Developing Lung** 3
Jichao Chen

2 **Epithelial–Mesenchymal Crosstalk in Lung Development** 17
David Warburton

3 **Lung Vasculogenesis and Angiogenesis** 25
Bernard Thébaud and Mervin C. Yoder

4 **Neural Regulation of Lung Development** 43
Alan J. Burns, Lucy J. Freem, and Jean-Marie Delalande

5 **Fetal and Neonatal Origins of Lung Disease** 63
Foula Sozo, Megan O'Reilly, and Richard Harding

Part II Adult Lung Stem and Progenitor Cells

6 **Stem and Progenitor Cells of the Trachea and Proximal Airways** 97
Ahmed E. Hegab, Tomoko Betsuyaku, and Brigitte N. Gomperts

7 **Stem Cells of the Distal Bronchiolar Airways** 113
Robert E. Hynds and Adam Giangreco

8 **Regenerative Cells in the Ageing Lung** 127
Andrew M. Hoffman

Part III Regulation of Lung Regeneration and Repair

9 **Stromal Regulation of Lung Regeneration and Repair** 149
Jonathan L. McQualter

10 **Immunomodulatory Regulation of Lung Regeneration and Repair** 159
Jonathan L. McQualter, Desiree Anthony, Ross Vlahos, and Steven Bozinovski

11 **Extracellular Matrix Specification of Regenerative Cells in the Adult Lung** 169
Janette K. Burgess and Gavin Tjin

12 **Role of Mechanical Stress in Lung Repair and Regeneration** 191
Connie C.W. Hsia and Priya Ravikumar

Part IV Cellular Therapies for Lung Disease

13 **Cellular Origins of Fibrotic Lung Diseases** 213
Vibha N. Lama

14 **Mesenchymal Stromal Cell-Based Therapies for Lung Disease** 225
Daniel Chambers

15 **Lung Vascular Regeneration and Repair** 243
Mervin C. Yoder and Bernard Thébaud

16 **Directed Differentiation of Human Pluripotent Stem Cells into Lung and Airway Epithelial Cells** 265
Hans-Willem Snoeck

Part V Bioengineering the Lung

17 **Regenerating the Respiratory Tract** 289
Mei Ling Lim, Alexandra B. Firsova, Neus Feliu, Elena V. Kuevda, Philipp Jungebluth, and Paolo Macchiarini

18 **Acellular Lung Scaffolds in Lung Bioengineering** 309
Darcy E. Wagner, Franziska E. Uhl, and Daniel J. Weiss

Part VI Afterword

19 **Afterword: Prospects and Challenges in Lung Regenerative Medicine** 351
Ivan Bertoncello

Index 359

Contributors

Desiree Anthony Lung Health Research Centre, Department of Pharmacology and Therapeutics, University of Melbourne, Melbourne, VIC, Australia

Ivan Bertoncello Lung Health Research Centre, Department of Pharmacology and Therapeutics, University of Melbourne, Melbourne, VIC, Australia

Tomoko Betsuyaku Division of Pulmonary Medicine, Department of Medicine, Keio University School of Medicine, Tokyo, Japan

Steven Bozinovski School of Health Sciences and Health Innovations Research Institute, RMIT University, Bundoora, VIC, Australia

Janette K. Burgess Woolcock Institute of Medical Research, The University of Sydney, Sydney, Australia

Discipline of Pharmacology, The University of Sydney, Sydney, Australia

Central Clinical School, The University of Sydney, Sydney, Australia

Alan J. Burns Stem Cells and Regenerative Medicine, Birth Defects Research Centre, UCL Institute of Child Health, London, UK

Department of Clinical Genetics, Erasmus University Medical Center, Rotterdam, The Netherlands

Daniel Chambers School of Medicine, The University of Queensland, Brisbane, QLD, Australia

Queensland Lung Transplant Service, The Prince Charles Hospital, Brisbane, QLD, Australia

Jichao Chen Department of Pulmonary Medicine, Center for Stem Cells and Developmental Biology, The University of Texas M. D. Anderson Cancer Center, Houston, TX, USA

Jean-Marie Delalande Stem Cells and Regenerative Medicine, Birth Defects Research Centre, UCL Institute of Child Health, London, UK

Centre for Digestive Diseases, Barts and The London School of Medicine and Dentistry, Queen Mary University of London, London, UK

Neus Feliu Advanced Center for Translational and Regenerative Medicine (ACTREM), Division of Ear, Nose, Throat, Department for Clinical Science, Intervention and Technology, Karolinska Institute, Stockholm, Sweden

Alexandra B. Firsova Advanced Center for Translational and Regenerative Medicine (ACTREM), Division of Ear, Nose, Throat, Department for Clinical Science, Intervention and Technology, Karolinska Institute, Stockholm, Sweden

Lucy J. Freem Stem Cells and Regenerative Medicine, Birth Defects Research Centre, UCL Institute of Child Health, London, UK

Division of Developmental Biology, The Roslin Institute, University of Edinburgh Easter Bush Campus, Midlothian, UK

Adam Giangreco Division of Medicine, Lungs for Living Research Centre, Rayne Institute, University College London, London, UK

Brigitte N. Gomperts Department of Pediatrics, Mattel Children's Hospital, University of California, Los Angeles, Los Angeles, CA, USA

Pulmonary Medicine, University of California, Los Angeles, CA, USA

Jonsson Comprehensive Cancer Center, University of California, Los Angeles, CA, USA

Eli and Edythe Broad Stem Cell Research Center, University of California, Los Angeles, CA, USA

Richard Harding Department of Anatomy and Developmental Biology, Monash University, Melbourne, VIC, Australia

Ahmed E. Hegab Division of Pulmonary Medicine, Department of Medicine, Keio University School of Medicine, Tokyo, Japan

Andrew M. Hoffman Regenerative Medicine Laboratory, Department of Clinical Sciences, Cummings School of Veterinary Medicine, Tufts University, North Grafton, MA, USA

Connie C.W. Hsia Department of Internal Medicine, University of Texas Southwestern Medical Center, Dallas, TX, USA

Robert E. Hynds Division of Medicine, Lungs for Living Research Centre, Rayne Institute, University College London, London, UK

Philipp Jungebluth Advanced Center for Translational and Regenerative Medicine, Division of Ear, Nose, Throat, Department for Clinical Science, Intervention and Technology, Karolinska Institute, Stockholm, Sweden

Elena V. Kuevda International Research, Clinical and Education Center of Regenerative Medicine, Kuban State Medical University, Krasnodar, Russian Federation

Vibha N. Lama Division of Pulmonary and Critical Care Medicine, Department of Internal Medicine, University of Michigan Medical School, Ann Arbor, MI, USA

Mei Ling Lim Advanced Center for Translational and Regenerative Medicine, Division of Ear, Nose, Throat, Department for Clinical Science, Intervention and Technology, Karolinska Institute, Stockholm, Sweden

Paolo Macchiarini International Research, Clinical and Education Center of Regenerative Medicine, Kuban State Medical University, Krasnodar, Russian Federation

Advanced Center for Translational and Regenerative Medicine, European Airway Institute, Division of Ear, Nose, Throat, Department for Clinical Science, Intervention and Technology, Karolinska Institute, Stockholm, Sweden

Jonathan L. McQualter School of Health Sciences and Health Innovations Research Institute, RMIT University, Bundoora, VIC, Australia

Megan O'Reilly Department of Pediatrics, Women and Children's Health Research Institute, University of Alberta, Edmonton, AB, Canada

Priya Ravikumar Department of Internal Medicine, University of Texas Southwestern Medical Center, Dallas, TX, USA

Hans-Willem Snoeck Columbia Center for Translational Immunology, Columbia University Medical Center, New York, NY, USA

Foula Sozo Department of Anatomy and Developmental Biology, Monash University, Melbourne, VIC, Australia

Bernard Thébaud Division of Neonatology, Department of Pediatrics, Sinclair Centre for Regenerative Medicine, Ottawa Hospital Research Institute, Children's Hospital of Eastern Ontario, University of Ottawa, Ottawa, ON, Canada

Gavin Tjin Woolcock Institute of Medical Research, The University of Sydney, Sydney, Australia

Central Clinical School, The University of Sydney, Sydney, Australia

Franziska E. Uhl University of Vermont College of Medicine, Burlington, VT, USA

Ross Vlahos School of Health Sciences and Health Innovations Research Institute, RMIT University, Bundoora, VIC, Australia

Darcy E. Wagner Comprehensive Pneumology Center, Munich, Germany

David Warburton Developmental Biology and Regenerative Medicine Program, Saban Research Institute, Children's Hospital Los Angeles, Keck School of

Medicine and Ostrow School of Dentistry, University of Southern California, Los Angeles, CA, USA

Daniel J. Weiss University of Vermont College of Medicine, Burlington, VT, USA

Mervin C. Yoder Department of Pediatrics, Wells Center for Pediatric Research, Indiana University School of Medicine, Indianapolis, IN, USA

Part I
Fetal Lung Development

Chapter 1
The Regulation of Branching Morphogenesis in the Developing Lung

Jichao Chen

Overview

Lung branching morphogenesis generates a tree-like structure and allows efficient airflow to millions of gas exchange units, the alveoli. The initiation, continuation, and termination of lung branching morphogenesis require precise control of the specification, maintenance, and depletion of lung epithelial progenitors. Here I will focus on these progenitors and describe the branch morphology across species and in comparison with other branching organs; the underlying cell biology including polarity, movement, and matrix interaction; and an initial gene regulatory network that can be expanded using genetic and genomic approaches. I will also describe when and how branching ends and its implications in premature birth and developmental plasticity.

Relationship Between Lung Epithelial Progenitors and Branching Morphogenesis

The air lumen from the proximal airway to the peripheral gas exchange compartment is bordered by a continuous monolayer of epithelial cells embedded in mesenchymal tissues. The developmental process forming the respiratory tree, termed branching morphogenesis, is not simply the sculpting of a homogeneous epithelial sheet, but rather concerted morphogenesis and differentiation of specialized epithelial progenitors. These progenitors are commonly defined by expression

J. Chen (✉)
Department of Pulmonary Medicine, Center for Stem Cells and Developmental Biology, The University of Texas M. D. Anderson Cancer Center, 2121 W. Holcombe Blvd, RM7.716, Houston, TX 77030, USA
e-mail: Jchen16@mdanderson.org

I. Bertoncello (ed.), *Stem Cells in the Lung*, Stem Cell Biology and Regenerative Medicine, DOI 10.1007/978-3-319-21082-7_1

of transcription factors *Sox9* and *Id2* [1, 2], and the expression of the former has been traced throughout development [3, 4]. The *Sox9* progenitors are specified in the embryonic foregut, build the entire respiratory tree including both the airway and gas exchange compartments, and are depleted after birth (Fig. 1.1). The temporal and spatial relationships between epithelial progenitors and branching morphogenesis are described below.

Temporal: The respiratory primordium forms in the ventral portion of the anterior foregut around embryonic day (E) 9 in mouse and gestation week 5 in human [1, 5]. A combination of *Wnt*, *Bmp*, and *Fgf* signalings is required to activate the respiratory lineage factor *Nkx2.1* [6–10]. The *Nkx2.1* respiratory primordium gives rise to both the trachea and the lung. A subset of *Nkx2.1* cells express *Sox9*, constituting the lung epithelial buds that will subsequently undergo branching morphogenesis (Fig. 1.1) [3].

After specification, *Sox9* progenitors undergo branching morphogenesis forming new branches at an exponential rate. The basic unit of a branch consists of a branch tip and a branch stalk. *Sox9* progenitors are always associated with branch tips throughout the branching process (Fig. 1.1) and thus must be precisely regulated in at least three aspects. First, the progenitors need to self-renew at a rate that matches the exponentially growing respiratory tree. Second, a controlled fraction of their progeny needs to exit the progenitor pool, forming branch stalks and differentiating into airway and alveolar cells [3, 4]. Third, depending on whether branch tips are split in a symmetric or asymmetric manner to form bifurcation or lateral branches, respectively, the progenitors need to be partitioned between new branch tips accordingly (Fig. 1.2).

Branching morphogenesis does not continue forever, and *Sox9* progenitors are depleted after birth (Fig. 1.1). A time course analysis following branch tips along the lobe edge shows that the branch tip-stalk structure disappears after birth as the stalks widen, presumably as a result of differentiation into alveolar ducts and air filling. Meanwhile, *Sox9* progenitors persist as clusters at branch tips as late as

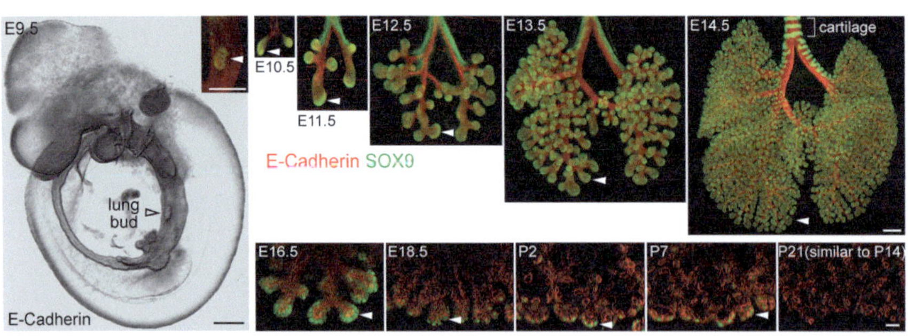

Fig. 1.1 A complete developmental history of SOX9 lung epithelial progenitors. Optical projection tomography (OPT, *top row*, scale: 250 um) and confocal images (*bottom row*, scale: 20 um) of immunostained whole embryo (E9.5), lungs (E10.5–E14.5) and lobe edges (E16.5–P21). SOX9 progenitors are indicated by *arrowheads*. *Open arrowhead*: lung bud forming on the foregut. Adapted from [3]

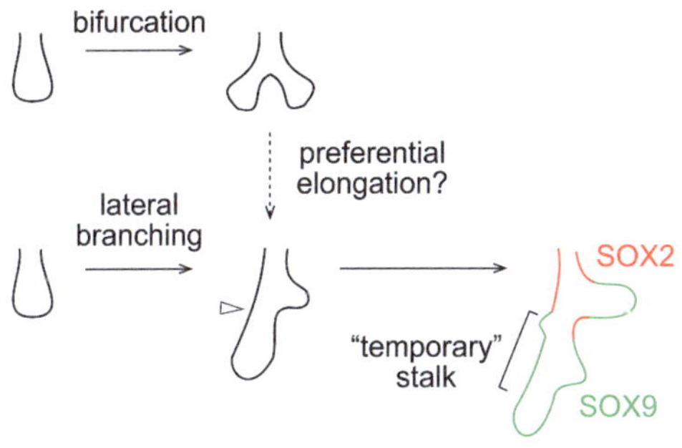

Fig. 1.2 Two modes of new branch formation. New branch tips form via bifurcation or lateral branching. One branch that is preferentially elongated after the symmetric bifurcation may appear as the main branch in the asymmetric lateral branching, suggesting that the two branching modes may be more similar than their final appearance. The main branch contains a "temporary" branch stalk that does not express SOX2 and can initiate new branches (*open arrowhead*)

postnatal day (P) 7 and potentially could continue to divide into smaller clusters, reminiscent of branch tip-splitting during branching morphogenesis. Given their location in the most distal portion of the respiratory tree and beyond alveolar ducts, these *Sox9* progenitor clusters presumably differentiate into primary saccules that will be further subdivided to form mature alveoli. Therefore, cessation of branching morphogenesis parallels depletion of *Sox9* progenitors.

Spatial: For bifurcation where branch tips form by splitting existing tips, it is intuitive that *Sox9* progenitors are located in branch tips, while their differentiating progeny, commonly marked by *Sox2* expression, are located in branch stalks. However, for lateral branching where branch tips form on existing stalks, I would like to introduce the concept of "temporary" branch stalks, in which a morphological branch stalk is molecularly a branch tip. Specifically, although "temporary" branch stalks are surrounded by airway smooth muscle [11], they are *Sox2* negative as supported by the absence of labeled distal epithelial cells in $Sox2^{CreER}$ lineage tracing experiments [4]. In other words, before *Sox9* progenitors become *Sox2*-positive "permanent" branch stalks, they remain capable of branching regardless of the tip-stalk morphology (Fig. 1.2). Therefore spatially, *Sox9* progenitors are located in the branching region of the developing respiratory tree.

The close temporal–spatial correlation between branching morphogenesis and *Sox9* epithelial progenitors suggests that regulation of branching morphogenesis is fundamentally regulation of *Sox9* progenitors by intrinsic (epithelial) and extrinsic (mesenchymal) signalings.

Morphological Features of Branching Morphogenesis

The building blocks, branch tips and stalks, form in two modes: bifurcation and lateral branching. Deployment of these branching modes must be regulated in a deterministic manner to form a respiratory tree of a highly reproducible shape and lobation pattern that is specific for each species. Meanwhile, space-filling must also play a role to ensure maximal space utilization without branch collision. This section will focus on the mouse lung whose branching has been studied in depth [12]; and this will be compared with lungs from other species and other branched organs.

Lobation: On a global level, branching morphogenesis is constrained by the size, shape, and arrangement of lobes termed lobation. The mouse lung consists of a single left lobe and four right lobes: cranial, middle, accessory, and caudal lobes. This asymmetry is regulated as part of the bilateral asymmetry of the body plan as situs inversus mutants, such as the *Dnahc11* mutant [12], have left-right inverted lungs as well as other internal organs. Closer examination of the branching structure suggests that the mouse lung might be more symmetric than its appearance with the exception of the accessory lobe (Fig. 1.3). Specifically, the caudal lobe can be considered the equivalent of the caudal portion of the left lobe (starting from L.L3) as both consist of a central medially curved lobar bronchus and multiple parallel lateral branches. The axis of the middle lobe parallels the lateral branches of the caudal lobe and thus may be the L.L2 equivalent, while the anterior portion of the cranial lobe is reminiscent of the L.L1.A1 branch. Thus, the main difference between left and right lobes is that individual right lobes are encased and thus separated by a single-layered mesothelium. Future studies of the mesothelium may shed light on mechanisms of normal lobation and abnormal lobe fusion.

The shape of individual lobes is apparently established early in development (Fig. 1.1), raising the interesting question of how branching morphogenesis at

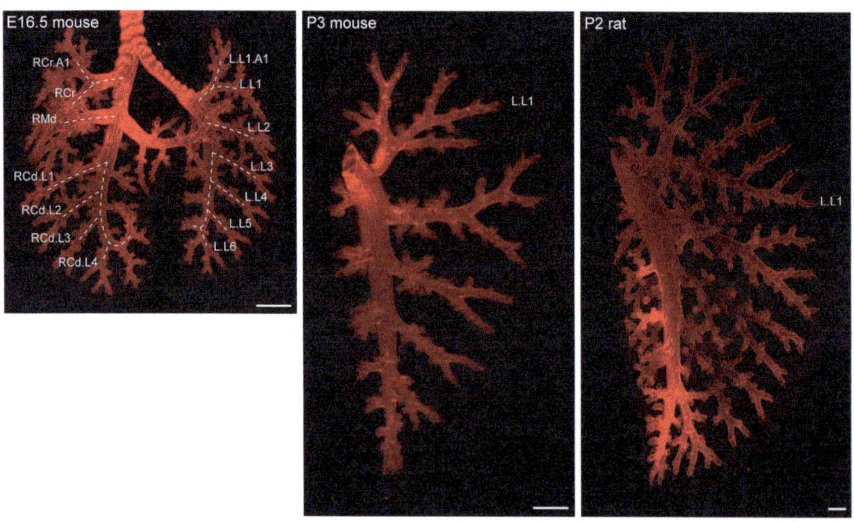

Fig. 1.3 Mouse and rat airways. OPT images of mouse and rat lung or left lobes immunostained for an airway marker SOX2. The mouse lung and left lobe are shown at the same magnification and the rat left lobe is half of that magnification. Scale: 500 um. The branch lineage is labeled to show symmetry between the left and right lobes

different locations advances the lobe surface in a concerted fashion to preserve lobe shape over time. Alternatively, lobe geometry is regulated by other means and feeds into the branching pattern control. Intriguingly, the rat lung is remarkably similar in lobe shape and branching pattern to the mouse lung, but is not simply a scaled-up version as it has more airway branch generations (Fig. 1.3).

Branching subroutines: Careful comparison of hundreds of developmental intermediates has revealed that branching in the mouse lung can be accounted for by three branching subroutines: domain branching, planar, and orthogonal bifurcation [12]. Furthermore, these three subroutines are deployed in three sequences, where orthogonal bifurcation represents a "dead-end" subroutine that will only form rosette-like branches and no longer switch to alternative subroutines [12]. It is interesting to note that, compared to orthogonal bifurcation, domain branching and planar bifurcation, by definition, form longer branch stalks that extend to the lobe edge and can initiate lateral branches. Thus, as discussed below, branch length (or elongation) may be an additional parameter other than branch angle that distinguishes the subroutines.

Dichotomous and monopodial branching: Branching in the mouse lung is often monopodial featuring minor daughter branches on the side of a major parent branch, while branching in the human lung is often dichotomous featuring symmetric daughter branches of comparable size [13]. Although this difference might reflect differential usage of lateral branching versus bifurcation across species, an additional contributing factor may be a difference in branch elongation after, but not during, branch formation. Specifically, differential elongation of one daughter branch after dichotomous branching can form an asymmetric structure reminiscent of the one formed via monopodial branching (Fig. 1.2). Consistent with this possible dichotomous-branching origin of monopodial branches, the major parent branch in monopodial branching is often curved (Fig. 1.3). Furthermore, no gene has been identified whose expression distinguishes dichotomous (bifurcation) and monopodial (lateral) branching, and computer simulation suggests a deterministic role of branch elongation in generating different branching modes [14]. The difference in branch elongation may be ultimately linked to lobe shape, which differs significantly between mouse and human.

Tip morphology in comparison with other branched organs: Compared to a bead-on-a-string design, branching morphogenesis packs a large number of terminal units without incurring a tortuous transport route. Thus, it is used in multiple organs, including the kidney, mammary, and salivary glands. However, the branch tip morphology among organs differs substantially. While branch tips in the lung are bulbous with obvious luminal space, the ureteric tree frequently contains trifurcating branch tips [15]; the salivary gland tips split via clefting due to replacement of cell–cell junctions with cell–matrix interactions [16]; and mammary gland branching requires epithelial-to-mesenchymal transition and migration of a bi-layered epithelium [17]. Although the Drosophila trachea system mediates respiratory function and requires the same fibroblast growth factor (*Fgf*) signaling as the lung, its branching appears more similar to vertebrate angiogenesis, where selected tip cells lead the rest toward a morphogen source [18]. It is worth noting that in the developing lung, the bulbous branch tip may appear on sections as a large diameter tube, but should not be mistaken as proximal airways, which do not have as large a lumen as those in the adult lung.

Cell Biology of Branching Morphogenesis

The cellular behaviors underlying lung branching morphogenesis are not well understood, in part because existing explant culture systems [19] are limited in recapitulating in vivo development. In particular, cultured lungs are constrained in size by the supporting filters, and the resulting limited branch outgrowth may disrupt the epithelium–mesenchyme cross talk that is essential for normal branching. Furthermore, cultured lungs flatten due to gravity; this disrupts branch shape, morphogen distribution, and potentially intraluminal fluid pressure, a mechanical stimulus possibly involved in branching [20]. In addition, given the exponential nature of branch tip number increase, a small difference in the initial tip number due to variation in developmental timing, coupled with ambiguity in branch tip counting due to altered morphology, may be misinterpreted as treatment effects. Nevertheless, genetic and culture experiments have implicated cell polarity, movement, and matrix interaction in lung branching.

Cell polarity: The developing lung epithelium is a monolayer polarized both along the apical–basal axis and within the monolayer plane. Epithelial-specific deletion of *Integrin beta-1* (*Itgb1*), a major mediator of cell–matrix interaction, completely abrogates branching [21]. The mutant epithelium frequently contains abnormal apical–basally oriented cell division and becomes multilayered with distinct gene expression among layers, reminiscent of a stratified epithelium such as the epidermis [22]. Overactivation of *Kras*, a small GTPase mediating receptor tyrosine kinase signaling, or loss of *Spry1* and *Spry2*, phosphatases limiting receptor tyrosine kinase signaling, shortens and widens branch tips in association with altered proportions of longitudinal versus circumferential cell divisions [23]. Although computer modeling suggests that abnormal cell division orientation largely explains the observed branch morphology phenotype, a similar mechanism has been proposed for formation of cystic renal tubes but may not be causal to their phenotype in some cases [24].

Cell movement: Although constrained by tight and adherent junctions, the lung epithelium is fluid with frequent interkinetic nuclear migration (INM) along the apical–basal axis and neighbor exchange via intercalation. Best known for its role in neurogenesis, INM positions the nucleus at the basal or apical side during the S or M phase of the cell cycle, respectively, and potentially exposes cells to location-specific signals [25, 26]. Asynchrony in INM among cells makes a monolayered epithelium appear pseudostratified and, depending on cell density, generates bottle-shaped cells with apical and basal thin extensions of varying length [27].

Real-time imaging of cultured lungs has confirmed the association between cell division orientation and branch growth in the longitudinal versus circumferential directions [28]. In addition, no migratory cellular extensions are observed, and instead epithelial cells move via neighbor exchange, a process reminiscent to intercalation during germ band elongation [29]. Unlike intercalation, the net apical surface area of lung epithelial cells can decrease locally via actomyosin contraction, leading to epithelium deformation and branch initiation, in culture and in a *Wnt* signaling mutant [27, 30]. Future genetic mosaic analyses are necessary to understand whether and how morphogenetic signals, including *Wnt* and *Fgf*, regulate cell shape in a cell-autonomous manner. Intriguingly, in the kidney, neighbor exchange occurs even during mitosis via a mitosis-associated cell dispersal process where apically dividing daughter cells move apart and insert back to the epithelium at distant sites [31]. If similar cell behavior also occurs in the lung, it would be interesting to determine how it might be integrated with oriented cell division to regulate branch morphology.

Extracellular matrix (*ECM*): The lung epithelium is surrounded by a basement membrane that contains fibronectin, collagen, nidogen, and laminin and is thicker around branch stalks than tips [32], suggesting a potential role of ECM in branching morphogenesis. Although such a role is demonstrated in cultured lungs [16, 33], genetic mutants either die before lung formation or exhibit little branching phenotype, presumably due to essential roles of ECM in cell survival or functional redundancy among related genes [34]. Interestingly, several ECM mutants, including those affecting ECM turnover, have abnormal perinatal lung maturation, highlighting the importance of ECM in supporting the air-filled alveolar region [34–37].

Genetic Control of Branching Morphogenesis

Defects in lung branching morphogenesis often lead to respiratory failure and death at birth, a characteristic phenotype that has allowed an effectively forward genetic screen and identification of a number of branching control genes and signalings [1, 38]. Future studies need to assemble the existing parts list into a gene regulatory network, especially with regard to their role in *Sox9* epithelial progenitors as they give rise to the entire epithelial tree. This section will summarize the major branching signalings focusing on the epithelial progenitors; highlight importance of nomenclature in phenotypic characterization; and describe available genetic and genomic tools to integrate existing knowledge toward a system-level understanding. Due to the aforementioned limitations of lung explant culture, only genetic mutants are included although caution is also warranted with the interpretation of these models because of the potential nonspecific toxicity of genetic tools [39–41].

Multiple signalings: The primary branching signal utilizes the *Fgf10-Fgfr2* pathway. *Fgf10* expression is dynamic and localized in the mesenchyme just ahead of growing branch tips [42], and both *Fgf10* and *Fgfr2* mutants completely lack branching [43, 44]. Epithelial-specific deletion of *Fgfr2* leads to a lung with only left and right main bronchi that are devoid of *Sox9* progenitors, while epithelial overactivation of *Kras*, a mediator of receptor tyrosine kinase including *Fgf* signalings, leads to expansion and persistence of *Sox9* progenitors [45]. Such gain-of-function and loss-of-function data support a dominant role of the *Fgf* signaling in progenitor maintenance.

The *Wnt* signaling in the developing lung is complex as multiple *Wnt* genes are present and canonical *Wnt* activity, depending on the reporter, is detected in either the epithelial or mesenchymal compartment [46]. *Wnt2/2b*-dependent canonical *Wnt* signaling controls respiratory fate specification in the foregut [6, 7]. The *Wnt7b* mutant has a significantly smaller lung with fewer albeit normal looking branches [47]. The *Wnt5a* mutant also has a smaller lung and a higher density of epithelial cells at late stages, although it is unknown if the total number of branches is increased [48]. Epithelial deletion of *Ctnnb1*, the mediator of canonical *Wnt* signaling, decreases the number of *Sox9* progenitors whose domain becomes disproportional to that of their *Sox2* expressing progeny [4, 49, 50].

Similar complexity exists for the *Bmp* and *Tgf* signalings. Although *Bmp4* is specifically expressed in the *Sox9* progenitors and antagonizes *Fgf10* in culture, its in vivo role is still debatable [4, 51, 52]. Similarly, although deletion of *Alk3* (*Bmpr1a*), a *Bmp* receptor, using *Sftpc* genetic drivers disrupts branching [52, 53], Shh^{Cre}-mediated deletion of *Alk3* has no detectable phenotype (unpublished observation). Only deletion of both *Bmp* receptors, *Alk3* and *Alk6*, disrupts respiratory fate specification in the foregut [8]. Such redundancy among homologous genes may also apply to the *Bmp* ligands. Epithelial deletion of *Alk5*, a *Tgf* receptor, leads to a smaller lung with fewer and dilated branches [54].

Therefore, existing evidence indicates that the epithelial progenitors receive inputs from *Fgf*, *Wnt*, and *Tgf* signalings. Although not discussed here, the mesenchyme is also precisely regulated by multiple signalings including *Fgf9*, *Wnt*, and *Shh*. Disruption of these will also in turn affect epithelial branching [55–57].

Phenotype nomenclature: Gross histological examination prompted by the characteristic neonatal lethality phenotype often leads to a description of the lung phenotype as hypoplastic or hypercellular, which likely corresponds to defective branching or alveolar differentiation, respectively. Further analyses of such mutants need to define phenotypes using an increasing number of molecular markers available for each process.

Another term commonly used is proximal–distal patterning, which requires clarification and discussion. First, patterning is classically defined as a onetime cell fate choice in a homogeneous field of naïve cells in response to a morphogen gradient, such as the anterior–posterior patterning of the Drosophila embryo [58]. Although the lung epithelium has differential gene expression proximal–distally (e.g., distal *Sox9* and proximal *Sox2*), the proximal region is not the equivalent of the distal region, but instead is made of the continuous influx of the progeny of the distally located *Sox9* progenitors. Therefore, proximal–distal patterning, or "proportion" to be exact in the context of branching, reflects cumulative, constant cell fate choices of *Sox9* progenitors between self-renewal and differentiation. Such a concept has been explored in recent studies [4, 59].

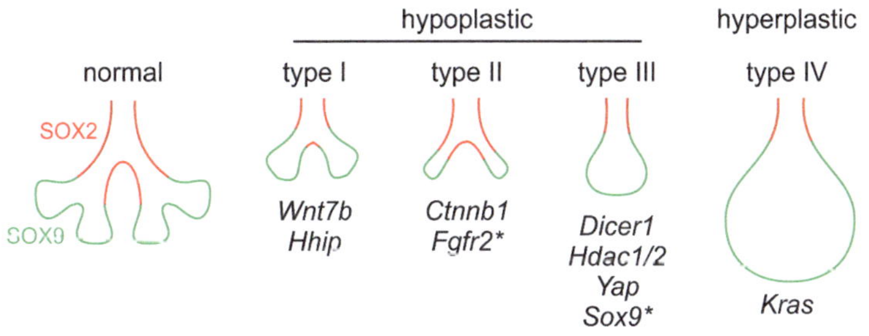

Fig. 1.4 Proximal–distal proportion in mutants with different branch morphology. Type I mutants [47, 57] have fewer normal-looking branches with a normal proximal–distal proportion. Type II mutants [4, 45, 49, 50] have fewer branching containing a smaller progenitor region. *Asterisk*: the *Fgfr2* mutant has no SOX9 progenitors. Type III mutants [45, 59, 60, 70] have fewer branches that are counterintuitively dilated as in the hyperplastic type IV mutant. Compared with the normal lung, type III mutants have a larger distal region with respect to individual branches, but a smaller distal region with respect to the whole lung. *Asterisk*: the *Sox9* mutant has a distal region that does not express SOX9. Type IV mutants [23, 45] have a larger lung with fewer and dilated branches

Second, definition of proximal–distal patterning may be confounded by branch morphology and depends on whether individual branches or the whole lung is compared between the control and mutant. As illustrated in Fig. 1.4, a simple slow branching mutant (type I) will have fewer branches with a normal proximal–distal proportion. Type II mutants can have a disproportionally small distal region as the result of decreased progenitor self-renewal with normal differentiation. Type III mutants are characterized by a smaller lung with fewer branches that are counterintuitively dilated. Due to this abnormal branch morphology, individual branches appear to have an expanded distal region while the total distal region in the whole lung is smaller. The apparent distal bias may be actually accompanied by decreased progenitor self-renewal as supported by the epithelial apoptosis phenotype in the *Dicer* mutant and the branch dilation phenotype from epithelial cell ablation in the *Sftpc-rtTA; tetO-DTA* mutant [4, 60]. Therefore, type II and type III mutants may be similar in having decreased progenitor self-renewal and total distal region in the whole lung, while their apparent difference in proximal–distal patterning with respect to individual branches may be confounded by branch dilation. Finally, Type IV mutants can have a disproportionally large distal region as the result of increased progenitor self-renewal with normal or decreased differentiation.

Last, proximal–distal patterning is often inferred from sections stained for distal and proximal markers, such as *Sox9* and *Sox2*, respectively. Caution should be taken in interpreting section-based results because sections near the lobe surface are expected to have more distal regions than those near the center. Furthermore, the proportion of distal versus proximal regions on sections reflects the average proximal–distal patterning in the whole lung instead of individual branches, an important difference for the type III mutants.

Therefore, while proximal–distal patterning is a useful phenotype descriptor, its interpretation may vary depending on mutant types (Fig. 1.4) and thus requires clarification, such as whether in the context of individual branches or the whole lung.

Gene regulatory network in epithelial progenitors: Despite the importance of tissue cross talk, branching morphogenesis ultimately is about self-renewal of *Sox9* progenitors to sustain branch formation and their differentiation into proximal airway progeny. Genetic and genomic tools are available to integrate known controlling signals into a gene regulatory network in the progenitors.

The relationship between genes in a network is classically studied using epistasis analysis, where the mutant phenotype of genes functioning downstream is expected to be dominant over, or epistatic to, that of upstream genes [61]. Epistasis analysis is essential in identifying causal instead of associative changes in gene expression, especially in systems with robust feedback regulation such as branching morphogenesis. Several genetic tools are available to specifically target the lung epithelium including epithelial progenitors. Shh^{Cre} is highly efficient due to its early activity in the foregut epithelium, but may be limited in studying branching morphogenesis if progenitor specification is blocked as in the *Ctnnb1* mutant [6, 7]. $Sox9^{CreER}$ and $Id2^{CreER}$ allow specific targeting of the progenitors at desired time points and efficiencies, and generation of mosaic mutants to study cell-autonomous effects [2, 4, 45]. Additional targeting tools include *Sftpc-rtTA; tetO-Cre* and $Nkx2.1^{CreER}$ [45, 62]. A possible epistasis analysis to integrate two major signalings, *Fgf* and *Wnt*, is to examine the phenotype of $Kras^{LSL\text{-}G12D/+}$; $Ctnnb1^{CKO/-}$; $Sox9^{CreER/+}$ double mutants, where the single mutants have an opposite effect on progenitor self-renewal (Fig. 1.4) [4].

Targeted and genomic screens have generated an increasing list of progenitor markers besides *Sox9* and *Id2* [3, 45, 63, 64]. A genome-level molecular definition of progenitors will allow comprehensive mutant analyses and gene network construction. For example, progenitor marker analyses in the *Sox9* mutant identify *Sox9*-dependent (e.g., *Clu* and *Mia1*) and independent (e.g., *Bmp4*, *Spry2*, and *Id2*) genes that are placed downstream or parallel (potentially upstream) to *Sox9* [45]. Furthermore, a gene controlling markers that are a subset of those controlled by another gene is likely to function downstream of the other gene. Such marker analyses can be extended to the whole genome by transcriptome profiling of mutant lungs or purified progenitors.

Termination of Branching Morphogenesis

Early in development, *Sox9* progenitors expand at least several hundred folds from less than a few hundred at specification to clusters of tens at each of the thousands of branch tips at the peak of branching morphogenesis (Fig. 1.1). This expansion is due to progenitor self-renewal, outpacing their differentiation into airway progeny. However, late in development, progenitor self-renewal must slow down, leading to eventual progenitor depletion and differentiation into their alveolar progeny. It is unclear when this transition in the balance between self-renewal and differentiation occurs and whether it corresponds to the transition from airway to alveolar differentiation. The timing of progenitor depletion and alveolar differentiation must be coupled with gestation periods to ensure a functional gas exchange region at birth. An intriguing candidate is glucocorticoid signaling as the maternal level of glucocorticoids peaks before birth and earlier administration of exogenous glucocorticoids activates alveolar differentiation prematurely. However, loss of the glucocorticoid receptor only causes a small delay in alveolar differentiation, suggesting the involvement of additional lung maturation factors [4, 65].

Additional insights may be obtained by examining lungs from species that do not have branching morphogenesis. Specifically, the frog lung consists of two main bronchi that do not branch but instead undergo alveolar differentiation after specification. Intriguingly, *Sox9*, although present in the frog genome, is not expressed at branch tips of the early frog lung, suggesting an antagonistic relationship between branching morphogenesis and alveolar differentiation [45, 66]. Consistent with this idea, loss of *Sox9* in the mouse lung impairs branching and allows premature expression of some alveolar markers, while overactivation of *Kras* leads to persistent *Sox9* progenitors and arrested alveolar differentiation [45]. Further cross species comparison using RNA sequencing may identify the molecular basis underlying lung evolution and specialization [67].

Given the exponential nature of branch number increase, the last few rounds of branching morphogenesis are expected to have a dominant effect on the final number of branches and consequently alveoli. This termination process may be sensitive to environmental perturbations such as oxygen and ventilation that are used to support premature birth and potentially contribute to bronchopulmonary dysplasia and chronic respiratory diseases. Even in clinically normal birth, suboptimal branching termination may lead to a reduced functional reserve or stressed stem cells that may manifest clinically upon injury or aging. This developmental plasticity matches epidemiological observations and has been explored experimentally in other organs [68, 69].

In summary, lung epithelial progenitors undergo branching morphogenesis to build both the airway and alveolar compartments. Cellular and genetic control of the progenitors via multiple signalings dictates branch number and morphology. Mechanistic understanding of progenitor specification, maintenance, and depletion has implications for clinical and subclinical respiratory diseases.

Acknowledgements I thank Edwin Ostrin for constructive comments and Jiang Chang for providing rat lungs. This work was supported by The University of Texas System Rising STARS award (JC) and The University of Texas MD Anderson Cancer Center Start-up Fund (JC). JC is a R. Lee Clark Fellow of The University of Texas MD Anderson Cancer Center.

The author declares no conflict of interest.

References

1. Morrisey EE, Hogan BL (2010) Preparing for the first breath: genetic and cellular mechanisms in lung development. Dev Cell 18(1):8–23
2. Rawlins EL et al (2009) The Id2+ distal tip lung epithelium contains individual multipotent embryonic progenitor cells. Development 136(22):3741–3745

3. Yang J, Chen J (2014) Developmental programs of lung epithelial progenitors: a balanced progenitor model. Wiley Interdiscip Rev Dev Biol 3(5):331–347
4. Alanis DM et al (2014) Two nested developmental waves demarcate a compartment boundary in the mouse lung. Nat Commun 5:3923
5. Warburton D et al (2010) Lung organogenesis. Curr Top Dev Biol 90:73–158
6. Goss AM et al (2009) Wnt2/2b and beta-catenin signaling are necessary and sufficient to specify lung progenitors in the foregut. Dev Cell 17(2):290–298
7. Harris-Johnson KS et al (2009) beta-Catenin promotes respiratory progenitor identity in mouse foregut. Proc Natl Acad Sci U S A 106(38):16287–16292
8. Domyan ET et al (2011) Signaling through BMP receptors promotes respiratory identity in the foregut via repression of Sox2. Development 138(5):971–981
9. Li Y et al (2008) Bmp4 is required for tracheal formation: a novel mouse model for tracheal agenesis. Dev Biol 322(1):145–155
10. Serls AE et al (2005) Different thresholds of fibroblast growth factors pattern the ventral foregut into liver and lung. Development 132(1):35–47
11. Kumar ME et al (2014) Mesenchymal cells. Defining a mesenchymal progenitor niche at single-cell resolution. Science 346(6211):1258810
12. Metzger RJ et al (2008) The branching programme of mouse lung development. Nature 453 (7196):745–750
13. Horsfield K, Cumming G (1968) Morphology of the bronchial tree in man. J Appl Physiol 24 (3):373–383
14. Menshykau D, Kraemer C, Iber D (2012) Branch mode selection during early lung development. PLoS Comput Biol 8(2), e1002377
15. Costantini F, Kopan R (2010) Patterning a complex organ: branching morphogenesis and nephron segmentation in kidney development. Dev Cell 18(5):698–712
16. Sakai T, Larsen M, Yamada KM (2003) Fibronectin requirement in branching morphogenesis. Nature 423(6942):876–881
17. Andrew DJ, Ewald AJ (2010) Morphogenesis of epithelial tubes: Insights into tube formation, elongation, and elaboration. Dev Biol 341(1):34–55
18. Gariano RF, Gardner TW (2005) Retinal angiogenesis in development and disease. Nature 438 (7070):960–966
19. Carraro G, del Moral PM, Warburton D (2010) Mouse embryonic lung culture, a system to evaluate the molecular mechanisms of branching. J Vis Exp 40(pi):2035
20. Nogawa H, Hasegawa Y (2002) Sucrose stimulates branching morphogenesis of embryonic mouse lung in vitro: a problem of osmotic balance between lumen fluid and culture medium. Dev Growth Differ 44(5):383–390
21. Chen J, Krasnow MA (2012) Integrin Beta 1 suppresses multilayering of a simple epithelium. PLoS One 7(12), e52886
22. Lechler T, Fuchs E (2005) Asymmetric cell divisions promote stratification and differentiation of mammalian skin. Nature 437(7056):275–280
23. Tang N et al (2011) Control of mitotic spindle angle by the RAS-regulated ERK1/2 pathway determines lung tube shape. Science 333(6040):342–345
24. Nishio S et al (2010) Loss of oriented cell division does not initiate cyst formation. J Am Soc Nephrol 21(2):295–302
25. Meyer EJ, Ikmi A, Gibson MC (2011) Interkinetic nuclear migration is a broadly conserved feature of cell division in pseudostratified epithelia. Curr Biol 21(6):485–491
26. Del Bene F et al (2008) Regulation of neurogenesis by interkinetic nuclear migration through an apical-basal notch gradient. Cell 134(6):1055–1065
27. Kadzik RS et al (2014) Wnt ligand/Frizzled 2 receptor signaling regulates tube shape and branch-point formation in the lung through control of epithelial cell shape. Proc Natl Acad Sci U S A 111(34):12444–12449
28. Schnatwinkel C, Niswander L (2013) Multiparametric image analysis of lung branching morphogenesis. Dev Dyn 242(6):622–637

29. Bertet C, Sulak L, Lecuit T (2004) Myosin-dependent junction remodelling controls planar cell intercalation and axis elongation. Nature 429(6992):667–671
30. Kim HY, Varner VD, Nelson CM (2013) Apical constriction initiates new bud formation during monopodial branching of the embryonic chicken lung. Development 140 (15):3146–3155
31. Packard A et al (2013) Luminal mitosis drives epithelial cell dispersal within the branching ureteric bud. Dev Cell 27(3):319–330
32. Mollard R, Dziadek M (1998) A correlation between epithelial proliferation rates, basement membrane component localization patterns, and morphogenetic potential in the embryonic mouse lung. Am J Respir Cell Mol Biol 19(1):71–82
33. Relan NK, Schuger L (1999) Basement membranes in development. Pediatr Dev Pathol 2 (2):103–118
34. Nguyen NM, Senior RM (2006) Laminin isoforms and lung development: all isoforms are not equal. Dev Biol 294(2):271–279
35. Oblander SA et al (2005) Distinctive functions of membrane type 1 matrix-metalloprotease (MT1-MMP or MMP-14) in lung and submandibular gland development are independent of its role in pro-MMP-2 activation. Dev Biol 277(1):255–269
36. Kheradmand F, Rishi K, Werb Z (2002) Signaling through the EGF receptor controls lung morphogenesis in part by regulating MT1-MMP-mediated activation of gelatinase A/MMP2. J Cell Sci 115(Pt 4):839–848
37. Gill SE et al (2003) A null mutation for tissue inhibitor of metalloproteinases-3 (Timp-3) impairs murine bronchiole branching morphogenesis. Dev Biol 261(2):313–323
38. Cardoso WV, Lu J (2006) Regulation of early lung morphogenesis: questions facts and controversies. Development 133(9):1611–1624
39. Morimoto M, Kopan R (2009) rtTA toxicity limits the usefulness of the SP-C-rtTA transgenic mouse. Dev Biol 325(1):171–178
40. Whitsett JA, Perl KA (2006) Conditional control of gene expression in the respiratory epithelium: a cautionary note. Am J Respir Cell Mol Biol 34(5):519–520
41. Sisson TH et al (2006) Expression of the reverse tetracycline-transactivator gene causes emphysema-like changes in mice. Am J Respir Cell Mol Biol 34(5):552–560
42. Bellusci S et al (1997) Fibroblast growth factor 10 (FGF10) and branching morphogenesis in the embryonic mouse lung. Development 124(23):4867–4878
43. De Moerlooze L et al (2000) An important role for the IIIb isoform of fibroblast growth factor receptor 2 (FGFR2) in mesenchymal-epithelial signalling during mouse organogenesis. Development 127(3):483–492
44. Sekine K et al (1999) Fgf10 is essential for limb and lung formation. Nat Genet 21(1):138–141
45. Chang DR et al (2013) Lung epithelial branching program antagonizes alveolar differentiation. Proc Natl Acad Sci U S A 110(45):18042–18051
46. Al Alam D et al (2011) Contrasting expression of canonical Wnt signaling reporters TOPGAL BATGAL and Axin2(LacZ) during murine lung development and repair. PLoS One 6(8), e23139
47. Rajagopal J et al (2008) Wnt7b stimulates embryonic lung growth by coordinately increasing the replication of epithelium and mesenchyme. Development 135(9):1625–1634
48. Li C et al (2002) Wnt5a participates in distal lung morphogenesis. Dev Biol 248(1):68–81
49. Mucenski ML et al (2003) beta-Catenin is required for specification of proximal/distal cell fate during lung morphogenesis. J Biol Chem 278(41):40231–40238
50. Shu W et al (2005) Wnt/beta-catenin signaling acts upstream of N-myc, BMP4, and FGF signaling to regulate proximal-distal patterning in the lung. Dev Biol 283(1):226–239
51. Weaver M, Dunn NR, Hogan BL (2000) Bmp4 and Fgf10 play opposing roles during lung bud morphogenesis. Development 127(12):2695–2704
52. Eblaghie MC et al (2006) Evidence that autocrine signaling through Bmpr1a regulates the proliferation, survival and morphogenetic behavior of distal lung epithelial cells. Dev Biol 291 (1):67–82

53. Sun J et al (2008) Prenatal lung epithelial cell-specific abrogation of Alk3-bone morphogenetic protein signaling causes neonatal respiratory distress by disrupting distal airway formation. Am J Pathol 172(3):571–582
54. Xing Y et al (2010) Signaling via Alk5 controls the ontogeny of lung Clara cells. Development 137(5):825–833
55. Yin Y, Wang F, Ornitz DN (2011) Mesothelial- and epithelial-derived FGF9 have distinct functions in the regulation of lung development. Development 138(15):3169–3177
56. De Langhe SP et al (2008) Formation and differentiation of multiple mesenchymal lineages during lung development is regulated by beta-catenin signaling. PLoS One 3(1), e1516
57. Chuang PT, Kawcak T, McMahon AP (2003) Feedback control of mammalian Hedgehog signaling by the Hedgehog-binding protein, Hip1, modulates Fgf signaling during branching morphogenesis of the lung. Genes Dev 17(3):342–347
58. St Johnston D, Nusslein-Volhard C (1992) The origin of pattern and polarity in the Drosophila embryo. Cell 68(2):201–219
59. Mahoney JE et al (2014) The hippo pathway effector Yap controls patterning and differentiation of airway epithelial progenitors. Dev Cell 30(2):137–150
60. Harris KS et al (2006) Dicer function is essential for lung epithelium morphogenesis. Proc Natl Acad Sci U S A 103(7):2208–2213
61. Phillips PC (2008) Epistasis–the essential role of gene interactions in the structure and evolution of genetic systems. Nat Rev Genet 9(11):855–867
62. Perl AK, Zhang L, Whitsett JA (2009) Conditional expression of genes in the respiratory epithelium in transgenic mice: cautionary notes and toward building a better mouse trap. Am J Respir Cell Mol Biol 40(1):1–3
63. Herriges JC et al (2012) Genome-scale study of transcription factor expression in the branching mouse lung. Dev Dyn 241(9):1432–1453
64. Diez-Roux G et al (2011) A high-resolution anatomical atlas of the transcriptome in the mouse embryo. PLoS Biol 9(1), e1000582
65. Smith BT (1979) Lung maturation in the fetal rat: acceleration by injection of fibroblast-pneumonocyte factor. Science 204(4397):1094–1095
66. Rankin SA et al (2014) A molecular atlas of Xenopus respiratory system development. Dev Dyn 244(1):69–85
67. Cieri RL et al (2014) New insight into the evolution of the vertebrate respiratory system and the discovery of unidirectional airflow in iguana lungs. Proc Natl Acad Sci U S A 111 (48):17218–17223
68. Cebrian C et al (2014) The number of fetal nephron progenitor cells limits ureteric branching and adult nephron endowment. Cell Rep 7(1):127–137
69. Stocks J, Sonnappa S (2013) Early life influences on the development of chronic obstructive pulmonary disease. Ther Adv Respir Dis 7(3):161–173
70. Wang Y et al (2013) Development and regeneration of Sox2+ endoderm progenitors are regulated by a HDAC1/2-Bmp4/Rb1 regulatory pathway. Dev Cell 24(4):345–358

Chapter 2
Epithelial–Mesenchymal Crosstalk in Lung Development

David Warburton

Mesenchyme Crosstalk to Epithelium

Alescio and Cassini [1] showed that ectopic branching of the embryonic mouse trachea can be induced by transplantation of distal lung mesenchyme to a more proximal location next to the tracheal epithelium, while conversely proximal epithelium can suppress distal branching. These classic experiments have been widely repeated by Shannon [2] and others so that the concept that distal lung mesenchyme can induce and control stereotypic branching morphogenesis of the airway epithelium has become well established.

Growth Factors in Epithelial–Mesenchymal Crosstalk

Subsequent work by ourselves and many other groups have further established that mesenchymal–epithelial induction of branching morphogenesis of the lung epithelium is mediated by peptide growth factors emanating from the mesenchyme and stimulating their cognate receptors upon the epithelium and vice versa. Among these factors BMP4, EGF, FGF7 and 10, HGF, PDGF, Shh, TGFβ, VEGF, and Wnt-β-catenin signaling have been found to play complex interlocking roles in this process (reviewed in [3–5]). FGF signaling is likewise a highly conserved branch induction mechanism, which is also inductive in fruit fly tracheal morphogenesis [6, 7].

D. Warburton (✉)
Developmental Biology and Regenerative Medicine Program, Saban Research Institute, Children's Hospital Los Angeles, Keck School of Medicine and Ostrow School of Dentistry, University of Southern California, 4661 Sunset Boulevard, Los Angeles, CA 90027, USA
e-mail: dwarburton@chla.usc.edu

I. Bertoncello (ed.), *Stem Cells in the Lung*, Stem Cell Biology and Regenerative Medicine, DOI 10.1007/978-3-319-21082-7_2

Actually, I tend to think of growth factor-mediated branching morphogenesis as a default mechanism of airway epithelium. For example, FGF10 is capable of inducing vigorous but randomly directional branching of isolated embryonic lung epithelium in the absence of lung mesenchyme [8]. Therefore, stereotypy must be what is conferred by the lung mesenchyme, suppressing branching where it is not needed.

There are several theories about how this might work. One is that Turing morphogenetic gradients must exist within the lung mesenchyme, that chemoattract epithelial tips up the gradient to extend and branch in the direction of the ascending morphogenetic gradient [9]. This is in fact quite likely to be the case for the first few rounds of morphogenetic branching, because pools of FGF10 expression in the right place and at the right time can be identified by in situ hybridization, and ectopic branching can be induced by FGF10-soaked beads [10]. Moreover, the rate of branching can be manipulated by Sprouty genes that function as inducible inhibitors of FGF signaling [7]. However, it is not really known exactly how the subsequent rounds of stereotypic branching (up to 9 in mouse and 16 in human), let alone the subsequent rounds of non-stereotypic, the so-called fractal branching (branches 9–11 in mouse, 16–21 in humans). Although we do know that EGF signaling must have something important to do with it, since EGF null mutation cuts the rate of branching in half [11]. Whereas, conversely increased luminal pressure brought on by tracheal obstruction increases branching via an FGF10-FGF receptor-dependent mechanism [12].

Epithelial–Mesenchymal Clocks

Another very interesting question is how does the epithelial branch timing clock work? We do now know from our recent experiments that rhythmic calcium waves within the epithelium that are driven from proximal bronchial epithelial pacemakers are important, because reversible inhibition of these waves with SERCA pump inhibitors can reversibly alter the rate of epithelial branching, thus determining the inter branch length [13]. Substances such as nicotine can also speed up branching morphogenesis and indeed overall lung maturation [14]. Certainly, there are important interspecies differences in the timing of stereotypic branching. In mouse, the branching clock strikes frequently so that interbranch lengths are short. Of course, the mature mouse lung is also relatively small. Interbranch lengths in human embryonic lung are much longer than mouse, which raises the question of why this might be and how interbranch lengths are matched to eventual respiratory surface area and body size. I wonder how these scaling rules work in whales and dinosaurs?

Mesenchyme is Not Boring But Busy

In the same era as Alescio and Cassini [1], when I was at medical school (1967–1973), the mesenchyme in general, we were taught, is regarded as rather a boring jelly-like substance in which not much was happening. Now, of course, we realize that the mesenchyme, particularly in the developing lung, is in fact a very busy place, with highly dynamic cell movements as well as comprising several essential tissues including blood vessels, lymphatics, immune cells, smooth muscle cells, and nerves. Our recent studies show that the tiny nerve axon projections that stereotypically accompany the developing airways in fact play an essential role in controlling branching morphogenesis, but not seemingly via parasympathetic neurotransmission [15]. It remains to be seen exactly how this system works. Likewise, we have shown that interfering with hemangioblast development into capillary networks by blocking VEGF signaling has important regional effects, not only on angio- and vasculogenesis, but also on epithelial branching morphogenesis in the embryonic lung. In particular, blockade of VEG signaling with a dominant negative decoy receptor approach impairs the induction of anterior epithelial branches from the major bronchi, which normally arise from populations of stem and/or progenitor cells expressing Sox9, which should arise at intervals along the anterior surface of the major bronchi [16, 17]. These condensations fail to appear in the presence of VEGF blockade and the absence of a correctly organized capillary vasculature.

Development of the smooth muscle cells within the mesenchyme of the lung also has important functional implications for the subsequent development of wheezing in childhood and indeed asthma. We have shown that some of the peripheral smooth muscle cells surrounding the bronchi actually arise from erstwhile FGF10 expressing cells, originally located within the peripheral mesenchyme, which translocate to surround and invest the distal bronchi [8]. Other populations of smooth muscle cells appear to originate from Sox9-positive mesenchymal progenitors that surround the distal trachea [18].

Pericytes are another class of mesenchymal cell that invest and maintain the capillary endothelium during embryonic development. We have recently discovered that pericytes arise from a distinct population of mesenchymal stem/progenitor cells that express Tbx4 and invade the lung from outside. Tbx4 expression also marks a subpopulation of smooth muscle cells that arrive in the embryonic lung [19].

Another important part of the lung where epithelial–mesenchymal crosstalk is key is the trachea. We and others recently reported that lung mesenchymal expression of Sox9 is essential for correct induction of the tracheal cartilages as well as for correct tracheal epithelial development [18].

An equally fundamental concept in epithelial–mesenchymal crosstalk in lung development is that timing and location are everything. This has been referred to as temporo-spatial control of inductive gene expression. Key examples of this include: retinoic acid signaling, in the absence of which no lung forms [20]; FGF10 signaling, in the absence of which no lung forms distal to the major bronchial origins [21]; and BMP4 signaling, which controls critical events at the time of transition to air breathing [22]. TGF beta signaling also plays important compartment and temporo-spatial specific roles in the epithelium versus the mesenchyme that may become manifest at later stages of lung development such as the alveolar phase [23].

Conclusion

In summary, temporo-spatial epithelial–mesenchymal crosstalk is essential for the correct stereotypic organization of the bronchial epithelial tree and subsequently, although not discussed in detail here, the alveoli. Within the mesenchyme, correct organization and function of the capillaries and nerves is likewise essential to achieve a fully functional lung (Fig. 2.1). Dysfunction of epithelial–mesenchymal crosstalk can lead to respiratory defects from absence of the lung through gross malformation of the lobes, through alveolar capillary dysplasia. Successful treatment of defects caused by defective epithelial–mesenchymal crosstalk has so far eluded small molecule approaches because of the exigency of repairing this complex signaling network. Surgical approaches such as controlled tracheal obstruction have achieved some mixed success, resulting in a large but grossly dysfunctional lung that is incapable of sustained gas exchange. Thus, significant challenges remain in fully parsing the mechanisms of epithelial–mesenchymal crosstalk in lung development. However, the effort may eventually pay off if we can re-entrain these developmental mechanisms to slow or reverse the decline in lung function that occurs over the advancing years of the full human life span.

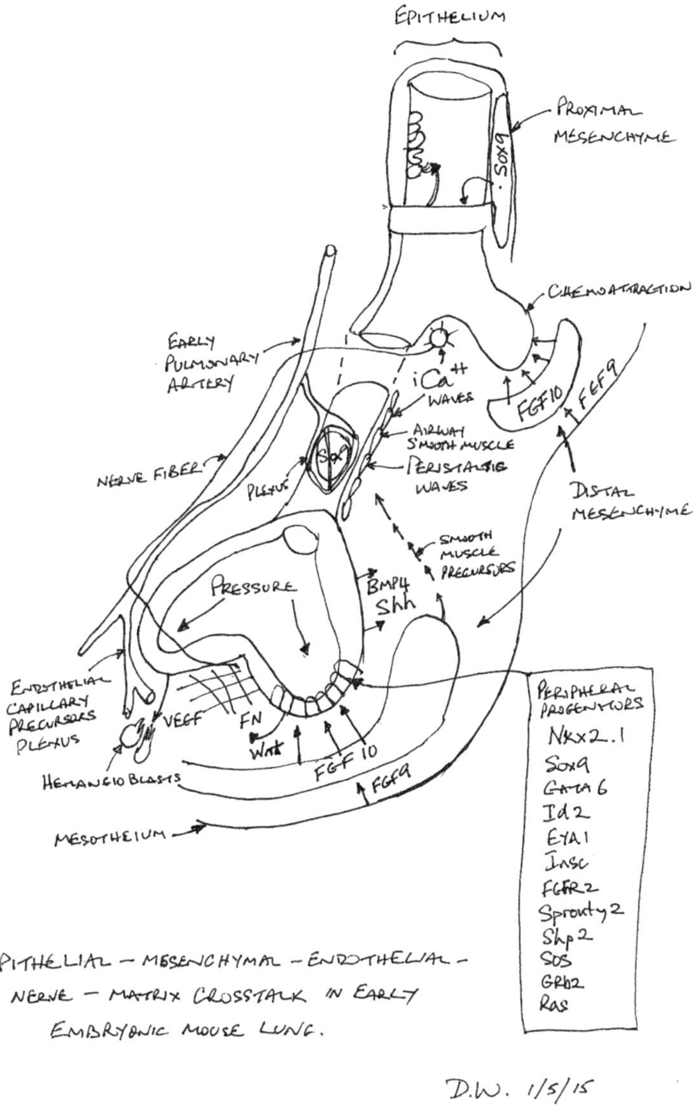

Fig. 2.1 Epithelial-mesenchymal-endothelial-nerve-matrix crosstalk in early mouse lung. This concept figure diagrams many of the known interactions between these compartments in lung morphogenesis. In the proximal lung, Sox9-positive mesenchymal progenitors play a key role in inducing tracheal cartilage, which in turn plays an inductive role on tracheal epithelial progenitors. In the first generation of airways, FGF10 signaling from the mesenchyme exerts an inductive and chemoattractive effect on the lung epithelium, which are both required for the subsequent induction of the distal epithelial progenitors as well as organization of all the distal lung structures. In the peripheral epithelium, FGF10 is still an important inductive signal for peripheral progenitors as well as for branching. FGF10 progenitors in the peripheral mesenchyme also act as progenitors for distal airway smooth muscle under the influence of Shh and BMP4 signaling. Wnt signaling from distal epithelium induces fibronectin to be laid down in potential distal clefts and this is essential for cleft formation. The early pulmonary arteries connect with early distal capillaries that form from hemangioblasts under the control of VEGF secreted by the epithelium. The formation of

Acknowledgements D.W. wishes to thank the NIH and, in particular, the US and UK taxpayers who over many years have provided continual support for his education, career development, and for the body of work summarized briefly here. As this is a short monograph, D.W. wishes to apologize for not having room to mention the many inspirational papers and important contributions by other colleagues in the field, but refers the reader to publications wherein he has reviewed as many as possible of them elsewhere.

References

1. Alescio T, Cassini A (1962) Induction in vitro of tracheal buds by pulmonary mesenchyme grafted on tracheal epithelium. J Exp Zool 150:83–94
2. Shannon JM (1994) Induction of alveolar type II cell differentiation in fetal tracheal epithelium by grafted distal lung mesenchyme. Dev Biol 166(2):600–614
3. Warburton D, Schwarz M, Tefft D, Flores-Delgado G, Anderson KD, Cardoso WV (2000) The molecular basis of lung morphogenesis. Mech Dev 92(1):55–81
4. Warburton D, Bellusci S, Del Moral PM, Kaartinen V, Lee M, Tefft D, Shi W (2003) Growth factor signaling in lung morphogenetic centers: automaticity, stereotypy and symmetry. Respir Res 4:5
5. Warburton D, El-Hashash A, Carraro G, Tiozzo C, Sala F, Rogers O et al (2010) Lung organogenesis. Curr Top Dev Biol 90:73–158
6. Tefft JD, Lee M, Smith S, Leinwand M, Zhao J, Bringas P Jr et al (1999) Conserved function of mSpry-2, a murine homolog of Drosophila sprouty, which negatively modulates respiratory organogenesis. Curr Biol 9(4):219–222
7. Mailleux AA, Tefft D, Ndiaye D, Itoh N, Thiery JP, Warburton D, Bellusci S (2001) Evidence that SPROUTY2 functions as an inhibitor of mouse embryonic lung growth and morphogenesis. Mech Dev 102(1–2):81–94
8. Ramasamy SK, Mailleux AA, Gupte VV, Mata F, Sala FG, Veltmaat JM et al (2007) Fgf10 dosage is critical for the amplification of epithelial cell progenitors and for the formation of multiple mesenchymal lineages during lung development. Dev Biol 307(2):237–247
9. Warburton D (2008) Developmental biology: order in the lung. Nature 453(7196):733–735
10. Park WY, Miranda B, Lebeche D, Hashimoto G, Cardoso WV (1998) FGF-10 is a chemotactic factor for distal epithelial buds during lung development. Dev Biol 201(2):125–134
11. Miettinen PJ, Warburton D, Bu D, Zhao JS, Berger JE, Minoo P et al (1997) Impaired lung branching morphogenesis in the absence of functional EGF receptor. Dev Biol 186 (2):224–236

Fig. 2.1 (continued) capillary plexi is likewise essential for the sprouting of side branches from condensation so Sox9-positive epithelial progenitor cells. Sympathetic nerve fibers arise from ganglia in the mediastinum, cross the proximal airway and early pulmonary arteries in a stereotypic manner and then follow the superior surface of the airway all the way to the periphery of the lung. Both nerve twigs and single capillary branches can be found between each lung epithelial cleft. FGF9 is expressed by the primitive mesothelium and induces FGF10 within the subjacent distal mesenchyme. Calcium ion waves arise from pacemakers in the mediastinum and proximal bronchi and pass down the bronchi toward the periphery, orchestrating distal-wards peristaltic waves that increase pressure within the peripheral airway tips. Peripheral tip progenitors can be identified by the co-expression of large number of characteristic transcriptional factors, growth factor receptors, and downstream signaling molecules that control their behavior

12. Unbekandt M, del Moral PM, Sala FG, Bellusci S, Warburton D, Fleury V (2008) Tracheal occlusion increases the rate of epithelial branching of embryonic mouse lung via the FGF10-FGFR2b-Sprouty2 pathway. Mech Dev 125(3–4):314–324
13. Jesudason EC, Keshet E, Warburton D (2010) Entrained pulmonary clocks: epithelium and vasculature keeping pace. Am J Physiol Lung Cell Mol Physiol 299(4):L453–L454
14. Wuenschell CW, Sunday ME, Singh G, Minoo P, Slavkin HC, Warburton D (1996) Embryonic mouse lung epithelial progenitor cells co-express immunohistochemical markers of diverse mature cell lineages. J Histochem Cytochem 44(2):113–123
15. Bower DV, Lee HK, Lansford R, Zinn K, Warburton D, Fraser SE, Jesudason EC (2014) Airway branching has conserved needs for local parasympathetic innervation but not neurotransmission. BMC Biol 12(1):92
16. Del Moral PM, Sala FG, Tefft D, Shi W, Keshet E, Bellusci S, Warburton D (2006) VEGF-A signaling through Flk-1 is a critical facilitator of early embryonic lung epithelial to endothelial crosstalk and branching morphogenesis. Dev Biol 290(1):177–188
17. Lazarus A, Del-Moral PM, Ilovich O, Mishani E, Warburton D, Keshet E (2011) A perfusion-independent role of blood vessels in determining branching stereotypy of lung airways. Development 138(11):2359–2368
18. Turcatel G, Rubin N, Menke DB, Martin G, Shi W, Warburton D (2013) Lung mesenchymal expression of Sox9 plays a critical role in tracheal development. BMC Biol 11:117
19. Zhang W, Menke DB, Jiang M, Chen H, Warburton D, Turcatel G et al (2013) Spatial-temporal targeting of lung-specific mesenchyme by a Tbx4 enhancer. BMC Biol 11:111
20. Chen F, Cao Y, Qian J, Shao F, Niederreither K, Cardoso WV (2010) A retinoic acid-dependent network in the foregut controls formation of the mouse lung primordium. J Clin Invest 120(6):2040–2048
21. Min H, Danilenko DM, Scully SA, Bolon B, Ring BD, Tarpley JE et al (1998) Fgf-10 is required for both limb and lung development and exhibits striking functional similarity to Drosophila branchless. Genes Dev 12(20):3156–3161
22. Sun J, Chen H, Chen C, Whitsett JA, Mishina Y, Bringas P Jr et al (2008) Prenatal lung epithelial cell-specific abrogation of Alk3-bone morphogenetic protein signaling causes neonatal respiratory distress by disrupting distal airway formation. Am J Pathol 172(3):571–582
23. Chen H, Zhuang F, Liu YH, Xu B, Del Moral P, Deng W et al (2008) TGF-beta receptor II in epithelia versus mesenchyme plays distinct roles in the developing lung. Eur Respir J 32 (2):285–295

Chapter 3
Lung Vasculogenesis and Angiogenesis

Bernard Thébaud and Mervin C. Yoder

Introduction

In 1959, Liebow observed that the alveolar septae in centrilobular emphysema were remarkably thin and almost avascular [1]. He postulated that a reduction in the blood supply of the small precapillary blood vessels might induce the disappearance of alveolar septae. Despite this early observation, pulmonary vessels were thought for many years, to be passive bystanders in lung development, following the branching pattern of the airways. At the other extreme of life, interrupted alveolar development in bronchopulmonary dysplasia (BPD), the chronic lung disease of prematurity, is also accompanied by a marked rarefaction of lung capillaries, suggesting a potential role of the developing lung vascular bed during normal alveolar development, injury, and repair. This review will summarize literature that has challenged old notions that the development of the blood vessels in the lung passively follows that of the airways. Evidence is presented that lung blood vessels actively promote alveolar growth during development and contribute to the maintenance of alveolar structures throughout postnatal life, opening new therapeutic avenues for lung diseases characterized by alveolar damage by modulation of vasculogenesis and angiogenesis.

B. Thébaud (✉)
Division of Neonatology, Department of Pediatrics, Sinclair Centre for Regenerative Medicine, Ottawa Hospital Research Institute, Children's Hospital of Eastern Ontario, University of Ottawa, 501 Smyth Road, Ottawa, ON, Canada, K1H 8L6
e-mail: bthebaud@ohri.ca

M.C. Yoder
Department of Pediatrics, Wells Center for Pediatric Research, Indiana University School of Medicine, Indianapolis, IN, USA

I. Bertoncello (ed.), *Stem Cells in the Lung*, Stem Cell Biology and Regenerative Medicine, DOI 10.1007/978-3-319-21082-7_3

Normal Lung Development in Brief

The classical stages of lung development: Lung development is classically subdivided into five overlapping stages in human and rodents, on the basis of gross histological features. The first four stages, termed the embryonic, pseudoglandular, canalicular, and saccular stages, occur during gestation. At the end of the saccular stage at about 36 weeks, lungs have formed alveolar ducts and air sacs. Alveolarization, the final stage of lung development, begins in the near-term lung prior to birth but primarily occurs postnatally, during the first 2–3 years of life, and may continue at a slower rate beyond childhood [2, 3]. The formation of alveoli occurs by the outgrowth of secondary septae that subdivide terminal saccules into anatomic alveoli. Premature infants at greatest risk for BPD in the post-surfactant era are born at 24–28 weeks, during the late canalicular or saccular stage of lung development just as the airways become juxtaposed to pulmonary vessels.

Lung angiogenesis: Formation of the pulmonary circulation has been primarily described as dependent on two basic processes: vasculogenesis and angiogenesis. Vasculogenesis is the de novo formation of blood vessels from angioblasts or endothelial precursor cells that migrate and differentiate in response to local cues to form vascular tubes. Angiogenesis is the formation of new blood vessels from preexisting ones. It has generally been accepted that the distal vasculature arises by vasculogenesis, and the proximal vasculature by angiogenesis, but this remains controversial [4]. De novo formation of blood vessels, i.e., vasculogenesis, from blood islands present within the mesenchyme have been described in the embryonic lung at embryonic day 9 (E9) in the mouse [5]. Angiogenesis starts around E12 when arteries and veins begin to sprout from the central pulmonary vascular trunks. Around E14, peripheral sinusoids and central vessel sprouts connect and establish a vascular network. This union of peripheral and central vascular structures is accompanied by extensive branching of the arteries, which follow the branching pattern of the airways. Again, the relative contributions of vasculogenesis and angiogenesis to lung vascular growth during each stage of lung development are controversial, and additional studies with appropriate experimental models are required to better define these underlying mechanisms. Studies in the developing chick embryo suggest that vasculogenesis is the main process by which the pulmonary vasculature forms [6]. Vascular fillings show that the lumen of the pulmonary arteries sprouted from the sixth pharyngeal arch arteries. However, serial sections and immunohistochemical localization of fetal liver kinase-1 protein (Flk-1), the receptor for vascular endothelial growth factor (VEGF) showed that the pulmonary arterial tree formed from endothelial cell precursors and coalescence of isolated blood vessels in the mediastinal splanchnic mesenchyme centrally to the developing lung tissue distally. Pulmonary veins grew from the left atrium to the developing lungs. Parera et al. suggested distal angiogenesis as a new mechanism for lung vascular morphogenesis, based on morphological analysis from the onset of lung development (E9.5) until the pseudoglandular stage (E13.5) in Tie2-LacZ transgenic mice [4]. In their model, capillary networks surrounding the terminal buds exist from the first morphological sign of lung development and then expand by formation of new capillaries from preexisting vessels as the lung bud grows [4]. The capacity of the primitive lung mesenchyme to differentiate into vascular endothelium was recently demonstrated in a new culture system and cells with the highest Flk-1 expression differentiated into endothelia more efficiently [7]. Studies in human fetal lung suggest that airways act as a template for pulmonary artery development, and that endothelial tubes form around the terminal buds of distal air spaces, suggesting an inductive influence of the epithelium [8]. Pereda et al. used immunolocalization and expression of CD31, CD34, FLT-1, KDR, and VEGF to show that at day 31 postfertilization (pf), a capillary plexus was already installed, and a few primitive erythroblasts were seen within the lumen of some blood vessels. Around day 45 pf, an increased amount of primitive erythroblasts was detected in the parenchyma surrounding the distal segment of the bronchial tree. The expression of FLT-1, KDR, CD31, and CD34 was observed in endothelial cells of the capillary plexus, and VEGF was detected in the endodermal epithelium. This indicated that the initial formation of the capillary plexus around the tip of the growing airway bud occurs by vasculogenesis likely regulated by VEGF. The increasing number of primitive erythroblasts from week 6 onward, associated with a change in the shape of the blood vessels, suggests a remodeling process and that the generation of new distal vessels at the tip of the lung bud occurs mainly by a process of angiogenesis. Clearly, new tools, such as lineage tracing studies routinely used in developmental biology, need to be applied to deepen our understanding about the formation of the pulmonary circulation [9].

The final important step of microvascular maturation overlaps the alveolar stage of lung development [10]. Capillaries, which are organized as double capillary layers in the immature gas-exchange region, later remodel to form a single capillary layer. This process is completed in the rat by about the third postnatal week. The alveolar wall thins and its cellular composition changes. In the rat, the thickness of the alveolar wall and the air–gas barrier (the distance between alveolar gas and capillary blood) decrease by 20–25 %. Between birth and adulthood, the alveolar and capillary surface areas expand nearly 20-fold and the capillary volume by 35-fold. Further expansion of the capillary network subsequently occurs via two angiogenic mechanisms: sprouting angiogenesis from preexisting vessels and intussusceptive growth [11]. Little is known about intussusceptive microvascular growth in the lung. This novel mode of blood vessel formation and remodeling occurs by internal division of the preexisting capillary plexus (insertion of transcapillary tissue pillars) without sprouting, which may underlie alveolar growth and remodeling throughout adult life and thus be amenable to therapeutic modulation for lung regeneration [12]. This is supported by recent findings suggesting that intussusceptive angiogenesis contributes to post-pneumonectomy lung growth. Much more needs to be learned about the anatomic events underlying lung vascular development and time-specific mechanisms that regulate growth and function at each stage. Nonetheless, current knowledge gained mostly from developmental biology and cancer literature about key angiogenic growth factors has been exploited to investigate the role of lung angiogenesis during normal alveolar development.

Evidence That Angiogenic Growth Factor-Driven Angiogenesis Promotes Normal Alveolar Development

VEGF is pivotal for the proper formation of blood vessels: VEGF is a highly specific mitogen and survival factor for vascular endothelial cells. VEGF binds to transmembrane tyrosine kinase receptors, VEGFR-1 (flt-1) and VEGFR-2 (flk-1/KDR), which are expressed on vascular endothelium [13]. The absolute requirement of VEGF for development of the embryonic vasculature in mice has been demonstrated by inactivation studies of VEGF alleles [14, 15] and knockouts of VEGFR-1 [16] and VEGFR-2 [17]. In each of these studies, inactivation of the target genes resulted in lethal phenotypes characterized by deficient organization of endothelial cells. Inducible Cre-loxP-mediated gene targeting or administration of a soluble VEGF receptor chimeric protein (mFlt (1-3)-IgG) to inactivate VEGF in early postnatal life results in increased mortality, stunted body growth, and impaired organ development [18]. VEGF inhibition resulted in less significant alterations as the animal matured, and the dependence on VEGF is lost around the fourth postnatal week. Interestingly, this period coincides with the end of alveolarization and microvasculature maturation in the lung.

VEGF promotes normal alveolar development: The spatial relationship between receptor and ligand suggests that VEGF plays a role in the development of the alveolar capillary bed. In addition, pharmacological and genetic inactivation of VEGF arrests alveolar development. VEGF mRNA and protein are localized to distal airway epithelial cells and the basement membrane subjacent to the airway epithelial cells [19]. This suggests that translocation of VEGF protein occurs after its synthesis in the epithelium. VEGFR-1 and VEGFR-2 mRNA expression also increases during normal mouse lung development [20, 21] and is localized to the pulmonary endothelial cells closely apposed to the developing epithelium [22]. $VEGF_{120}$, $VEGF_{164}$, and $VEGF_{188}$ are present in alveolar type II cells in the developing mouse lung, and their expression peaks during the canalicular stage, when most of the vessel growth occurs in the lung, then decreases towards until day 10 postnatal (P10) when it increases to levels that are maintained through adulthood [22].

Targeted exon deletion of the VEGF gene reveals that mice that lack the heparin-binding isoforms $VEGF_{164}$ and $VEGF_{188}$ display a variety of vascular defects, including a significant reduction in the formation of air spaces and capillaries, resulting in distended and underdeveloped alveoli [23]. Likewise, pharmacological and genetic VEGF inhibition during alveolar development decreases alveolarization and pulmonary arterial density, features encountered in clinical BPD [23–26]. Chronic treatment of adult rats with the VEGFR-1 and -2 blocker

SU5416 leads to enlargement of the air spaces, indicative of emphysema [27], suggesting that VEGF is required for the formation, but also the maintenance of the pulmonary vasculature and alveolar structures throughout adulthood. Conversely, lung overexpression of VEGF during normal lung development disrupts the lung architecture suggesting a tight regulation of VEGF-driven angiogenesis to insure proper lung development [28, 29]. Pharmacological VEGF inhibition in lung renal capsule grafts [26] and genetic VEGF inactivation [30] show that selective inactivation of VEGF in respiratory epithelium results in an almost complete absence of pulmonary capillaries, demonstrating the dependence of pulmonary capillary development on epithelium-derived Vegf-A. Deficient capillary formation in Vegf-A deficient lungs was associated with a defect in primary septae formation, coupled with suppression of epithelial cell proliferation and decreased hepatocyte growth factor (HGF) expression. Lung endothelial cells express HGF, and selective deletion of the HGF receptor gene in respiratory epithelium phenocopies the malformation of septae, confirming the requirement for epithelial HGF signaling in normal septae formation and suggesting that HGF serves as an endothelium-derived factor that signals to the epithelium. In summary, these observations suggest that inhibition of vascular growth itself may directly impair alveolarization.

Perturbation of nitric oxide (NO) is associated with arrested alveolar and lung vascular growth: While the role of the endothelium-derived relaxing factor NO in the regulation of the pulmonary vascular tone in the perinatal period is well established [31], little was known about its potential role in the structural development of the pulmonary vasculature. Studies suggest that VEGF-induced lung angiogenesis is in part mediated by NO. SU5416 (VEGF inhibitor)-induced arrested alveolar and vascular growth in newborn rats is associated with decreased lung eNOS protein expression and NO production; treatment with inhaled NO improves vascular and alveolar growth in this model [32]. Lungs of late fetal and neonatal eNOS-deficient mice have a paucity of distal arteries and reduced alveolarization [33] and are more susceptible for failed vascular and alveolar growth after exposure to mild hypoxia and hyperoxia [34].

Hypoxia inducible factors (HIFs): Despite its crucial role in O_2 homeostasis, the role of HIFs during alveolar development was relatively unexplored until recently. HIF, a master regulator of O_2 homeostasis discovered in 1992 by Dr. Semenza [35], regulates the expression of a range of angiogenic factors. HIFs are heterodimers that comprise one of three O_2-sensitive α-subunits (HIF-1, HIF-2, or HIF-3) and a constitutively expressed β-subunit (also known as aryl hydrocarbon receptor nuclear translocator, ARNT) [36]. Under normoxic conditions, the α-subunit is rapidly hydroxylated by prolyl hydroxylase domain enzymes (PHD1-3) [37, 38] allowing recognition of the α-subunit by the von Hippel–Lindau (VHL) E3 ubiquitin ligase complex and subsequent proteosomal degradation [39].

Not surprisingly, HIFs contribute to normal lung development [40]. In the human fetal lung, all members of the HIF signaling pathway are expressed in the epithelium [41, 42]. HIF-2α and ARNT proteins are also localized in the mesenchyme [41, 42]. In vitro, HIF inhibition during early lung development, using antisense knockdown, decreases vascular development and lung branching morphogenesis in lung explants; exogenous VEGF only partially overcomes this arrest in lung development [43]. Conversely, conditional overexpression of HIF-1α in lung epithelium disrupts branching morphogenesis, impairs lung maturation, and causes pulmonary vascular abnormalities, including hemorrhage and increased lymphangiogenesis [44]. In contrast to the role of HIF during early lung development, there is relatively little information about HIF's function during postnatal alveolar growth, mainly because HIF deficiency is embryonically [45, 46] or immediately postnatally [47, 48] lethal. Cre-mediated deletion of HIF-1α in the developing lung epithelium results in neonatal respiratory distress associated with decreased surfactant production [48]. HIF-2α$^{-/-}$ mice die at birth due to respiratory distress and can be rescued by VEGF treatment [47]. These observations suggest a role for HIFs during lung maturation and surfactant synthesis. The temporal expression pattern of HIF-1α mRNA and HIF-2α mRNA peaking during the alveolar stage of rat lung development provides circumstantial evidence for HIF's role during postnatal lung development [49, 50]. HIF inhibition during the crucial period of alveolar development, by intratracheal dominant-negative adenovirus (dnHIF-1α)-mediated gene transfer or chetomin, a small molecule HIF inhibitor, efficiently decreased lung HIF-1α, HIF-2α, and VEGF expression and led to air space enlargement and arrested lung vascular growth [50].

PECAM-1: Administration of an anti-PECAM-1 antibody that inhibits endothelial cell migration, but not proliferation or survival in vitro, also impairs alveolarization in neonatal rats, without reducing endothelial cell content [51]. These data suggest that the loss of PECAM-1 function compromises postnatal lung development and provides evidence that inhibition of endothelial cell function, in contrast to loss of viable endothelial cells, inhibits alveolarization.

Evidence That Angiogenesis is Disrupted in Experimental Models of Arrested Alveolarization and in Human BPD and Emphysema

Link between angiogenesis and arrested alveolar growth/loss of alveoli: The proposed link between alveolarization and angiogenesis is suggested by the secondary abnormalities that occur in one process when the other is primarily affected. Alveolar hypoplasia and dysmorphic changes of the lung vasculature are consistent findings in experimental and human BPD. Already in 1959, Liebow had made similar observations in adult patients with centrilobular emphysema: alveolar septae were thin and nearly avascular, suggesting that a reduction in the blood supply of the small precapillary blood vessels might induce the disappearance of alveolar septae [1]. Lung vascular development is disrupted in all animal models mimicking BPD [52–55]. The first evidence that abnormal vascular development may contribute to neonatal lung disease came from autopsy studies showing reduced pulmonary microvascularization in infants dying from BPD [56]. A more recent postmortem study of newborns dying after short and prolonged durations of mechanical ventilation also quantified lung microvascular growth [57]. This study confirmed the reduction in vascular branching arteries, but interestingly lung PECAM-1 protein content (a marker of endothelial cells) was decreased in infants dying after brief ventilation, but was increased after prolonged ventilation [57]. These findings suggest a transient decrease in endothelial proliferation, followed by a brisk proliferative response, despite a reduction in vessel number. This observation suggests that dysmorphic lung vascular growth in BPD may not necessarily result simply from a reduction in the number of endothelial cells, emphasizing the need for more extensive studies in human and in animal models of BPD to better define the mechanisms that underlie early and the time-dependent sequence of events that precede the development of impaired distal lung structure.

Decreased VEGF and HIF signaling in BPD: Various animal models of impaired alveolar development also display abnormal lung vascular development [58–62]. Accordingly, animal and human studies show decreased expression of VEGF and its receptors in chronic newborn lung injury. Decreased VEGF expression is found in alveolar type II cells of newborn rabbits [63] or in newborn rat lungs [25] exposed to 100 % O_2. In the preterm baboon model of BPD, arrested platelet endothelial cellular adhesion molecule-1 (PECAM-1, a distal lung endothelial cell marker) expression and reduced capillary density are associated with lower VEGF and VEGFR-1 mRNA and protein expression seen at 125 vs. 140 day term animals [64]. Similar observations exist in chronically ventilated premature sheep [65] and in term mice [66]. Premature sheep with antenatal endotoxin exposure (mimicking chorioamnionitis, another risk factor for BPD) also display decreased lung VEGF expression [67]. In humans, similar observations have been made in some [68–70], but not all [71] studies. Infants developing BPD have lower VEGF levels than those surviving without BPD [70]. VEGF may participate in pulmonary repair after acute lung injury. In lung tissue from infants who died from BPD, the typical patterns of alveolar simplification with "dysmorphic" microvasculature are associated with reduced lung VEGF and VEGFR-1 mRNA and protein expression [68]. Likewise, there are lower VEGF levels in the tracheal aspirates of preterm infants dying from severe respiratory distress syndrome than survivors and in infants subsequently developing BPD, as compared to premature infants surviving without pulmonary complications [68].

Hypoxia is a major stimulus of VEGF expression [72]. Premature exposure of the developing lung to a hyperoxic environment may downregulate VEGF expression. Even ambient O_2 levels (21 %), i.e., premature birth per se, may interfere with normal lung vascular development [73]. Accordingly, HIF signaling is impaired in the lung of hyperoxic newborn rats [49, 50] and in large newborn animal models of ventilation-induced lung injury [74, 75] mimicking BPD.

Perturbation of other angiogenic growth factors provides further evidence for a link between disrupted angiogenesis and BPD: Despite its central role in vascular formation, VEGF works in concert with other factors, notably Angiopoietins (Ang) to stabilize the vascular wall. Unlike mouse embryos lacking VEGF or VEGFR-2, embryos lacking Ang1 or it receptor Tie2 develop a rather normal primary vasculature [76–78]. However, endothelial cells fail to associate appropriately with underlying support cells, which are the cells that provide the Ang1 protein that acts on endothelial Tie2 receptors [79]. While high concentrations of VEGF lead to immature, leaky, and hemorrhagic vessels, Ang1 generates vessels that are resistant to leak suggesting that Ang1 maximizes interactions between endothelial cells and their surrounding support cells and matrix [80]. These findings indicate that Ang1 is complementary to VEGF during vessel formation, acting at a later stage of angiogenesis to elicit vessel maturation and integrity. However, little is known about the role of Ang during normal alveolarization. In the baboon lung, Ang1 is mainly expressed in the septal mesenchymal cells, and Ang1 and its receptor Tie2 increase during lung development. Conversely, Tie-2 expression is decreased in lungs of ventilated baboons [64] and humans with BPD [81, 82]. The role of Ang2 during angiogenesis is less clear. Ang2 displays similarly high affinity for Tie2 and may act as a Tie2 antagonist. In lung epithelial cells, Ang2 expression is induced during hyperoxia [83]. Hyperoxia-induced oxidant injury, cell death, inflammation, permeability, and mortality are ameliorated in Ang2 knockout mice and in Ang2 siRNA-treated mice. Finally, Ang2 tracheal aspirate levels are increased in newborns that develop BPD. Likewise, chronically ventilated preterm lambs [84] and baboons [85] have decreased lung eNOS expression, suggesting that NO deficiency may contribute to the decreased alveolarization seen in these models of BPD.

Counter intuitively NF-κB, a transcription factor traditionally associated with inflammation, plays a unique protective role in the neonatal lung. Constitutive NF-κB expression is higher in neonatal lung and in primary pulmonary endothelial cells compared with the adult [86]. Inhibiting constitutive NF-κB activity in the neonatal pulmonary endothelial cells with either pharmacological inhibitors or RNA interference blocked pulmonary endothelial cells' survival, decreased proliferation, and impaired in vitro angiogenesis. NF-κB was found to be a direct regulator of VEGFR2. Accordingly, blocking NF-κB activity during the alveolar stage of lung development in mice induced the alveolar simplification reminiscent of BPD and significantly reduced pulmonary capillary density. This data suggest that disruption of NF-κB signaling may contribute to the pathogenesis of BPD.

The Forkhead Box (Fox) family of transcription factors plays important roles in regulating expression of genes involved in cellular proliferation and differentiation. Newborn foxf1(+/−) mice with diminished Foxf1 levels displayed defects in alveolarization and lung capillaries and died postnatally [21]. Interestingly, surviving newborn foxf1(+/−) mice exhibited increased pulmonary Foxf1 levels and normal adult lung morphology, suggesting that wild-type Foxf1 levels are required for lung development and function. However, when challenged with butylated hydroxytoluene-induced lung injury, foxf1(+/−) mice died from severe lung hemorrhage and this was associated with decreased pulmonary Foxf1 expression and increased alveolar endothelial cell apoptosis that disrupted capillary integrity [87]. Thus, sustained expression of Foxf1 is essential for normal lung repair and endothelial cell survival. FOXF1's crucial role in the formation of the vasculature was recently confirmed in mice with endothelial specific deletion of Foxf1 [88]. The clinical implication is that in humans, FOXF1 haploinsufficiency is associated with alveolar capillary dysplasia with misalignment of pulmonary veins, a rare, lethal developmental disorder of the lung [89].

Axonal guidance cues (*AGC*): are molecules that regulate neural network formation in the nervous system and the outgrowth of axons by acting as attractants, guiding neurons to their target or as repellents, creating exclusion zones that neurons avoid [90]. AGC are also involved in angiogenesis, cell migration, and early lung branching morphogenesis. Thus, AGC are appealing candidates in guiding also the outgrowth of secondary crests during alveolar development and repair [91]. Mice homozygous for the hypomorphic knockin allele ephrinB2$^{\Delta V/\Delta V}$ - encoding mutant ephrinB2 with a disrupted C-terminal PDZ interaction motif - die by 2 weeks of age and have enlarged air spaces, suggesting a crucial role for EphrinB2 during alveolar development [92]. EphrinB2 knockdown using intranasal siRNA during the postnatal stage of alveolar development in rats arrested alveolar and vascular growth. Conversely, hyperoxic exposed rats with enlarged air spaces and arrested lung vascular growth have decreased lung EphrinB2 and receptor EphB4 expression [93]. Similar observations have been obtained with semaphorins, another AGC family of molecules that play important neural and vascular roles [93–96].

Antiangiogenic factors: Endothelial monocyte activating polypeptide II (EMAP-II) and endostatin are potent natural angiogenesis inhibitors. The role of EMAP-II has been explored only at earlier stages of lung development, but its activation decreases neovascularization, interrupts lung branching morphogenesis, and decreases lung surfactant [97]. Endostatin was recently measured in tracheal aspirate fluid, and higher endostatin concentrations correlated with parameters reflecting lower lung maturity [98]. Endostatin may also contribute to aberrant epithelial repair in patients with pulmonary fibrosis that display elevated bronchoalveolar lavage endostatin levels [99]. These correlated with a number of elevated cytokines and impaired lung function in IPF.

It is very likely that other pro- and antiangiogenic growth factors play a role during lung angiogenesis and the balance between these growth factors is crucial for the coordinated assembly and remodeling of blood vessels during alveolar development. For example, hyperoxia increases the potent angiostatin, Pigment Epithelium-derived Factor (PEDF), in neonatal mice [100]. Conversely, hyperoxia reduced alveolarization in wilt-type mice, but not in PEDF (−/−) mice. Interestingly, VEGF stimulation of proliferation, migration, and capillary tube formation of MFLM-91U (a fetal mouse lung endothelial cell line) was inhibited by PEDF. MFLM-91U cells exposed to conditioned medium from E17 fetal mouse lung alveolar type II cells cultured in hyperoxia formed fewer capillary tubes than conditioned medium from type II cells cultured in room air. This effect was abolished by PEDF inhibition. These data support that interrupted alveolar development in BPD likely results from an altered balance between pro- and antiangiogenic factors.

Endoglin (CD105) is a hypoxia-inducible transforming growth factor-beta (TGF-β) co-receptor. Excessive lung TGF-β expression contributes to arrested alveolarization [101]. High levels of endoglin expression have been described in vascular endothelial cells in tissues undergoing active angiogenesis, such as in regenerating and inflamed tissues or tumors [102]. Lungs of short-term ventilated preterm infants showed significant upregulation of endoglin mRNA and protein levels, immunolocalized to the microvasculature compared to age-matched nonventilated control lungs [103]. Similar but more variable endoglin upregulation was noted in lungs of long-term ventilated infants with BPD. This was associated with decreased mRNA levels of VEGF, Ang1, and their respective receptors [103].

In summary, these data provide strong evidence that angiogenesis is necessary for alveolarization during normal lung development, and that an imbalance in angiogenic growth factors during a critical period of growth contributes to the late sequelae of BPD. These observations also provide the rationale for testing the therapeutic potential of pro-angiogenic growth factors to promote alveolar development.

Proof of Principle Experiments Indicating a Therapeutic Potential of Angiogenic Growth Factors to Prevent Arrested Alveolar Growth

Recombinant human VEGF (rhVEGF) treatment of newborn rats during or after exposure to hyperoxia enhances vessel growth and improves alveolarization [104, 105]. Likewise, postnatal intratracheal adenovirus-mediated VEGF gene therapy improves survival, promotes lung capillary formation, preserves alveolar development, and regenerates new alveoli in this same model of irreversible lung injury [25]. In both animal studies, constitutive VEGF expression induced immature and leaky capillaries and lung edema. However, combining lung VEGF and Ang1 (which promotes vascular maturation) gene transfer preserves alveolarization and enhances angiogenesis with more mature capillaries that are less permeable, reducing the vascular leakage seen in VEGF-induced capillaries [25].

These observations highlight the tightly orchestrated process of angiogenesis and points towards the need to closely recapitulate this process to warrant efficient and safe angiogenesis. The transcription factor HIF regulates several angiogenic growth factors. Accordingly, HIF activation via inhibition of prolyl hydroxylase domain-containing proteins prevents lung injury in the premature baboon model of BPD and supports further a potential role for angiogenic growth factor in promoting alveolar development [106]. Similarly, airway delivery of adenovirus-mediated HIF-1α therapy to chronic hyperoxic exposed neonatal rats, restored HIF-1α, eNOS, VEGF, VEGFR2, and Tie2 expression, and preserved and rescued alveolar growth and mature lung capillary formation [50].

Given that VEGF-induced angiogenesis is in part mediated by NO, some of these findings may explain the beneficial effects of early and prolonged low-dose inhaled NO seen in some clinical trials to prevent BPD [107–109], but not others

[110, 111]. Members of the NO-signaling pathway or other pulmonary vasodilators may exert similar beneficial effects [112–116].

Likewise, EphrinB2 preserved alveolar epithelial cell viability in O_2, decreased O_2-induced alveolar epithelial cell apoptosis, accelerated alveolar epithelial cell wound healing, and maintained lung microvascular endothelial cell viability, and proliferation and vascular network formation in vitro [117]. In vivo, treatment with intranasal EphrinB2 decreased alveolar epithelial and endothelial cell apoptosis, preserved alveolar and vascular growth in hyperoxic rats, and attenuated pulmonary hypertension [117].

Conclusions and Future Directions

These observations provide proof of concept for the crucial role of the lung vasculature in what is traditionally thought of as an air space disease and open new therapeutic avenues to protect or regenerate new alveoli. However, much more needs to be learned about the morphological changes and the mechanisms that regulate lung vascular development. For example, intussusceptive microvascular growth, a relatively recent described mode of blood vessel formation in the lung [11], may promote alveolar development and remodeling throughout adult life and thus be amenable to therapeutic modulation for lung regeneration if we can unravel its regulatory mechanisms. Recapitulating the tightly orchestrated process of angiogenesis, as demonstrated by therapeutic interventions with VEGF and HIF, will be crucial if this strategy is to be exploited for treatment purposes. In this respect, recent insight into stem cell biology has enabled the identification of endothelial progenitor cells and their reparative potential. Indeed, if angiogenic growth factors contribute to alveolar homeostasis, then vascular progenitor cells emerge as appealing candidates likely to be involved in the same mechanisms. The exciting perspectives of using vascular repair cells to treat lung diseases are explored in Chap. 17.

Acknowledgments B.T. is supported by the Canadian Institutes for Health Research (CIHR), the Ontario Thoracic Society, the Canadian Stem Cell Network, the Children's Hospital of Eastern Ontario Research Institute and the Ottawa Hospital Research Institute. M.Y. is supported in part by the Riley Children's Foundation.

References

1. Liebow AA (1959) Pulmonary emphysema with special reference to vascular changes. Am Rev Respir Dis 80:67–93
2. Burri PH, Dbaly J, Weibel ER (1974) The postnatal growth of the rat lung. I. Morphometry. Anat Rec 178:711–730

3. Burri PH, Moschopulos M (1992) Structural analysis of fetal rat lung development. Anat Rec 234:399–418
4. Parera MC, van Dooren M, van Kempen M, de Krijger R, Grosveld F, Tibboel D, Rottier R (2005) Distal angiogenesis: a new concept for lung vascular morphogenesis. Am J Physiol Lung Cell Mol Physiol 288:L141–L149
5. deMello DE, Sawyer D, Galvin N, Reid LM (1997) Early fetal development of lung vasculature. Am J Respir Cell Mol Biol 16:568–581
6. Anderson-Berry A, O'Brien EA, Bleyl SB, Lawson A, Gundersen N, Ryssman D et al (2005) Vasculogenesis drives pulmonary vascular growth in the developing chick embryo. Dev Dyn 233:145–153
7. Yamamoto Y, Baldwin HS, Prince LS (2012) Endothelial differentiation by multipotent fetal mouse lung mesenchymal cells. Stem Cells Dev 21:1455–1465
8. Hall SM, Hislop AA, Pierce CM, Haworth SG (2000) Prenatal origins of human intrapulmonary arteries: formation and smooth muscle maturation. Am J Respir Cell Mol Biol 23:194–203
9. Peng T, Morrisey EE (2013) Development of the pulmonary vasculature: current understanding and concepts for the future. Pulm Circ 3:176–178
10. Burri P (1999) Lung development and pulmonary angiogenesis. In: Gaultier C, Bourbon JR, Post M (eds) Lung Development. Oxford University Press, New York, pp 122–151
11. Djonov V, Schmid M, Tschanz SA, Burri PH (2000) Intussusceptive angiogenesis: its role embryonic vascular network formation. Circ Res 86:286–292
12. Ackermann M, Houdek JP, Gibney BC, Ysasi A, Wagner W, Belle J et al (2014) Sprouting and intussusceptive angiogenesis in postpneumonectomy lung growth: mechanisms of alveolar neovascularization. Angiogenesis 17:541–551
13. Neufeld G, Cohen T, Gengrinovitch S, Poltorak Z (1999) Vascular endothelial growth factor (VEGF) and its receptors. Faseb J 13:9–22
14. Ferrara N, Carver-Moore K, Chen H, Dowd M, Lu L, O'Shea KS et al (1996) Heterozygous embryonic lethality induced by targeted inactivation of the VEGF gene. Nature 380:439–442
15. Carmeliet P, Ferreira V, Breier G, Pollefeyt S, Kieckens L, Gertsenstein M et al (1996) Abnormal blood vessel development and lethality in embryos lacking a single VEGF allele. Nature 380:435–439
16. Fong GH, Rossant J, Gertsenstein M, Breitman ML (1995) Role of the Flt-1 receptor tyrosine kinase in regulating the assembly of vascular endothelium. Nature 376:66–70
17. Shalaby F, Rossant J, Yamaguchi TP, Gertsenstein M, Wu XF, Breitman ML, Schuh AC (1995) Failure of blood-island formation and vasculogenesis in Flk-1-deficient mice. Nature 376:62–66
18. Gerber HP, Hillan KJ, Ryan AM, Kowalski J, Keller GA, Rangell L et al (1999) VEGF is required for growth and survival in neonatal mice. Development 126:1149–1159
19. Healy AM, Morgenthau L, Zhu X, Farber HW, Cardoso WV (2000) VEGF is deposited in the subepithelial matrix at the leading edge of branching airways and stimulates neovascularization in the murine embryonic lung. Dev Dyn 219:341–352
20. Gebb SA, Shannon JM (2000) Tissue interactions mediate early events in pulmonary vasculogenesis. Dev Dyn 217:159–169
21. Kalinichenko VV, Lim L, Stolz DB, Shin B, Rausa FM, Clark J et al (2001) Defects in pulmonary vasculature and perinatal lung hemorrhage in mice heterozygous null for the Forkhead Box f1 transcription factor. Dev Biol 235:489–506
22. Ng YS, Rohan R, Sunday ME, Demello DE, D'Amore PA (2001) Differential expression of VEGF isoforms in mouse during development and in the adult. Dev Dyn 220:112–121
23. Galambos C, Ng YS, Ali A, Noguchi A, Lovejoy S, D'Amore PA, DeMello DE (2002) Defective pulmonary development in the absence of heparin-binding vascular endothelial growth factor isoforms. Am J Respir Cell Mol Biol 27:194–203

24. Jakkula M, Le Cras TD, Gebb S, Hirth KP, Tuder RM, Voelkel NF, Abman SH (2000) Inhibition of angiogenesis decreases alveolarization in the developing rat lung. Am J Physiol Lung Cell Mol Physiol 279:L600–L607
25. Thebaud B, Ladha F, Michelakis ED, Sawicka M, Thurston G, Eaton F et al (2005) Vascular endothelial growth factor gene therapy increases survival, promotes lung angiogenesis, and prevents alveolar damage in hyperoxia-induced lung injury: evidence that angiogenesis participates in alveolarization. Circulation 112:2477–2486
26. Zhao L, Wang K, Ferrara N, Vu TH (2005) Vascular endothelial growth factor co-ordinates proper development of lung epithelium and vasculature. Mech Dev 122:877–886
27. Kasahara Y, Tuder RM, Taraseviciene-Stewart L, Le Cras TD, Abman S, Hirth PK et al (2000) Inhibition of VEGF receptors causes lung cell apoptosis and emphysema. J Clin Invest 106:1311–1319
28. Akeson AL, Cameron JE, Le Cras TD, Whitsett JA, Greenberg JM (2005) Vascular endothelial growth factor-A induces prenatal neovascularization and alters bronchial development in mice. Pediatr Res 57:82–88
29. Le Cras TD, Spitzmiller RE, Albertine KH, Greenberg JM, Whitsett JA, Akeson AL (2004) VEGF causes pulmonary hemorrhage, hemosiderosis, and air space enlargement in neonatal mice. Am J Physiol Lung Cell Mol Physiol 287:L134–L142
30. Yamamoto H, Yun EJ, Gerber HP, Ferrara N, Whitsett JA, Vu TH (2007) Epithelial-vascular cross talk mediated by VEGF-A and HGF signaling directs primary septae formation during distal lung morphogenesis. Dev Biol 308:44–53
31. Abman S, Kinsella J, Mercier J (1999) Nitric oxide and endothelin in the developing pulmonary circulation: physiologic and clinical implications. In: Gaultier C, Bourbon JR, Post M (eds) Lung Development. Oxford University Press, New York, pp 196–202
32. Balasubramaniam V, Maxey AM, Morgan DB, Markham NE, Abman SH (2006) Inhaled NO restores lung structure in eNOS-deficient mice recovering from neonatal hypoxia. Am J Physiol Lung Cell Mol Physiol 291:L119–L127
33. Han RN, Babaei S, Robb M, Lee T, Ridsdale R, Ackerley C et al (2004) Defective lung vascular development and fatal respiratory distress in endothelial NO synthase-deficient mice: a model of alveolar capillary dysplasia? Circ Res 94:1115–1123
34. Balasubramaniam V, Tang JR, Maxey A, Plopper CG, Abman SH (2003) Mild hypoxia impairs alveolarization in the endothelial nitric oxide synthase (eNOS) deficient mouse. Am J Physiol Lung Cell Mol Physiol 284:L964–L971
35. Semenza GL, Wang GL (1992) A nuclear factor induced by hypoxia via de novo protein synthesis binds to the human erythropoietin gene enhancer at a site required for transcriptional activation. Mol Cell Biol 12:5447–5454
36. Semenza GL (2012) Hypoxia-inducible factors in physiology and medicine. Cell 148:399–408
37. Bruick RK, McKnight SL (2001) A conserved family of prolyl-4-hydroxylases that modify HIF. Science 294:1337–1340
38. Majmundar AJ, Wong WJ, Simon MC (2010) Hypoxia-inducible factors and the response to hypoxic stress. Mol Cell 40:294–309
39. Maxwell PH, Wiesener MS, Chang GW, Clifford SC, Vaux EC, Cockman ME et al (1999) The tumour suppressor protein VHL targets hypoxia-inducible factors for oxygen-dependent proteolysis. Nature 399:271–275
40. Shimoda LA, Semenza GL (2011) HIF and the lung role of hypoxia-inducible factors in pulmonary development and disease. Am J Respir Crit Care Med 183:152–156
41. Groenman F, Rutter M, Caniggia I, Tibboel D, Post M (2007) Hypoxia-inducible factors in the first trimester human lung. J Histochem Cytochem 55:355–363
42. Rajatapiti P, van der Horst IW, de Rooij JD, Tran MG, Maxwell PH, Tibboel D et al (2008) Expression of hypoxia-inducible factors in normal human lung development. Pediatr Dev Pathol 11:193–199

43. van Tuyl M, Liu J, Wang J, Kuliszewski M, Tibboel D, Post M (2005) Role of oxygen and vascular development in epithelial branching morphogenesis of the developing mouse lung. Am J Physiol Lung Cell Mol Physiol 288:L167–L178
44. Bridges JP, Lin S, Ikegami M, Shannon JM (2012) Conditional hypoxia inducible factor-1alpha induction in embryonic pulmonary epithelium impairs maturation and augments lymphangiogenesis. Dev Biol 362:24–41
45. Iyer NV, Kotch LE, Agani F, Leung SW, Laughner E, Wenger RH et al (1998) Cellular and developmental control of O_2 homeostasis by hypoxia- inducible factor 1 alpha. Genes Dev 12:149–162
46. Ryan HE, Lo J, Johnson RS (1998) HIF-1 alpha is required for solid tumor formation and embryonic vascularization. EMBO J 17:3005–3015
47. Compernolle V, Brusselmans K, Acker T, Hoet P, Tjwa M, Beck H et al (2002) Loss of HIF-2alpha and inhibition of VEGF impair fetal lung maturation, whereas treatment with VEGF prevents fatal respiratory distress in premature mice. Nat Med 8:702–710
48. Saini Y, Harkema JR, LaPres JJ (2008) HIF1alpha is essential for normal intrauterine differentiation of alveolar epithelium and surfactant production in the newborn lung of mice. J Biol Chem 283:33650–33657
49. Hosford GE, Olson DM (2003) Effects of hyperoxia on VEGF, its receptors, and HIF-2alpha in the newborn rat lung. Am J Physiol Lung Cell Mol Physiol 285:L161–L168
50. Vadivel A, Alphonse RS, Etches N, van Haaften T, Collins JJ, O'Reilly M et al (2014) Hypoxia-inducible factors promote alveolar development and regeneration. Am J Respir Cell Mol Biol 50:96–105
51. DeLisser HM, Helmke BP, Cao G, Egan PM, Taichman D, Fehrenbach M et al (2006) Loss of PECAM-1 function impairs alveolarization. J Biol Chem 281:8724–8731
52. D'Angio CT, Ryan RM (2014) Animal models of bronchopulmonary dysplasia. The preterm and term rabbit models. Am J Physiol Lung Cell Mol Physiol 307:L959–L969
53. Berger J, Bhandari V (2014) Animal models of bronchopulmonary dysplasia. The term mouse models. Am J Physiol Lung Cell Mol Physiol 307:L936–L947
54. O'Reilly M, Thebaud B (2014) Animal models of bronchopulmonary dysplasia. The term rat models. Am J Physiol Lung Cell Mol Physiol 307:L948–L958
55. Yoder BA, Coalson JJ (2014) Animal models of bronchopulmonary dysplasia. The preterm baboon models. Am J Physiol Lung Cell Mol Physiol 307:L970–L977
56. Tomashefski JF Jr, Oppermann HC, Vawter GF, Reid LM (1984) Bronchopulmonary dysplasia: a morphometric study with emphasis on the pulmonary vasculature. Pediatr Pathol 2:469–487
57. De Paepe ME, Mao Q, Powell J, Rubin SE, DeKoninck P, Appel N et al (2006) Growth of pulmonary microvasculature in ventilated preterm infants. Am J Respir Crit Care Med 173:204–211
58. Randell SH, Mercer RR, Young SL (1990) Neonatal hyperoxia alters the pulmonary alveolar and capillary structure of 40-day-old rats. Am J Pathol 136:1259–1266
59. Roberts RJ, Weesner KM, Bucher JR (1983) Oxygen-induced alterations in lung vascular development in the newborn rat. Pediatr Res 17:368–375
60. Wilson WL, Mullen M, Olley PM, Rabinovitch M (1985) Hyperoxia-induced pulmonary vascular and lung abnormalities in young rats and potential for recovery. Pediatr Res 19:1059–1067
61. Shaffer SG, O'Neill D, Bradt SK, Thibeault DW (1987) Chronic vascular pulmonary dysplasia associated with neonatal hyperoxia exposure in the rat. Pediatr Res 21:14–20
62. Han RN, Buch S, Tseu I, Young J, Christie NA, Frndova H et al (1996) Changes in structure, mechanics, and insulin-like growth factor-related gene expression in the lungs of newborn rats exposed to air or 60% oxygen. Pediatr Res 39:921–929
63. Maniscalco WM, Watkins RH, D'Angio CT, Ryan RM (1997) Hyperoxic injury decreases alveolar epithelial cell expression of vascular endothelial growth factor (VEGF) in neonatal rabbit lung. Am J Respir Cell Mol Biol 16:557–567

64. Maniscalco WM, Watkins RH, Pryhuber GS, Bhatt A, Shea C, Huyck H (2002) Angiogenic factors and alveolar vasculature: development and alterations by injury in very premature baboons. Am J Physiol Lung Cell Mol Physiol 282:L811–L823
65. Bland RD, Xu L, Ertsey R, Rabinovitch M, Albertine KH, Wynn KA et al (2007) Dysregulation of pulmonary elastin synthesis and assembly in preterm lambs with chronic lung disease. Am J Physiol Lung Cell Mol Physiol 292:L1370–L1384
66. Bland RD, Mokres LM, Ertsey R, Jacobson BE, Jiang S, Rabinovitch M et al (2007) Mechanical ventilation with 40% oxygen reduces pulmonary expression of genes that regulate lung development and impairs alveolar septation in newborn mice. Am J Physiol Lung Cell Mol Physiol 293:L1099–L1110
67. Kallapur SG, Bachurski CJ, Le Cras TD, Joshi SN, Ikegami M, Jobe AH (2004) Vascular changes after intra-amniotic endotoxin in preterm lamb lungs. Am J Physiol Lung Cell Mol Physiol 287:L1178–L1185
68. Lassus P, Turanlahti M, Heikkila P, Andersson LC, Nupponen I, Sarnesto A, Andersson S (2001) Pulmonary vascular endothelial growth factor and Flt-1 in fetuses, in acute and chronic lung disease, and in persistent pulmonary hypertension of the newborn. Am J Respir Crit Care Med 164:1981–1987
69. Lassus P, Ristimaki A, Ylikorkala O, Viinikka L, Andersson S (1999) Vascular endothelial growth factor in human preterm lung. Am J Respir Crit Care Med 159:1429–1433
70. Bhatt AJ, Pryhuber GS, Huyck H, Watkins RH, Metlay LA, Maniscalco WM (2001) Disrupted pulmonary vasculature and decreased vascular endothelial growth factor, Flt-1, and TIE-2 in human infants dying with bronchopulmonary dysplasia. Am J Respir Crit Care Med 164:1971–1980
71. D'Angio CT, Maniscalco WM, Ryan RM, Avissar NE, Basavegowda K, Sinkin RA (1999) Vascular endothelial growth factor in pulmonary lavage fluid from premature infants: effects of age and postnatal dexamethasone. Biol Neonate 76:266–273
72. Shweiki D, Itin A, Soffer D, Keshet E (1992) Vascular endothelial growth factor induced by hypoxia may mediate hypoxia-initiated angiogenesis. Nature 359:843–845
73. Hjalmarson O, Sandberg K (2002) Abnormal lung function in healthy preterm infants. Am J Respir Crit Care Med 165:83–87
74. Asikainen TM, Ahmad A, Schneider BK, White CW (2005) Effect of preterm birth on hypoxia-inducible factors and vascular endothelial growth factor in primate lungs. Pediatr Pulmonol 40:538–546
75. Grover TR, Asikainen TM, Kinsella JP, Abman SH, White CW (2007) Hypoxia-inducible factors HIF-1alpha and HIF-2alpha are decreased in an experimental model of severe respiratory distress syndrome in preterm lambs. Am J Physiol Lung Cell Mol Physiol 292:L1345–L1351
76. Sato TN, Tozawa Y, Deutsch U, Wolburg-Buchholz K, Fujiwara Y, Gendron-Maguire M et al (1995) Distinct roles of the receptor tyrosine kinases Tie-1 and Tie-2 in blood vessel formation. Nature 376:70–74
77. Suri C, Jones PF, Patan S, Bartunkova S, Maisonpierre PC, Davis S et al (1996) Requisite role of angiopoietin-1, a ligand for the TIE2 receptor, during embryonic angiogenesis. Cell 87:1171–1180
78. Dumont DJ, Gradwohl G, Fong GH, Puri MC, Gertsenstein M, Auerbach A, Breitman ML (1994) Dominant-negative and targeted null mutations in the endothelial receptor tyrosine kinase, tek, reveal a critical role in vasculogenesis of the embryo. Genes Dev 8:1897–1909
79. Suri C, McClain J, Thurston G, McDonald DM, Zhou H, Oldmixon EH et al (1998) Increased vascularization in mice overexpressing angiopoietin-1. Science 282:468–471
80. Thurston G, Suri C, Smith K, McClain J, Sato TN, Yancopoulos GD, McDonald DM (1999) Leakage-resistant blood vessels in mice transgenically overexpressing angiopoietin-1. Science 286:2511–2514
81. Bhatt AJ, Amin SB, Chess PR, Watkins RH, Maniscalco WM (2000) Expression of vascular endothelial growth factor and Flk-1 in developing and glucocorticoid-treated mouse lung. Pediatr Res 47:606–613

82. Thomas W, Seidenspinner S, Kramer BW, Kawczynska-Leda N, Chmielnicka-Kopaczyk M, Marx A et al (2009) Airway concentrations of angiopoietin-1 and endostatin in ventilated extremely premature infants are decreased after funisitis and unbalanced with bronchopulmonary dysplasia/death. Pediatr Res 65:468–473
83. Bhandari V, Choo-Wing R, Lee CG, Zhu Z, Nedrelow JH, Chupp GL et al (2006) Hyperoxia causes angiopoietin 2-mediated acute lung injury and necrotic cell death. Nat Med 12:1286–1293
84. MacRitchie AN, Albertine KH, Sun J, Lei PS, Jensen SC, Freestone AA et al (2001) Reduced endothelial nitric oxide synthase in lungs of chronically ventilated preterm lambs. Am J Physiol Lung Cell Mol Physiol 281:L1011–L1020
85. Afshar S, Gibson LL, Yuhanna IS, Sherman TS, Kerecman JD, Grubb PH et al (2003) Pulmonary NO synthase expression is attenuated in a fetal baboon model of chronic lung disease. Am J Physiol Lung Cell Mol Physiol 284:L749–L758
86. Iosef C, Alastalo TP, Hou Y, Chen C, Adams ES, Lyu SC et al (2012) Inhibiting NF-kappaB in the developing lung disrupts angiogenesis and alveolarization. Am J Physiol Lung Cell Mol Physiol 302:L1023–L1036
87. Kalinichenko VV, Zhou Y, Shin B, Stolz DB, Watkins SC, Whitsett JA, Costa RH (2002) Wild-type levels of the mouse Forkhead Box f1 gene are essential for lung repair. Am J Physiol Lung Cell Mol Physiol 282:L1253–L1265
88. Ren X, Ustiyan V, Pradhan A, Cai Y, Havrilak JA, Bolte CS et al (2014) FOXF1 transcription factor is required for formation of embryonic vasculature by regulating VEGF signaling in endothelial cells. Circ Res 115:709–720
89. Stankiewicz P, Sen P, Bhatt SS, Storer M, Xia Z, Bejjani BA et al (2009) Genomic and genic deletions of the FOX gene cluster on 16q24.1 and inactivating mutations of FOXF1 cause alveolar capillary dysplasia and other malformations. Am J Hum Genet 84:780–791
90. Carmeliet P, Tessier-Lavigne M (2005) Common mechanisms of nerve and blood vessel wiring. Nature 436:193–200
91. Prodhan P, Kinane TB (2002) Developmental paradigms in terminal lung development. Bioessays 24:1052–1059
92. Wilkinson GA, Schittny JC, Reinhardt DP, Klein R (2008) Role for ephrinB2 in postnatal lung alveolar development and elastic matrix integrity. Dev Dyn 237:2220–2234
93. Vadivel A, Alphonse RS, Collins JJ, van Haaften T, O'Reilly M, Eaton F, Thebaud B (2013) The axonal guidance cue semaphorin 3C contributes to alveolar growth and repair. PLoS One 8, e67225
94. Becker PM, Tran TS, Delannoy MJ, He C, Shannon JM, McGrath-Morrow S (2011) Semaphorin 3A contributes to distal pulmonary epithelial cell differentiation and lung morphogenesis. PLoS One 6, e27449
95. Joza S, Wang J, Fox E, Hillman V, Ackerley C, Post M (2012) Loss of semaphorin-neuropilin-1 signaling causes dysmorphic vascularization reminiscent of alveolar capillary dysplasia. Am J Pathol 181:2003–2017
96. Joza S, Wang J, Tseu I, Ackerley C, Post M (2013) Fetal, but not postnatal, deletion of semaphorin-neuropilin-1 signaling affects murine alveolar development. Am J Respir Cell Mol Biol 49:627–636
97. Schwarz MA, Zhang F, Gebb S, Starnes V, Warburton D (2000) Endothelial monocyte activating polypeptide II inhibits lung neovascularization and airway epithelial morphogenesis. Mech Dev 95:123–132
98. Janer J, Andersson S, Haglund C, Lassus P (2007) Pulmonary endostatin perinatally and in lung injury of the newborn infant. Pediatrics 119:e241–e246
99. Richter AG, McKeown S, Rathinam S, Harper L, Rajesh P, McAuley DF et al (2009) Soluble endostatin is a novel inhibitor of epithelial repair in idiopathic pulmonary fibrosis. Thorax 64:156–161

100. Chetty A, Bennett M, Dang L, Nakamura D, Cao GJ, Mujahid S et al (2015) Pigment epithelium-derived factor mediates impaired lung vascular development in neonatal hyperoxia. Am J Respir Cell Mol Biol 52:295–303
101. Nakanishi H, Sugiura T, Streisand JB, Lonning SM, Roberts JD Jr (2007) TGF-beta-neutralizing antibodies improve pulmonary alveologenesis and vasculogenesis in the injured newborn lung. Am J Physiol Lung Cell Mol Physiol 293:L151–L161
102. ten Dijke P, Goumans MJ, Pardali E (2008) Endoglin in angiogenesis and vascular diseases. Angiogenesis 11:79–89
103. De Paepe ME, Patel C, Tsai A, Gundavarapu S, Mao Q (2008) Endoglin (CD105) up-regulation in pulmonary microvasculature of ventilated preterm infants. Am J Respir Crit Care Med 178:180–187
104. Kunig A, Balasubramaniam V, Markham NE, Seedorf G, Gien J, Abman SH (2006) Recombinant human VEGF treatment transiently increases lung edema but enhances lung structure after neonatal hyperoxia. Am J Physiol Lung Cell Mol Physiol 289:L529–L535
105. Kunig AM, Balasubramaniam V, Markham NE, Morgan D, Montgomery G, Grover TR, Abman SH (2005) Recombinant human VEGF treatment enhances alveolarization after hyperoxic lung injury in neonatal rats. Am J Physiol Lung Cell Mol Physiol 289:L529–L535
106. Asikainen TM, Chang LY, Coalson JJ, Schneider BK, Waleh NS, Ikegami M et al (2006) Improved lung growth and function through hypoxia-inducible factor in primate chronic lung disease of prematurity. FASEB J 20:1698–1700
107. Ballard RA, Truog WE, Cnaan A, Martin RJ, Ballard PL, Merrill JD et al (2006) Inhaled nitric oxide in preterm infants undergoing mechanical ventilation. N Engl J Med 355:343–353
108. Kinsella JP, Cutter GR, Walsh WF, Gerstmann DR, Bose CL, Hart C et al (2006) Early inhaled nitric oxide therapy in premature newborns with respiratory failure. N Engl J Med 355:354–364
109. Schreiber MD, Gin-Mestan K, Marks JD, Huo D, Lee G, Srisuparp P (2003) Inhaled nitric oxide in premature infants with the respiratory distress syndrome. N Engl J Med 349:2099–2107
110. Kinsella JP, Cutter GR, Steinhorn RH, Nelin LD, Walsh WF, Finer NN, Abman SH (2014) Noninvasive inhaled nitric oxide does not prevent bronchopulmonary dysplasia in premature newborns. J Pediatr 165:1104–1108, e1101
111. Mercier JC, Hummler H, Durrmeyer X, Sanchez-Luna M, Carnielli V, Field D et al (2010) Inhaled nitric oxide for prevention of bronchopulmonary dysplasia in premature babies (EUNO): a randomised controlled trial. Lancet 376:346–354
112. Luong C, Rey-Perra J, Vadivel A, Gilmour G, Sauve Y, Koonen D et al (2011) Antenatal sildenafil treatment attenuates pulmonary hypertension in experimental congenital diaphragmatic hernia. Circulation 123:2120–2131
113. Vadivel A, Abozaid S, van Haaften T, Sawicka M, Eaton F, Chen M, Thebaud B (2010) Adrenomedullin promotes lung angiogenesis, alveolar development, and repair. Am J Respir Cell Mol Biol 43:152–160
114. Vadivel A, Alphonse RS, Ionescu L, Machado DS, O'Reilly M, Eaton F et al (2014) Exogenous hydrogen sulfide (H2S) protects alveolar growth in experimental O_2-induced neonatal lung injury. PLoS One 9, e90965
115. Vadivel A, Aschner JL, Rey-Parra GJ, Magarik J, Zeng H, Summar M et al (2010) L-citrulline attenuates arrested alveolar growth and pulmonary hypertension in oxygen-induced lung injury in newborn rats. Pediatr Res 68:519–525
116. Ladha F, Bonnet S, Eaton F, Hashimoto K, Korbutt G, Thebaud B (2005) Sildenafil improves alveolar growth and pulmonary hypertension in hyperoxia-induced lung injury. Am J Respir Crit Care Med 172:750–756
117. Vadivel A, van Haaften T, Alphonse RS, Rey-Parra GJ, Ionescu L, Haromy (2012) Critical role of the axonal guidance cue EphrinB2 in lung growth, angiogenesis, and repair. Am J Respir Crit Care Med 185:564–574

Chapter 4
Neural Regulation of Lung Development

Alan J. Burns, Lucy J. Freem, and Jean-Marie Delalande

Early Lung Development

The trachea and lungs are endoderm-derived organs that develop as outgrowths of the foregut. Early in development the foregut is patterned into regions with different fates by a number of morphogenic signals that are expressed in anterior–posterior and dorsoventral gradients in order to set up overlapping regions of transcription factor expression (reviewed in [1–4]). For example, Nkx2.1 is a transcription factor expressed in the ventral endoderm and is necessary for lung endoderm specification. The Nkx2.1 domain of expression overlaps with that of the rostrally expressed transcription factor Sox2 and with that of the caudally expressed transcription factor Hex1 [5], so this combination of transcription factors characterizes the region that will become the lungs. In addition to transcription factors, several morphogenic signals, including retinoic acid and Sonic hedgehog, ensure correct patterning of

A.J. Burns, Ph.D. (✉)
Stem Cells and Regenerative Medicine, Birth Defects Research Centre, UCL Institute of Child Health, 30 Guilford Street, London WC1N 1EH, UK

Department of Clinical Genetics, Erasmus University Medical Center, Rotterdam, The Netherlands
e-mail: alan.burns@ucl.ac.uk

L.J. Freem
Stem Cells and Regenerative Medicine, Birth Defects Research Centre, UCL Institute of Child Health, 30 Guilford Street, London WC1N 1EH, UK

Division of Developmental Biology, The Roslin Institute, University of Edinburgh, Easter Bush Campus, Midlothian, UK

J.-M. Delalande
Stem Cells and Regenerative Medicine, Birth Defects Research Centre, UCL Institute of Child Health, 30 Guilford Street, London WC1N 1EH, UK

Centre for Digestive Diseases, Barts and The London School of Medicine and Dentistry, Queen Mary University of London, London, UK

I. Bertoncello (ed.), *Stem Cells in the Lung*, Stem Cell Biology and Regenerative Medicine, DOI 10.1007/978-3-319-21082-7_4

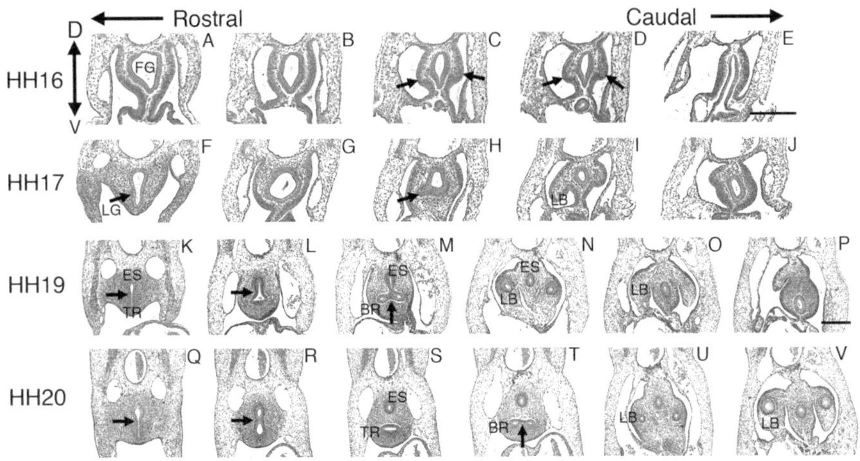

Fig. 4.1 Development of the lung buds in the early chick embryo. At HH stage 17 (day 2.5) the laryngotracheal groove forms in the ventral foregut (*F*) and deepens over time (*K*), eventually separating from the gut to form the trachea (*S*). The trachea bifurcates caudally to produce the bronchi (*M*, *T*). The lung buds initially form as swellings in the wall of the foregut (*D*), which grow outwards over time (*I*, *N*, *V*). *FG* foregut, *LG* Laryngotracheal groove, *ES* esophagus, *BR* bronchi, *LB* lung buds. From [11]

lung development during organogenesis [6–8]. The morphogen Sonic hedgehog (Shh), its receptor Patched (Ptc) and the downstream zinc-finger transcription factors Gli1, 2, and 3 are all members of a pathway necessary for the correct dorsoventral patterning of the trachea and lung buds. Shh signalling is required for the partitioning of the esophagus from the trachea and bronchi [9] and for normal lung mesenchyme proliferation [7, 10].

During early embryogenesis (embryonic day 9, E9, in mice, E2.5 in chick and week 5 in human development) [11, 12], the trachea and lung buds arise by evaginating from a presumptive respiratory territory in the ventral wall of the primitive foregut (reviewed in [13]). Following initial evagination, a laryngotracheal groove forms in the ventral anterior foregut and deepens to separate from the gut and form the trachea (E9.5–E12.5 in mouse). The trachea then bifurcates caudally to produce the bronchi. Two swellings in the gut wall develop, forming the early lung buds. The lumen of the trachea then separates from that of the esophagus by the narrowing and pinching together of the two sides of the groove (see Fig. 4.1). This process occurs in a caudal to rostral direction. The bronchi are formed as the epithelial tubules extend out into the lung buds. After separation of the trachea and bronchi from the gut, the epithelial tubules within the mammalian lung that will form the airways undergo successive branching events to generate the respiratory tree (E12.5–E16.5 in mouse). These epithelial tubules are surrounded by a network of capillaries, the early vascular plexus. The larger, anterior, conducting airways branch successively to give rise to smaller airways and eventually join with the mesenchyme at the alveoli, the site of respiratory exchange. The lung mesenchyme that makes up the lung buds grows and divides into lobes as the lung develops [4].

The branched network that forms during lung development eventually comprises millions of airways arranged in intricate patterns, maximizing the surface exchange necessary for respiration. The branching pattern of the airway epithelium into the surrounding mesenchyme is complex but remarkably stereotyped, and is generated by three geometrically simple modes of branching (domain branching, planar and orthogonal bifurcation), that occur under the influence of local parameters [14]. The genetic basis of the branching program is still not elucidated, although it has been shown that multiple signaling pathways, including the Wnt, bone morphogenetic protein (Bmp), and fibroblast growth factor (Fgf) pathways, are implicated in its regulation [4, 14].

Shortly after the airway branching process begins, differentiation of the airway smooth muscle (ASM) and cartilage, that will give both the rigidity and elasticity critical for effective air exchange in the lungs, commences. ASM is one of the main targets for lung innervation (see section below) and is first detected at E10.5 in mouse [15, 16] and by week 6 of human development [12]. In E10.5 mouse embryos, ASM differentiation is first detected in the mesenchyme overlying the carina (the point where the trachea splits into the main bronchi). While differentiating during the same time window, ASM and cartilage are initially localized to separate domains, with the cartilage differentiating in a much wider domain along the length of the trachea and main bronchi [16]. Recent studies show that, despite previous assumptions, stalk mesenchyme of the epithelial bud does not contribute to ASM, but that ASM progenitors are located in a niche just ahead of branch tips [17, 18]. These cells subsequently migrate along the tip to reach the stalk and begin differentiating into ASM [18]. By E11.5 in the mouse, the ASM domain expands proximally along the trachea and distally along the main bronchi. At later stages, ASM and cartilage develop a complementary spatial pattern of differentiation. Interestingly, a loss of either ASM (by conditional deletion of the smooth muscle differentiation gene *Srf*) or cartilage (by conditional deletion of the cartilage differentiation gene *Sox9*) results in an increase in cell number and in the domain of the remaining tissue lineage, suggesting a communication feedback mechanism between ASM and cartilage [16].

Development of Lung Innervation

During the early development of the lungs, a complex neuronal network is laid out. The neurons that contribute to this network are generally divided into two categories: extrinsic neurons, which have their cell bodies outside the respiratory tract (i.e., parasympathetic innervation from the nodose and jugular ganglia in the brain stem, and sympathetic innervation from dorsal root and thoracic ganglia), and the intrinsic innervation composed of local parasympathetic ganglia, as well as true intrinsic ganglia [15, 19–21]. This nerve network is formed from an intricate combination of migrating neural crest cells that differentiate into neurons, and extrinsic axonal outgrowths, the true complexity of which is yet to be fully elucidated. As outlined below, intrinsic neurons have been implicated in important roles in early stages of lung development, whereas extrinsic nerves represent the main regulatory neural influence at later stages, in particular, after birth.

Development of Intrinsic Lung Innervation

Intrinsic lung neurons, whose cell bodies reside within the lungs, are derived from neural crest cells (NCC), a highly migratory, transitory cell population that gives rise to a wide diversity of cell types throughout the embryo including all neurons and glia of the peripheral nervous system [22]. Their development begins with the migration of NCC from the dorsal aspect of the neural tube into and along the embryonic foregut. Fate mapping studies in chick embryos demonstrated that NCC arising in the vagal (hindbrain) region of the neural tube, adjacent to somites 1–7, migrate in a rostrocaudal direction along the entire length of the gut to form the enteric nervous system (ENS), the intrinsic innervation of the gastrointestinal tract [23], and reviewed in [24–26]. As these vagal NCC migrate along the foregut, a subpopulation of cells moves tangentially from the foregut into the adjacent lung buds, a process revealed by NCC tracing in the chick [11] and mouse [19], and by antibody labelling of NCC in the human embryo [12]. This subpopulation of NCC therefore appears to undergo a two-step migration from one organ to another: first from the vagal region of the neural tube into and along the gut, and then tangentially from the (fore)gut into the lungs (Fig. 4.2). Concomitant with this early vagal NCC migration, the trachea and lung buds are in the process of separating from the foregut, so the lung buds and gut endoderm are physically connected, facilitating the direct migration of NCC into the lung buds. At these early stages of lung bud development, NCC are present in relatively high numbers compared to the size of the buds (Fig. 4.2f–i). However, this appears to be a transitory situation, as the lung buds subsequently increase in size dramatically, decreasing this ratio.

Within the chick lung, vagal NCC have been shown to differentiate to form neurons that are present adjacent to the future airways and are closely associated with ASM [11]. These intrinsic lung neurons appear to aggregate and form the airway parasympathetic ganglia. Lung innervation studies in human [12, 29], pig [30] and mouse [15, 20] demonstrated that intrinsic lung innervation is closely associated with ASM in the distal lung. Surrounding the future airways, a neuronal plexus lies on top of the ASM and extends into the distal lung over time along the growing airways [15]. Intrinsic airway ganglia are often found at conducting airway branching points and act as sites of convergence for several different classes of pulmonary innervation [31]. They receive input from extrinsic and other intrinsic neurons, and may act as sites of signal integration before signaling to downstream targets of innervation in the ASM [32].

Although the mechanisms and signaling cues that direct the tangential migration of NCC from the gut into the lungs have not yet been elucidated, perhaps the most plausible explanation for this behavior is a cell autonomous (chemotactic) response

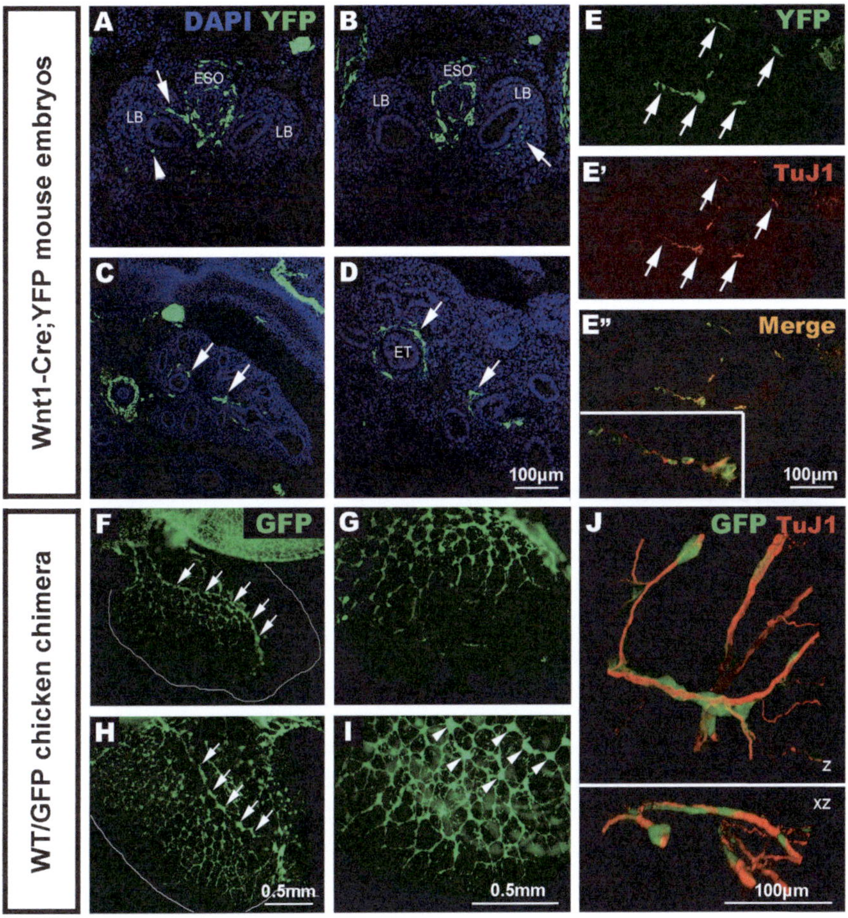

Fig. 4.2 Early development of lung innervation from neural crest cells and extrinsic nerve fibers. (**a**–**e**) Sections of *Wnt1-Cre; YFP* embryos immunostained using anti-GFP. (**a**, **b**) E10.5 embryo sections showing positively labelled neural crest cells within the esophagus, trachea, and lung buds. (**c**, **d**) E13.5 embryo sections showing numerous YFP cells (*arrows*) within the developing lungs closely associated with the epithelial tubules. (**e**) The majority of YFP+ tissue (**e**) is positive for the neural marker, TuJ1 (**e**') and clearly overlaps (**e**" and *inset*). (**f**–**i**) Wholemount preparations of foregut and lungs from chicken embryos that received a vagal GFP graft to specifically label neural crest cells. (**f**, **g**) At E7.5 the network of GFP+ NCC within the lungs is extensive, and an interconnected chain of GFP+ cells runs along the length of the lung (**f**, *arrows*), with progressively finer GFP+ projections extending towards the periphery (**g**). At E8.5 (**h**, **i**), numerous GFP+ cells and interconnected ganglia (**h**, *arrows*) are present within the lung. High magnification reveals interconnected GFP+ ganglia (**i**, *arrowheads*). (**j**) Association of migrating neural crest cells and extrinsic innervation during chicken lung development. Confocal 3D reconstruction of the developing chick lung at E6, showing the close association of the migrating lung neural crest cells (*green*) with the extrinsic nerve fibers (*red*). *Z*: 3D reconstruction of high magnification confocal image stack. *XZ*: 90° rotation of the same 3D reconstruction around the *X* axis. *ESO* esophagus, *LB* lung bud, *ET* epithelial tubule. From [19, 27, 28]

to a lung-specific signal. Such mechanisms, involving the guidance proteins Netrins (netrin-1 and -3) and netrin receptors (DCC, deleted in colorectal cancer, and the adenosine A2b receptor) have been reported to be involved in the perpendicular "secondary" migration of NCC from the outer (myenteric) to the inner (submucosal) layers of the gut and also, interestingly, out of the foregut into the developing pancreas [33], where NCC also form ganglia [34]. However, to date no specific molecular signature, or unique expression of cell surface receptors, has been identified in the NCC that colonize the lung. In fact, in chick embryo studies addressing this question, lung NCC were found to express many genes in common with enteric NCC, including Sox10, EdnrB, and Ret [11]. A large body of evidence has demonstrated the importance of the RET signaling pathway in ENS development, and activation of Ret receptor and Gfra1 co-receptor on migrating enteric NCC by the GDNF ligand is essential for NCC migration, survival, proliferation, and differentiation (reviewed in [35]). Due to this key role for RET in NCC development, and that $Ret^{-/-}$ mice have a depressed ventilatory response to inhaled CO_2 [36] the potential role of RET signaling in the development of lung innervation has been investigated by a number of groups. Our group showed that Ret-expressing lung NCC are attracted to GDNF in cultured lungs [19]. Further, studies by Langsdorf et al. [20] demonstrated that in $Ret^{-/-}$ mice the number of intrinsic lung neurons is reduced by 80 % in the trachea and primary bronchi, but these effects were not evident in $Gdnf^{-/-}$ and $NRTN^{-/-}$ mice suggesting redundancy in signaling through the Gdnf family ligands, Gdnf, NRTN, artemin and persephin [37]. However, in conflicting studies, when Freem et al. examined the lungs of $Ret^{-/-}$ and $Gfra1^{-/-}$ mice, no major differences in the extent of lung innervation were observed [19]. Nevertheless, Langsdorf et al. reported that in Ret mutant embryos intrinsic lung ganglia were reduced in size, further pointing to a possible role for the RET signaling pathway in survival and/or proliferation of lung NCC [20]. Thus although some studies have implicated signal(s) such as Ret in NCC colonization of the lung, other studies raise alternate possibilities. Freem et al. demonstrated that NCC following the enteric migration route along the gut can equally re-migrate to the lungs after back-transplantation from the gut to their initial site of migration within the vagal neural tube [27]. This suggests that vagal NCC are not prespecified to colonize either the gut or the lung and therefore they may migrate into the lung passively or by responding to (as yet unknown) signaling cues expressed by the lung and/or the gut. How NCC discriminate between signals such as Ret/Gdnf in the gut and lung remains to be determined but differing levels of Ret expression by these cell types may be important, as has previously been shown for vagal and sacral NCC that differ in their ability to colonize the gastrointestinal tract [38].

Another possible mechanism for facilitating migration of NCC from the gut into the lung is via the vascular system, which has been suggested to act as a substrate for the directed migration of NCC in the gut [39]. In support of this idea, the vascular system begins to form within the developing lung buds concomitant with the migration of vagal NCC into the lung [28]. However, our recent studies in chick embryos do not support this idea, as it was found that in the lung the innervation and

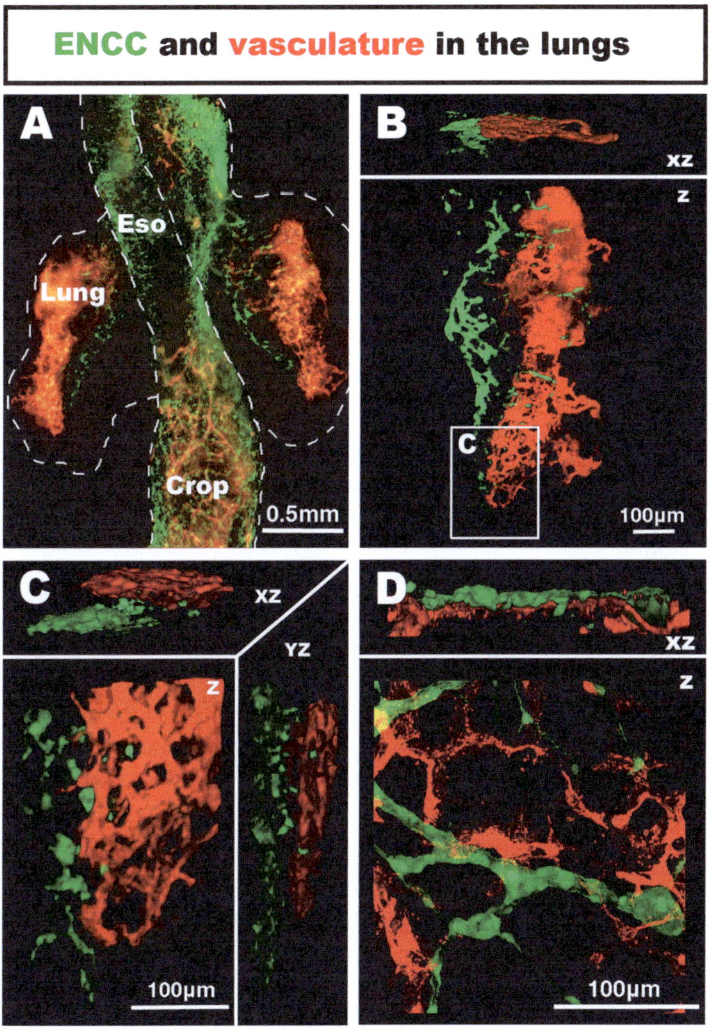

Fig. 4.3 Co-development of neural crest-derived innervation and associated vasculature in the embryonic chicken lungs. (**a**–**c**) Wholemount preparation of the lungs at E5.5. NCC are located within the mediodorsal region of the lungs, whereas the vascular system is located on the lateral side. (**d**) High magnification confocal image of the lungs at E11.5 (HH37–38). NCC form a thin ganglionated network intertwined with the vascular system. From [28]

the vascular system initially develop on opposite sides of the lung buds, clearly establishing two distinct early spatiotemporal patterns [28] (Fig. 4.3a–c). Interestingly though, the nerve and vascular networks are found in very close proximity at later stages of development, suggesting that neurovascular units within the lungs are functionally important (Fig. 4.3d) [28].

Throughout their early development, lung NCC are closely associated with the developing vagus nerves (see Fig. 4.2). Tollet et al. using a $p75^{NTR}$ antibody to label neural crest-derived cells in mouse embryos, revealed NCC in the vagus nerve and in processes extending from the vagus to the lung [15]. This suggests that NCC may also migrate along these paired extrinsic nerves to gain access to the lung. Indeed, following preestablished nerve fibers has already been recognized as a migratory behavior for another population of NCC, the sacral NCC, which contributes to the hindgut ENS. Studies on chick [40, 41] and mouse [42] have shown that the sacral NCC, which are known to share molecular signatures with vagal NCC [38], follow already established nerve projections from the extrinsic nerve of Remak in the chick, or pelvic plexus in the mouse, as a route for gaining entry to the hindgut. Moreover, this nerve-associated migratory behavior seems to be more common than previously recognized. In particular, a recent study from the Brunet laboratory showed that migrating cranial NCC can only form parasympathetic ganglia after being guided to the site of ganglion formation by nerves laying down fibers in "preganglionic" territories [43]. Interestingly, it can also be noted that an organ like the liver, which is only innervated by extrinsic nerve fibers after the initial wave of NCC migration along the gut has passed, does not contain any intrinsic ganglia [44]. Together, these studies suggest that preestablished neural projections offer an additional pathway for migrating NCC to colonize the lungs and that an absence of projecting nerve fibers could even prevent NCC entering an organ, such as the liver, during early development. However, detailed analysis of the interrelationships between intrinsic and extrinsic innervation is hampered by the fact that there are no known developmental models with a selective deficiency in intrinsic lung neurons, and by a lack of molecular markers that distinguish intrinsic from extrinsic neurons.

Development of Extrinsic Lung Innervation

As discussed above, extrinsic pulmonary innervation begins to develop during embryonic stages concomitant with intrinsic lung innervation. The extrinsic innervation of the lung, which will allow direct bronchopulmonary communication with the central nervous system (CNS), arises from the vagal and sympathetic nerve trunks of the autonomic nervous system. A proportion of extrinsic lung innervation, the vagus nerve fibers from the vagal nodose ganglia, is derived from the nodose epibranchial placode [45]. Other sensory innervation, including that from the dorsal root ganglia, vagal jugular ganglia, and cervical thoracic ganglia, is neural crest-derived [46]. The vagal fibers that innervate the mouse lungs have been shown to contain a mix of placode-derived and neural crest-derived fibers [47].

Parasympathetic innervation from the brainstem, nodose ganglia, and jugular ganglia enters the lung via the vagus nerve, while sympathetic pulmonary innervation comes from the dorsal root ganglia and cervical thoracic sympathetic ganglia [48]. In contrast to intrinsic lung neurons, whose numbers have been reported to be reduced in *Ret* mutant mice [20], ASM innervation in the distal lung remains unaffected in these mutants, suggesting that extrinsic lung innervation may require a different neurogenic signal [21, 49]. Studies of E13.5 mouse embryos have shown that extrinsic axons extend into the distal lung and are present in the vicinity of developing ASM [15]. This association between ASM and growing axons is maintained during later embryonic stages and elaborates into a complex neural network closely associated with the smooth muscle [15]. This temporal and spatial interrelationship between axon outgrowth and ASM expression led to the investigation of potential ASM-derived neurotrophic factor(s) coordinating these two aspects of lung development [49, 50]. Xingbin's group showed that, as Shh signaling induces the specification of ASM, it also represses an inhibitory microRNA (miR-206), which leads to the activation of Brain-derived-neurotrophic-factor (BDNF) expression in the smooth muscle. BDNF is then used as an ASM-derived neurotrophic signal by the extrinsic neurons to extend their axons into the distal lung [49]. In accordance with this result, a study using TrkB (the receptor for BDNF) null mice, showed that the lungs of these mutants had thinner bronchial epithelium, a larger airway luminal diameter, and larger alveolar spaces than their wild-type siblings. This suggests that BDNF/TrkB is important for cell proliferation and/or migration in the lung and ultimately for adequate development of epithelium and ASM [50, 51].

Again, as for the NCC that migrate from the gut into the lung, the guidance cues that direct axons into and within the lungs are elusive, although parallels can be made with signals guiding extrinsic axons innervating the gut. In this organ vagal sensory axons have been reported to respond to chemoattractive netrin signaling cues mediated through the receptor Deleted in Colon Cancer (DCC) as shown in in vitro experiments. Vagal axons in the gut terminate their response to netrin in the presence of laminin, expressed in the extracellular matrix of the gut, suggesting a possible mechanism for vagus nerve terminal positioning [52].

Developmental Interactions Between the Airway Innervation and Its Cellular Targets

As already mentioned, ASM is one of the main targets for lung innervation. ASM has been suggested to have a potential role in the antenatal development and growth of the lung [53]. The idea is that the peristaltic contraction of ASM moves fluid through the developing lung [54] and that the pressure of the fluid against the lung tissue may provide a mechanical stimulus that further promotes lung growth [55]. Evidence for this comes from studies showing that increasing intraluminal pressure within the lung stimulates epithelial branching in culture and may increase lung growth in vivo [56]. In the absence of positive internal pressure, lung growth is decreased [29]. In further support of the possible link between embryonic ASM peristalsis and lung growth, ASM dysfunction has been linked to hypoplastic lung growth [57]. Also, ASM dysfunction occurs in, and may cause, lung hypoplasia in a congenital diaphragmatic hernia (CDH) rat model, where lung hypoplasia precedes CDH [58]. This peristaltic movement of ASM appears to be initiated and maintained independently of neuronal input [59]. Intrinsic calcium channel activity is important in maintaining ASM contractility prenatally [60] and tone postnatally [59], but the role of lung innervation in driving ASM peristalsis has not been investigated. ASM contraction is mediated by intrinsic calcium currents and appears to be modulated by neuronal input [61].

Other, (trophic) interaction(s) between the developing lung innervation and cell types such as the ASM have not been studied extensively during development, but an interesting parallel can again be made with the gastrointestinal tract. In particular, the innervation of the stomach develops in a similar manner to that of the lungs with a contribution from vagal NCC that give rise to intrinsic ganglia, as well as extrinsic innervation from the vagus nerves [62]. Recent work by Faure et al. [63] highlighted the importance of enteric NCC in regulating smooth muscle patterning and differentiation in the stomach. Specifically, results suggested that a crucial number of enteric NCC is required for maintenance of smooth muscle identity and differentiation. These authors demonstrated that experimental reduction of the numbers of enteric NCC induced sustained activation of the BMP and Notch pathways in the stomach mesenchyme, which impaired smooth muscle development [63]. Considering the importance of the Notch pathway in multiple aspects of lung development (as reviewed in [64]), this type of muscle developmental regulation by a neural component could also be important for ASM development in the lungs, but as yet this remains only a tantalizing hypothesis.

Another important target of lung innervation is the neuroepithelial bodies (NEB). Pulmonary neuroendocrine cells (PNECs), also called neuroepithelial endocrine cells, are hypoxia-sensitive chemoreceptors found in the lung airway epithelium adjacent to the lumen of the airway (reviewed in [65–68]). They express the calcitonin neurotransmitter marker CGRP [69] and unlike intrinsic pulmonary neurons are not derived from the neural crest, instead arising from the endoderm that forms the airway epithelium [70]. Some, but not all, PNECs aggregate with other PNECs to form specialized structures innervated by multiple extrinsic nerves, and are termed NEBs [66]. During lung epithelium development, different assemblies of Notch receptors coordinate the number and distribution of NEBs in the conducting airways [71]. NEBs are also sites of airway epithelial stem cell reservoirs [72]. NEBs are innervated by vagal afferents, which transmit information to the CNS, as well as dorsal root ganglion afferents that may modulate intrinsic neuron activity via axon reflexes. Intrinsic neurons are found in close proximity with NEBs and may influence their responses by releasing nitric oxide (NO) [68]. In the mature lungs, the role of NEBs is to respond to intraluminal stimuli, including pO_2 and pCO_2, pH of the airway surface fluid, and environmental stimuli such as nicotine [68]. An increase in the density of NEBs has been reported in lungs from cases of sudden infant death syndrome (SIDS) [73, 74]. One of the possible consequences of this reported increased density of NEBs is increased sensitivity to hypoxia. The control of proliferation of NEBs and PNECs is thus potentially important in early respiratory function and, as we will see in the following paragraph, is likely influenced by the co-development of the neuronal network.

Only a few studies have investigated potential interactions between the developing extrinsic innervation with other cells type during lung development. Most noteworthy is a recent paper, using the 2-photon laser ablation technique targeting the local parasympathetic ganglia in organ culture, which demonstrated that elimination of these ganglia abolishes lung branching and diminishes endothelial and epithelial cell proliferation [75]. Using drosophila, Bower et al. were also able to demonstrate that this neural control of airway morphogenesis is conserved between invertebrates and mammals, highlighting the importance of this developmental interaction. Prior to this finding, other studies have also suggested that parasympathetic nerves are a vital component of the progenitor cell niche during development, maintaining a pool of progenitors for organogenesis [76, 77]. Using keratin-5 as a marker for epithelial progenitor cells, Knox et al. demonstrated that the removal of the parasympathetic ganglion decreased the number and morphogenesis of progenitor cells in mouse embryonic submandibular gland culture. Moreover, the study also showed that this regulation of epithelial development is controlled via the acetylcholine-signaling pathway [76]. This is significant as embryonic submandibular glands share many of the features of lung development, including branching morphogenesis, NCC colonization and development of an extensive parasympathetic and sympathetic extrinsic innervation [78, 79].

Insight from Developmental Defects Affecting Airway Innervation

As mentioned above, the biggest hurdles preventing a better understanding of the role of lung innervation during development are (1) that there are no known developmental (mouse) models with a selective deficiency in intrinsic lung neurons and/or extrinsic innervation, and (2) that there is a lack of molecular markers that distinguish intrinsic from extrinsic neurons. One possible route for exploring these processes is analyzing human conditions or congenital abnormalities that impair lung function and where there is at least some understanding of the genes and signaling pathways underlying the defects. Some of these conditions are briefly discussed and listed below.

Central Congenital Hypoventilation Syndrome and Phox2b

Central congenital hypoventilation syndrome (CCHS) is a disorder that results in an absence of adequate autonomic control of respiration with decreased sensitivity to hypoxia and hypercapnia. In CCHS respiration slows dramatically during sleep and does not increase in response to hypoxia, causing suffocation. CCHS is primarily considered a disorder of central respiratory pattern generation and can cause death before the age of 2 years (reviewed in [80]). Mutations in *PHOX2B* are strongly associated with CCHS in humans [80–83] and have a similar effect on respiration in mouse models [84].

Phox2b is a paired homeobox transcription factor expressed by neurons in the central and peripheral nervous system [85]. It is necessary for the development of much of the autonomic nervous system, including neural crest-derived peripheral autonomic innervation and neural crest-derived intrinsic enteric neurons [86, 87]. Phox2b is also important in the development of neurons in the brainstem that regulate ventilation in response to carbon dioxide levels [88]. There also exists a weaker association between CCHS and *RET* mutations, in which *RET* mutations increase the risk of intestinal aganglionosis when accompanying *PHOX2B* mutations [89]. Phox2b has been reported to regulate the downstream expression of *Ret* and *Gdnf* [86]. As $Ret^{-/-}$ mice have aganglionosis due to failure of NCC to colonize the entire gut and also a depressed ventilatory response to inhaled CO_2 [36], there may be an, as yet unidentified, link between NCC development and breathing control.

In CCHS the brainstem control of ventilation exhibits a reduced response to hypoxia. It is possible that a defect in oxygen sensing by NEBs or in transmission of information by airway intrinsic ganglia could contribute to deficient brainstem control of ventilation. Although the non-neural crest-derived brainstem neurons involved in respiratory control that are affected by *PHOX2B* mutation have been studied in the context of CCHS, the role of neural crest-derived intrinsic lung innervation in CCHS has not been investigated to date.

Waardenburg Syndrome Type IV and Sox10

Sox10 is a member of the SoxE transcription factor gene family, expressed in delaminating and migrating NCC, including the NCC that migrate into the lungs [11, 90]. Sox10 regulates neural crest development, along with its earlier expressed relatives *Sox8* and *Sox9*, and maintains NCC pluripotency and survival. Sox10 promotes neurogenesis in autonomic lineages by switching on the proneural gene *Mash1*. It also controls the transcription of a number of downstream PNS markers, including Ret, which are important in the development of the neural crest-derived ENS (reviewed in [90, 91]). Later in development Sox10 is involved in specification of neural crest-derived glia [90].

In humans, mutations in *SOX10* cause neurocristopathies of varying severity. In Waardenburg Syndrome Type IV (WS-IV), patients have deficiencies in neural crest-derived melanocytes, resulting in hypopigmentation, sensorineural hearing loss, and defects in enteric ganglia formation [92]. Waardenburg syndrome type IV-C (OMIM ID 613266), also called Waardenburg-Shah syndrome [93], is caused by heterozygous *SOX10* mutation. WS-IV-A (OMIM 277580) and B (OMIM 613265) have similar features but are caused by *EDNRB* and *EDN-3* mutations respectively. Defects in glial cells causing hypomyelination, resulting in neuropathy, are sometimes later consequences of *SOX10* mutation and when accompanying the symptoms of WS-IV are part of PCWH (Peripheral demyelinating neuropathy, central dysmyelinating leukodystrophy, Waardenburg syndrome and Hirschsprung disease) (OMIM ID 609136). Importantly though, no respiratory symptoms have been described in WS-IV patients.

The *Sox10*^{Dom} mutation, which is a mouse model for WS-IV [94], affects skin pigmentation, and PNS and enteric ganglia formation. *Sox10*^{Dom} heterozygous mutant mice have deficient NCC migration along the gut [95], producing a variable-length aganglionic distal bowel phenotype [94, 96]. Embryos homozygous for the mutation die before birth, usually before E15, which unfortunately does not allow for investigation of potential deficits in breathing due to a deficit of NCC-derived lung neurons. NCC appear to be absent from *Sox10*^{Dom} E14.5 homozygous mutant mouse lungs but not from heterozygous mutant lungs, indicating that the aganglionic gut phentotype seen in heterozygous mutants does not extend to the lungs [97]. To our knowledge, potential neural crest-derived neural defects in E15 (or earlier) mouse lungs have not been definitively investigated to date.

DiGeorge Syndrome and Tbx1

Tbx1 is a member of the T-box transcription factor family. It is necessary for the normal development of the NCC-derived pharyngeal arch structures and for the migration of cardiac NCC within the arches [98]. *Tbx1* is expressed in the pharyngeal arch mesenchyme and in the developing lung epithelium [99]. Tbx1 is necessary for the release of NCC guidance molecule Gbx2 from the pharyngeal arches, guiding cardiac NCC towards the heart [100]. Such non-cell-autonomous gene mutations can therefore disrupt NCC migration if they disrupt extrinsic factors or structures required for NCC migration.

DiGeorge syndrome, also called velocardiofacial syndrome, results from a heterozygous deletion in chromosome 22q11 [101]. The *TBX1* locus is within this region. Characteristics are highly variable and may include congenital heart defects, palate defects, craniofacial abnormalities, and learning disabilities, most of which appear to stem from problems in pharyngeal arch development. Laryngotracheal abnormalities are observed in a proportion of patients, including tracheoesophageal fistulae, shortened trachea, missing or softened cartilage in the upper airways and tracheal bronchus, or blind-ending bronchus. These factors can cause restrictive lung disease, where the amount of air that can be inhaled is physically restricted [102].

Tbx1 heterozygous mutant mice show most aspects of the DiGeorge syndrome phenotype, most notably defects in cardiac outflow tract development [98, 103]. The role of Tbx1 in neural crest migration, particularly in cardiac neural crest migration, has been linked to DiGeorge syndrome [104]. *Tbx1* mouse mutants display disruption of cranial nerve ganglia and migration [105] and homozygous *Tbx1* mutants have been shown to have decreased extrinsic vagal innervation of the stomach [106] and of the lungs [97]. The effect of homozygous *Tbx1* mutation on the development of the vagus nerve is interesting when considering the interaction between intrinsic NCC-derived lung neurons and extrinsic lung innervation from the vagus nerve. NCC in E14.5 Tbx1 mutant lungs do not migrate into the distal lung tissue, possibly because extrinsic innervation does not extend into the distal lung in these mutants [97]. It is thus possible that Tbx1 may be involved in the guidance of NCC and/or extrinsic innervation into the lung, but again this remains to be definitively determined.

Summary of the Final Layout of Airway Innervation

Ultimately, the lungs develop both extrinsic and intrinsic innervation with both sensory and effector functions (Fig. 4.4). Extrinsic vagal and sympathetic nerve branches innervate the lung. Bronchial branches of the vagus nerve join nerve fibers

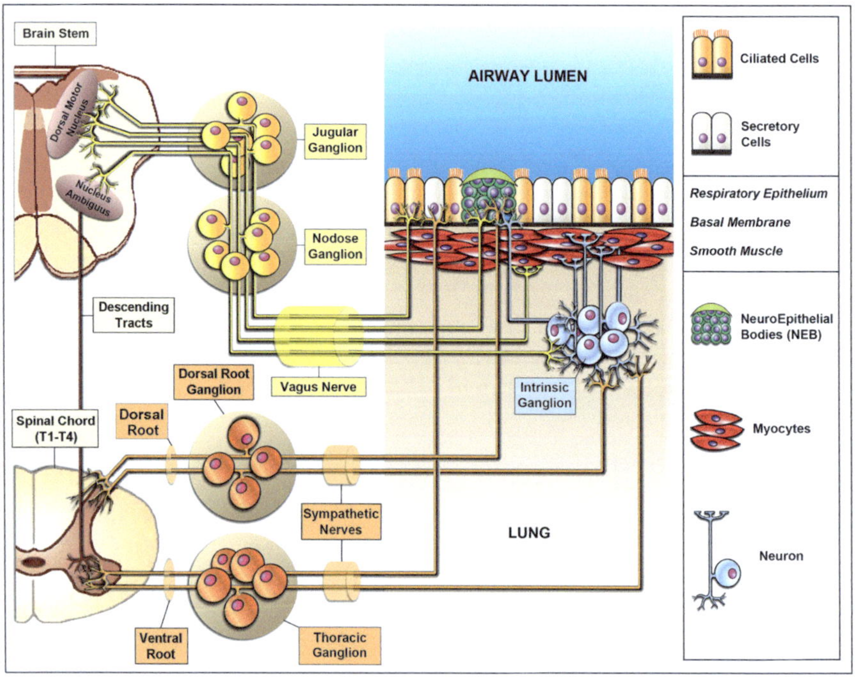

Fig. 4.4 Schematic representation of the extrinsic and intrinsic innervation of the respiratory tract. Extrinsic neurons have their cell bodies in (1) the jugular and nodose ganglia, projecting anteriorly to the dorsal motor nucleus and the nucleus ambiguous of the brain stem and extending their axons posteriorly, via the vagus nerve, to the goblet cells, Club cells and neuroepithelial bodies within the airway epithelium, the airway smooth muscle, as well as the intrinsic ganglia; (2) the dorsal root ganglia and thoracic ganglia projecting anteriorly to the spinal chord at the T1–T4 level via the dorsal and the ventral root respectively. These neurons project to the lung via the sympathetic nerves to connect to the airway epithelium and neuroepithelial bodies. Neural crest cell-derived intrinsic neurons (intrinsic ganglia) connect to the airway smooth muscle and NEBS

from the sympathetic trunk before entering the lung, forming mixed nerve trunks. Extrinsic projections can innervate intrinsic pulmonary neurons or directly innervate sensory or effector structures. Extrinsic and intrinsic neuronal fibers form a network of nerves around the airways, called the pulmonary plexus.

Efferent nerves regulate ASM tone and mucus secretion from submucosal glands, goblet cells, and Club cells in the airway epithelium. Vascular permeability and blood flow in pulmonary blood vessels is also regulated by pulmonary innervation. Sensory nerves provide homeostatic feedback on lung inflation, and detect irritant chemicals and oxygen concentration within the lung either directly or through sensory cells.

Concluding Remarks

Despite numerous investigations on the development of lung innervation stretching back over 20 years, many questions remain unanswered. More studies are required to characterize the genes and signaling pathways involved in the development of intrinsic and extrinsic lung innervation and to better understand how this intricate neural network is laid out and interconnected with other cell types within the developing lung. In particular, little is known about how the developing neural tissue interacts and influences other cell types during functional lung formation. Additionally, the developmental and functional roles of intrinsic neurons within this network have not been fully elucidated. It is essential that such studies are performed, as they will be key in unraveling the role that alterations or defects in lung innervation play in health and disease before and after birth.

References

1. Morrisey EE, Hogan BLM (2010) Preparing for the first breath: genetic and cellular mechanisms in lung development. Dev Cell 18(1):8–23
2. Ornitz DM, Yin Y (2012) Signaling networks regulating development of the lower respiratory tract. Cold Spring Harb Perspect Biol 2012:4(5)
3. Rankin SA, Zorn AM (2014) Gene regulatory networks governing lung specification. J Cell Biochem 115(8):1343–1350
4. Herriges M, Morrisey EE (2014) Lung development: orchestrating the generation and regeneration of a complex organ. Development 141(3):502–513
5. Grapin-Botton A, Melton (2000) Endoderm development: from patterning to organogenesis. Trends Genet 16(3):124–130
6. Warburton D et al (2000) The molecular basis of lung morphogenesis. Mech Dev 92 (1):55–81
7. Kugler MC et al (2015) Sonic Hedgehog signaling in the lung—from development to disease. Am J Respir Cell Mol Biol 52(1):1–13

8. Warburton D et al (2005) Molecular mechanisms of early lung specification and branching morphogenesis. Pediatr Res 57(5 Pt 2):26R–37R
9. Litingtung Y et al (1998) Sonic hedgehog is essential to foregut development. Nat Genet 20 (1):58–61
10. Bellusci S et al (1997) Involvement of Sonic hedgehog (Shh) in mouse embryonic lung growth and morphogenesis. Development 124(1):53–63
11. Burns AJ, Delalande JM (2005) Neural crest cell origin for intrinsic ganglia of the developing chicken lung. Dev Biol 277(1):63–79
12. Burns AJ, Thapar N, Barlow AJ (2008) Development of the neural crest-derived intrinsic innervation of the human lung. Am J Respir Cell Mol Biol 38(3):269–275
13. Billmyre KK, Hutson M, Klingensmith J (2015) One shall become two: separation of the esophagus and trachea from the common foregut tube. Dev Dyn 244(3):277–288
14. Metzger RJ et al (2008) The branching programme of mouse lung development. Nature 453 (7196):745–750
15. Tollet J, Everett AW, Sparrow MP (2001) Spatial and temporal distribution of nerves, ganglia, and smooth muscle during the early pseudoglandular stage of fetal mouse lung development. Dev Dyn 221(1):48–60
16. Hines EA et al (2013) Establishment of smooth muscle and cartilage juxtaposition in the developing mouse upper airways. Proc Natl Acad Sci U S A 110(48):19444–19449
17. Weaver M, Batts L, Hogan BL (2003) Tissue interactions pattern the mesenchyme of the embryonic mouse lung. Dev Biol 258(1):169–184
18. Kumar ME et al (2014) Mesenchymal cells. Defining a mesenchymal progenitor niche at single-cell resolution. Science 346(6211), 1258810
19. Freem LJ et al (2010) The intrinsic innervation of the lung is derived from neural crest cells as shown by optical projection tomography in Wnt1-Cre; YFP reporter mice. J Anat 217 (6):651–664
20. Langsdorf A et al (2011) Neural crest cell origin and signals for intrinsic neurogenesis in the mammalian respiratory tract. Am J Respir Cell Mol Biol 44(3):293–301
21. Aven L, Ai X (2013) Mechanisms of respiratory innervation during embryonic development. Organogenesis 9(3):194–198
22. Kalcheim C, Le Douarin NM (1999) The neural crest developmental and cell biology series: 36, 2nd edn. Cambridge University Press, Cambridge
23. Le Douarin NM, Teillet MA (1973) The migration of neural crest cells to the wall of the digestive tract in avian embryo. J Embryol Exp Morphol 30(1):31–48
24. Goldstein AM, Hofstra RMW, Burns AJ (2013) Building a brain in the gut: development of the enteric nervous system. Clin Genet 83(4):307–316
25. Sasselli V, Pachnis V, Burns AJ (2012) The enteric nervous system. Dev Biol 366(1):64–73
26. Avetisyan M, Schill EM, Heuckeroth RO (2015) Building a second brain in the bowel. J Clin Invest 125(3):899–907
27. Freem LJ et al (2012) Lack of organ specific commitment of vagal neural crest cell derivatives as shown by back-transplantation of GFP chicken tissues. Int J Dev Biol 56 (4):245–254
28. Delalande JM et al (2014) Vascularisation is not necessary for gut colonisation by enteric neural crest cells. Dev Biol 385(2):220–229
29. Sparrow MP, Weichselbaum M, McCray PB (1999) Development of the innervation and airway smooth muscle in human fetal lung. Am J Respir Cell Mol Biol 20(4):550–560
30. Sparrow MP, Warwick SP, Everet AW (1995) Innervation and function of the distal airways in the developing bronchial tree of fetal pig lung. Am J Respir Cell Mol Biol 13(5):518–525
31. Brouns I et al (2009) Neurochemical pattern of the complex innervation of neuroepithelial bodies in mouse lungs. Histochem Cell Biol 131(1):55–74
32. Myers AC (2001) Transmission in autonomic ganglia. Respir Physiol 125(1–2):99–111
33. Jiang Y, Liu MT, Gershon MD (2003) Netrins and DCC in the guidance of migrating neural crest-derived cells in the developing bowel and pancreas. Dev Biol 258(2):364–384

34. Kirchgessner AL, Gershon MD (1990) Innervation of the pancreas by neurons in the gut. J Neurosci 10(5):1626–1642
35. Sasselli V et al (2012) Evaluation of ES-derived neural progenitors as a potential source for cell replacement therapy in the gut. BMC Gastroenterol 12:81
36. Burton MD et al (1997) RET proto-oncogene is important for the development of respiratory CO_2 sensitivity. J Auton Nerv Syst 63(3):137–143
37. Baloh RH et al (2000) The GDNF family ligands and receptors—implications for neural development. Curr Opin Neurobiol 10(1):103–110
38. Delalande JM et al (2008) The receptor tyrosine kinase RET regulates hindgut colonization by sacral neural crest cells. Dev Biol 313(1):279–292
39. Nagy N et al (2009) Endothelial cells promote migration and proliferation of enteric neural crest cells via beta1 integrin signaling. Dev Biol 330(2):263–272
40. Burns AJ, Le Douarin NM (1998) The sacral neural crest contributes neurons and glia to the post- umbilical gut: spatiotemporal analysis of the development of the enteric nervous system. Development 125(21):4335–4347
41. Burns AJ, Le Douarin NM (2001) Enteric nervous system development: analysis of the selective developmental potentialities of vagal and sacral neural crest cells using quail-chick chimeras. Anat Rec 262(1):16–28
42. Wang X et al (2011) Analysis of the sacral neural crest cell contribution to the hindgut enteric nervous system in the mouse embryo. Gastroenterology 141(3):992–1002; e1–6
43. Espinosa-Medina I et al (2014) Neurodevelopment. Parasympathetic ganglia derive from Schwann cell precursors. Science 345(6192):87–90
44. Delalande JM, Milla PJ, Burns AJ (2004) Hepatic nervous system development. Anat Rec 280(1):848–853
45. Baker CV, Bronner-Fraser M (2001) Vertebrate cranial placodes I. Embryonic induction. Dev Biol 232(1):1–61
46. Kwong K et al (2008) P2X2 receptors differentiate placodal vs. neural crest C-fiber phenotypes innervating guinea pig lungs and esophagus. Am J Physiol Lung Cell Mol Physiol 295 (5):L858–L865
47. Nassenstein C et al (2010) Phenotypic distinctions between neural crest and placodal derived vagal C-fibres in mouse lungs. J Physiol 588(23):4769–4783
48. Kummer W et al (1992) The sensory and sympathetic innervation of guinea-pig lung and trachea as studied by retrograde neuronal tracing and double-labelling immunohistochemistry. Neuroscience 49(3):715–737
49. Radzikinas K et al (2011) A Shh/miR-206/BDNF cascade coordinates innervation and formation of airway smooth muscle. J Neurosci 31(43):15407–15415
50. Prakash YS, Martin RJ (2014) Brain-derived neurotrophic factor in the airways. Pharmacol Ther 143(1):74–86
51. García-Suárez O et al (2009) TrkB is necessary for the normal development of the lung. Respir Physiol Neurobiol 167(3):281–291
52. Ratcliffe E, D'Autreaux F, Gershon MD (2008) Laminin terminates the Netrin/DCC mediated attraction of vagal sensory axons. Dev Neurobiol 68(7):960–971
53. Jesudason EC (2009) Airway smooth muscle: an architect of the lung? Thorax 64(6):541–545
54. Schittny JC, Miserocchi G, Sparrow MP (2000) Spontaneous peristaltic airway contractions propel lung liquid through the bronchial tree of intact and fetal lung explants. Am J Respir Cell Mol Biol 23(1):11–18
55. Sparrow M, Warwick SP, Mitchell HW (1994) Foetal airway motor tone in prenatal lung development of the pig. Eur Respir J 7(8):1416–1424
56. Unbekandt M et al (2008) Tracheal occlusion increases the rate of epithelial branching of embryonic mouse lung via the FGF10-FGFR2b-Sprouty2 pathway. Mech Dev 125 (3–4):314–324

57. Jesudason EC et al (2006) Peristalsis of airway smooth muscle is developmentally regulated and uncoupled from hypoplastic lung growth. Am J Physiol Lung Cell Mol Physiol 291(4): L559–L565
58. Featherstone NC et al (2006) Airway smooth muscle dysfunction precedes teratogenic congenital diaphragmatic hernia and may contribute to hypoplastic lung morphogenesis. Am J Respir Cell Mol Biol 35(5):571–578
59. Sanderson MJ et al (2008) Regulation of airway smooth muscle cell contractility by Ca2+ signaling and sensitivity. Proc Am Thorac Soc 5(1):23–31
60. Bai Y, Zhang M, Sanderson MJ (2007) Contractility and Ca2+ Signaling of Smooth Muscle Cells in Different Generations of Mouse Airways. Am J Respir Cell Mol Biol 36(1):122–130
61. Berridge MJ (2008) Smooth muscle cell calcium activation mechanisms. J Physiol 586 (21):5047–5061
62. Ratcliffe EM, Farrar NR, Fox EA (2011) Development of the vagal innervation of the gut: steering the wandering nerve. Neurogastroenterol Motil 23(10):898–911
63. Faure S et al (2015) Enteric neural crest cells regulate vertebrate stomach patterning and differentiation. Development 142(2):331–342
64. Xu K, Moghal N, Egan SE (2012) Notch signaling in lung development and disease. Adv Exp Med Biol 727:89–98
65. Cutz E, Jackson A (1999) Neuroepithelial bodies as airway oxygen sensors. Respir Physiol 115(2):201–214
66. Van Lommel A (2001) Pulmonary neuroendocrine cells (PNEC) and neuroepithelial bodies (NEB): chemoreceptors and regulators of lung development. Paediatr Respir Rev 2 (2):171–176
67. Adriaensen D et al (2006) Evidence for a role of neuroepithelial bodies as complex airway sensors: comparison with smooth muscle-associated airway receptors. J Appl Physiol 101 (3):960–970
68. Cutz E et al (2013) Recent advances and controversies on the role of pulmonary neuroepithelial bodies as airway sensors. Semin Cell Dev Biol 24(1):40–50
69. Keith IM (1991) Calcitonin gene-related peptide and its mRNA in pulmonary neuroendocrine cells and ganglia. Histochemistry 96(4):311–315
70. Sorokin SP, Hoyt RF Jr, Shaffer MJ (1997) Ontogeny of neuroepithelial bodies: correlations with mitogenesis and innervation. Microsc Res Tech 37(1):43–61
71. Morimoto M et al (2012) Different assemblies of Notch receptors coordinate the distribution of the major bronchial Clara, ciliated and neuroendocrine cells. Development 139 (23):4365–4373
72. Reynolds SD et al (2000) Neuroepithelial bodies of pulmonary airways serve as a reservoir of progenitor cells capable of epithelial regeneration. Am J Pathol 156(1):269–278
73. Gillan JE et al (1989) Abnormal patterns of pulmonary neuroendocrine cells in victims of sudden infant death syndrome. Pediatrics 84(5):828–834
74. Cutz E et al (2007) Pulmonary neuroendocrine cells and neuroepithelial bodies in sudden infant death syndrome: potential markers of airway chemoreceptor dysfunction. Pediatr Dev Pathol 10(2):106–116
75. Bower DV et al (2014) Airway branching has conserved needs for local parasympathetic innervation but not neurotransmission. BMC Biol 12(1):92
76. Knox SM et al (2010) Parasympathetic innervation maintains epithelial progenitor cells during salivary organogenesis. Science 329(5999):1645–1647
77. Knox SM et al (2013) Parasympathetic stimulation improves epithelial organ regeneration. Nat Commun 4:1494
78. Melnick M, Jaskoll T (2000) Mouse submandibular gland morphogenesis: a paradigm for embryonic signal processing. Crit Rev Oral Biol Med 11(2):199–215
79. Patel VN, Rebustini IT, Hoffman MP (2006) Salivary gland branching morphogenesis. Differentiation 74(7):349–364

80. Rand CM, Carroll MS, Weese-Mayer DE (2014) Congenital central hypoventilation syndrome: a neurocristopathy with disordered respiratory control and autonomic regulation. Clin Chest Med 35(3):535–545
81. Berry-Kravis EM et al (2006) Congenital central hypoventilation syndrome: PHOX2B mutations and phenotype. Am J Respir Crit Care Med 174(10):1139–1144
82. Weese-Mayer DE et al (2009) Congenital central hypoventilation syndrome from past to future: model for translational and transitional autonomic medicine. Pediatr Pulmonol 44 (6):521–535
83. Weese-Mayer DE, Berry-Kravis EM (2004) Genetics of congenital central hypoventilation syndrome: lessons from a seemingly orphan disease. Am J Respir Crit Care Med 170 (1):16–21
84. Amiel J et al (2003) Polyalanine expansion and frameshift mutations of the paired-like homeobox gene PHOX2B in congenital central hypoventilation syndrome. Nat Genet 33 (4):459–461
85. Pattyn A et al (1997) Expression and interactions of the two closely related homeobox genes Phox2a and Phox2b during neurogenesis. Development 124(20):4065–4075
86. Pattyn A et al (1999) The homeobox gene Phox2b is essential for the development of autonomic neural crest derivatives. Nature 399(6734):366–370
87. Dauger S et al (2003) Phox2b controls the development of peripheral chemoreceptors and afferent visceral pathways. Development 130(26):6635–6642
88. Onimaru H, Ikeda K, Kawakami K (2008) CO_2-sensitive preinspiratory neurons of the parafacial respiratory group express Phox2b in the neonatal rat. J Neurosci 28 (48):12845–12850
89. de Pontual L et al (2006) Mutations of the RET gene in isolated and syndromic Hirschsprung's disease in human disclose major and modifier alleles at a single locus. J Med Genet 43(5):419–423
90. Kipanyula MJ et al (2014) Signaling pathways bridging fate determination of neural crest cells to glial lineages in the developing peripheral nervous system. Cell Signal 26(4):673–682
91. Sauka-Spengler T, Bronner-Fraser M (2008) A gene regulatory network orchestrates neural crest formation. Nat Rev Mol Cell Biol 9(7):557–568
92. Badner JA, Chakravarti A (1990) Waardenburg syndrome and Hirschsprung disease: evidence for pleiotropic effects of a single dominant gene. Am J Med Genet 35(1):100–104
93. Shah KN et al (1981) White forelock, pigmentary disorder of irides, and long segment Hirschsprung disease: possible variant of Waardenburg syndrome. J Pediatr 99(3):432–435
94. Hakami RM et al (2006) Genetic evidence does not support direct regulation of EDNRB by SOX10 in migratory neural crest and the melanocyte lineage. Mech Dev 123(2):124–134
95. Southard-Smith EM, Kos L, Pavan WJ (1998) Sox10 mutation disrupts neural crest development in Dom Hirschsprung mouse model. Nat Genet 18(1):60–64
96. Walters LC et al (2010) Genetic background impacts developmental potential of enteric neural crest-derived progenitors in the Sox10Dom model of Hirschsprung disease. Human Mol Genet 19(22):4353–4372
97. Freem LJ (2011) The development of the neural crest-derived intrinsic innervation of the lung. University College, London
98. Lindsay EA et al (2001) Tbx1 haploinsufficiency in the DiGeorge syndrome region causes aortic arch defects in mice. Nature 410(6824):97–101
99. Chapman DL et al (1996) Expression of the T-box family genes, Tbx1-Tbx5, during early mouse development. Dev Dyn 206(4):379–390
100. Calmont A et al (2009) Tbx1 controls cardiac neural crest cell migration during arch artery development by regulating Gbx2 expression in the pharyngeal ectoderm. Development 136 (18):3173–3183
101. Epstein JA (2001) Developing models of DiGeorge syndrome. Trends Genet 17(10):S13–S17
102. Huang RY, Shapiro NL (2000) Structural airway anomalies in patients with DiGeorge syndrome: a current review. Am J Otolaryngol 21(5):326–330

103. Jerome LA, Papaioannou VE (2001) DiGeorge syndrome phenotype in mice mutant for the T-box gene, Tbx1. Nat Genet 27(3):286–291
104. Kochilas L et al (2002) The role of neural crest during cardiac development in a mouse model of DiGeorge syndrome. Dev Biol 251(1):157–166
105. Vitelli F et al (2002) Tbx1 mutation causes multiple cardiovascular defects and disrupts neural crest and cranial nerve migratory pathways. Hum Mol Genet 11(8):915–922
106. Calmont A et al (2010) Absence of the vagus nerve in the stomach of Tbx1 −/− mutant mice. Neurogastroenterol Motil 23(2):125–130

Chapter 5
Fetal and Neonatal Origins of Lung Disease

Foula Sozo, Megan O'Reilly, and Richard Harding

Evidence for Developmental Programming: Epidemiological and Experimental Evidence

Like many organs, the lung can be affected during its period of development in such a way that its later structure and function are altered. This can occur as a result of altered structural development or altered cellular function as a result of genetic or epigenetic alterations. Many such alterations can impair lung function for much or all of the life span and can also increase the risk of later pulmonary illness. Developmental insults or challenges that have the potential to induce persistent alterations in lung structure or function include physical and metabolic changes in the developmental environment. As the lung undergoes development during both prenatal and early postnatal life, its development can be altered by environmental and/or epigenetic factors that affect the intrauterine environment or the environment during infancy and childhood. Altered development can include gross structural alterations such as lung hypoplasia caused by congenital diaphragmatic hernia (CDH) or oligohydramnios, for example, or more subtle changes that may lead to persistent oxidative stress or inflammation of lung tissue.

In this chapter, we review current knowledge of factors commonly operating during development that can persistently affect the structure and function of the lung, and the underlying processes. It has long been known that impaired lung function during early life is associated with impaired lung function later in life; this is known as "tracking" of lung function [1]. More recently, it has become accepted

F. Sozo • R. Harding, Ph.D., D.Sc. (✉)
Department of Anatomy and Developmental Biology, Monash University, 19 Innovation Walk, Melbourne, VIC 3800, Australia
e-mail: richard.harding@monash.edu

M. O'Reilly
Department of Pediatrics, Women and Children's Health Research Institute, University of Alberta, Edmonton, AB, Canada T6G 2E1

I. Bertoncello (ed.), *Stem Cells in the Lung*, Stem Cell Biology and Regenerative Medicine, DOI 10.1007/978-3-319-21082-7_5

that impaired lung function during early life can be caused by environmental factors operating during fetal and early postnatal life, and that these can consequently lead to impaired lung function and increased disease risk much later in life: this has become known as "developmental programming" [2, 3]. These two concepts (tracking and developmental programming) are both manifestations of the same phenomenon, as both are caused by persistent changes in the structure and/or function of lung tissue. They are both likely a result of lung development proceeding along a genetically determined program, such that alterations in development at one stage cannot be corrected at a subsequent stage.

Evidence for developmental programming of the lung in humans has been derived from numerous epidemiological and clinical studies and in recent years has become supported and enhanced by an abundance of experimental evidence. Lung diseases of infancy and childhood thought to have fetal or neonatal origins include chronic lung disease of prematurity, lung hypoplasia, persistent pulmonary hypertension (PPH), and asthma. Pulmonary diseases of later life, such as chronic obstructive pulmonary disease (COPD), are thought to have their origins in early life [4]. Similarly, it is likely that accelerated lung aging may be a consequence of altered lung development, such as impaired alveolarization.

Fetal Causes of Altered Lung Development and Later Pulmonary Illness

Fetal Nutrition and Growth

Restricted fetal growth affects around 8–10 % of live births and is recognized as a significant cause of perinatal morbidity [5]. Multiple organs are known to be developmentally affected by fetal growth restriction (FGR), in that their structure and function can be persistently altered. One of the first studies to show a link between FGR and later pulmonary health was an epidemiological study of adult men in the UK [6]; that study showed statistical links between low birthweight (and weight at 1 year) on later lung function and death from respiratory illness. Since that study was published, similar studies from several countries have confirmed that FGR can have a persistent adverse effect on the developing lung [7]. These observations have led to numerous investigations of the underlying processes. The most common causes of FGR are placental insufficiency and a maternal diet that is restricted in calories or protein. Placental insufficiency usually occurs as a result of maternal vascular disease (e.g., hypertension or pre-eclampsia), multiple gestation or maternal tobacco smoking. In the presence of placental insufficiency and FGR, the fetus is likely to be hypoglycemic, hypolipidemic, acidemic, and hypoxemic [8, 9]. Thus, the etiology of altered lung development in the presence of FGR is likely to be multifactorial, and to date it has been difficult to isolate any given factor as being a cause of altered lung development.

Respiratory Effects of FGR Observed During Infancy and Childhood

During infancy, FGR has been associated with respiratory insufficiency and poor lung function [10–13]. It is now clear that this deficit in lung function persists into childhood and beyond, supporting the concept of "tracking" of lung function. School-age children who experienced FGR have poorer lung function, as evidenced by reduced expiratory flow rates and greater airway reactivity, compared to children of the same age born with normal birthweight [14–16]. Many infants with FGR are born preterm, and such infants are found to have even poorer lung function than FGR infants born at term [17]. In twin pairs of children and adolescents, it is apparent that the twin with the lower birthweight, indicative of FGR, has poorer lung function and increased airway reactivity [18, 19]. Such studies of twins are important because they eliminate many maternal and genetic factors.

Respiratory Effects of FGR Observed During Adult Life

Early studies on the relation between fetal growth and adult lung function only had access to data on birthweight. Therefore, such data could have included subjects who were of low birthweight as a result of preterm birth. Later studies have avoided the potentially confounding effects of preterm birth by only including subjects born at term [20]; such studies have in general confirmed a significant relationship between birthweight at term and respiratory illness during adulthood [21–24]. Nine studies of lung function in male and female subjects, performed in several different countries, formed the basis of a meta-analysis of birthweight effects on adult lung function [7]; the analysis revealed a small but significant reduction in adult forced expiratory volume in 1 s (FEV_1) with decreasing birthweight. Taken together with lung function data from infants and children, the above studies provide strong evidence that FGR causes changes in lung development that impair lung function throughout postnatal life and may contribute to COPD in adult subjects of low birthweight [25, 26]. Although there are apparently no morphometric analyses of the lungs of adults exposed to FGR, data from animal studies suggest that impaired alveolarization and structural alterations to small conducting airways (see below) play an important mechanistic role.

Experimental Studies

Numerous animal studies using a variety of species have been performed in an attempt to elucidate the mechanisms by which FGR alters fetal lung development [10]. Experimentally, FGR has been induced by maternal food restriction (often protein restriction), mechanical constriction of the uterine or umbilical arteries, maternal hyperthermia or pre-pregnancy ablation of placentation sites.

Using sheep, a series of studies has shown that FGR, experimentally induced by placental insufficiency, has a profound effect on fetal lung development. In such models, the major factors leading to changes in organ development are likely to be fetal nutrient restriction and hypoxemia, as well as possible elevations in circulating corticosteroids. FGR was induced to coincide with the saccular and early alveolar stages of lung development, which occur before birth in sheep as in humans [27]. The effects of FGR on lung architecture were studied in the near-term fetus, 8-week-old lamb (childhood equivalent), and 2-year-old (adult) sheep. Interestingly, some changes not present before term became apparent with postnatal development. At 8 weeks of age, there were 31 % fewer alveoli in FGR lambs, compared to controls, and this reduction was still evident in adult sheep [28, 29]. In accordance with fewer alveoli, the alveoli of FGR animals were larger at both 8 weeks and 2 years. Thus, FGR induced by late-gestational nutrient restriction led to a permanent reduction in alveolarization, which resulted in a reduction in alveolar surface area relative to both lung volume and body weight [29]. It was also found that the inter-alveolar walls were significantly thicker than in control animals at 8 weeks (73 % thicker) and 2 years (47 % thicker) [29] due to an increased extracellular matrix; as alveolar walls normally become progressively thinner with lung development, thicker alveolar walls suggest persistently inhibited development. Importantly, the blood–air barrier was also thicker: in adults, it was 43 % thicker after FGR than in control animals [29]. The combination of fewer alveoli, thicker alveolar walls, and a thicker blood–air barrier would be likely to reduce pulmonary compliance and diffusing capacity, as found at 8 weeks [30]. The persistence of such alterations in lung architecture into adulthood would be expected to impair adult lung function and to increase the rate of the age-related decline in lung function.

The findings in sheep are largely consistent with findings in other laboratory species following nutrient restriction coincident with lung development [31–33]. Using rodent models of FGR, it has been shown that the deposition of extracellular matrix proteins, including elastin, is impaired by FGR [32, 34], and that the impairment of alveolarization is likely to be mediated by alterations in retinoic acid receptors [35]. In addition to alterations in alveolar structure, pulmonary surfactant has been reported to be affected by FGR, although findings are not consistent between models [10].

The conducting airways may also be affected by FGR, although few structural studies have been undertaken. In near-term fetal sheep subjected to late-gestation FGR, it was found that airway walls were proportionally thinner than in controls (less cartilage and epithelium); however, these changes had resolved by 8 weeks after birth [36]. There was also evidence that airway submucosal glands were attenuated in the FGR animals after birth [36].

The mechanisms by which FGR alters lung development, with some changes persisting into adult life, are still under investigation. In most cases, FGR results from placental insufficiency, which induces a complex and variable mix of physiological alterations in the fetus, including reduced nutrient supply, altered levels of growth factors, hypoxemia, and endocrine alterations, including elevated levels of circulating corticosteroids [37]. Each of these factors has the potential to alter aspects of prenatal lung development, including alveolarization. Owing to the postnatal lung's limited ability to recover from structural deficits that occur during development, many of the alterations induced during fetal (and neonatal) life are likely to persist throughout life. There is now strong evidence that FGR induces epigenetic alterations in genes that are critical to lung development [38].

Amniotic Fluid Volume

The fetal lung develops in an expanded state; indeed, the volume of luminal liquid within the lung during fetal life (fetal lung liquid) is greater than functional residual capacity (FRC) in the postnatal air-filled lung [39]. A remarkable feature of fetal lung development is that the degree of lung expansion by fetal lung liquid has a profound, direct effect on lung growth. This is clearly demonstrated experimentally by the removal of lung liquid leading to cessation of lung growth and lung hypoplasia; conversely, chronically increasing the degree of fetal lung expansion (e.g., by tracheal occlusion) leads to augmented growth of lung tissue [39].

An inadequate amount of amniotic fluid surrounding the fetus (oligohydramnios) is a common cause of fetal lung hypoplasia. Clinically, oligohydramnios is associated with respiratory insufficiency at birth and increased hospitalization during childhood [40]. Oligohydramnios can be caused by reduced production of amniotic fluid (usually due to fetal renal abnormalities) or by premature rupture of the fetal membranes. The lack of amniotic fluid allows the elastic uterine wall to "force" the fetal trunk into a state of increased flexion; this exaggerated trunk flexion increases intra-abdominal pressure, elevating the fetal diaphragm and compressing the fetal lung, as observed in the sheep and human fetus [41, 42]. Thoracic dimensions are reversibly reduced by experimental oligohydramnios [43]. The degree of lung hypoplasia at the time of birth is related to the duration and severity of the oligohydramnios. Oligohydramnios caused by reduced fetal urine production occurs from earlier in gestation than oligohydramnios resulting from membrane rupture; hence, the effects on the developing lung are likely to be more severe.

Attempts have been made to restore amniotic fluid volume, and hence fetal lung expansion, by infusing an appropriate fluid into the amniotic sac. In an ovine model of oligohydramnios, serial infusions of artificial amniotic fluid led to a reduction in lung hypoplasia and normalized airspace fraction [44]. A recent review of five randomized controlled trials suggests that amnioinfusion leads to a reduction in pulmonary hypoplasia [45]. It is likely that amnioinfusion improves fetal lung development by reducing the degree of fetal trunk flexion, allowing the fetal lungs to expand; it may also improve placental perfusion, resulting in improved oxygenation and nutrient delivery to the fetus [45].

Congenital Diaphragmatic Hernia

A less common cause of fetal lung hypoplasia is CDH, in which a defect in closure of the diaphragm (unilateral or more rarely bilateral) allows the lung to recoil and abdominal contents to enter the chest. The defect normally occurs during embryogenesis, resulting in the lung being unexpanded from an early stage in its development. CDH affects 1–4 per thousand live births, and usually has a poor prognosis; it is fatal in about 30 % of cases.

Fetal lung hypoplasia, if severe, can lead to respiratory distress or death at birth. The hypoplastic lung has a reduced number of airway generations, fewer alveoli, an underdeveloped vascular network, and hence a reduced capacity for gas exchange. The extent to which the hypoplastic lung can recover structurally after birth is unclear, as few studies have examined the potential for catch-up lung growth following intrauterine compromise. Given the limited ability of the postnatal lung to form new alveoli, it is likely that the lungs of survivors of CDH have reduced gas exchange, which may in turn limit exercise capacity. A recent study of almost 100 survivors of CDH aged up to 3 years showed that lung function was impaired; the extent of impairment was related to the degree of lung hypoplasia at birth and the duration of mechanical ventilation [46].

One of the major long-term complications of CDH is PPH, which is thought to be a result of poorly developed vasculature in the lung, and impaired vasodilator mechanisms [47]. PPH is evident in infants and children with CDH, and is likely to persist into adulthood, potentially contributing to adult lung disease; however, current survivors of CDH are still young and therefore there is little information available on the long-term outcomes of CDH survivors.

The fetal lungs can be expanded by occlusion of the trachea, due to the continued production of fetal lung liquid by the distal lung epithelium [39]. In animal studies, tracheal occlusion has been shown to increase pulmonary luminal volume, thereby rapidly accelerating lung tissue growth and alveolarization [48]. Currently, there is an ongoing clinical trial of tracheal occlusion performed endoscopically under local anesthesia in fetuses diagnosed with CDH [49].

Fetal Exposures

Fetal exposure to an increasing number of chemicals is now known to alter lung development, with long-term consequences. These include the products of maternal tobacco smoking, alcohol ingested by the mother, antenatal corticosteroids, and pro-inflammatory cytokines. Each of these exposures has been reported to alter lung development in the fetus, with potential long-term consequences for lung function and respiratory health.

Maternal Tobacco Smoking and Nicotine

Tobacco smoking during pregnancy has long been known to impact upon placental function, fetal lung development, and postnatal respiratory health in exposed offspring. Fetal exposure to the products of maternal smoking leads to an increased postnatal risk of pulmonary infections, wheezing, reduced lung function (low FEV_1), and asthma [50–52]. The products of maternal smoking that are the most important causal factors in altering fetal lung development are likely to be nicotine and carbon monoxide, both of which readily cross the placenta, although many other chemicals may also play a role. Numerous experimental studies have been performed to identify the mechanisms by which maternal smoking alters lung development: these have largely focused on the role of whole smoke or nicotine administered to the mother [53]. Such studies have shown that fetal nicotine exposure permanently impairs alveolarization and reduces the ability of exposed offspring to protect themselves against environmental stressors. Nicotine has been shown to be genotoxic, to increase the production of reactive oxygen species (ROS), and to reduce the antioxidant capacity of the lung [54]. Thus, the fetal lungs develop in an environment of oxidant–antioxidant imbalance, with concomitant adverse effects of oxidants and nicotine on cell integrity [54]. As a result of fetal disturbances to lung development, exposed offspring are more prone to develop respiratory diseases such as asthma and emphysema later in life [53, 54]. Recent studies have shown that retinoic signalling in the fetal lung may be perturbed by maternal smoking [50], and that collagen deposition may be increased via upregulated transforming growth factor (TGF)-β signalling [55]. It is now considered likely that maternal smoking is capable of inducing alterations in the fetal epigenome that could result in life-long alterations in lung function [56].

Maternal Alcohol Consumption

Recent studies suggest that a substantial number of women consume alcohol during pregnancy, with 30 % of women in the USA reporting consumption of alcohol at some stage during their pregnancy and 8.3 % reporting consumption of four or more alcoholic drinks on one occasion at least once during pregnancy [57]. This is cause for concern, as experimental studies indicate that ethanol readily crosses the placenta [58]. Indeed, chronic maternal alcohol consumption can lead to a spectrum of developmental defects in the offspring, collectively termed fetal alcohol spectrum disorders (FASD). The incidence of FASD is estimated to be 1 in 100 live births, although the prevalence could be as high as 2–5 % [59, 60]. Fetal alcohol syndrome is the most severe form of FASD and is characterized by growth restriction, craniofacial dysmorphology, and central nervous system dysfunction. Although most research into the effects of fetal ethanol exposure has been focused on the developing brain, it has also been associated with congenital abnormalities of the respiratory tract [61] and an increased susceptibility to respiratory infections in childhood [62]. In animal models, exposure to high levels of ethanol throughout gestation or at mid-gestation causes reduced fetal lung growth [63, 64]; however, moderate exposure in late gestation has no effect [65]. Despite a lack of effect on lung growth or alveolarization, moderate ethanol exposure in late gestation increases collagen deposition in the ovine fetal lung, decreases surfactant protein (SP) levels, and alters surfactant phospholipid composition [65–67]; such changes could compromise lung function after birth. Evidence also suggests that prenatal ethanol exposure may reduce the ability of the lungs to clear pathogens, as indicated by decreases in ciliary beat frequency [66] and pro-inflammatory cytokine gene expression [65], an increase in apoptosis of alveolar macrophages and a decrease in alveolar macrophage function both under normal conditions and in response to infection [68–70]. A recent study of prepubescent sheep exposed to ethanol in late gestation, however, reported no significant differences in lung growth, lung architecture, pro-inflammatory cytokine gene expression, and surfactant phospholipid composition, although *SP-D* mRNA levels were increased [67]. As ethanol exposure did not continue to birth or after birth, this study suggests that the developing lung has some capacity to undergo repair in the absence of further exposure. As SP-D has immunoregulatory roles and roles in surfactant phospholipid homeostasis [39], the observed increase in *SP-D* mRNA levels following ethanol exposure suggests that innate pulmonary immunity and lung function may be persistently altered. Further studies investigating the long-term effects of fetal ethanol exposure on the lung are warranted.

Antenatal Exposure to Corticosteroids

During the last several decades, corticosteroids have been routinely administered to pregnant women who show signs of preterm delivery (8–12 % of pregnancies). These agents, primarily betamethasone and dexamethasone, are administered during preterm labor to enhance lung maturation in the preterm fetus, thereby increasing its chances of survival. The principal beneficial effect of corticosteroids for preterm survival is the stimulation of surfactant production and release by the immature lung; this occurs as a result of an increase in the number and maturity of alveolar epithelial type II cells [71]. It has long been recognized that, as well as stimulating surfactant production, corticosteroids inhibit secondary septation and alveolarization in the immature lung when administered before or during the process of alveolarization [72]. The effects of corticosteroids during development can be long lasting in many tissues [73], depending on the timing and duration of corticosteroid exposure, and the lung is also likely to be permanently affected, potentially as a result of epigenetic modifications [74]. This persistent effect has been clearly demonstrated in rats exposed to corticosteroids during early postnatal life [75] and is thought to result from a persistent imbalance between cell proliferation and cell death in secondary septa [76]. Because such a high proportion of infants who are born preterm (i.e., birth prior to 37 completed weeks of gestation) are exposed to corticosteroids, the adverse effects of corticosteroids on structural development of the lung, and hence on later lung function, are likely to be a significant factor in the etiology of respiratory illness and reduced lung function during all stages of life.

Intrauterine Inflammation (Chorioamnionitis)

In many pregnancies, the fetus is exposed to inflammation of the surrounding membranes (chorion and amnion) as a result of bacterial infection; this is often referred to as chorioamnionitis. Intrauterine inflammation is present in up to 60 % of pregnancies that result in preterm birth, and it is now considered that inflammation plays a causal role in the etiology of preterm birth [77]. In such pregnancies, the fetus is exposed to inflammatory mediators over prolonged periods, and it is now clear that these mediators can enter the fetal lungs [78]. Since the first demonstration that intrauterine inflammation improves respiratory function in the preterm neonate [79], numerous studies have investigated the impact of intrauterine inflammation on fetal lung development. Such studies have shown that lipopolysaccharide (LPS), a component of the Gram-negative bacterial cell wall, and pro-inflammatory cytokines accelerate the maturation of alveolar epithelial type II cells in the preterm fetus, such that they produce more surfactant; in addition, pulmonary vascular development is enhanced [80]. It has also been reported that intrauterine inflammation inhibits pulmonary elastin deposition and alveolarization [81]; these effects may be transient, as other studies have not shown a persistent effect of intrauterine inflammation on lung morphometry [82, 83]. The long-term effects of intrauterine inflammation on the lungs have received very little attention. However, a recent study in sheep has shown that it can lead to persistent changes in gene expression of pro-inflammatory cytokines in the postnatal lung, as well as reductions in protein expression of SP-A and -C [82]; together, these changes may impair innate immunity of the postnatal lung. It is clear that more research is needed regarding the pulmonary programming effects of intrauterine inflammation.

Neonatal Causes of Lung Disease or Altered Lung Development

Early postnatal environments that interfere with normal lung development and maturation may result in persistent alterations in lung structure and function, increasing the risk for respiratory disease in adult life. The first two years of life are an important time for lung development, and many studies have shown that a large proportion of adult respiratory diseases stem from events that occur during infancy and early childhood. In the sections below, we provide an overview of some of the factors in the neonatal period that can alter lung development and may be key contributors to adult lung disease.

Preterm Birth

Preterm birth, defined as birth prior to 37 completed weeks of gestation, affects approximately 11 % of pregnancies worldwide [84]. Preterm birth can be categorized into very preterm birth (<32 weeks' gestation) and extremely preterm birth (<28 weeks' gestation). As lung development occurs throughout gestation, and critical stages of lung development occur late in gestation, babies born preterm are born with structurally and functionally immature lungs. Very preterm infants are born during the saccular stage of lung development, which is prior to the formation of definitive alveoli. Extremely preterm babies can be born during the canalicular (pre-saccular) stage of lung development and therefore are born with a thick lung interstitium and prior to the production of surfactant. Owing to lung immaturity, infants born more than 6–8 weeks before term usually experience respiratory insufficiency at birth, referred to as the respiratory distress syndrome (RDS). Therefore, infants who are born very preterm or extremely preterm usually require respiratory support to ensure their survival; the degree of support and the length of stay in hospital are proportional to the degree of prematurity [85]. Respiratory support of preterm infants can involve supplemental oxygen therapy and/or assisted ventilation, including continuous positive airway pressure (CPAP), nasal intermittent positive pressure ventilation (nIPPV), or mechanical ventilation via an endotracheal tube. Improvements in the respiratory care of preterm infants, especially the introduction of exogenous surfactant therapy, more gentle ventilation strategies and more cautious use of supplemental oxygen, have resulted in an increase in their survival, in particular those who are born at earlier gestational ages. Despite the necessity of this respiratory intervention, the incidence of chronic lung disease or bronchopulmonary dysplasia (BPD) in this population remains high [86]. BPD is characterized by disrupted lung development, in particular alveolar simplification (fewer and larger alveoli), variable interstitial fibrosis, and increased airway smooth muscle (see Fig. 5.1) [87, 88].

Although BPD is a multifactorial disease, major factors in its etiology are thought to be inflammation, oxidative stress, and inappropriate repair mechanisms due to prolonged exposure to high oxygen concentrations (hyperoxia) during supplemental oxygen therapy or mechanical ventilation [89–91]. Mechanical ventilation can contribute to lung injury by inducing inflammation as a result of repetitive tissue stretch (shear stress) and over-inflation (barotrauma) of incompliant lung tissue; particularly damaging is repetitive opening of collapsed alveoli in atelectatic regions of the surfactant deficient lung [92]. Numerous other fetal and neonatal factors associated with preterm birth can contribute to the pathogenesis of BPD or alterations in lung development; such factors include the severity of prematurity, exposure to antenatal or postnatal corticosteroids, prenatal infection or inflammation, postnatal infection, intrauterine growth restriction, and inadequate postnatal nutrition (see Fig. 5.2) [91, 93]. Regardless of etiology, it is now apparent that persistent changes within the lungs of preterm infants, especially those who developed BPD, lead to an increased risk of poor lung function in childhood and

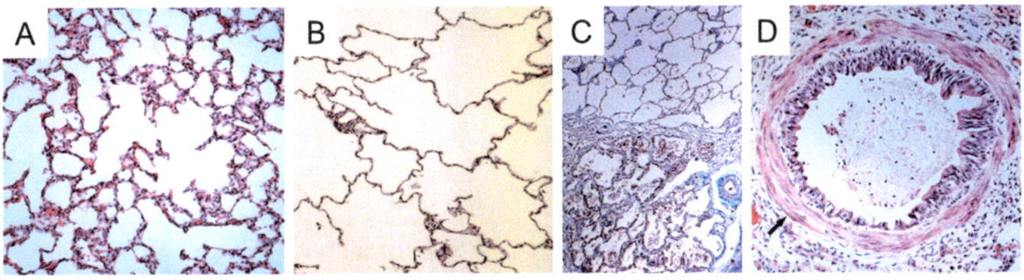

Fig. 5.1 Histological representation of the alveolar region of a lung from (**a**; ×25 magnification) a term infant, and (**b**; ×50 magnification) a surfactant-treated preterm infant with BPD. Photomicrographs of a preterm infant with "healed" BPD, demonstrating (**c**; ×30 magnification) diffuse alveolar septal fibrosis (*bottom*) with enlarged alveoli (*top*), and (**d**; ×100 magnification) hyperplastic smooth muscle (*arrow*) surrounding a bronchiole (modified from [196])

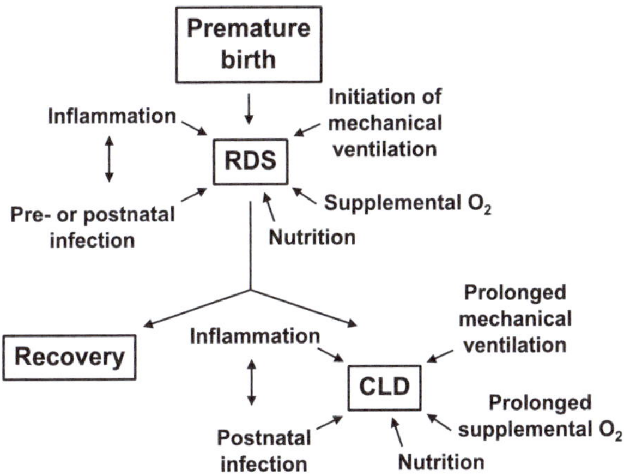

Fig. 5.2 Factors that are likely to contribute to respiratory distress syndrome (RDS) in infants who are born preterm, and factors that may lead to the subsequent development of chronic lung disease (CLD) [93]

late adolescence, as indicated by reductions in FEV_1, forced mid-expiratory flow ($FEF_{25–75\ \%}$), peak expiratory flow (PEF), transfer across the lung of carbon monoxide (TL_{CO}), and increased airway resistance [94–99]. Individuals who were born preterm also have an increased risk of reduced exercise capacity [97, 98], asthma [94, 99–101], COPD [102], and respiratory infections [103–105] later in life. The contribution of hyperoxia, mechanical ventilation, and corticosteroids to altered lung development, and later lung disease following preterm birth will be discussed in more detail in the following sections.

Hyperoxia

Preterm infants are often administered hyperoxic gas to increase blood oxygen levels. Oxygen is added to inhaled gas to reach a target arterial oxygen saturation (SaO_2) of 85–99 %, with the aim of preventing tissue hypoxia. In the past, the use of high oxygen concentrations (up to 100 % O_2) was common, although concentrations were reduced once it became apparent that high concentrations could lead to tissue injury [106–108]. Currently, preterm babies are normally resuscitated using room air (21 % O_2) and then the fraction of inspired oxygen (FiO_2) is increased until adequate blood oxygenation levels are achieved [109]. However, as fetuses normally have an arterial partial pressure of oxygen (PaO_2) of approximately 25–30 mmHg and a SaO_2 of 70–80 %, exposure of an immature lung with an immature antioxidant system to even low supplemental oxygen concentrations may be detrimental. It is difficult to determine the contribution of neonatal hyperoxia alone to the long-term pulmonary outcomes of preterm birth and BPD. However, oxygen supplementation during the neonatal period in infants born very preterm has been identified as an independent risk factor for asthma in childhood [101]; further, prolonged oxygen therapy is associated with low FEV_1 and the use of inhaled bronchodilators in very-low-birth-weight (preterm) children [96], suggesting an impact on the conducting airways.

Numerous experimental studies have shown that neonatal inhalation of hyperoxic gas can persistently alter the development of alveoli, conducting airways and pulmonary vasculature. Specifically, neonatal hyperoxia leads to fewer and larger alveoli (reviewed in [110]), altered alveolar epithelial cell differentiation [111], altered cellular composition of the bronchiolar epithelium [112–114], altered surfactant composition [111, 115], increased airway smooth muscle [116–119], decreased bronchiolar collagen [112], decreased pulmonary capillary density [120], and remodelling of the lung extracellular matrix [111, 121] and small pulmonary arteries [120]. Consequently, neonatal hyperoxia is associated with later reductions in lung function, including altered airway resistance and reactivity, dynamic lung compliance, tissue elastance and tissue damping, and pulmonary hypertension, as evidenced by right ventricular hypertrophy [111, 112, 120, 122]. Importantly, some of the effects of neonatal hyperoxia on the lung are long-lasting, with alterations in lung structure and function observed in early and advanced adulthood [111, 112, 118, 122–125]. Interestingly, recent studies and reviews have indicated that the oxygen concentration used and the duration of oxygen exposure are important determinants of lung injury [110, 111, 119].

Lung injury and alterations to lung structure that are induced by neonatal exposure to hyperoxia are thought to be due to oxidative stress and a subsequent pro-inflammatory response (see Fig. 5.3) (reviewed in [110]). Experimentally, neonatal treatments aimed at minimizing ROS and exploiting antioxidant defense, as well as treatments that target inflammatory cell influx into the lung and reduce the degree of lung inflammation, have shown promise in improving lung structure following neonatal hyperoxia [110]. Due to the importance of pulmonary vascular

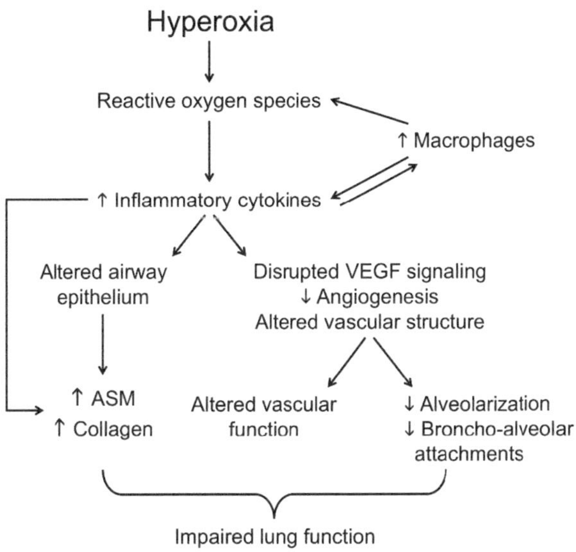

Fig. 5.3 Potential pathways whereby exposure of the developing lung to hyperoxia leads to alterations in the structure and function of the airways, pulmonary vasculature and alveoli, thus contributing to impaired lung function later in life. *ASM* airway smooth muscle, *VEGF* vascular endothelial growth factor [110]

development in regulating alveolarization, treatments promoting angiogenesis following neonatal hyperoxia have also received much attention [110]. Recently, the use of stem/progenitor cells to repair or prevent lung injury has also been trialled in experimental models, and clinical trials for their use in preterm infants are currently underway, as discussed below.

Mechanical Ventilation

Owing to lung immaturity, preterm infants usually require assisted ventilation. Different ventilation strategies are employed depending on the requirement of the infant; these range from noninvasive strategies, including those involving constant airway pressure (i.e., CPAP using a constant or variable flow, or nasal cannula with low or high flow) or variable airway pressure (i.e., nIPPV, nasal positive airway pressure or nasal high-frequency ventilation), to invasive positive pressure ventilation requiring endotracheal intubation, including patient-triggered ventilation, volume target ventilation, and high-frequency oscillation [126]. It is now recognized that invasive ventilation strategies can increase the incidence of BPD and lung injury, which can increase the risk for persistent lung disease in infants born preterm. Furthermore, intubation is associated with trauma at the site of intubation (i.e., larynx and upper trachea), as well as bacterial colonization, and sepsis [126]. Noninvasive ventilation strategies have been shown to reduce these adverse effects associated with mechanical ventilation, improve gas exchange, and reduce the incidence of BPD [127]. Thus, not only are noninvasive ventilation strategies now used for weaning infants off mechanical ventilation to reduce extubation failure, they are increasingly being used as primary methods of respiratory support for preterm infants with respiratory insufficiency. However, even noninvasive techniques have been associated with adverse effects, including pneumothorax, pulmonary over-inflation, increased work of breathing, decreased cardiac output, as well as nasal complications and gastrointestinal distension and perforation [126]. Large randomized controlled trials are required to determine the beneficial effects of different modes of noninvasive ventilation on long-term respiratory and non-respiratory outcomes.

Ventilator-induced lung injury is believed to be a result of either volutrauma or barotrauma, which is injury related to over-expansion of the lungs (due to high volumes or pressures), atelectrauma, which is injury caused by repeated opening of collapsed alveoli, or biotrauma, which is injury caused by up-regulation of pro-inflammatory responses to mechanical ventilation, or a combination of these factors [126]. In the clinical setting, it is difficult to determine the precise cause of lung injury in preterm infants due to the multifactorial and individualized nature of their care, both before and after birth. Experimentally, it has been shown in rodents and sheep that even brief mechanical ventilation (2–24 h) of the developing lung alone (i.e., without other confounding factors) leads to impaired formation of alveoli and lung microvessels, as indicated by decreases in secondary crest density, alveolar number, and CD31 staining, and increases in airspace and alveolar size; these changes in lung structure are likely due to disruption of elastin deposition in the lung [92, 128–130] and altered expression of other genes/proteins that regulate alveolar formation and angiogenesis, including *platelet derived growth factor-A* (*PDGF-A*), *tenascin-C*, and vascular endothelial growth factor receptor 2 (VEGF-R2) (see Fig. 5.4) [92, 131]. Studies have also identified numerous transcription factors and genes that are dysregulated by mechanical ventilation, contributing to lung injury and altered lung development; likely factors include those involved in TGF-β signalling, of which elastin is a target, mechanotransduction and the immune response [132–134]. Furthermore, recent studies have shown that invasive mechanical ventilation in preterm lambs causes an increase in histone deacetylatase (HDAC) 1 and subsequent genome-wide histone hypoacetylation in the lung, indicating alterations in the epigenetic regulation of gene expression [135]. Insulin-like growth factor-1 (IGF-1) is one gene whose expression is increased in mechanically ventilated preterm lambs and in infants who have died from RDS or BPD, and which appears to be epigenetically modified by mechanical ventilation [135]. Further studies are required to identify other genes that are modified by mechanical ventilation and the specific pathways leading to their alteration [93]. A greater understanding of the mechanisms by which mechanical ventilation alters gene expression and leads to altered lung development will enable targeted interventions to be developed. In this regard, histone acetylation is increased and alveolar formation is improved when preterm lambs that are mechanically ventilated are treated with HDAC inhibitors [135].

In addition to altering alveolar structure, a short period (2–12 h) of positive pressure ventilation of the immature (prenatal) sheep lung induces injury and remodelling of the small conducting airways (i.e., bronchioles), including epithelial cell detachment, alterations in the areas of epithelium and collagen in intact airways, and an increase in airway smooth muscle [129, 130, 136]. There is

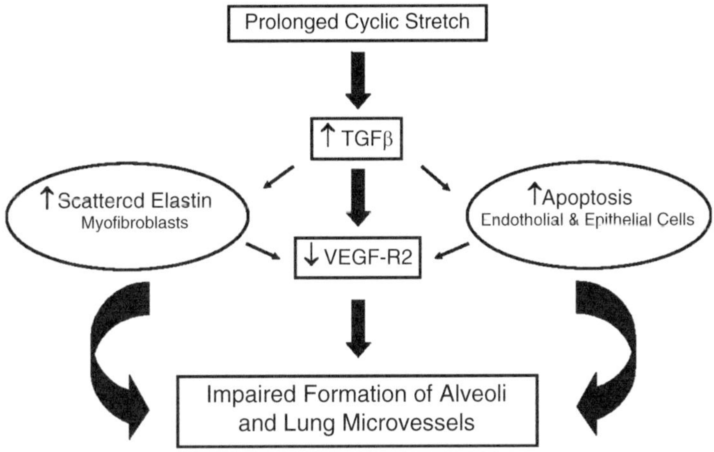

Fig. 5.4 Potential pathways whereby prolonged cyclic stretch of the lung caused by positive-pressure mechanical ventilation alters the formation of the pulmonary vasculature and alveoli. *TGF-β* transforming growth factor-β, *VEGF-R2* vascular endothelial growth factor receptor 2 [92]

evidence that airway remodelling persists in the absence of further ventilation, with changes in epithelial area, epithelial cell proliferation and apoptosis, airway smooth muscle and collagen, and a decrease in the number of bronchiolar-alveolar attachments [129, 130, 136]. There is also evidence, however, that the fetal lung is capable of spontaneous repair following ventilation-induced injury, with normalization of alveolar structure in the absence of continued ventilation [129, 130], although this recovery could potentially be restricted to the fetal lung. Changes in alveolarization and airway wall structure following mechanical ventilation could contribute to the increased risk of poor lung function and obstructive lung disease reported in individuals who were born preterm. Indeed, airway reactivity is increased 48 h after brief mechanical ventilation of rat pups [137]. Further studies investigating the long-term effects of mechanical ventilation of the immature lung on respiratory health are needed.

Corticosteroids

Corticosteroids have been used in preterm infants with evolving or established BPD to reduce lung inflammation and improve gas exchange; corticosteroids are used to reduce the need for oxygen supplementation [91] and to decrease the risk of retinopathy of prematurity and death [138]. Postnatal corticosteroids, predominantly dexamethasone, are generally administered systemically and can be given early ($\leq$7 days after birth) or late ($>$7 days). Although systemic corticosteroid therapy can have short-term benefits, it has been linked to adverse long-term outcomes, including detrimental effects on lung structure, neurodevelopment, cardiovascular function, and the gastrointestinal system [91, 138]. Although many of these effects have been linked to the prolonged use of high-dose corticosteroids (especially dexamethasone), studies have indicated that even low-dose corticosteroids may have adverse effects [91]. Inhaled corticosteroids have been trialled in preterm infants with the aim of reducing the side effects associated with systemic corticosteroid use, although these have had limited success [91, 138]. Studies investigating the optimal timing, dose, and type of corticosteroids given to preterm infants to improve short-term outcomes and reduce adverse long-term outcomes are warranted.

Nutrition

Adequate nutrition plays a key role in normal lung development. Experimentally, reduced nutrition during lung development before and after birth can affect alveolarization and bronchiolar structure [28, 31, 139]. The effects of undernutrition during fetal life on lung development have been described above. Lung development continues after birth and it has been shown in sheep that, at maturity, animals that experienced postnatal growth restriction have a smaller lung volume and a smaller pulmonary surface area, relative to lung and body size, than animals that grow normally [140]. Undernutrition during the early postnatal period has been shown to disrupt lung development, especially when combined with exposure to hyperoxia [112, 141, 142]. This is of particular importance in the context of very preterm infants who form most of their alveoli after birth. In addition, very preterm infants have a high risk of developing BPD, and it is known that malnutrition represents one of the many factors contributing to BPD [143]. Maintaining a careful balance of nutritional support for very and extremely preterm infants represents a considerable challenge. Interventions that are necessary to support the survival of these preterm infants, combined with developmental complications they may encounter (e.g., prolonged orotracheal intubation, medications that alter gastrointestinal motility, altered palatal groove formation, irritability due to neurological status, gastrointestinal immaturity) make feeding by mouth a problematic task; therefore, nutritional support is often provided via parenteral or enteral routes [144]. However, most extremely preterm infants receive minimal enteral feeding during the first two weeks of life, due to the abovementioned problems, and their nutritional requirements need to be supplemented with parenteral feeding. Recently, it was reported that a critical minimal amount of enteral feeding is required to prevent the development of BPD [145]. Preterm infants who developed BPD received less enteral feeding during the first two weeks of life, even though it was compensated by parenteral nutrition [145]. Furthermore, it has been reported that, compared to preterm infants without BPD, those with BPD have a greater energy expenditure (up to 25 % above total caloric needs), likely due to their increased work of breathing, stress, inflammation, medications, and the perceived need for catch-up growth [143, 144].

Another aspect of nutritional management of preterm infants that has been associated with an increased risk for BPD is fluid intake. It has been reported that high fluid intake during the first days after birth may increase the risk of developing BPD, through the persistence of a patent ductus arteriosus and a higher fluid content in the pulmonary interstitial tissue; these complications could contribute to decreased lung compliance and increase the need for respiratory support [143, 146]. Overall, it is suggested that adequate nutritional management of preterm infants with evolving or established BPD can be maintained by adopting enteral nutrition as soon as possible, avoiding excess fluid administration during the first days after birth, and progressively increasing energy intake [147].

The use of specific micronutrients, such as vitamins A, C, and E, inositol, iron, selenium, glutamine, cysteine, and methionine, may be beneficial in the nutritional management of preterm infants to prevent or lessen the severity of BPD (reviewed in [144]). Indeed, optimal feeding and growth in preterm infants remains a challenge, and overnutrition of the preterm infant to achieve catch-up growth may actually be detrimental to long-term health [148]. Although accelerated growth of the preterm infant has beneficial effects on neurodevelopment, it has also been associated with many metabolic and cardiovascular implications that present in childhood and adulthood [148]. Further research is needed in order to identify the optimal method of managing preterm nutrition with the goal of enhancing lung growth, preventing BPD, and avoiding long-term, potentially harmful programming effects of catch-up growth.

Although infants born preterm are faced with a vulnerable period during which nutrition can greatly affect their lung development and subsequent lung health, nutritional status can also impact upon the mature lung. Emphysema is commonly observed in patients with anorexia nervosa and was also noted at autopsies of humans who died from starvation [149]. These findings support the concept that lung structure/function is dependent upon nutritional status.

Respiratory Infections

The link between early-life respiratory infections, particularly viral infections, and subsequent increased risk of recurrent wheezing and asthma is being increasingly investigated. Viral respiratory infections during infancy have been shown to contribute to the development of wheezing in childhood. In particular, the occurrence of symptomatic rhinovirus infection during infancy (<12 months of age) has been identified as the most significant risk factor for the development of wheezing 1 year after infection [150] and into early childhood [151]. Infection with respiratory syncytial virus (RSV) in infancy has also been linked with wheezing in pre-school children, albeit with a weaker relationship compared to rhinovirus infections [151]. A later study confirmed these findings, showing that bronchiolitis caused by viruses other than RSV was associated with a greater risk of recurrent wheezing (within the first 2 and 3 years after hospitalization for bronchiolitis) compared to RSV-associated wheezing [152]. Not only has rhinovirus infection been associated with recurrent wheezing, but rhinovirus infection during the first

3 years of life is also significantly associated with reduced lung function (lower FEV_1, percent predicted FEV_1, $FEV_{0.5}$, $FEF_{25-75\ \%}$, FEV_1/forced vital capacity (FVC), and $FEV_{0.5}$/FVC) at 8 years of age compared to children with non-wheezing rhinovirus illnesses and no rhinovirus illnesses [153]. Furthermore, this study showed that children with RSV-induced wheezing did not have significant alterations in any of the lung function measurements, indicating that infections in early childhood caused by rhinovirus, rather than RSV, are significant predictors of decreased lung function in childhood [153]. Not only are viral respiratory infections during early postnatal life associated with subsequent wheezing, but they often precede the development of childhood asthma. In a cohort of children that was followed up to 6 years of age after specific viral respiratory wheezing illnesses in early childhood (within the first 3 years of life), rhinovirus-associated wheezing illnesses were the most significant predictors of the subsequent development of asthma [154]. This, however, does not exclude RSV as a potential risk factor for asthma. Recent studies showed that severe RSV-induced bronchiolitis during infancy (<12 months of age) is associated with an increased risk of asthma at 6 and 7 years of age [155, 156]. Furthermore, RSV infection in infancy was shown to be associated with a 3.2-fold increase in the risk of wheezing and impaired lung function (lower FEV_1, FVC, and FEV_1/FVC, and increased respiratory system resistance) at age 6 [155]. Another recent study reported on a study population at 15–18 years of age, and compared patients hospitalized for bronchiolitis in infancy (<24 months of age) caused by RSV or rhinovirus to a control population [157]. This study showed that patients hospitalized in infancy for bronchiolitis had an increased risk of asthma in adolescence; of the bronchiolitis patients, those who had rhinovirus infection had a higher risk of self-reported asthma, but not doctor-diagnosed asthma, than those who had RSV infection [157]. Studies that have extended further into adulthood have found that hospitalization for bronchiolitis in infancy (<24 months of age) is associated with an increased risk of asthma, and an increased use of asthma medication at 28–31 years [158]. In that study, doctor-diagnosed asthma was present in adulthood in one-third of former bronchiolitis patients, whereas asthma was present in only 10 % of controls. An earlier study of the same cohort demonstrated that the prevalence of asthma at 28–31 years was similar to that recorded 10 years earlier at 18–21 years of age [159], indicating the early-onset and persistence of long-term sequelae following early-life respiratory infection. However, these studies did not differentiate between RSV and rhinovirus bronchiolitis. The link between early-life respiratory infections and subsequent long-term respiratory morbidity has also been highlighted in a follow-up study that showed an association between RSV bronchiolitis in the first year of life and increased prevalence of asthma or recurrent wheeze at 18 years of age; 39 % of RSV subjects vs. 9 % of control subjects [160]. A reduction in spirometric airway function was reported in the RSV cohort compared to controls, regardless of whether the RSV subjects presented with or without asthma or recurrent wheeze [160].

Bacterial colonization of the hypopharynx with *S. pneumonia*, *H. influenza*, or *M. catarrhalis* in healthy neonates at 4 weeks of age has been associated with an increased risk of pneumonia and bronchiolitis in early life (up to 3 years) independent of asthma [161]; this suggests that an altered microbiome of the neonatal airway may affect the maturation of the immune system, leading to an increased risk of pneumonia and bronchiolitis. Furthermore, radiologically confirmed pneumonia in the first 2 years of life has been associated with asthma or asthma-like symptoms in pre-school children (4–5 years of age) [162]. This is confirmed in a meta-analysis that showed an overall risk of major long-term respiratory sequelae following childhood pneumonia (before 5 years of age) in non-hospitalized children of 5.5 %; this risk was almost three times greater in children who had been hospitalized, with adenovirus pneumonia having the highest risk (approximately 55 %) [163]. Major respiratory sequelae were defined as restrictive lung disease, obstructive lung disease, and/or bronchiectasis, and the most common type of major outcome reported was restrictive lung disease [163]. These findings indicate that children diagnosed with pneumonia should be considered at risk of long-term respiratory sequelae, even after acute symptoms have resolved.

Environmental Exposures

Many studies have highlighted the deleterious effects of a suboptimal environment on lung development (reviewed in [164]). Harmful environmental exposures include both outdoor pollutants (e.g., industrial and motor vehicle emissions) and indoor pollutants (e.g., environmental tobacco smoke (ETS) and fungal spores). In particular, ETS exposure in early life has been identified as a significant risk factor for poor pulmonary outcome in later life. A recent meta-analysis indicated that passive exposure to ETS (in both pre- and postnatal life) increases the incidence of wheeze and asthma in children and adolescents by at least 20 % [165]. Maternal smoking was associated with a 52 % increase in the risk of wheeze and a 20 % increase in the risk of asthma at 5–18 years of age [165]. Not only has ETS exposure during infancy been associated with increased risk of wheeze and asthma, but a meta-analysis has also shown it to increase the risk of lower respiratory infections, particularly bronchiolitis [166]. All types of passive ETS exposure (pre- and postnatal maternal, paternal, both parents, any household member smoking), but particularly maternal smoking, led to a significant increase in the risk of infants developing lower respiratory infections in the first 2 years of life [166]. This meta-analysis highlights the importance of limiting ETS exposure during both gestation and childhood to prevent the onset of wheezing, asthma, and respiratory infections. In contrast to the large number of reports on effects of ETS exposure, there are few reports on the effects of exposure to traffic-related air pollution during infancy and childhood and its effects on later respiratory morbidity. A recent systematic review and meta-analysis showed that early childhood exposure to traffic-related air pollution was associated with the development of asthma throughout childhood up to 12 years of age, with the magnitude of risk increasing with age

[167]. However, further studies are needed to investigate the relation between effects of exposure at specific ages and adult outcomes.

Potential Targets for Cell-Based Therapies: Neonatal Treatments for Preterm Infants

The use of cell-based therapies to prevent or repair neonatal lung injury in very preterm infants has now evolved from experimental studies using animal models to phase I and II clinical trials in very preterm infants at risk of BPD. The rationale for the current clinical trials is based on knowledge gained from studies examining the perturbation of pulmonary stem cells in neonatal lung injury, as well as experimental studies in animal models that found stem cell-based therapies to be beneficial. Studies in animal models and in humans have shown that altered function or depletion of resident stem cells within the developing lung likely contributes to the pathogenesis of BPD. In rodent models of BPD, in which newborn pups are exposed to high concentrations of oxygen, lung epithelial cells, mesenchymal stromal cells (MSCs), and endothelial progenitor cells (EPCs) were shown to be significantly reduced in number and also displayed altered differentiation potential [123, 168–170]. Similar perturbations to resident stem cells have been reported in tracheal aspirates and umbilical cord blood samples taken from preterm infants [171–175]. The presence of MSCs in tracheal aspirates has been reported as an indicator of BPD development, and these MSCs were shown to demonstrate a lung-specific gene expression and to secrete pro-inflammatory cytokines; this suggests that the MSCs in the tracheal aspirates originated from the lungs and have the capacity to contribute to lung injury [171, 172]. In samples of umbilical cord blood from preterm infants, endothelial colony forming cells (ECFCs; a subset of EPCs) were shown to have an increased susceptibility to in vitro exposure to hyperoxic gas, and were found in reduced numbers among BPD infants compared to non-BPD preterm infants [173–175]. This suggests that a reduction in the levels of angiogenic progenitor cells may contribute to disrupted vasculogenic mechanisms during lung development and lead to impaired vascular and alveolar growth. In response to the findings of stem cell perturbation in BPD, experimental studies have focused on their therapeutic replacement with exogenous-derived stem cells (reviewed in [176]). One of the most commonly researched cell types is the MSC, which can be easily sourced from bone marrow, adipose tissue, muscle, peripheral blood, umbilical cord blood, umbilical cord Wharton's jelly, and placenta. Numerous components of neonatal lung injury have been attenuated by the administration of MSCs, including inflammation, vascular damage, impaired alveolarization, and fibrosis [170, 177–188]. Furthermore, MSCs have been reported to have low engraftment and differentiation into the lung, suggesting a paracrine-mediated therapeutic effect. This has been supported by studies showing similar protective properties of cell-free conditioned media from MSCs in experimental models of neonatal lung injury. Other cell types (ECFCs, angiogenic cells, and amnion epithelial cells) have also been examined in experimental studies and have shown promise in their therapeutic benefit for neonatal lung injury [189–194].

The overall aim of experimental studies on neonatal lung regeneration has been to generate evidence to create a rationale for translating experimental findings into the clinic. Encouragingly, there has recently been translation of findings from animal studies into clinical trials in preterm infants, with the results of the first phase I clinical trial (NCT01297205) of stem cell therapy in preterm infants being published [195]. The authors assessed the safety and feasibility of human umbilical cord blood-derived MSCs, administered to nine very preterm infants who were at risk for developing BPD. The infants were born at 24–26 weeks of gestation (630–1030 g) and needed continuous ventilatory support that could not be decreased due to significant respiratory distress prior to enrolment into the study. Three patients were given a low dose of MSCs (1×10^7 cells/kg) via the trachea, and six were given a high dose (2×10^7 cells/kg). Treatment-related serious adverse effects were assessed for 84 days following transplantation of the MSCs. The authors reported no serious adverse effects or dose-limiting toxicity associated with transplantation of the MSCs. There was no significant difference in serious adverse effects between the low-dose and high-dose MSC group or between the MSC group and a matched-comparison group, except in BPD severity; the MSC group had significantly lower BPD severity than the matched-comparison group. Furthermore, the inflammatory markers matrix metalloproteinase (MMP)-9, interleukin (IL)-6, IL-8, tumor necrosis factor (TNF)-α, and TGF-β were reduced in tracheal aspirate fluid after MSC transplantation. The findings of this study are important in establishing the feasibility, safety, and efficacy of MSC treatment for BPD in preterm infants. Follow-up studies are currently underway to assess the long-term safety of MSC transplantation in the same preterm infants reported in the phase I clinical trial, at approximately 2 years of age (NCT01632475) and 5 years of age (NCT02023788). Furthermore, a randomized, double-blind, multicenter, phase II clinical trial (NCT0182957) has commenced that will investigate the efficacy and safety of a low dose of MSCs versus a control group for the treatment of BPD in preterm infants; this trial will be followed up with another study (NCT01897987) investigating the long-term safety and efficacy of the phase II trial. It is anticipated that findings from these current clinical trials will pave the way for the future treatment of neonatal lung diseases using stem cell-based therapies.

Conclusions

The lung develops during prenatal and early postnatal life, and is therefore vulnerable to adverse environmental conditions during the embryonic, fetal, and postnatal stages of lung development. The peripheral lung, including alveoli, the pulmonary microcirculation, and small conducting airways, is especially vulnerable to injury or altered development. Owing to the lung's limited capacity for repair, lung injury, or altered lung development caused by a suboptimal fetal or early postnatal environment can persist into later life, adversely affecting adult lung function and increasing the risk of respiratory illness. Changes in lung structure can be due to a direct developmental effect or to changes in gene and protein expression caused by alterations in the nutritional, physical/mechanical, oxidative, or immunological environments. In addition, a range of environmental factors may induce epigenetic alterations in key genes that regulate lung development to cause persistent alterations in lung structure and/or function. Although there are currently no treatments capable of repairing the injured lung, there is now encouraging experimental evidence on the effectiveness of cell-based therapies to prevent or repair injury to the immature lungs, with clinical trials now underway in infants who are born very preterm and are at risk of developing chronic lung disease.

References

1. Martinez FD (2009) The origins of asthma and chronic obstructive pulmonary disease in early life. Proc Am Thorac Soc 6:272–277
2. Barker DJ, Thornburg KL (2013) Placental programming of chronic diseases, cancer and lifespan: a review. Placenta 34:841–845
3. Vickers MH (2014) Early life nutrition, epigenetics and programming of later life disease. Nutrients 6:2165–2178
4. Narang I, Bush A (2012) Early origins of chronic obstructive pulmonary disease. Semin Fetal Neonatal Med 17:112–118
5. Pallotto EK, Kilbride HW (2006) Perinatal outcome and later implications of intrauterine growth restriction. Clin Obstet Gynecol 49:257–269
6. Barker DJ, Godfrey KM, Fall C, Osmond C, Winter PD, Shaheen SO (1991) Relation of birth weight and childhood respiratory infection to adult lung function and death from chronic obstructive airways disease. BMJ 303:671–675
7. Lawlor DA, Ebrahim S, Davey Smith G (2005) Association of birth weight with adult lung function: findings from the British Women's Heart and Health Study and a meta-analysis. Thorax 60:851–858
8. Cetin I, Alvino G (2009) Intrauterine growth restriction: implications for placental metabolism and transport. A review. Placenta 30:S77–S82
9. Pecks U, Brieger M, Schiessl B, Bauerschlag DO, Piroth D, Bruno B et al (2012) Maternal and fetal cord blood lipids in intrauterine growth restriction. J Perinat Med 40:287–296
10. Briana DD, Malamitsi-Puchner A (2013) Small for gestational age birth weight: impact on lung structure and function. Paediatr Respir Rev 14:256–262
11. Dezateux C, Lum S, Hoo AF, Hawdon J, Costeloe K, Stocks J (2004) Low birth weight for gestation and airway function in infancy: exploring the fetal origins hypothesis. Thorax 59:60–66
12. Minior VK, Divon MY (1998) Fetal growth restriction at term: myth or reality? Obstet Gynecol 92:57–60
13. Tyson JE, Kennedy K, Broyles S, Rosenfeld CR (1995) The small for gestational age infant: accelerated or delayed pulmonary maturation? Increased or decreased survival? Pediatrics 95:534–538
14. Kotecha SJ, Watkins WJ, Heron J, Henderson J, Dunstan FD, Kotecha S (2010) Spirometric lung function in school-age children: effect of intrauterine growth retardation and catch-up growth. Am J Respir Crit Care Med 181:969–974

15. Rona RJ, Gulliford MC, Chinn S (1993) Effects of prematurity and intrauterine growth on respiratory health and lung function in childhood. BMJ 306:817–820
16. Wjst M, Popescu M, Trepka MJ, Heinrich J, Wichmann HE (1998) Pulmonary function in children with initial low birth weight. Pediatr Allergy Immunol 9:80–90
17. Morsing E, Gustafsson P, Brodszki J (2012) Lung function in children born after foetal growth restriction and very preterm birth. Acta Paediatr 101:48–54
18. Nikolajev K, Heinonen K, Hakulinen A, Lansimies E (1998) Effects of intrauterine growth retardation and prematurity on spirometric flow values and lung volumes at school age in twin pairs. Pediatr Pulmonol 25:367–370
19. Nikolajev K, Koskela H, Korppi M (2008) Birth weight and adult lung function: a within-pair analysis of twins followed up from birth. World J Pediatr 4:222–226
20. Edwards CA, Osman LM, Godden DJ, Campbell DM, Douglas JG (2003) Relationship between birth weight and adult lung function: controlling for maternal factors. Thorax 58:1061–1065
21. Boezen HM, Vonk JM, van Aalderen WM, Brand PL, Gerritsen J, Schouten JP et al (2002) Perinatal predictors of respiratory symptoms and lung function at a young adult age. Eur Respir J 20:383–390
22. Canoy D, Pekkanen J, Elliott P, Pouta A, Laitinen J, Hartikainen AL et al (2007) Early growth and adult respiratory function in men and women followed from the fetal period to adulthood. Thorax 62:396–402
23. Narang I, Rosenthal M, Cremonesini D, Silverman M, Bush A (2008) Longitudinal evaluation of airway function 21 years after preterm birth. Am J Respir Crit Care Med 178:74–80
24. Rona RJ, Smeeton NC, Bustos P, Amigo H, Diaz PV (2005) The early origins hypothesis with an emphasis on growth rate in the first year of life and asthma: a prospective study in Chile. Thorax 60:549–554
25. Bush A (2008) COPD: a pediatric disease. COPD 5:53–67
26. Stocks J, Hislop A, Sonnappa S (2013) Early lung development: lifelong effect on respiratory health and disease. Lancet Respir Med 1:728–742
27. Alcorn DG, Adamson TM, Maloney JE, Robinson PM (1981) A morphologic and morphometric analysis of fetal lung development in the sheep. Anat Rec 201:655–667
28. Maritz GS, Cock ML, Louey S, Joyce BJ, Albuquerque CA, Harding R (2001) Effects of fetal growth restriction on lung development before and after birth: a morphometric analysis. Pediatr Pulmonol 32:201–210
29. Maritz GS, Cock ML, Louey S, Suzuki K, Harding R (2004) Fetal growth restriction has long-term effects on postnatal lung structure in sheep. Pediatr Res 55:287–295
30. Joyce BJ, Louey S, Davey MG, Cock ML, Hooper SB, Harding R (2001) Compromised respiratory function in postnatal lambs after placental insufficiency and intrauterine growth restriction. Pediatr Res 50:641–649
31. Das RM (1984) The effects of intermittent starvation on lung development in suckling rats. Am J Pathol 117:326–332
32. Joss-Moore LA, Wang Y, Yu X, Campbell MS, Callaway CW, McKnight RA et al (2011) IUGR decreases elastin mRNA expression in the developing rat lung and alters elastin content and lung compliance in the mature rat lung. Physiol Genomics 43:499–505
33. Lechner AJ (1985) Perinatal age determines the severity of retarded lung development induced by starvation. Am Rev Respir Dis 131:638–643
34. Rehan VK, Sakurai R, Li Y, Karadag A, Corral J, Bellusci S et al (2012) Effects of maternal food restriction on offspring lung extracellular matrix deposition and long term pulmonary function in an experimental rat model. Pediatr Pulmonol 47:162–171
35. Joss-Moore L, Carroll T, Yang Y, Fitzhugh M, Metcalfe D, Oman J et al (2013) Intrauterine growth restriction transiently delays alveolar formation and disrupts retinoic acid receptor expression in the lung of female rat pups. Pediatr Res 73:612–620
36. Wignarajah D, Cock ML, Pinkerton KE, Harding R (2002) Influence of intrauterine growth restriction on airway development in fetal and postnatal sheep. Pediatr Res 51:681–688

37. Thorn SR, Rozance PJ, Brown LD, Hay WW Jr (2011) The intrauterine growth restriction phenotype: fetal adaptations and potential implications for later life insulin resistance and diabetes. Semin Reprod Med 29:225–236
38. Joss-Moore LA, Albertine KH, Lane RH (2011) Epigenetics and the developmental origins of lung disease. Mol Genet Metab 104:61–66
39. Harding R, Hooper SB (1996) Regulation of lung expansion and lung growth before birth. J Appl Physiol 81:209–224
40. Chien LN, Chiou HY, Wang CW, Yeh TF, Chen CM (2014) Oligohydramnios increases the risk of respiratory hospitalization in childhood: a population-based study. Pediatr Res 75:576–581
41. Albuquerque CA, Smith KR, Saywers TE, Johnson C, Cock ML, Harding R (2002) Relation between oligohydramnios and spinal flexion in the human fetus. Early Hum Dev 68:119–126
42. Harding R, Hooper SB, Dickson KA (1990) A mechanism leading to reduced lung expansion and lung hypoplasia in fetal sheep during oligohydramnios. Am J Obstet Gynecol 163:1904–1913
43. Harding R, Liggins GC (1991) The influence of oligohydramnios on thoracic dimensions of fetal sheep. J Dev Physiol 16:355–361
44. Nicksa GA, Yu DC, Kalish BT, Klein JD, Turner CG, Zurakowski D et al (2011) Serial amnioinfusions prevent fetal pulmonary hypoplasia in a large animal model of oligohydramnios. J Pediatr Surg 46:67–71
45. Hofmeyr GJ, Eke AC, Lawrie TA (2014) Amnioinfusion for third trimester preterm premature rupture of membranes. Cochrane Database Syst Rev 3, CD000942
46. Panitch HB, Weiner DJ, Feng R, Perez MR, Healy F, McDonough JM et al (2014) Lung function over the first 3 years of life in children with congenital diaphragmatic hernia. Pediatr Pulmonol doi:10.1002/ppul.23082
47. Pierro M, Thebaud B (2014) Understanding and treating pulmonary hypertension in congenital diaphragmatic hernia. Semin Fetal Neonatal Med 19:357–363
48. Nardo L, Maritz G, Harding R, Hooper SB (2000) Changes in lung structure and cellular division induced by tracheal obstruction in fetal sheep. Exp Lung Res 26:105–119
49. Deprest J, Brady P, Nicolaides K, Benachi A, Berg C, Vermeesch J et al (2014) Prenatal management of the fetus with isolated congenital diaphragmatic hernia in the era of the TOTAL trial. Semin Fetal Neonatal Med 19:338–348
50. Manoli SE, Smith LA, Vyhlidal CA, An CH, Porrata Y, Cardoso WV et al (2012) Maternal smoking and the retinoid pathway in the developing lung. Respir Res 13:42
51. Stocks J, Dezateux C (2003) The effect of parental smoking on lung function and development during infancy. Respirology 8:266–285
52. Stocks J, Sonnappa S (2013) Early life influences on the development of chronic obstructive pulmonary disease. Ther Adv Respir Dis 7:161–173
53. Maritz GS, Harding R (2011) Life-long programming implications of exposure to tobacco smoking and nicotine before and soon after birth: evidence for altered lung development. Int J Environ Res Public Health 8:875–898
54. Maritz GS (2013) Perinatal exposure to nicotine and implications for subsequent obstructive lung disease. Paediatr Respir Rev 14:3–8
55. Dasgupta C, Xiao D, Xu Z, Yang S, Zhang L (2012) Developmental nicotine exposure results in programming of alveolar simplification and interstitial pulmonary fibrosis in adult male rats. Reprod Toxicol 34:370–377
56. Suter MA, Anders AM, Aagaard KM (2013) Maternal smoking as a model for environmental epigenetic changes affecting birthweight and fetal programming. Mol Hum Reprod 19:1–6
57. Ethen MK, Ramadhani TA, Scheuerle AE, Canfield MA, Wyszynski DF, Druschel CM et al (2009) Alcohol consumption by women before and during pregnancy. Matern Child Health J 13:274–285

58. Kenna K, Sozo F, De Matteo R, Hanita T, Gray SP, Tare M et al (2012) Alcohol exposure during late gestation: multiple developmental outcomes in sheep. J Dev Org Health Dis 3:224–236
59. May PA, Gossage JP, Kalberg WO, Robinson LK, Buckley D, Manning M et al (2009) Prevalence and epidemiologic characteristics of FASD from various research methods with an emphasis on recent in-school studies. Dev Disabil Res Rev 15:176–192
60. Public, editor (2005) Fetal alcohol spectrum disorder (FASD): a framework for action, 2003: National Library of Canada Cataloguing in Publication
61. Baumann P, Schild C, Hume RF, Sokol RJ (2006) Alcohol abuse—a persistent preventable risk for congenital anomalies. Int J Gynaecol Obstet 95:66–72
62. Johnson S, Knight R, Marmer DJ, Steele RW (1981) Immune deficiency in fetal alcohol syndrome. Pediatr Res 15:908–911
63. Inselman LS, Fisher SE, Spencer H, Atkinson M (1985) Effect of intrauterine ethanol exposure on fetal lung growth. Pediatr Res 19:12–14
64. Wang X, Gomutputra P, Wolgemuth DJ, Baxi L (2007) Effects of acute alcohol intoxication in the second trimester of pregnancy on development of the murine fetal lung. Am J Obstet Gynecol 197(269):e261–e264
65. Sozo F, O'Day L, Maritz G, Kenna K, Stacy V, Brew N et al (2009) Repeated ethanol exposure during late gestation alters the maturation and innate immune status of the ovine fetal lung. Am J Physiol Lung Cell Mol Physiol 296:L510–L518
66. Lazic T, Wyatt TA, Matic M, Meyerholz DK, Grubor B, Gallup JM et al (2007) Maternal alcohol ingestion reduces surfactant protein A expression by preterm fetal lung epithelia. Alcohol 41:347–355
67. Sozo F, Vela M, Stokes V, Kenna K, Meikle PJ, De Matteo R et al (2011) Effects of prenatal ethanol exposure on the lungs of postnatal lambs. Am J Physiol Lung Cell Mol Physiol 300: L139–L147
68. Gauthier TW, Ping XD, Gabelaia L, Brown LA (2010) Delayed neonatal lung macrophage differentiation in a mouse model of in utero ethanol exposure. Am J Physiol Lung Cell Mol Physiol 299:L8–L16
69. Gauthier TW, Ping XD, Harris FL, Wong M, Elbahesh H, Brown LA (2005) Fetal alcohol exposure impairs alveolar macrophage function via decreased glutathione availability. Pediatr Res 57:76–81
70. Gauthier TW, Young PA, Gabelaia L, Tang SM, Ping XD, Harris FL et al (2009) In utero ethanol exposure impairs defenses against experimental group B streptococcus in the term Guinea pig lung. Alcohol Clin Exp Res 33:300–306
71. Orgeig S, Morrison JL, Sullivan LC, Daniels CB (2015) The development of the pulmonary surfactant system. In: Harding R, Pinkerton KE (eds) The lung: development, aging and the environment, 2nd edn. Elsevier, London, pp 183–209
72. Tschanz SA, Damke BM, Burri PH (1995) Influence of postnatally administered glucocorticoids on rat lung growth. Biol Neonate 68:229–245
73. Seckl JR (2004) Prenatal glucocorticoids and long-term programming. Eur J Endocrinol 151 (Suppl 3):U49–U62
74. Yu HR, Kuo HC, Chen CC, Sheen JM, Tiao MM, Chen YC et al (2014) Prenatal dexamethasone exposure in rats results in long-term epigenetic histone modifications and tumour necrosis factor-alpha production decrease. Immunology 143:651–660
75. Tschanz SA, Haenni B, Burri PH (2002) Glucocorticoid induced impairment of lung structure assessed by digital image analysis. Eur J Pediatr 161:26–30
76. Luyet C, Burri PH, Schittny JC (2002) Suppression of cell proliferation and programmed cell death by dexamethasone during postnatal lung development. Am J Physiol Lung Cell Mol Physiol 282:L477–L483
77. Goldenberg RL, Hauth JC, Andrews WW (2000) Intrauterine infection and preterm delivery. N Engl J Med 342:1500–1507

78. Moss TJ, Nitsos I, Kramer BW, Ikegami M, Newnham JP, Jobe AH (2002) Intra-amniotic endotoxin induces lung maturation by direct effects on the developing respiratory tract in preterm sheep. Am J Obstet Gynecol 187:1059–1065
79. Watterberg KL, Demers LM, Scott SM, Murphy S (1996) Chorioamnionitis and early lung inflammation in infants in whom bronchopulmonary dysplasia develops. Pediatrics 97:210–215
80. Galinsky R, Polglase GR, Hooper SB, Black MJ, Moss TJ (2013) The consequences of chorioamnionitis: preterm birth and effects on development. J Pregnancy 2013:412831
81. Kramer BW, Kallapur SG, Moss TJ, Nitsos I, Polglase GP, Newnham JP et al (2010) Modulation of fetal inflammatory response on exposure to lipopolysaccharide by chorioamnion, lung, or gut in sheep. Am J Obstet Gynecol 202(77):e71–e79
82. Atik A, Sozo F, Orgeig S, Suri L, Hanita T, Harding R et al (2012) Long-term pulmonary effects of intrauterine exposure to endotoxin following preterm birth in sheep. Reprod Sci 19:1352–1364
83. Kallapur SG, Nitsos I, Moss TJ, Kramer BW, Newnham JP, Ikegami M et al (2005) Chronic endotoxin exposure does not cause sustained structural abnormalities in the fetal sheep lungs. Am J Physiol Lung Cell Mol Physiol 288:L966–L974
84. Blencowe H, Cousens S, Oestergaard MZ, Chou D, Moller AB, Narwal R et al (2012) National, regional, and worldwide estimates of preterm birth rates in the year 2010 with time trends since 1990 for selected countries: a systematic analysis and implications. Lancet 379:2162–2172
85. Hintz SR, Bann CM, Ambalavanan N, Cotten CM, Das A, Higgins RD (2010) Predicting time to hospital discharge for extremely preterm infants. Pediatrics 125:e146–e154
86. Jensen EA, Schmidt B (2014) Epidemiology of bronchopulmonary dysplasia. Birth Defects Res A Clin Mol Teratol 100:145–157
87. Coalson JJ (2003) Pathology of new bronchopulmonary dysplasia. Semin Neonatol 8:73–81
88. Husain AN, Siddiqui NH, Stocker JT (1998) Pathology of arrested acinar development in postsurfactant bronchopulmonary dysplasia. Hum Pathol 29:710–717
89. Jobe AJ (1999) The new BPD: an arrest of lung development. Pediatr Res 46:641–643
90. Saugstad OD (2010) Oxygen and oxidative stress in bronchopulmonary dysplasia. J Perinat Med 38:571–577
91. Kinsella JP, Greenough A, Abman SH (2006) Bronchopulmonary dysplasia. Lancet 367:1421–1431
92. Mokres LM, Parai K, Hilgendorff A, Ertsey R, Alvira CM, Rabinovitch M et al (2010) Prolonged mechanical ventilation with air induces apoptosis and causes failure of alveolar septation and angiogenesis in lungs of newborn mice. Am J Physiol Lung Cell Mol Physiol 298:L23–L35
93. Albertine KH, Yoder BA (2015) Pulmonary consequences of preterm birth. In: Harding R, Pinkerton KE (eds) The lung: development, aging and the environment, 2nd edn. Elsevier, London, pp 311–327
94. Doyle LW (2006) Respiratory function at age 8–9 years in extremely low birthweight/very preterm children born in Victoria in 1991–1992. Pediatr Pulmonol 41:570–576
95. Doyle LW, Faber B, Callanan C, Freezer N, Ford GW, Davis NM (2006) Bronchopulmonary dysplasia in very low birth weight subjects and lung function in late adolescence. Pediatrics 118:108–113
96. Korhonen P, Laitinen J, Hyodynmaa E, Tammela O (2004) Respiratory outcome in school-aged, very-low-birth-weight children in the surfactant era. Acta Paediatr 93:316–321
97. Smith LJ, van Asperen PP, McKay KO, Selvadurai H, Fitzgerald DA (2008) Reduced exercise capacity in children born very preterm. Pediatrics 122:e287–e293
98. Kilbride HW, Gelatt MC, Sabath RJ (2003) Pulmonary function and exercise capacity for ELBW survivors in preadolescence: effect of neonatal chronic lung disease. J Pediatr 143:488–493

99. Gross SJ, Iannuzzi DM, Kveselis DA, Anbar RD (1998) Effect of preterm birth on pulmonary function at school age: a prospective controlled study. J Pediatr 133:188–192
100. Gessner BD, Chimonas MA (2007) Asthma is associated with preterm birth but not with small for gestational age status among a population-based cohort of Medicaid-enrolled children <10 years of age. Thorax 62:231–236
101. Mai XM, Gaddlin PO, Nilsson L, Finnstrom O, Bjorksten B, Jenmalm MC et al (2003) Asthma, lung function and allergy in 12-year-old children with very low birth weight: a prospective study. Pediatr Allergy Immunol 14:184–192
102. Brostrom EB, Akre O, Katz-Salamon M, Jaraj D, Kaijser M (2013) Obstructive pulmonary disease in old age among individuals born preterm. Eur J Epidemiol 28:79–85
103. Stoll BJ, Gordon T, Korones SB, Shankaran S, Tyson JE, Bauer CR et al (1996) Late-onset sepsis in very low birth weight neonates: a report from the National Institute of Child Health and Human Development Neonatal Research Network. J Pediatr 129:63–71
104. Webber S, Wilkinson AR, Lindsell D, Hope PL, Dobson SR, Isaacs D (1990) Neonatal pneumonia. Arch Dis Child 65:207–211
105. Doyle LW, Anderson PJ (2009) Long-term outcomes of bronchopulmonary dysplasia. Semin Fetal Neonatal Med 14:391–395
106. Campbell K (1951) Intensive oxygen therapy as a possible cause of retrolental fibroplasia; a clinical approach. Med J Aust 2:48–50
107. Northway WH Jr, Rosan RC, Porter DY (1967) Pulmonary disease following respirator therapy of hyaline-membrane disease. Bronchopulmonary dysplasia. N Engl J Med 276:357–368
108. Ryan H (1952) Retrolental fibroplasia; a clinicopathologic study. Am J Ophthalmol 35:329–342
109. Richmond S, Wyllie J (2010) European Resuscitation Council Guidelines for Resuscitation 2010 Section 7. Resuscitation of babies at birth. Resuscitation 81:1389–1399
110. Sozo F, O'Reilly M (2015) The effects of neonatal hyperoxia on lung development. In: Harding R, Pinkerton KE (eds) The lung: development, aging and the environment, 2nd edn. Elsevier, London, pp 329–349
111. Yee M, Chess PR, McGrath-Morrow SA, Wang Z, Gelein R, Zhou R et al (2009) Neonatal oxygen adversely affects lung function in adult mice without altering surfactant composition or activity. Am J Physiol Lung Cell Mol Physiol 297:L641–L649
112. O'Reilly M, Hansbro PM, Horvat JC, Beckett EL, Harding R, Sozo F (2014) Bronchiolar remodeling in adult mice following neonatal exposure to hyperoxia: relation to growth. Anat Rec (Hoboken) 297:758–769
113. Massaro GD, McCoy L, Massaro D (1986) Hyperoxia reversibly suppresses development of bronchiolar epithelium. Am J Physiol 251:R1045–R1050
114. Massaro GD, McCoy L, Massaro D (1988) Development of bronchiolar epithelium: time course of response to oxygen and recovery. Am J Physiol 254:R755–R760
115. White CW, Greene KE, Allen CB, Shannon JM (2001) Elevated expression of surfactant proteins in newborn rats during adaptation to hyperoxia. Am J Respir Cell Mol Biol 25:51–59
116. Belik J, Jankov RP, Pan J, Tanswell AK (2003) Chronic O2 exposure enhances vascular and airway smooth muscle contraction in the newborn but not adult rat. J Appl Physiol 94:2303–2312
117. Denis D, Fayon MJ, Berger P, Molimard M, De Lara MT, Roux E et al (2001) Prolonged moderate hyperoxia induces hyperresponsiveness and airway inflammation in newborn rats. Pediatr Res 50:515–519
118. O'Reilly M, Harding R, Sozo F (2014) Altered small airways in aged mice following neonatal exposure to hyperoxic gas. Neonatology 105:39–45
119. Hartman WR, Smelter DF, Sathish V, Karass M, Kim S, Aravamudan B et al (2012) Oxygen dose responsiveness of human fetal airway smooth muscle cells. Am J Physiol Lung Cell Mol Physiol 303:L711–L719

120. Ladha F, Bonnet S, Eaton F, Hashimoto K, Korbutt G, Thebaud B (2005) Sildenafil improves alveolar growth and pulmonary hypertension in hyperoxia-induced lung injury. Am J Respir Crit Care Med 172:750–756
121. Chetty A, Cao GJ, Severgnini M, Simon A, Warburton R, Nielsen HC (2008) Role of matrix metalloprotease-9 in hyperoxic injury in developing lung. Am J Physiol Lung Cell Mol Physiol 295:L584–L592
122. Dauger S, Ferkdadji L, Saumon G, Vardon G, Peuchmaur M, Gaultier C et al (2003) Neonatal exposure to 65% oxygen durably impairs lung architecture and breathing pattern in adult mice. Chest 123:530–538
123. Yee M, Vitiello PF, Roper JM, Staversky RJ, Wright TW, McGrath-Morrow SA et al (2006) Type II epithelial cells are critical target for hyperoxia-mediated impairment of postnatal lung development. Am J Physiol Lung Cell Mol Physiol 291:L1101–L1111
124. Yee M, White RJ, Awad HA, Bates WA, McGrath-Morrow SA, O'Reilly MA (2011) Neonatal hyperoxia causes pulmonary vascular disease and shortens life span in aging mice. Am J Pathol 178:2601–2610
125. Thibeault DW, Mabry S, Rezaiekhaligh M (1990) Neonatal pulmonary oxygen toxicity in the rat and lung changes with aging. Pediatr Pulmonol 9:96–108
126. Mahmoud RA, Roehr CC, Schmalisch G (2011) Current methods of non-invasive ventilatory support for neonates. Paediatr Respir Rev 12:196–205
127. Gupta S, Sinha SK, Donn SM (2009) Ventilatory management and bronchopulmonary dysplasia in preterm infants. Semin Fetal Neonatal Med 14:367–373
128. Allison BJ, Crossley KJ, Flecknoe SJ, Davis PG, Morley CJ, Harding R et al (2008) Ventilation of the very immature lung in utero induces injury and BPD-like changes in lung structure in fetal sheep. Pediatr Res 64:387–392
129. Brew N, Hooper SB, Allison BJ, Wallace MJ, Harding R (2011) Injury and repair in the very immature lung following brief mechanical ventilation. Am J Physiol Lung Cell Mol Physiol 301:L917–L926
130. Brew N, Hooper SB, Zahra V, Wallace M, Harding R (2013) Mechanical ventilation injury and repair in extremely and very preterm lungs. PLoS One 8, e63905
131. Bland RD, Mokres LM, Ertsey R, Jacobson BE, Jiang S, Rabinovitch M et al (2007) Mechanical ventilation with 40% oxygen reduces pulmonary expression of genes that regulate lung development and impairs alveolar septation in newborn mice. Am J Physiol Lung Cell Mol Physiol 293:L1099–L1110
132. Kompass KS, Deslee G, Moore C, McCurnin D, Pierce RA (2010) Highly conserved transcriptional responses to mechanical ventilation of the lung. Physiol Genomics 42:384–396
133. Hillman NH, Moss TJ, Kallapur SG, Bachurski C, Pillow JJ, Polglase GR et al (2007) Brief, large tidal volume ventilation initiates lung injury and a systemic response in fetal sheep. Am J Respir Crit Care Med 176:575–581
134. Wallace MJ, Probyn ME, Zahra VA, Crossley K, Cole TJ, Davis PG et al (2009) Early biomarkers and potential mediators of ventilation-induced lung injury in very preterm lambs. Respir Res 10:19
135. Joss-Moore LA, Lane RH, Albertine KH (2015) Epigenetics and the developmental origins of lung disease. In: Harding R, Pinkerton KE (eds) The lung: development, aging and the environment, 2nd edn. Elsevier, London, pp 299–308
136. O'Reilly M, Hooper SB, Allison BJ, Flecknoe SJ, Snibson K, Harding R et al (2009) Persistent bronchiolar remodeling following brief ventilation of the very immature ovine lung. Am J Physiol Lung Cell Mol Physiol 297:L992–L1001
137. Iben SC, Haxhiu MA, Farver CF, Miller MJ, Martin RJ (2006) Short-term mechanical ventilation increases airway reactivity in rat pups. Pediatr Res 60:136–140
138. Schulzke SM, Pillow JJ (2010) The management of evolving bronchopulmonary dysplasia. Paediatr Respir Rev 11:143–148

139. Massaro GD, McCoy L, Massaro D (1988) Postnatal undernutrition slows development of bronchiolar epithelium in rats. Am J Physiol 255:R521–R526
140. Maritz G, Probyn M, DeMatteo R, Snibson K, Harding R (2008) Lung parenchyma at maturity is influenced by postnatal growth but not by moderate preterm birth in sheep. Neonatology 93:28–35
141. Mataloun MM, Leone CR, Mascaretti RS, Dohlnikoff M, Rebello CM (2009) Effect of postnatal malnutrition on hyperoxia-induced newborn lung development. Braz J Med Biol Res 42:606–613
142. Mataloun MM, Rebello CM, Mascaretti RS, Dohlnikoff M, Leone CR (2006) Pulmonary responses to nutritional restriction and hyperoxia in premature rabbits. J Pediatr (Rio J) 82:179–185
143. Dani C, Poggi C (2012) Nutrition and bronchopulmonary dysplasia. J Matern Fetal Neonatal Med 25(Suppl 3):37–40
144. Biniwale MA, Ehrenkranz RA (2006) The role of nutrition in the prevention and management of bronchopulmonary dysplasia. Semin Perinatol 30:200–208
145. Wemhoner A, Ortner D, Tschirch E, Strasak A, Rudiger M (2011) Nutrition of preterm infants in relation to bronchopulmonary dysplasia. BMC Pulm Med 11:7
146. Guo MM, Chung CH, Chen FS, Chen CC, Huang HC, Chung MY (2015) Severe bronchopulmonary dysplasia is associated with higher fluid intake in very Low-birth-weight infants: a retrospective study. Am J Perinatol 30:155–162
147. Moya F (2014) Preterm nutrition and the lung. World Rev Nutr Diet 110:239–252
148. Harding JE, Derraik JG, Berry MJ, Jaquiery AL, Alsweiler JM, Cormack BE et al (2013) Optimum feeding and growth in preterm neonates. J Dev Org Health Dis 4:215–222
149. Jobe AH (2006) Let's feed the preterm lung. J Pediatr (Rio J) 82:165–166
150. Midulla F, Pierangeli A, Cangiano G, Bonci E, Salvadei S, Scagnolari C et al (2012) Rhinovirus bronchiolitis and recurrent wheezing: 1-year follow-up. Eur Respir J 39:396–402
151. Lemanske RF Jr, Jackson DJ, Gangnon RE, Evans MD, Li Z, Shult PA et al (2005) Rhinovirus illnesses during infancy predict subsequent childhood wheezing. J Allergy Clin Immunol 116:571–577
152. Valkonen H, Waris M, Ruohola A, Ruuskanen O, Heikkinen T (2009) Recurrent wheezing after respiratory syncytial virus or non-respiratory syncytial virus bronchiolitis in infancy: a 3-year follow-up. Allergy 64:1359–1365
153. Guilbert TW, Singh AM, Danov Z, Evans MD, Jackson DJ, Burton R et al (2011) Decreased lung function after preschool wheezing rhinovirus illnesses in children at risk to develop asthma. J Allerg Clin Immunol 128:532–U147
154. Jackson DJ, Gangnon RE, Evans MD, Roberg KA, Anderson EL, Pappas TE et al (2008) Wheezing rhinovirus illnesses in early life predict asthma development in high-risk children. Am J Respir Crit Care Med 178:667–672
155. Zomer-Kooijker K, van der Ent CK, Ermers MJ, Uiterwaal CS, Rovers MM, Bont LJ et al (2014) Increased risk of wheeze and decreased lung function after respiratory syncytial virus infection. PLoS One 9, e87162
156. Bacharier LB, Cohen R, Schweiger T, Yin-Declue H, Christie C, Zheng J et al (2012) Determinants of asthma after severe respiratory syncytial virus bronchiolitis. J Allergy Clin Immunol 130(91-100), e103
157. Ruotsalainen M, Hyvarinen MK, Piippo-Savolainen E, Korppi M (2013) Adolescent asthma after rhinovirus and respiratory syncytial virus bronchiolitis. Pediatr Pulmonol 48:633–639
158. Backman K, Piippo-Savolainen E, Ollikainen H, Koskela H, Korppi M (2014) Increased asthma risk and impaired quality of life after bronchiolitis or pneumonia in infancy. Pediatr Pulmonol 49:318–325
159. Piippo-Savolainen E, Remes S, Kannisto S, Korhonen K, Korppi M (2004) Asthma and lung function 20 years after wheezing in infancy: results from a prospective follow-up study. Arch Pediatr Adolesc Med 158:1070–1076

160. Sigurs N, Aljassim F, Kjellman B, Robinson PD, Sigurbergsson F, Bjarnason R et al (2010) Asthma and allergy patterns over 18 years after severe RSV bronchiolitis in the first year of life. Thorax 65:1045–1052
161. Vissing NH, Chawes BLK, Bisgaard H (2013) Increased risk of pneumonia and bronchiolitis after bacterial colonization of the airways as neonates. Am J Respir Crit Care Med 188:1246–1252
162. Santos JC, Zhang L, Menegatti PK, Guasselli CS, Celso Filho CM, Maito LR et al (2011) Pneumonia in the first 2 years of life, and asthma in preschool-age children. Pediatr Int 53:576–580
163. Edmond K, Scott S, Korczak V, Ward C, Sanderson C, Theodoratou E et al (2012) Long term sequelae from childhood pneumonia; systematic review and meta-analysis. PLoS One 7, e31239
164. La Grutta S, Indinnimeo L, di Coste A, Ferrante G, Landi M, Pelosi U et al (2013) Environmental risk factors and lung diseases in children: from guidelines to health effects. Early Hum Dev 89(Suppl 3):S59–S62
165. Burke H, Leonardi-Bee J, Hashim A, Pine-Abata H, Chen Y, Cook DG et al (2012) Prenatal and passive smoke exposure and incidence of asthma and wheeze: systematic review and meta-analysis. Pediatrics 129:735–744
166. Jones LL, Hashim A, McKeever T, Cook DG, Britton J, Leonardi-Bee J (2011) Parental and household smoking and the increased risk of bronchitis, bronchiolitis and other lower respiratory infections in infancy: systematic review and meta-analysis. Respir Res 12:5
167. Bowatte G, Lodge C, Lowe AJ, Erbas B, Perret J, Abramson MJ et al (2014) The influence of childhood traffic-related air pollution exposure on asthma, allergy and sensitization: a systematic review and a meta-analysis of birth cohort studies. Allergy 70:245–256
168. Balasubramaniam V, Mervis CF, Maxey AM, Markham NE, Abman SH (2007) Hyperoxia reduces bone marrow, circulating, and lung endothelial progenitor cells in the developing lung: implications for the pathogenesis of bronchopulmonary dysplasia. Am J Physiol Lung Cell Mol Physiol 292:L1073–L1084
169. Irwin D, Helm K, Campbell N, Imamura M, Fagan K, Harral J et al (2007) Neonatal lung side population cells demonstrate endothelial potential and are altered in response to hyperoxia-induced lung simplification. Am J Physiol Lung Cell Mol Physiol 293:L941–L951
170. van Haaften T, Byrne R, Bonnet S, Rochefort GY, Akabutu J, Bouchentouf M et al (2009) Airway delivery of mesenchymal stem cells prevents arrested alveolar growth in neonatal lung injury in rats. Am J Respir Crit Care Med 180:1131–1142
171. Popova AP, Bozyk PD, Bentley JK, Linn MJ, Goldsmith AM, Schumacher RE et al (2010) Isolation of tracheal aspirate mesenchymal stromal cells predicts bronchopulmonary dysplasia. Pediatrics 126:e1127–e1133
172. Bozyk PD, Popova AP, Bentley JK, Goldsmith AM, Linn MJ, Weiss DJ et al (2011) Mesenchymal stromal cells from neonatal tracheal aspirates demonstrate a pattern of lung-specific gene expression. Stem Cells Dev 20:1995–2007
173. Borghesi A, Massa M, Campanelli R, Bollani L, Tzialla C, Figar TA et al (2009) Circulating endothelial progenitor cells in preterm infants with bronchopulmonary dysplasia. Am J Respir Crit Care Med 180:540–546
174. Baker CD, Ryan SL, Ingram DA, Seedorf GJ, Abman SH, Balasubramaniam V (2009) Endothelial colony-forming cells from preterm infants are increased and more susceptible to hyperoxia. Am J Respir Crit Care Med 180:454–461
175. Baker CD, Balasubramaniam V, Mourani PM, Sontag MK, Black CP, Ryan SL et al (2012) Cord blood angiogenic progenitor cells are decreased in bronchopulmonary dysplasia. Eur Respir J 40:1516–1522
176. O'Reilly M, Thebaud B (2012) Cell-based strategies to reconstitute lung function in infants with severe bronchopulmonary dysplasia. Clin Perinatol 39:703–725
177. Ahn SY, Chang YS, Kim SY, Sung DK, Kim ES, Rime SY et al (2013) Long-term (postnatal day 70) outcome and safety of intratracheal transplantation of human umbilical cord blood-derived mesenchymal stem cells in neonatal hyperoxic lung injury. Yonsei Med J 54:416–424

178. Aslam M, Baveja R, Liang OD, Fernandez-Gonzalez A, Lee C, Mitsialis SA et al (2009) Bone marrow stromal cells attenuate lung injury in a murine model of neonatal chronic lung disease. Am J Respir Crit Care Med 180:1122–1130
179. Chang YS, Choi SJ, Ahn SY, Sung DK, Sung SI, Yoo HS et al (2013) Timing of umbilical cord blood derived mesenchymal stem cells transplantation determines therapeutic efficacy in the neonatal hyperoxic lung injury. PLoS One 8, e52419
180. Chang YS, Choi SJ, Sung DK, Kim SY, Oh W, Yang YS et al (2011) Intratracheal transplantation of human umbilical cord blood-derived mesenchymal stem cells dose-dependently attenuates hyperoxia-induced lung injury in neonatal rats. Cell Transplant 20:1843–1854
181. Chang YS, Oh W, Choi SJ, Sung DK, Kim SY, Choi EY et al (2009) Human umbilical cord blood-derived mesenchymal stem cells attenuate hyperoxia-induced lung injury in neonatal rats. Cell Transplant 18:869–886
182. Hansmann G, Fernandez-Gonzalez A, Aslam M, Vitali SH, Martin T, Mitsialis SA et al (2012) Mesenchymal stem cell-mediated reversal of bronchopulmonary dysplasia and associated pulmonary hypertension. Pulm Circ 2:170–181
183. Sutsko RP, Young KC, Ribeiro A, Torres E, Rodriguez M, Hehre D et al (2013) Long-term reparative effects of mesenchymal stem cell therapy following neonatal hyperoxia-induced lung injury. Pediatr Res 73:46–53
184. Tropea KA, Leder E, Aslam M, Lau AN, Raiser DM, Lee JH et al (2012) Bronchioalveolar stem cells increase after mesenchymal stromal cell treatment in a mouse model of bronchopulmonary dysplasia. Am J Physiol Lung Cell Mol Physiol 302:L829–L837
185. Waszak P, Alphonse R, Vadivel A, Ionescu L, Eaton F, Thebaud B (2012) Preconditioning enhances the paracrine effect of mesenchymal stem cells in preventing oxygen-induced neonatal lung injury in rats. Stem Cells Dev 21:2789–2797
186. Zhang H, Fang J, Su H, Yang M, Lai W, Mai Y et al (2012) Bone marrow mesenchymal stem cells attenuate lung inflammation of hyperoxic newborn rats. Pediatr Transplant 16:589–598
187. Zhang H, Fang J, Wu Y, Mai Y, Lai W, Su H (2013) Mesenchymal stem cells protect against neonatal rat hyperoxic lung injury. Expert Opin Biol Ther 13:817–829
188. Zhang X, Wang H, Shi Y, Peng W, Zhang S, Zhang W et al (2012) Role of bone marrow-derived mesenchymal stem cells in the prevention of hyperoxia-induced lung injury in newborn mice. Cell Biol Int 36:589–594
189. Alphonse RS, Vadivel A, Fung M, Shelley WC, Critser PJ, Ionescu L et al (2014) Existence, functional impairment, and lung repair potential of endothelial colony-forming cells in oxygen-induced arrested alveolar growth. Circulation 129:2144–2157
190. Baker CD, Seedorf GJ, Wisniewski BL, Black CP, Ryan SL, Balasubramaniam V et al (2013) Endothelial colony-forming cell conditioned media promote angiogenesis in vitro and prevent pulmonary hypertension in experimental bronchopulmonary dysplasia. Am J Physiol Lung Cell Mol Physiol 305:L73–L81
191. Balasubramaniam V, Ryan SL, Seedorf GJ, Roth EV, Heumann TR, Yoder MC et al (2010) Bone marrow-derived angiogenic cells restore lung alveolar and vascular structure after neonatal hyperoxia in infant mice. Am J Physiol Lung Cell Mol Physiol 298:L315–L323
192. Hodges RJ, Jenkin G, Hooper SB, Allison B, Lim R, Dickinson H et al (2012) Human amnion epithelial cells reduce ventilation-induced preterm lung injury in fetal sheep. Am J Obstet Gynecol 206(448):e448–e455
193. Vosdoganes P, Hodges RJ, Lim R, Westover AJ, Acharya RY, Wallace EM et al (2011) Human amnion epithelial cells as a treatment for inflammation-induced fetal lung injury in sheep. Am J Obstet Gynecol 205(2):156; e26–33
194. Vosdoganes P, Lim R, Koulaeva E, Chan ST, Acharya R, Moss TJ et al (2013) Human amnion epithelial cells modulate hyperoxia-induced neonatal lung injury in mice. Cytotherapy 15:1021–1029

195. Chang YS, Ahn SY, Yoo HS, Sung SI, Choi SJ, Oh WI et al (2014) Mesenchymal stem cells for bronchopulmonary dysplasia: phase 1 dose-escalation clinical trial. J Pediatr 164 (966-972), e966
196. Agrons GA, Courtney SE, Stocker JT, Markowitz RI (2005) From the archives of the AFIP: lung disease in premature neonates: radiologic-pathologic correlation. Radiographics 25:1047–1073

Part II
Adult Lung Stem and Progenitor Cells

Chapter 6
Stem and Progenitor Cells of the Trachea and Proximal Airways

Ahmed E. Hegab, Tomoko Betsuyaku, and Brigitte N. Gomperts

Introduction

The lung is a complex organ in which the airways and blood vessels form a multigenerational dichotomous branching tree-like structure which is structurally and functionally divided into two main regions: the conducting airways and the air exchange units. The conducting airways are further subdivided into the trachea, bronchi, and bronchioles, while the air exchange units of the distal lung consist of alveoli that are lined by Type I and Type II alveolar epithelial cells.

The more than 42 diverse cell types distributed in different regional micro-environments along the branching airways is an important feature of lung complexity: each cell type designed to perform a specific function. The large conducting airways, which are the subject of this chapter, comprise the trachea and bronchi and are lined by a pseudostratified columnar mucociliary epithelium that functions in host defense. Below this epithelial lining lie the submucosal glands (SMG) which produce the majority of the serous and mucus secretions in the airway. Because these proximal airways are constantly exposed to an external environment which is a potential source of injury caused by challenges including temperature changes,

A.E. Hegab (✉) • T. Betsuyaku
Division of Pulmonary Medicine, Department of Medicine, Keio University School of Medicine, 35 Shinanomachi, Shinjuku-Ku, Tokyo 160-8582, Japan
e-mail: ahegab@a2.keio.jp

B.N. Gomperts
Department of Pediatrics, Mattel Children's Hospital, University of California, Los Angeles, CA, USA

Pulmonary Medicine, University of California, Los Angeles, CA, USA

Jonsson Comprehensive Cancer Center, University of California, Los Angeles, CA, USA

Eli and Edythe Broad Stem Cell Research Center, University of California, Los Angeles, CA, USA

I. Bertoncello (ed.), *Stem Cells in the Lung*, Stem Cell Biology and Regenerative Medicine, DOI 10.1007/978-3-319-21082-7_6

airborne particle inhalation and pathogens, the airway epithelium needs an efficient repair mechanism. It is now well established that these epithelial cell lineages are maintained during homeostasis, and repaired after injury, by regulation of the proliferation and differentiation of stem cells. In this chapter, we describe the significant progress made in recent years in understanding the proximal airway epithelium, its stem and progenitor cell subpopulations, their niche and the mechanisms regulating its repair, while reconciling important differences in the organization of regenerative cells and their regional microenvironments in the human and rodent lung.

Specifically, the cartilaginous trachea and bronchi in the human lung are characterized by the presence of SMG throughout their submucosa, unlike the bronchioles which are the more distal branches of airways that lack cartilage and glands. In rodents however, the distinction between bronchi and bronchioles is less evident as SMG are detected mainly around the first tracheal cartilaginous C-shaped ring (C1) with smaller rudimentary glands being visible down to only C4–6. Rodents also have no cartilage in their intrapulmonary airways. There are also major differences in the epithelial cellular lining of human and rodent airways. In humans, the pseudostratified columnar epithelium consists of basal, ciliated, and secretory cells (mainly goblet cells with a few Club cells—previously known as Clara cells) which line the trachea and bronchi. Whereas in rodents, the pseudostratified columnar epithelium consists of basal, ciliated, and secretory cells (mainly Club and serous cells with almost no goblet cells) which line only the trachea. The intrapulmonary airways down to the alveoli are lined by a simple columnar epithelium that consists mainly of Club and ciliated cells with a few neuroendocrine cells [1–3].

Evidence for Both "Dedicated" and "Facultative" Stem Cells in Proximal Airways

Different organs and tissues have stem and progenitor cells with different proliferative and reparative capacities. Tissues like the skin, intestine, and hematopoietic system have a constant turnover and highly efficient repair after injury. On the other hand, tissues like the brain and heart undergo little turnover and repair poorly after injury, while the lung is characterized by very slow but continuous turnover and variable reparative capacity that depends on the severity, anatomical site, repetition, and nature of injury [4, 5].

Traditionally, quiescence was considered to be an essential property of tissue stem cells. This idea was based on the notion that stem cells are few cells that cycle rarely and slowly in order to protect their genome from developing mutations [6]. Accordingly, a stem cell of a tissue should be able to "retain" a mitotic label for prolonged periods. Label-retaining cells (LRCs) were identified by administering BrdU to animals for long periods, followed by a long washout time. Cells that were identified by immunostaining to retain the BrdU label were considered the stem cells of that tissue. Several studies have identified LRCs in several locations in the proximal airways including the tracheal SMG ducts and dispersed foci within the surface epithelium (probably the inter-cartilaginous regions) of the trachea, as well as Club cells associated with the neuroepithelial bodies [7, 8]. Because lung epithelial turnover is so slow, these studies had to inflict an injury on the lung to induce a state of repair in order to identify LRCs. A recent study suggests that, proximal airway epithelial cells turn over every 6 months, which is much longer than the previous estimates [9, 10]. However, it is possible that the LRCs detected after injury represent a different stem cell subpopulation than the stem cell responsible for maintaining the airways during very slow homeostatic turnover. Indeed, recent findings in the skin and intestine have confirmed that LRCs are not the only stem cells in these tissues, demonstrating that quiescence and label retention are not inherent properties of all tissue stem cells [11].

Similar findings from several tissues have led to the classification of tissue stem cells as either "dedicated" or "facultative." Dedicated stem cells are thought to be few in number, more undifferentiated, and their self-renewal to be slow (quiescent) and capable of differentiation into more committed progenitors or specialized cell types. Facultative stem cells are thought to represent a larger stem cell pool, which exhibit differentiated features when in the quiescent state but have the capacity to proliferate, self-renew, and give rise to differentiated progeny in response to injury (Fig. 6.1) [12].

Different Stem Cell Subtypes Within the Airway Basal Cell Pool

Basal cells in several organs and tissues express Trp-63 (p63), cytokeratins (K)5 and K14 [13]. Airway basal cells were thought to constitutively express K14 [14] like basal cells of other epithelia, but this was later shown not to be the case [15, 16]. Naïve mouse tracheal basal cells express K14 in only 10–20 % of cells [16, 17] while in apparently healthy human bronchi K14 is expressed in only 1–2 % of cells [18]. K14 is also expressed in the basal and myoepithelial cells (MEC) of the SMG of both human and mouse [17, 18], as discussed below. However, after various injuries and/or infection and with chronic inflammation K14 expression is upregulated in many basal cells [17–19].

The functional role of K14 upregulation after injury and in pathological conditions is not clear. A recent study of stratified epithelium showed that loss of K14 expression was associated with reduced cell proliferation and delay in cell cycle progression [20]. As naïve airway basal cells proliferate very slowly, the need to upregulate K14 expression would be understandable in the context of "activation" of an otherwise-dormant basal stem cell to quickly proliferate in response to injury (Fig. 6.1). Whether K5+/K14− and K5+/K14+ represent two different subtypes of

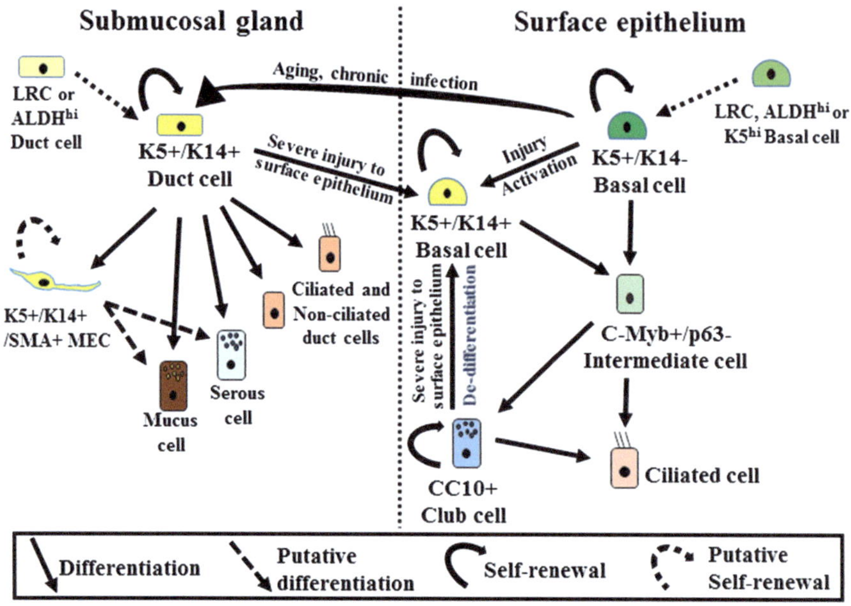

Fig. 6.1 Schematic representation of our current understanding of stem and progenitor cells responsible for maintenance and repair of the proximal airway surface epithelium and submucosal glands. *LRC* label retaining cells, *ALDH* aldehyde dehydrogenase, *MEC* myoepithelial cells, *K* keratin, *SMA* smooth muscle actin

basal cell with some degree of plasticity, or whether they are the same cell in a different state of activation is still controversial [16].

Evidence that basal cells are the "facultative" stem cells of the proximal airways and that they maintain and repair the proximal airways after injury has accumulated over many years [21, 22]. The most concrete characterization of their behavior as stem cells comes from lineage tracing studies. Engelhardt et al. performed a thorough study of the ability of the human proximal airway epithelium to reconstitute a denuded tracheal xenograft [23]. Epithelial cells were color-labeled using retroviral vectors so that the resulting clone sizes and differentiation potential could be followed. They observed that approximately 45 % of clones were multipotent, giving rise to basal, intermediate, secretory, and ciliated cells [23]. This finding is consistent with later rigorous lineage tracing experiments using a *K5-CreER* transgenic mouse model to follow the fate of K5+ tracheal basal cells in vivo during early postnatal growth, adult tracheal homeostasis, and tracheal repair after injury [24]. This study showed that basal cells self-renew and differentiate into both Club and ciliated cells, and that a long observation period of up to 14 weeks did not lead to loss of the label from basal cells indicating that they are the genuine stem cells in the trachea [24]. By contrast, lineage labeling of tracheal CC10+ (Club) cells showed that while labeled tracheal Club cells do proliferate and differentiate into ciliated cells they are gradually replaced over time by unlabeled cells, presumably the progeny of basal cells [25]. These experiments also indicated that while the differentiative potential of Club cells in basal cell-containing airways was more restricted than that of basal cells during homeostasis or after limited injury, they seemed to be able to generate basal cells after severe epithelial injury (albeit with very low efficiency) [25]. This ability of Club cells to "de-differentiate" into basal cells with a greater developmental potential was recently extensively studied by combining lineage tracing of Club cells with cellular ablation of basal cells (Fig. 6.1) [26]. Hegab et al. used the syngeneic heterotopic transplantation of mouse trachea as a model for severe hypoxic/ischemic injury for airway epithelium [27]. Soon after the injury, most cells of the tracheal surface epithelium underwent apoptosis and sloughed, with only a few surviving cells that were mostly K5+/K14− basal cells. Over the following days, these surviving basal cells, proliferated, acquired K14 expression, and started to differentiate gradually into secretory cells followed later by ciliated cells. BrdU labeling confirmed that most of the proliferating cells through the early repair phase were K5+ basal cells while in the later repair and remodeling period; some secretory cells were also BrdU labeled (Fig. 6.1) [27].

Collectively, it seems that during homeostasis and after different forms of injury, many basal cells throughout the proximal airways possess the ability to self-renew and differentiate towards the secretory and ciliated lineages. Club cells, in the presence of basal cells, have limited proliferative capacity but if basal cells were extensively damaged, they may contribute to the repair process by enhanced proliferation and possibly (de-) differentiation.

One interesting finding from basal cell lineage tracing is that basal cells responded to injury induced by SO_2 inhalation by differentiating more towards a ciliated than Club cell fate, while during homeostasis the reverse was true [24]. This suggests that the fate of the basal cell progeny is dependent on interactions with its surrounding environment, namely, the niche. However, it may also indicate the presence of two or more subtypes of basal cells: one that is active during homeostasis and tends to differentiate more into Club cells, while the other subtype is activated in response to injury, which in the case of SO_2 inhalation injury produces more ciliated cells. It is also possible that a different subtype of basal cell could respond to different types of injury. When Ghosh et al. lineage traced basal cells after inducing a selective injury to Club cells by naphthalene administration, they detected four different types of clones at 40 days post-injury. 50 % of these clones contained only basal cells, 25 % contained basal and secretory cells, and 19 % of clones had basal, secretory, and ciliated cells, while only 6 % contained basal and ciliated cell clones [16].

Indeed, there is accumulating evidence that there is a subset of basal cells, which has more stem cell characteristics than the rest of the basal cells. When basal cells with higher expression of K5 were sorted from a *K5-GFP* transgenic mouse, they produced more colonies that were also larger in size compared to basal cells with lower K5 expression [28]. Recently, sorting basal cells based on their aldehyde dehydrogenase activity (a commonly used marker of normal tissue and cancer stem cells [29]), enriched for a subset of basal cells that possessed more efficient two-dimensional (2D) [30] and three-dimensional (3D) colony/sphere formation ability (Fig. 6.1) [19, 31]. In the 3D colony/sphere formation assay, unfractionated basal cell spheres consistently formed spheres with a wide range of sizes [17, 24]. Further analysis of these spheres showed that they also had variable wall thicknesses and ratios of basal to differentiated cells (Hegab AE unpublished data).

Collectively, these data suggest the presence of different types of basal cells that have different responses to injury and possess different proliferative and differentiation abilities. Further studies are warranted to detect markers that can identify these different types in order to perform individualized characterization of their behavior both in vivo and in vitro.

Signaling Pathways Governing Basal Cell Proliferation and Differentiation

It is only recently that we have started to understand the factors regulating basal cell self-renewal versus differentiation during homeostasis and in response to injury. All basal cells in different epithelial tissues are typically marked by the transcription factor p63 (TP63, Trp63). P63 and basal cells seem to be essential for proper function of many organs. For example, loss of p63 from mammary gland basal cells leads to dramatic defects in luminal cell proliferation and differentiation, resulting in lactation failure [32], and loss of p63 in basal cells in the prostate results in disturbed ductal integrity and abnormal differentiation of luminal cells [33]. On the other hand, the trachea of P63 null mice contains no basal cells but shows a well-arranged columnar ciliated epithelium [34].

It is now well established that many stem cells can divide symmetrically, especially during organ development or in response to injury, when cell number expansion becomes a priority to establish an epithelial barrier, and this is also true of airway basal stem cells [35]. Asymmetric cell division (where daughter cells resulting from a cell division adopt divergent fates) is considered as a defining characteristic of stem cell differentiation and is associated with a significant drop in symmetric cell division [35]. Notch has been found to play two temporally distinct roles in airway epithelial repair—firstly, in self-renewal of basal cells and secondly in guiding the differentiation of basal cells towards ciliated or secretory lineages. Tsao et al. used a mouse model with conditional disruption of epithelial (but not mesenchymal) Notch. Inhibition of Notch in the developing lung of this model resulted in intrapulmonary airways that were entirely lined by ciliated cells but did not show any effect on the numbers of tracheal basal cells [36].

Notch has also been found to be important in adult airway epithelium homeostasis and during repair after injury. Using an in vivo reporter allele and genetic and pharmacologic gain- and loss-of-function experiments to demonstrate the role of Notch in cell fate determination of the proximal airway epithelium, Rock et al. found that Notch signaling drove basal cell differentiation to the secretory lineage, while the absence of Notch shifted the balance towards producing more ciliated cells [37]. They also speculated that Notch signaling is not required for self-renewal of basal cells [37]. However, further in-depth studies revealed that different combinations of Notch receptor activation and ligand binding resulted in completely different airway epithelial differentiation profiles [38]. It seems that components of the Notch pathway play complicated and tightly regulated roles in directing basal cell destiny. Paul et al. examined Notch pathway components both in vitro and in vivo to determine their role in controlling basal cell proliferation in response to injury [39]. Wild type mice received intratracheal polidocanol in order to slough off the airway surface epithelium and then the repairing airway epithelium was examined for Notch pathway components. Several components were upregulated during the repair process and were weakly expressed under steady-state conditions, suggesting that some Notch pathway components are active during basal cells proliferation.

Paul et al. also showed that dynamic changes in the level of reactive oxygen species (ROS) directly influence the ability of basal cells to self-renew and proliferate and that this is a tightly regulated process. This intriguing finding also suggests that exogenous ROS from air pollutants and cigarette smoke may influence airway epithelial cell proliferation and differentiation by altering the balance between the self-renewal and differentiation of basal stem cells. Thus, homeostasis is achieved by a regulatory loop in which the rise in ROS levels activates NRF2, the master transcription factor regulating the antioxidant response and NRF2, in its turn, has a direct regulating role on the Notch-mediated role in basal cell proliferation [39].

The epidermal growth factor receptor (EGFR) pathway in the lung epithelium regulates several key cellular processes including self-renewal, proliferation, repair, and differentiation. Aberrant EGFR signaling can cause basal cell hyperplasia, goblet cell hyperplasia, and induction of mucus production and/or promote fibrosis and collagen deposition [40]. An activating mutation in EGFR is associated with non-small cell lung cancer (NSCLC) and drugs that specifically target the EGFR pathway are now being used to treat NSCLC [41]. EGFR was found to have a role in controlling basal cell proliferation in homeostasis and during repair [42]. The EGFR inhibitor, AG1478, decreased basal cell proliferation and mucin production in experimental models [42, 43]. The Wnt-β-catenin pathway has also been shown to play a role in basal cell proliferation and differentiation both in vivo and in vitro [44–46].

One additional aspect of basal cell regulation was delineated with the identification of Myb as a key transcription factor that is expressed in post-mitotic epithelial cells that are destined to become multi-ciliated cells. Myb was shown to act downstream of Notch and upstream of Foxj1 and thus to be an essential regulator of basal cell differentiation towards ciliated lineages [47]. Another study demonstrated that Myb not only marked intermediate cells of the airway epithelium but was also critical in controlling their differentiation to both the ciliated and secretory lineages. Inhibiting Myb in airway epithelium in vitro, and *Myb* gene deletion in vivo, resulted in p63-negative cells which could not proceed to differentiate into ciliated or secretory cell lineages [48].

The Stem Cell Niche of the Airway Epithelium

It is well established that transcriptional regulation of a stem cell's decision to proliferate or differentiate towards a specific lineage is markedly affected by mutual signaling with various components in the stem cell niche [49]. Tadokoro et al. have recently described a novel interaction between IL-6, a niche component, and Stat3, a transcription factor. They showed that IL-6 originating from cells in the nearby stroma activate basal cell Stat3-dependent signaling resulting in a shift of basal cell differentiation towards a ciliated cell fate. This regulation was found to be both at the level of ciliogenesis-related genes and transcription factors like the Multicilin gene and FoxJ1, as well as by inhibition of the Notch pathway [50].

Culturing human bronchial and mouse tracheal epithelial cells (MTEC) at an air–liquid interface (ALI) has been the standard assay for studying airway epithelium and its niche in vitro for many years [51, 52]. In this assay, a mixture of airway epithelial cells is seeded on a collagen-coated membrane in a transwell. Cells are submerged in medium and allowed to form "cobblestones" for several days before the medium is removed from the upper chamber to generate an ALI. A pseudostratified epithelium containing basal, Club, ciliated, and goblet cells is usually apparent starting after 5 days at the ALI. Various experiments, treatments, and interventions can be applied to the ALI and the effect on cells can be examined at varying time points.

More recently, the sphere/colony formation assay in a 3D culture system has emerged as a useful method to study and characterize various stem cell properties, such as proliferation, differentiation, and niche interactions [53, 54]. In this assay, the size and number of spheres reflects the proliferation capacity of stem and progenitor cells, while the ability to serially propagate spheres with subsequent passages is considered a measure of self-renewal potential of the basal cells. The sphere assay has several advantages for studying stem cells over the ALI culture system. It allows for direct observation of the proliferation and self-renewal potential of basal cells with various treatments or genetic alterations. Mixing of different types of cells or differently labeled cells, allows for identification of the spatial relationship between putative stem cells and putative niche cells and/or their differentiated progeny. There are an increasing number of studies exploiting this sphere/colony formation assay to identify various properties of airway basal cells and niche interactions [17, 19, 24, 31, 39]. However, as with any in vitro assay, the interpretation of findings about the behavior of basal cells from the sphere/colony formation assay needs to be evaluated carefully and further validated in vivo.

Intermediate Cells

Using specific markers, cell types populating the airway pseudostratified epithelium were identified as basal (30 % of cells), ciliated (30 %), secretory (goblet, serous and/or Club cells) (30 %), and rare neuroendocrine cells (1–2 %) [1, 3]. However, about 8–10 % of cells do not stain for any of these known markers and are known as "indeterminate" cells or "intermediate" cells which are thought to be in a state of transition/differentiation [1, 55]. They were also thought of as early short-lived progenitor cells because they were observed to arise from basal cells that survived SO_2 injury before eventually differentiating into ciliated or secretory cell lineages. During this transition, they lost their basal cell markers and had not yet acquired ciliated or Club cell markers [37]. A disturbance in basal cell proliferation and differentiation programs, as might be seen in smokers and patients with COPD, characterized by basal cell hyperplasia or metaplasia is associated with an increase in incidence of both basal and intermediate cells [48]. Likewise, there is an increased incidence of intermediate cells and goblet cells in goblet cell hyperplasia [48]. The recent identification of the transcription factor, Myb, as a marker of these intermediate or early progenitor cells in both mouse and human (Fig. 6.1) [48] will likely pave the way for their further characterization by lineage tracing and exploration of their role during homeostasis and potential contribution to diseases of the lungs.

The Submucosal Glands

SMGs are complex branching structures located in the submucosal tissue of the human trachea and bronchi, but only in the most upper part of the trachea in mice. The SMG ducts open into the epithelium lining the surface of the airways. The duct branches into several collecting ducts which collect secretions from multiple mucus and serous tubules that form the glandular acini. Serous and mucus tubules are lined with polyhedral serous and mucus cells. The SMG tubules are surrounded by elongated MEC which express K5 and K14 in addition to α-smooth muscle actin. The SMG ducts are lined with a thin layer of stratified/pseudostratified epithelium. In the proximal part of the ducts, the luminal cells are ciliated. Both ciliated and basal cells of the SMG ducts seem to be in continuity with the ciliated and basal cells on the surface epithelium but basal cells of the SMG ducts differ from surface epithelial basal cells in that they are more flattened and express K14 [17, 18, 27]. Serous cells of the SMGs secret electrolyte-rich watery secretions, lysozyme, neutral mucins, and immunoglobulins. Mucus cells secret acidic mucins. The rate and nature of secretions from the SMGs is regulated by the cholinergic and non-cholinergic nervous systems and is affected by interactions with inflammatory cells, chemokines and allergens [56, 57]. Hypertrophic SMGs with enhanced secretion volume and a change in secretion constituents are important components of several lung diseases such as chronic bronchitis [58], cystic fibrosis [59], and asthma [60].

Stem Cells of the SMG

Much of the research on SMG stem cells has been conducted using animal models. It is therefore important to consider the structural and cellular differences between rodents and human airways and SMG [61], also highlighted in the introduction, when extrapolating these studies to humans. SMGs develop from the so-called primordial glandular placodes (PGPs), which are clusters of basally positioned epithelial progenitor cells within the developing airway epithelium. These PGPs start to invade into the submucosa and branch in response to signaling cues from the surrounding mesenchyme, eventually forming the branched networks of serous and mucus tubules of the SMGs [61].

Interestingly, when labeled human proximal airway epithelium was used to reconstitute a denuded tracheal xenograft, epithelial cells not only reconstituted the denuded tracheal surface but also invaded the submucosa and formed labeled SMGs, though infrequently [23]. Recently, the gradual appearance of age-related, gland-like structures (ARGLS) has also been described [62]. Interestingly, these ARGLS also seemed to arise by budding from the surface epithelium, similar to what is seen during SMG development and in the xenograft model (Fig. 6.1).

Collectively, it seems that some cells within the proximal airway surface epithelium, presumably some of the basal cells, are responsible for SMG embryonic development and they seem to preserve the potential to "bud" into the submucosa in a transplantation model and with aging.

SMG Duct Cells Are a SMG Stem Cell

Being in a relatively protected environment compared to the airway surface epithelium, SMG cells are expected to cycle even more slowly than the surface epithelium. Ciliated cells in the ducts of the SMG have a turnover rate of more than 18 months, much longer than the 6 months suggested for tracheal ciliated cells [10]. As mentioned above, LRCs were identified in SMG ducts after repeated injury to the surface epithelium [7]. In the same study, the ability of mouse SMG cells to contribute to the tracheal epithelial lining was suggested because when an enzyme-denuded trachea was transplanted subcutaneously in a nude mouse and examined after 28 days, the submucosal region showed cystic spaces lined by a "bronchiolized" ciliated epithelium [7]. These results suggest the potential participation of SMG duct stem cells in repairing injured surface epithelium. However, there is currently no injury model or cell ablation technique available that will specifically injure the SMG to help identify the stem cells maintaining and/or repairing the glands themselves.

Hegab et al. performed several studies to characterize the stem cells in the SMG ducts. In one study, they examined the chronological persistence or reappearance of epithelial cells in a tracheal graft after severe acute hypoxic/ischemic injury [27]. At day 1 after injury, all cells in the SMG bed were apoptotic and only a few K5+/K14+ cells in the gland ducts survived. These surviving duct cells were presumed to be the cells responsible for the ensuing repair [27]. Over the following days, these cells proliferated and gradually differentiated into secretory cells and MEC that invaded deeply into the submucosa eventually forming tubules that appeared to be functional. BrdU labeling confirmed that most proliferating cells during the early repair phase were K5+ duct cells while MECs were also seen to be proliferating during the later stages of repair and remodeling (Fig. 6.1) [27]. This study also confirmed the previous findings of Borthwick et al., [7] by demonstrating the ability of SMG cells to reconstitute the tracheal epithelium; enzymatically denuded prior to transplantation. Hegab et al. showed that SMGs could regenerate the denuded surface epithelium but this process was much slower than when surviving basal cells repaired the surface epithelium (Fig. 6.1) [27]. In another study, an alternative strategy was used to enrich SMG cells based on the finding that SMG duct cells, but not tubular cells, express Trop2 [17]. A 2-step enzyme digestion protocol was used to firstly detach the surface epithelium, and then digest SMG and other stromal cells in a second step prior to FACS sorting SMG duct cells based on their differential expression of Trop2 [17]. In human bronchi, a similar 2-step enzyme digestion protocol was used to separate surface epithelium from SMG cells and then sort basal and SMG duct cells on the basis of their differential expression of CD44 and CD166, respectively [19]. In the mouse, the self-renewal and differentiation potential of sorted SMG duct cells was demonstrated using both in vivo ectopic transplantation and in vitro colony/sphere formation assays, as well as lineage tracing of cells expressing K14 during repair after transplantation.

As mentioned above, in the naïve mouse trachea, K14 is expressed in basal cells of SMG ducts and in MECs but K14+ basal cells only represent about 10 % of the tracheal surface epithelial cells. They used *K14CrePR1* mice crossed with *ROSA26-floxSTOP-YFP* mice in order to express YFP only in K14+ cells on administration of RU486. To induce a proliferative state in the SMGs that turn over extremely slowly, the lineage tracing was performed using the hypoxic-ischemic injury model [17]. When the lineage marker was turned on with RU486 after airway injury, YFP expression was found in the SMGs, SMG ducts, and the SE adjacent to and overlying the SMG ducts but not in the surface epithelium of the distal trachea where SMGs are not present. The detection of this pattern of YFP+ cells strongly suggested that K14+ SMG duct cells contributed to repair of the SMG gland and participated in repair of the surface epithelium together with the basal cells that survived the injury (Fig. 6.1) [17]. As was seen with mouse basal cells [31], colony/sphere forming cells within SMG duct cells were enriched in the ALDH-expressing cell population [19].

SMG Myoepithelial Cells

MECs are flat, thin cells that wrap around the glandular acini, such as mammary, salivary, and airway SMG. Their function is to squeeze acini to help push their secretions into the ducts. They express both smooth muscle actin and keratins. In mammary glands, MECs were found to be the main stem cell of the gland [63] while in salivary glands it was found that the duct cells, rather than MECs, possessed stem cell characteristics [64]. On the other hand, both duct and MEC of the sweat glands seemed to harbor cells with stem cell properties [65, 66]. No study to date has examined the stem cell potential of airway SMG MECs even though they are a potentially clinically relevant cell type as their hypertrophy is commonly associated with fatal asthma [67].

Occasional YFP positive acinar cells were seen shortly after activation of the *K14/GFP* transgene in the *K14CrePR1x ROSA26-floxSTOP-YFP* mice by administration of RU486 [Hegab AE unpublished data]. These cells were so distal in the SMGs that it is more likely that they descended from their immediate neighboring K14+ MECs rather than from the more proximally positioned K14+ duct cells. During SMG repair from the severe acute hypoxic/ischemic injury described above, MECs were the first differentiated cell type to appear in addition to the surviving duct cells. BrdU labeling was seen in duct and MECs but not in secretory cells (Fig. 6.1) [27 and Hegab AE unpublished data]. More studies are required to characterize SMG MECs both in vitro and in vivo and to identify their possible contribution to various diseases that involve SMGs.

As the K5/K14-expressing SMG duct cells seem to be the anatomical continuum of the K5/K14-expressing MECs at the junction between the SMG ducts and tubules [17, 19, 27], it is possible that SMGs are hierarchically organized as follows: SMG duct cells have the potential to self-renew and differentiate into both MECs and acinar secretory cells, MECs self-renew and differentiate into acinar secretory cells, while acinar secretory cells are terminally differentiated cells. This is more or less similar to the tracheal cell hierarchy of basal, Club, and ciliated cells, respectively (Fig. 6.1).

Conclusion

In conclusion, much progress has been made over the last 5 years in increasing our understanding of the mechanisms of airway basal stem cell proliferation and differentiation to produce the mucociliary airway epithelium. However, there is still much to be learned about the development and repair of the proximal airway epithelium, its niche and the SMGs. It appears that there are multiple transcriptional pathways that are regulating these processes and the timing of signaling in each is critical for the initial proliferation followed by differentiation in normal airway repair. There is a large amount of overlap of these pathways with those that are active in adult stem cells of other epithelia, and there is much to be learned in studies from organs such as the skin, intestines, breast, and prostate and from other model organisms such as *Drosophila melanogaster* and *Chlamydomonas*. Many lung diseases arise from poor mucociliary clearance due to abnormal repair, and premalignant lesions of the large airways arise due to excessive self-renewal and a block in differentiation of basal cells. Thus, there is a great clinical need for the development of new therapies to improve mucociliary clearance and promote the normal pseudostratified columnar airway epithelium. Understanding the airway epithelial cell populations, their stem cell subpopulations, their niche, and the mechanisms involved in their repair will ultimately lead to novel therapies to promote repair of the normal airway epithelium, and a functional mucociliary escalator and thereby improve lung health.

References

1. Mercer RR, Russell ML, Roggli VL, Crapo JD (1994) Cell number and distribution in human and rat airways. Am J Respir Cell Mol Biol 10(6):613–624
2. Metzger RJ, Klein OD, Martin GR, Krasnow MA (2008) The branching programme of mouse lung development. Nature 453(7196):745–750
3. Hogan BL, Barkauskas CE, Chapman HA, Epstein JA, Jain R, Hsia CC et al (2014) Repair and regeneration of the respiratory system: complexity, plasticity, and mechanisms of lung stem cell function. Cell Stem Cell 15(2):123–138
4. Weiss DJ, Bertoncello I, Borok Z, Kim C, Panoskaltsis-Mortari A, Reynolds S et al (2011) Stem cells and cell therapies in lung biology and lung diseases. Proc Am Thorac Soc 8(3): 223–272
5. Grompe M (2012) Tissue stem cells: new tools and functional diversity. Cell Stem Cell 10(6): 685–689
6. Mackenzie IC, Bickenbach JR (1985) Label-retaining keratinocytes and Langerhans cells in mouse epithelia. Cell Tissue Res 242(3):551–556
7. Borthwick DW, Shahbazian M, Krantz QT, Dorin JR, Randell SH (2001) Evidence for stem-cell niches in the tracheal epithelium. Am J Respir Cell Mol Biol 24(6):662–670
8. Hong KU, Reynolds SD, Giangreco A, Hurley CM, Stripp BR (2001) Clara cell secretory protein-expressing cells of the airway neuroepithelial body microenvironment include a label-retaining subset and are critical for epithelial renewal after progenitor cell depletion. Am J Respir Cell Mol Biol 24(6):671–681
9. Bowden DH (1983) Cell turnover in the lung. Am Rev Respir Dis 128(2 Pt 2):S46–S48
10. Rawlins EL, Hogan BL (2008) Ciliated epithelial cell lifespan in the mouse trachea and lung. Am J Physiol Lung Cell Mol Physiol 295(1):L231–L234
11. Li L, Clevers H (2010) Coexistence of quiescent and active adult stem cells in mammals. Science 327(5965):542–545
12. Rawlins EL, Hogan BL (2006) Epithelial stem cells of the lung: privileged few or opportunities for many? Development 133(13):2455–2465
13. Hutton E, Paladini RD, Yu QC, Yen M, Coulombe PA, Fuchs E (1998) Functional differences between keratins of stratified and simple epithelia. J Cell Biol 143(2):487–499
14. Hong KU, Reynolds SD, Watkins S, Fuchs E, Stripp BR (2004) Basal cells are a multipotent progenitor capable of renewing the bronchial epithelium. Am J Pathol 164(2):577–588
15. Cole BB, Smith RW, Jenkins KM, Graham BB, Reynolds PR, Reynolds SD (2010) Tracheal Basal cells: a facultative progenitor cell pool. Am J Pathol 177(1):362–376

16. Ghosh M, Brechbuhl HM, Smith RW, Li B, Hicks DA, Titchner T et al (2011) Context-dependent differentiation of multipotential keratin 14-expressing tracheal basal cells. Am J Respir Cell Mol Biol 45(2):403–410
17. Hegab AE, Ha VL, Gilbert JL, Zhang KX, Malkoski SP, Chon AT et al (2011) Novel stem/progenitor cell population from murine tracheal submucosal gland ducts with multipotent regenerative potential. Stem Cells 29(8):1283–1293
18. Ooi AT, Mah V, Nickerson DW, Gilbert JL, Ha VL, Hegab AE (2010) Presence of a putative tumor-initiating progenitor cell population predicts poor prognosis in smokers with non-small cell lung cancer. Cancer Res 70(16):6639–6648
19. Hegab AE, Ha VL, Darmawan DO, Gilbert JL, Ooi AT, Attiga YS et al (2012) Isolation and in vitro characterization of basal and submucosal gland duct stem/progenitor cells from human proximal airways. Stem Cells Transl Med 1(10):719–724
20. Alam H, Sehgal L, Kundu ST, Dalal SN, Vaidya MM (2011) Novel function of keratins 5 and 14 in proliferation and differentiation of stratified epithelial cells. Mol Biol Cell 22(21):4068–4078
21. Ford JR, Terzaghi-Howe M (1992) Basal cells are the progenitors of primary tracheal epithelial cell cultures. Exp Cell Res 198(1):69–77
22. Zepeda ML, Chinoy MR, Wilson JM (1995) Characterization of stem cells in human airway capable of reconstituting a fully differentiated bronchial epithelium. Somat Cell Mol Genet 21(1):61–73
23. Engelhardt JF, Schlossberg H, Yankaskas JR, Dudus L (1995) Progenitor cells of the adult human airway involved in submucosal gland development. Development 121(7):2031–2046
24. Rock JR, Onaitis MW, Rawlins EL, Lu Y, Clark CP, Xue Y (2009) Basal cells as stem cells of the mouse trachea and human airway epithelium. Proc Natl Acad Sci U S A 106(31):12771–12775
25. Rawlins EL, Okubo T, Xue Y, Brass DM, Auten RL, Hasegawa H et al (2009) The role of Scgb1a1+ Clara cells in the long-term maintenance and repair of lung airway, but not alveolar, epithelium. Cell Stem Cell 4(6):525–534
26. Tata PR, Mou H, Pardo-Saganta A, Zhao R, Prabhu M, Law BM et al (2013) Dedifferentiation of committed epithelial cells into stem cells in vivo. Nature 503(7475):218–223
27. Hegab AE, Nickerson DW, Ha VL, Darmawan DO, Gomperts BN (2012) Repair and regeneration of tracheal surface epithelium and submucosal glands in a mouse model of hypoxic-ischemic injury. Respirology 17(7):1101–1113
28. Schoch KG, Lori A, Burns KA, Eldred T, Olsen JC, Randell SH (2004) A subset of mouse tracheal epithelial basal cells generates large colonies in vitro. Am J Physiol Lung Cell Mol Physiol 286(4):L631–L642
29. Muzio G, Maggiora M, Paiuzzi E, Oraldi M, Canuto RA (2012) Aldehyde dehydrogenases and cell proliferation. Free Radic Biol Med 52(4):735–746
30. Ghosh M, Helm KM, Smith RW, Giordanengo MS, Li B, Shen H et al (2011) A single cell functions as a tissue-specific stem cell and the in vitro niche-forming cell. Am J Respir Cell Mol Biol 45(3):459–469
31. Hegab AE, Ha VL, Bisht B, Darmawan DO, Ooi AT, Zhang KX (2014) Aldehyde dehydrogenase activity enriches for proximal airway basal stem cells and promotes their proliferation. Stem Cells Dev 23(6):664–675
32. Forster N, Saladi SV, van Bragt M, Sfondouris ME, Jones FE, Li Z (2014) Basal cell signaling by p63 controls luminal progenitor function and lactation via NRG1. Dev Cell 28(2):147–160
33. Kurita T, Medina RT, Mills AA, Cunha GR (2004) Role of p63 and basal cells in the prostate. Development 131(20):4955–4964
34. Daniely Y, Liao G, Dixon D, Linnoila RI, Lori A, Randell SH (2004) Critical role of p63 in the development of a normal esophageal and tracheobronchial epithelium. Am J Physiol Cell Physiol 287(1):C171–C181
35. Morrison SJ, Kimble J (2006) Asymmetric and symmetric stem-cell divisions in development and cancer. Nature 441(7097):1068–1074

36. Tsao PN, Vasconcelos M, Izvolsky KI, Qian J, Lu J, Cardoso WV (2009) Notch signaling controls the balance of ciliated and secretory cell fates in developing airways. Development 136(13):2297–2307
37. Rock JR, Gao X, Xue Y, Randell SH, Kong YY, Hogan BL (2011) Notch-dependent differentiation of adult airway basal stem cells. Cell Stem Cell 8(6):639–648
38. Morimoto M, Nishinakamura R, Saga Y, Kopan R (2012) Different assemblies of Notch receptors coordinate the distribution of the major bronchial Clara, ciliated and neuroendocrine cells. Development 139(23):4365–4373
39. Paul MK, Bisht B, Darmawan DO, Chiou R, Ha VL, Wallace WD et al (2014) Dynamic changes in intracellular ROS levels regulate airway basal stem cell homeostasis through Nrf2-dependent Notch signaling. Cell Stem Cell 15(2):199–214
40. Vallath S, Hynds RE, Succony L, Janes SM, Giangreco A (2014) Targeting EGFR signalling in chronic lung disease: therapeutic challenges and opportunities. Eur Respir J 44(2):513–522
41. Taron M, Ichinose Y, Rosell R, Mok T, Massuti B, Zamora L (2005) Activating mutations in the tyrosine kinase domain of the epidermal growth factor receptor are associated with improved survival in gefitinib-treated chemorefractory lung adenocarcinomas. Clin Cancer Res 11(16):5878–5885
42. Brechbuhl HM, Li B, Smith RW, Reynolds SD (2014) Epidermal growth factor receptor activity is necessary for mouse basal cell proliferation. Am J Physiol Lung Cell Mol Physiol 307(10):L800–L810
43. Hegab AE, Sakamoto T, Nomura A, Ishii Y, Morishima Y, Iizuka T (2007) Niflumic acid and AG-1478 reduce cigarette smoke-induced mucin synthesis: the role of hCLCA1. Chest 131(4): 1149–1156
44. Li C, Li A, Li M, Xing Y, Chen H, Hu L et al (2009) Stabilized beta-catenin in lung epithelial cells changes cell fate and leads to tracheal and bronchial polyposis. Dev Biol 334(1):97–108
45. Smith MK, Koch PJ, Reynolds SD (2012) Direct and indirect roles for β-catenin in facultative basal progenitor cell differentiation. Am J Physiol Lung Cell Mol Physiol 302(6):L580–L594
46. Brechbuhl HM, Ghosh M, Smith MK, Smith RW, Li B, Hicks DA (2011) β-catenin dosage is a critical determinant of tracheal basal cell fate determination. Am J Pathol 179(1):367–379
47. Tan FE, Vladar EK, Ma L, Fuentealba LC, Hoh R, Espinoza FH (2013) Myb promotes centriole amplification and later steps of the multiciliogenesis program. Development 140(20):4277–4286
48. Pan JH, Adair-Kirk TL, Patel AC, Huang T, Yozamp NS, Xu J et al (2014) Myb permits multilineage airway epithelial cell differentiation. Stem Cells 32(12):3245–3256
49. Wagers AJ (2012) The stem cell niche in regenerative medicine. Cell Stem Cell 10(4):362–369
50. Tadokoro T, Wang Y, Barak LS, Bai Y, Randell SH, Hogan BL (2014) IL-6/STAT3 promotes regeneration of airway ciliated cells from basal stem cells. Proc Natl Acad Sci U S A 111(35): E3641–E3649
51. You Y, Richer EJ, Huang T, Brody SL (2002) Growth and differentiation of mouse tracheal epithelial cells: selection of a proliferative population. Am J Physiol Lung Cell Mol Physiol 283(6):L1315–L1321
52. Gray TE, Guzman K, Davis CW, Abdullah LH, Nettesheim P (1996) Mucociliary differentiation of serially passaged normal human tracheobronchial epithelial cells. Am J Respir Cell Mol Biol 14(1):104–112
53. Pastrana E, Silva-Vargas V, Doetsch F (2011) Eyes wide open: a critical review of sphere-formation as an assay for stem cells. Cell Stem Cell 8(5):486–498
54. Sato T, van Es JH, Snippert HJ, Stange DE, Vries RG, van den Born M et al (2011) Paneth cells constitute the niche for Lgr5 stem cells in intestinal crypts. Nature 469(7330):415–418
55. Rock JR, Randell SH, Hogan BL (2010) Airway basal stem cells: a perspective on their roles in epithelial homeostasis and remodeling. Dis Model Mech 3(9–10):545–556
56. Marin MG, Culp DJ (1986) Isolation and culture of submucosal gland cells. Clin Chest Med 7(2):239–245

57. Ballard ST, Spadafora D (2007) Fluid secretion by submucosal glands of the tracheobronchial airways. Respir Physiol Neurobiol 159(3):271–277
58. Snider GL (1981) Pathogenesis of emphysema and chronic bronchitis. Med Clin North Am 65(3):647–665
59. Wood RE, Boat TF, Doershuk CF (1976) Cystic fibrosis. Am Rev Respir Dis 113(6):833–878
60. Fahy JV (2001) Remodeling of the airway epithelium in asthma. Am J Respir Crit Care Med 164(10 Pt 2):S46–S51
61. Liu X, Driskell RR, Engelhardt JF (2004) Airway glandular development and stem cells. Curr Top Dev Biol 64:33–56
62. Wansleeben C, Bowie E, Hotten DF, Yu YR, Hogan BL (2014) Age-related changes in the cellular composition and epithelial organization of the mouse trachea. PLoS One 9(3):e93496
63. Prater MD, Petit V, Alasdair Russell I, Giraddi RR, Shehata M, Menon S et al (2014) Mammary stem cells have myoepithelial cell properties. Nat Cell Biol 16(10):942–950
64. Pringle S, Van Os R, Coppes RP (2013) Concise review: adult salivary gland stem cells and a potential therapy for xerostomia. Stem Cells 31(4):613–619
65. Kurata R, Futaki S, Nakano I, Tanemura A, Murota H, Katayama I et al (2014) Isolation and characterization of sweat gland myoepithelial cells from human skin. Cell Struct Funct 39(2): 101–112
66. Lu CP, Polak L, Rocha AS, Pasolli HA, Chen SC, Sharma N et al (2012) Identification of stem cell populations in sweat glands and ducts reveals roles in homeostasis and wound repair. Cell 150(1):136–150
67. Green FH, Williams DJ, James A, McPhee LJ, Mitchell I, Mauad T (2010) Increased myoepithelial cells of bronchial submucosal glands in fatal asthma. Thorax 65(1):32–38

Chapter 7
Stem Cells of the Distal Bronchiolar Airways

Robert E. Hynds and Adam Giangreco

Introduction

The lungs represent a significant interface between the body and the outside world, with a total surface area of 40–50 m^2 in adult humans [1]. Within the lungs, a branched network of conducting airways includes the trachea and mainstem bronchi, distal bronchiolar airways, respiratory and terminal bronchioles, and alveolar epithelium. The primary functions of airways are to increase the lung surface area and act as a physical barrier to inhaled toxicants including fine and ultrafine particulate matter, viral and bacterial pathogens [2]. Conducting airways achieve this by producing antimicrobial secretory proteins and eliminating both pathogens and particulates through the continuous movement of a mucociliary escalator [3]. Club cells (formerly Clara cells), along with goblet cells, are responsible for the secretory functions of the airway epithelium [4]. The prominent mucins secreted include MUC5AC and MUC5B, although a broad range of large structurally related gel-forming mucin glycoproteins are additionally expressed in human airways [5]. A second important defence mechanism is the production of antimicrobial peptides and proteins, such as lysozyme and lactoferrin, to prevent microbial colonization of the lungs [6].

The proximal trachea and mainstem bronchi contain a pseudostratified epithelium comprising several cell types that can be broadly divided into basal keratinocytes and non-basal luminal cells. Basal cells line the basement membrane and in normal circumstances are not significantly exposed to the airway surface [7, 8]. In contrast, luminal cells, including abundant ciliated and mucous-secreting cells, maintain a direct interface with the apical lumen. In the distal conducting airway, the epithelium retains both ciliated and secretory characteristics but

R.E. Hynds • A. Giangreco (✉)
Lungs for Living Research Centre, Division of Medicine, University College London,
5 University Street, London, WC1E 6JF UK
e-mail: a.giangreco@ucl.ac.uk

I. Bertoncello (ed.), *Stem Cells in the Lung*, Stem Cell Biology and Regenerative Medicine, DOI 10.1007/978-3-319-21082-7_7

exhibits a simple cuboidal architecture [9]. Finally, the alveolar epithelium consists of type I and type II cells and normally lacks the ciliated, secretory, and basal cell types observed in the more proximal conducting airways [10]. It is also important to note that anatomical differences exist between the lungs of various mammalian species. For example, human airways exhibit a pseudostratified, keratinized epithelium with abundant basal cells, submucosal glands, and cartilaginous rings throughout the trachea, bronchi, and some bronchioles [11]. In contrast, only the trachea and mainstem bronchi of murine airways contain basal epithelial cells under homeostatic conditions [8]. These inter-species distinctions most likely result from differences in the respiratory demands placed upon human and murine lungs [12].

As a consequence of continuous pathogen and particulate exposure within lungs, all conducting airways undergo a slow but continuous renewal [13]. This process of constant regeneration results in the complete turnover of the human bronchiolar epithelium every 100–300 days. Normally, this is accomplished by endogenous stem cells with both long-term self-renewal and multipotent differentiation capacities. In addition to their role in homeostasis, a growing body of literature suggests that stem cells of the distal bronchiolar airways may contribute to chronic obstructive pulmonary disease (COPD) and lung cancer aetiology [14]. These chronic lung diseases cause approximately one in ten deaths worldwide, with COPD expected to become the third most common cause of death by 2020 [15]. Unfortunately, although treatments to reduce the symptoms of chronic lung disease have improved in recent years, there remain no effective cures. Thus, in order to develop new, more effective therapies for lung cancer and COPD it is necessary to understand the specific cellular, biochemical, and molecular properties of stem cells in lung health and disease [16].

In this chapter, we review current knowledge regarding the identity and characterization of distal bronchiolar stem cells. In particular, we focus on the role of these cells in airway homeostasis, repair, and regeneration. We describe recent methods to investigate distal bronchiolar stem cells in vivo and ex vivo and discuss key features of the distal bronchiolar niche, or microenvironment, that supports an endogenous lung stem cell's capacity for repair, regeneration, and differentiation.

In Vivo Airway Homeostasis and Repair

Under normal conditions, less than 0.5 % of bronchiolar epithelial cells undergo proliferation in any given day [17]. Therefore, most historical studies of distal bronchiolar stem cells have involved rodent models of airway injury that increase lung cell proliferation [18–21]. These included tobacco smoke, oxidant gas, or other chemicals administered via inhalation or injection. In the 1970s, secretory Club cells were identified as the predominant mitotic cell population in distal bronchiolar airways [22]. In mice and rats, oxidant exposure caused Club cell dedifferentiation to morphologically variant “Type A” cells that accounted for over 70 % of cell proliferation with damaged bronchioles [22]. Pulse-chase experiments involving tritiated thymidine nucleoside incorporation subsequently established that "Type A" Club cells were capable of multipotent differentiation into both Club and ciliated cell types [22, 23]. Importantly, Club cell proliferation was readily observed throughout the epithelium following mild to moderate airway damage, demonstrating the broad distribution of this stem cell population.

In recent years, genetically modified mouse models have demonstrated that Secretoglobin 1A1 (Club cell secretory protein, CCSP)-expressing cells are indeed a stem cell population that maintains distal bronchiolar homeostasis in murine lungs. Specifically, aggregation chimaera and Scgb1a1 lineage tracing models demonstrated that large numbers of clonally derived cell patches with a wide distribution of sizes were present in distal bronchioles in the absence of epithelial injury [24, 25]. On average, these airway cell clones were small, randomly distributed, and exhibited multipotent differentiation to both Club and ciliated cell lineages [24, 25]. The abundance, size, and distribution of these Scgb1a1-derived clones were consistent with a stochastic process of homeostasis common to multiple epithelial organs [26]. Overall, the results of these studies suggested that in murine airways an abundant population of Club cells functions as stem cells that maintain distal bronchiolar homeostasis (Fig. 7.1a).

In addition to their contribution to airway homeostasis, previous severe injury, and repair studies also identified subpopulations of Club cells that contribute to lung regeneration. The most frequently used severe bronchiolar injury model involves intraperitoneal or aerosolized delivery of naphthalene, a derivative of coal tar [27–30]. Naphthalene administration causes significant, Club cell-specific toxicity in murine airways [31]. The extent and severity of this injury can be experimentally determined through both dose and route of administration [29, 30]. The cause of Club cells' unique sensitivity to cytotoxic injury is their expression of cytochrome P450 monooxygenases; this family of metabolic enzymes acts to oxidize potentially damaging exogenous compounds rendering them more water soluble. Specifically, airway Club cells express cytochrome P450 isozyme 2 F2 (CYP2F2) which produces the toxic metabolite 1R, 2S-naphthalene oxide upon naphthalene exposure [28, 32].

Following naphthalene-mediated Club cell ablation, a small number of naphthalene-resistant Club cells (termed variant CCSP-expressing cells; vCE cells) survive by virtue of their reduced expression of CYP2F2. These vCE cells subsequently proliferate and regenerate all differentiated distal bronchiolar cell types [33, 34]. These phenotypically variant stem cells are typically found in association with both neuroepithelial bodies (NEBs) and at bronchiolar duct junctions [25, 33, 35]. In addition, a population of bronchiolar duct junction-associated vCE cells has been shown to express both CCSP and surfactant protein-C (SPC) [36]. Lineage tracing experiments involving these cells, termed bronchioalveolar stem cells or BASCs, demonstrate that they are capable of differentiation into both bronchiolar and alveolar epithelial cell types following certain types of lung injury [24, 37].

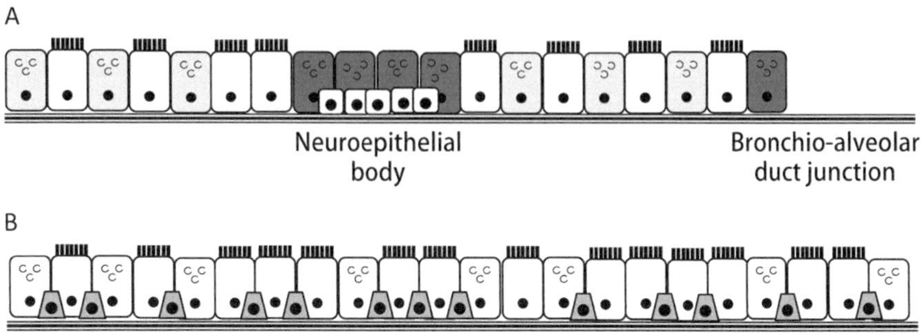

Fig. 7.1 Stem cells of the distal bronchiolar airways. (**a**) Murine distal airways are primarily composed of ciliated, secretory (Club), and neuroendocrine cells. Lineage tracing studies using genetically modified mouse models have shown that Club cells function as the predominant stem cell population that maintains normal airway homeostasis (*light grey cells*). NEB- and BADJ-associated, variant Club stem cells (NASCs and BASCs) additionally exhibit enhanced resistance to airway damage and contribute to lung regeneration following severe injury (*dark grey cells*). (**b**) Human distal airways contain abundant secretory (Club and goblet), ciliated, and basal epithelial cell types. In contrast to murine lungs, ex vivo and in vivo evidence supports a likely role for basal cells as human distal bronchiolar stem cells. Here, basal cells are thought to maintain both airway homeostasis and repair/regeneration capacities (*light grey cells*)

In addition to normal and variant Club cells, recent studies identified a population of p63-expressing basal-like cells that contribute to distal lung repair following influenza infection [38]. At the peak of influenza infection, the abundance of these p63+ basal cells increased dramatically and resulted in restoration of a phenotypically normal airway with abundant Club cells [38]. Subsequent work by this same group demonstrated that p63-expressing cells were in fact an essential component of normal distal airway regeneration post-influenza infection [39]. These studies, along with recent complementary evidence of dedifferentiation of CCSP+ cells into upper airway basal stem cells, suggest a previously unappreciated lineage plasticity among cells that survive epithelial injury [40, 41].

In contrast to murine airways, CCSP-expressing cells are significantly less abundant in human airways [9, 42] and a combination of both Club and basal cells are present in human distal bronchioles [43]. This suggests that either CCSP-expressing cells or basal cells might function as stem cells in human bronchioles [44]. In vitro studies using human cells support the hypothesis that keratin 5 and/or keratin 14-expressing basal cells might function as bronchiolar stem cells in human lungs [38]. Further studies by Teixeira et al. provide quantitative analysis of distal bronchiolar cell types within intact human airways [43]. These results suggest that basal cells are abundant and widely distributed human airway stem cells (Fig. 7.1b). It is as yet unclear whether these basal stem cells responsible for human airway homeostasis and regeneration are equivalent to the keratin-5 and keratin-14-expressing p63+ cells associated with post-influenza lung regeneration in mice [38, 39]. The relative contribution of CCSP-expressing Club cells and keratin 5-expressing basal cells to human distal lung homeostasis and repair also remains to be determined.

Recently, expression of Wnt-responsive genes such as leucine-rich repeat-containing G-protein coupled receptor 5 (Lgr5) and Lgr6 have been found in rare stem cells in a broad range of epithelial organs [45], including the lung [46]. Following these discoveries, a discrete population of E-Cadherin+/Lgr6+ cells was isolated from human small bronchioles [47]. Reminiscent of murine BASCs, single cells were capable of significant expansion in vitro and generated differentiated bronchioalveolar cell types when injected under the kidney capsule of immune-deficient mice [47]. The relationship of these cells to other known stem cell populations in either mouse or human lungs and their relative importance to in vivo homeostasis and repair will be an important question in future studies.

Ex Vivo Analyses of Distal Bronchiolar Stem Cells

Investigation of the growth and differentiation properties of prospectively isolated airway cells ex vivo represents an alternative strategy for the identification of distal bronchiolar stem cells [48, 49]. These studies establish the value of ex vivo lung stem cell cultures, with high correlation between in vitro stem cell characteristics and the endogenous behaviour of these cells in human and murine airways [50, 51]. The continued improvement of these protocols for lung stem cell isolation, expansion, and differentiation will not only reduce reliance on animal models but also provide a novel means to interrogate the relevance of stem cells in human lung disease.

Initially, murine lung stem cells were isolated from intact lungs using the cell surface markers Sca-1 and CD34 [36]. These cells expressed CCSP and SPC, were capable of in vitro self-renewal and exhibited multipotent differentiation into Club-like cells as well as type I and II alveolar epithelial cells. Thus, dual Sca1 and CD34-positive cells appear to represent an ex vivo equivalent of terminal bronchiole-associated BASCS [36]. Despite this, the purity and homogeneity of these prospectively isolated cells has been contested, in that the cell surface markers Sca-1 and CD34 additionally isolate non-BASC cell types [52–54]. It is similarly unclear how these findings might inform the prospective isolation of human distal bronchiolar stem cells given the lack of a human homologue for Sca-1 [55].

In separate studies, McQualter and colleagues determined that a population of $CD24^{low}$/integrin $\alpha6\beta4$+/EpCam+ cells were capable of multilineage differentiation [56]. These cells represent a non-ciliated population of airway cells with properties consistent with CCSP+ stem cells. Specifically, co-culture of isolated integrin $\alpha6\beta4$+/$CD24^{low}$ cells with lung-derived mesenchymal cells produces differentiated epithelial organoids with both bronchiolar Club and ciliated cell populations [56]. Here, mesenchyme-derived FGF10 was identified as an important mediator of epithelial stem cell expansion in vitro [56, 57]. This was consistent with evidence that parabronchial smooth muscle-derived FGF10 maintains vCE cell populations in vivo [58, 59]. As above, it remains to be determined whether human lungs contain an equivalent cell type with characteristics similar to these integrin $\alpha6\beta4+/CD24^{low}$ cells.

The Distal Bronchiolar Stem Cell Niche

A detailed understanding of the mechanisms that regulate airway stem cell behaviour will be necessary in order to use these cells for improving human lung health. In vivo, stem cells exist in a complex three-dimensional (3D) microenvironment, the "stem cell niche," which is finely tuned to generate the signals required to balance self-renewal and differentiation [60–62]. Thus, the stem cell niche functions as a dynamic system that integrates specific extracellular components, growth factors, signalling molecules, biomechanical properties, and spatial constraints that direct and determine cellular fate and phenotype [63]. While the niche of some stem cells has been studied extensively (such as that of intestinal or hematopoietic stem cells [64, 65]), the specific signals that regulate distal bronchiolar stem cells are less well understood.

In murine airways, vCE cells located adjacent to neuroepithelial bodies (NEBs) and bronchioalveolar duct junctions (BADJs) regenerate the airways following injury making these anatomical locations putative airway stem cell niches [33, 35, 36]. In support of this, epithelial cells in close proximity to pulmonary neuroendocrine cells and neuroepithelial bodies (NEBs) can exhibit increased proliferation [66, 67]. Similarly, in adult mice exposed to naphthalene, surviving vCE cells are co-localized with neuroendocrine cells located at airway branch points [35, 68]. Separately, the bronchioalveolar duct junction represents a second stem cell niche identified in distal bronchiolar airways [33, 36]. Here, neuroendocrine cells are largely absent, suggesting that other cell populations regulate BASC activation [33]. It is as yet unclear whether neuroepithelial bodies or terminal bronchioles serve a similar role in maintaining populations of injury-resilient stem cells in human conducting airways.

In recent years, studies have determined that the regulation of stem cells within their niche is coordinated through both intrinsic and extrinsic mechanisms [63]. Specific components of niche–stem cell coordination include stem cell–basement membrane interactions (governed by extracellular matrix stiffness and composition), cell–cell interactions (including direct communication and contact inhibition-mediated cell proliferation and signalling), and regulation via local and systemic secreted components, inflammatory and immune cells, and pathogenic stimuli (Fig. 7.2, [61, 69–71]). These are likely to be coordinated through pattern recognition receptors (such as toll-like receptors), cell adhesion molecules normally involved in regulation of adherens and tight junctions (such as cadherins, junctional adhesion molecules (JAMs) and other molecules including nectins and nectin-like molecules), receptor tyrosine kinases (RTKs), G-protein coupled receptors (GPCRs), and cytokine receptors. In turn, this coordinated signalling is thought to regulate cell homeostasis, proliferation, apoptosis, senescence, and motility via

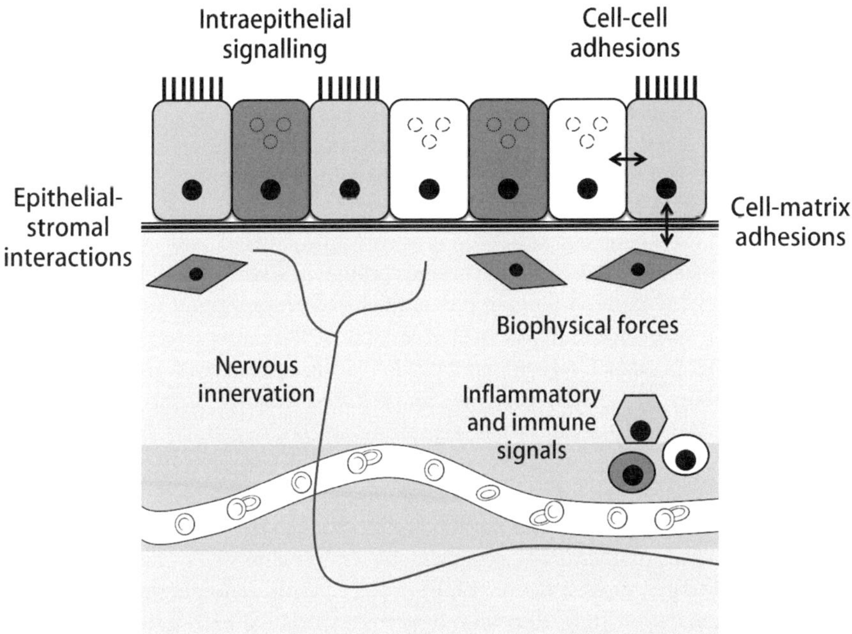

Fig. 7.2 Components of an epithelial stem cell niche. The niche, or local microenvironment where stem cells reside, functions as a dynamic system that integrates local, systemic, and cell-intrinsic signals to determine cellular fate and phenotype [60]. Local signals include interactions among and between cells and their extracellular matrix as well as intra-epithelial and bidirectional epithelial-stromal signalling cascades [63]. In turn, these signals both influence and respond to biophysical components of the niche including matrix stiffness, composition, local tension, and microenvironment shape. In addition, signals from nearby inflammatory and immune cells, nervous innervation, and the local circulation are known to influence stem cell phenotypes

several downstream networks. Although incompletely characterized, these are known to include the PI3K/Akt, Erk/MAPK, canonical and non-canonical beta-catenin, Notch-Delta, Jak/Stat, and Hedgehog signalling pathways [72–74].

In vivo, bronchiolar stem cell behaviour is determined by multiple niche-derived signalling networks [75, 76]. In particular, it was shown that stabilization of beta-catenin activity in bronchiolar stem cells promotes generation of a hyperplastic epithelium and expansion of naphthalene-resistant stem-like cells [77]. Similarly, ablation of Gata6, a negative regulator of the Wnt receptor Frizzled2, causes loss of epithelial differentiation and increased BASC abundance [78]. In each case, bronchiolar stem cell beta-catenin activity was governed by stromal cell-derived local Wnt and/or Fgf secretion [58, 59, 77, 78].

Notch signalling also impacts upon the differentiation status of the distal airway. Activation of canonical Notch signalling in developing murine airways results in goblet cell metaplasia and reduced numbers of ciliated cells [79]. Loss of Notch signalling in Hes1 −/− mice reduces the number of Club cells due to an increase in the proportion of cells that differentiate as pulmonary neuroendocrine cells during lung development [80]. In addition, naphthalene-mediated Club cell ablation is fatal in Notch1 mutant mice due to an inability to efficiently repair the damaged epithelium [81]. Loss of Notch signalling in adult mice through conditional inactivation of Pofut1 reduces Club cell number and increases the abundance of ciliated cells [82]. The depletion of Club cells in the absence of Notch signalling may arise as a consequence of lateral inhibition from ciliated cells that prevents neighbouring cells from adopting a Club cell phenotype [83]. Overall, these studies indicate that low levels of Notch favour differentiation towards ciliated and neuroendocrine lineages whereas elevated Notch signalling promotes Club and goblet cell differentiation via SAM-pointed domain containing ETS transcription factor (SPDEF) activation [84, 85]. Interestingly, misexpression of the active Notch intracellular domain in the proximal airway epithelium causes goblet cell metaplasia and reduced numbers of ciliated cells in the proximal but not distal airways [79], suggesting that the effects of Notch signalling are both spatially and temporally dependent.

In murine models, the vasculature is also important in driving signalling pathways that control lung regeneration. In a unilateral pneumonectomy model, matrix metalloproteinase 14 (MMP14) produced by endothelial cells causes increased release of epidermal growth factor (EGF) protein family ectodomains to stimulate epithelial regeneration [86]. Recently, a novel endothelial cell-derived signalling axis was found to influence the fate of BASCs [71]. Primary murine lung endothelial cells were able to support multiple passages of BASCs with bronchiolar and alveolar differentiation capacity in 3D culture [71]. Thrombospondin1 (TSP1)-deficient endothelial cells tip the balance of BASC differentiation in favour of bronchiolar cell types, suggesting that TSP1 is an endothelial-derived factor which promotes alveolar differentiation of BASCs [71]. Addition of recombinant BMP4 to BASC cultures produced the reverse effect in a TSP1-dependent manner, favouring alveolar differentiation [71]. Thus BMP4, likely produced by epithelial cells [87], induces TSP1 expression in endothelial cells via the calcineurin-NFAT pathway, which in turn promotes alveolar differentiation of murine distal lung stem cells.

Unfortunately, the signalling pathways and factors that regulate stem cell activity in human lungs remain only partially characterized. Despite this, studies of prospectively isolated Lgr6+ human lung stem cells injected under the murine kidney capsule have identified endogenous cells and factors that are involved in lung stem cell growth and differentiation [47]. Expression of the cytokine stromal cell-derived factor 1 (SDF-1) in transplanted stem cells activates and recruits stromal fibroblasts [88]. These fibroblasts secrete tumour necrosis factor α (TNFα), which provides an activating signal for lung stem cells to produce more TGFβ and consequently more SDF-1. Furthermore, endothelial cells were recruited to the transplantation site in a process dependent on secretion of interleukin 8 (IL-8) and vascular endothelial growth factor (VEGF) by activated fibroblasts [88]. While the relevance of these ex vivo findings to the native human lung stem cell niche remains unclear, these data support the aforementioned studies indicating the importance of endothelial signals in the murine BASC niche [71].

Distal Stem Cells in Lung Cancer

As well as their role in homeostasis, repair, and regeneration, there is now agreement that many types of epithelial tumours originate from single, dysfunctional stem cells [89]. This is consistent with the unique properties of tissue stem cells including their capacity for long-term self-renewal and generation of extensive progeny [90–92]. In the lungs, there is a good correlation between stem cell niches and the sites at which lung tumors are most commonly observed [93].

In distal airways, alveolar type II cells have traditionally been considered a likely cell of origin for lung adenocarcinomas based on the observation that this tumor type often expresses abundant surfactant protein C [94]. Indeed, in murine models of K-Ras-induced adenocarcinoma, recombination with SPC-CreER or CCSP-CreER was observed in both type II cells and BASCs but tumors were only observed in the alveoli [94, 95]. However, the BASC subset expressing both CCSP and SPC expands during tumorigenesis, suggesting that this population of cells may represent an alternative cancer initiating cell [36]. Recent attempts to clarify this problem have used CCSP-Cre and SPC-Cre adenoviruses to initiate K-Ras-dependent tumorigenesis [96] or induced recombination from more wide ranging promoters such as *polr2a* [97]. The former study showed CCSP-Cre adenovirus administration caused tumorigenesis in both the BADJ and the alveoli but infrequently in the bronchioles [96]. The latter study showed that K-Ras was activated in a range of epithelial cell types including CCSP+, SPC+, and CCSP+/SPC+ populations but that the development of adenocarcinomas occurred exclusively in the alveoli [97]. More of these BADJ lesions, some of which progressed to invasive cancer, occurred when intratracheal administration of Cre adenovirus was used to initiate recombination. Altogether, these studies suggest that multiple lung cell types may be capable of giving rise to K-Ras-induced tumours. This finding is consistent with the extensive plasticity observed following severe lung injury as well as the growing appreciation that environmental factors (in particular, immune surveillance via PD-1/PD-L1 signalling) play a dominant role in driving lung cancer initiation and progression [98, 99].

Conclusions and Perspectives

Substantial progress has been made in recent years in improving our understanding of the mechanisms of murine airway homeostasis, repair, and regeneration. The locations and identities of various endogenous airway stem cells are now well established and markers identified to facilitate the prospective isolation and characterization of these various populations. We are also beginning to understand the molecular mechanisms and signalling pathways that govern airway stem cell behaviour in lung health and disease. Despite this, an improved appreciation of how the local microenvironment or niche responds to and directs lung injury and repair remains a key research priority. It is also important to improve our understanding of the identity, location, plasticity, and functional significance of human lung stem cells in health and disease. Answering these questions in a convincing manner is likely to influence the development of new therapies for lung disease and subsequently improve human lung health.

References

1. Hasleton PS (1972) The internal surface area of the adult human lung. J Anat 112 (Pt 3):391–400
2. Rezaee F, Georas SN (2014) Breaking barriers. New insights into airway epithelial barrier function in health and disease. Am J Respir Cell Mol Biol 50(5):857–869
3. Houtmeyers E, Gosselink R, Gayan-Ramirez G, Decramer M (1999) Regulation of mucociliary clearance in health and disease. Eur Respir J 13(5):1177–1188
4. Massaro GD, Singh G, Mason R, Plopper CG, Malkinson AM, Gail DB (1994) Biology of the Clara cell. Am J Physiol 266(1 Pt 1):L101–L106
5. Bernacki SH, Nelson AL, Abdullah L, Sheehan JK, Harris A, Davis CW, Randell SH (1999) Mucin gene expression during differentiation of human airway epithelia in vitro. Muc4 and muc5b are strongly induced. Am J Respir Cell Mol Biol 20(4):595–604
6. Parker D, Prince A (2011) Innate immunity in the respiratory epithelium. Am J Respir Cell Mol Biol 45(2):189–201
7. Rock JR, Onaitis MW, Rawlins EL, Lu Y, Clark CP, Xue Y et al (2009) Basal cells as stem cells of the mouse trachea and human airway epithelium. Proc Natl Acad Sci U S A 106 (31):12771–12775
8. Rock JR, Randell SH, Hogan BL (2010) Airway basal stem cells: a perspective on their roles in epithelial homeostasis and remodeling. Dis Model Mech 3(9–10):545–556
9. Boers JE, Ambergen AW, Thunnissen FB (1999) Number and proliferation of clara cells in normal human airway epithelium. Am J Respir Crit Care Med 159(5 Pt 1):1585–1591
10. Rackley CR, Stripp BR (2012) Building and maintaining the epithelium of the lung. J Clin Invest 122(8):2724–2730
11. Tyler WS (1983) Comparative subgross anatomy of lungs. Pleuras, interlobular septa, and distal airways. Am Rev Respir Dis 128(2 Pt 2):S32–S36
12. Irvin CG, Bates JH (2003) Measuring the lung function in the mouse: the challenge of size. Respir Res 4:4
13. Rawlins EL, Hogan BL (2008) Ciliated epithelial cell lifespan in the mouse trachea and lung. Am J Physiol Lung Cell Mol Physiol 295(1):L231–L234
14. Gomperts BN, Strieter RM (2007) Stem cells and chronic lung disease. Annu Rev Med 58:285–298
15. Chapman KR, Mannino DM, Soriano JB, Vermeire PA, Buist AS, Thun MJ et al (2006) Epidemiology and costs of chronic obstructive pulmonary disease. Eur Respir J 27(1):188–207
16. Weiss DJ (2014) Concise review: current status of stem cells and regenerative medicine in lung biology and diseases. Stem Cells 32(1):16–25
17. Ayers MM, Jeffery PK (1988) Proliferation and differentiation in mammalian airway epithelium. Eur Respir J 1(1):58–80
18. Evans MJ, Hackney JD (1972) Cell proliferation in lungs of mice exposed to elevated concentrations of oxygen. Aerosp Med 43(6):620–622
19. Stripp BR, Maxson K, Mera R, Singh G (1995) Plasticity of airway cell proliferation and gene expression after acute naphthalene injury. Am J Physiol 269(6 Pt 1):L791–L799
20. van Rongen E, Thames HD Jr, Travis EL (1993) Recovery from radiation damage in mouse lung: interpretation in terms of two rates of repair. Radiat Res 133(2):225–233

21. Paul MK, Bisht B, Darmawan DO, Chiou R, Ha VL, Wallace WD et al (2014) Dynamic changes in intracellular ROS levels regulate airway basal stem cell homeostasis through Nrf2-dependent Notch signaling. Cell Stem Cell 15(2):199–214
22. Evans MJ, Cabral-Anderson LJ, Freeman G (1978) Role of the Clara cell in renewal of the bronchiolar epithelium. Lab Invest 38(6):648–653
23. Fanucchi MV, Murphy ME, Buckpitt AR, Philpot RM, Plopper CG (1997) Pulmonary cytochrome P450 monooxygenase and Clara cell differentiation in mice. Am J Respir Cell Mol Biol 17(3):302–314
24. Rawlins EL, Okubo T, Xue Y, Brass DM, Auten RL, Hasegawa H et al (2009) The role of Scgb1a1+ Clara cells in the long-term maintenance and repair of lung airway, but not alveolar, epithelium. Cell Stem Cell 4(6):525–534
25. Giangreco A, Arwert EN, Rosewell IR, Snyder J, Watt FM, Stripp BR (2009) Stem cells are dispensable for lung homeostasis but restore airways after injury. Proc Natl Acad Sci U S A 106(23):9286–9291
26. Simons BD, Clevers H (2011) Strategies for homeostatic stem cell self-renewal in adult tissues. Cell 145(6):851–862
27. O'Brien KA, Suverkropp C, Kanekal S, Plopper CG, Buckpitt AR (1989) Tolerance to multiple doses of the pulmonary toxicant, naphthalene. Toxicol Appl Pharmacol 99 (3):487–500
28. Plopper CG, Macklin J, Nishio SJ, Hyde DM, Buckpitt AR (1992) Relationship of cytochrome P-450 activity to Clara cell cytotoxicity. III. Morphometric comparison of changes in the epithelial populations of terminal bronchioles and lobar bronchi in mice, hamsters, and rats after parenteral administration of naphthalene. Lab Invest 67(5):553–565
29. Van Winkle LS, Buckpitt AR, Nishio SJ, Isaac JM, Plopper CG (1995) Cellular response in naphthalene-induced Clara cell injury and bronchiolar epithelial repair in mice. Am J Physiol 269(6 Pt 1):L800–L818
30. West JA, Pakehham G, Morin D, Fleschner CA, Buckpitt AR, Plopper CG (2001) Inhaled naphthalene causes dose dependent Clara cell cytotoxicity in mice but not in rats. Toxicol Appl Pharmacol 173(2):114–119
31. Mahvi D, Bank H, Harley R (1997) Morphology of a naphthalene-induced bronchiolar lesion. Am J Pathol 86(3):558–572
32. Ritter JK, Owens IS, Negishi M, Nagata K, Sheen YY, Gillette JR, Sasame H (1991) Mouse pulmonary cytochrome P-450 naphthalene hydroxylase: cDNA cloning, sequence, and expression in Saccharomyces cerevisiae. Biochemistry 30(48):11430–11437
33. Giangreco A, Reynolds SD, Stripp BR (2002) Terminal bronchioles harbor a unique airway stem cell population that localizes to the bronchoalveolar duct junction. Am J Pathol 161 (1):173–182
34. Hong KU, Reynolds SD, Giangreco A, Hurley CM, Stripp BR (2001) Clara cell secretory protein-expressing cells of the airway neuroepithelial body microenvironment include a label-retaining subset and are critical for epithelial renewal after progenitor cell depletion. Am J Respir Cell Mol Biol 24(6):671–681
35. Reynolds SD, Giangreco A, Power JH, Stripp BR (2000) Neuroepithelial bodies of pulmonary airways serve as a reservoir of progenitor cells capable of epithelial regeneration. Am J Pathol 156(1):269–278
36. Kim CF, Jackson EL, Woolfenden AE, Lawrence S, Babar I, Vogel S et al (2005) Identification of bronchioalveolar stem cells in normal lung and lung cancer. Cell 121(6):823–835
37. Tropea KA, Leder E, Aslam M, Lau AN, Raiser DM, Lee JH et al (2012) Bronchioalveolar stem cells increase after mesenchymal stromal cell treatment in a mouse model of bronchopulmonary dysplasia. Am J Physiol Lung Cell Mol Physiol 302(9):L829–L837
38. Kumar PA, Hu Y, Yamamoto Y, Hoe NB, Wei TS, Mu D et al (2011) Distal airway stem cells yield alveoli in vitro and during lung regeneration following H1N1 influenza infection. Cell 147(3):525–538

39. Zuo W, Zhang T, Wu DZ, Guan SP, Liew AA, Yamamoto Y et al (2014) p63Krt5 distal airway stem cells are essential for lung regeneration. Nature 517(7536):616–620
40. Tata PR, Mou H, Pardo-Saganta A, Zhao R, Prabhu M, Law BM et al (2013) Dedifferentiation of committed epithelial cells into stem cells in vivo. Nature 503(7475):218–223
41. Desai TJ, Krasnow MA (2013) Stem cells: differentiated cells in a back-up role. Nature 503 (7475):204–205
42. Kotton DN, Morrisey EE (2014) Lung regeneration: mechanisms, applications and emerging stem cell populations. Nat Med 20(8):822–832
43. Teixeira VH, Nadarajan P, Graham TA, Pipinikas CP, Brown JM, Falzon M et al (2013) Stochastic homeostasis in human airway epithelium is achieved by neutral competition of basal cell progenitors. Elife 2, e00966
44. Hogan BL, Barkauskas CE, Chapman HA, Epstein JA, Jain R, Hsia CC et al (2014) Repair and regeneration of the respiratory system: complexity, plasticity, and mechanisms of lung stem cell function. Cell Stem Cell 15(2):123–138
45. Koo BK, Clevers H (2014) Stem cells marked by the R-spondin receptor LGR5. Gastroenterology 147(2):289–302
46. Snippert HJ, Haegebarth A, Kasper M, Jaks V, van Es JH, Barker N et al (2010) Lgr6 marks stem cells in the hair follicle that generate all cell lineages of the skin. Science 327 (5971):1385–1389
47. Oeztuerk-Winder F, Guinot A, Ochalek A, Ventura JJ (2012) Regulation of human lung alveolar multipotent cells by a novel p38alpha MAPK/miR-17-92 axis. EMBO J 31 (16):3431–3441
48. Bertoncello I, McQualter JL (2013) Lung stem cells: do they exist? Respirology 18 (4):587–595
49. Leeman KT, Fillmore CM, Kim CF (2014) Lung stem and progenitor cells in tissue homeostasis and disease. Curr Top Dev Biol 107:207–233
50. McQualter JL, Bertoncello I (2015) Clonal culture of adult mouse lung epithelial stem/ progenitor cells. Methods Mol Biol 1235:231–241
51. Bertoncello I, McQualter J (2011) Isolation and clonal assay of adult lung epithelial stem/ progenitor cells. Curr Protoc Stem Cell Biol. Chapter 2: Unit 2G 1
52. Teisanu RM, Lagasse E, Whitesides JF, Stripp BR (2009) Prospective isolation of bronchiolar stem cells based upon immunophenotypic and autofluorescence characteristics. Stem Cells 27 (3):612–622
53. Raiser DM, Kim CF (2009) Commentary: Sca-1 and cells of the lung: a matter of different sorts. Stem Cells 27(3):606–611
54. Teisanu RM, Chen H, Matsumoto K, McQualter JL, Potts E, Foster WM et al (2011) Functional analysis of two distinct bronchiolar progenitors during lung injury and repair. Am J Respir Cell Mol Biol 44(6):794–803
55. Holmes C, Stanford WL (2007) Concise review: stem cell antigen-1: expression, function, and enigma. Stem Cells 25(6):1339–1347
56. McQualter JL, Yuen K, Williams B, Bertoncello I (2010) Evidence of an epithelial stem/ progenitor cell hierarchy in the adult mouse lung. Proc Natl Acad Sci U S A 107(4):1414–1419
57. McQualter JL, McCarty RC, Van der Velden J, O'Donoghue RJ, Asselin-Labat ML, Bozinovski S, Bertoncello I (2013) TGF-beta signaling in stromal cells acts upstream of FGF-10 to regulate epithelial stem cell growth in the adult lung. Stem Cell Res 11 (3):1222–1233
58. Volckaert T, Dill E, Campbell A, Tiozzo C, Majka S, Bellusci S, De Langhe SP (2011) Parabronchial smooth muscle constitutes an airway epithelial stem cell niche in the mouse lung after injury. J Clin Invest 121(11):4409–4419
59. Volckaert T, Campbell A, De Langhe S (2013) c-Myc regulates proliferation and Fgf10 expression in airway smooth muscle after airway epithelial injury in mouse. PLoS One 8(8): e71426
60. Moore KA, Lemischka IR (2006) Stem cells and their niches. Science 311(5769):1880–1885

61. Watt FM, Hogan BL (2000) Out of Eden: stem cells and their niches. Science 287 (5457):1427–1430
62. Jones DL, Wagers AJ (2008) No place like home: anatomy and function of the stem cell niche. Nat Rev Mol Cell Biol 9(1):11–21
63. Roeder I, Loeffler M, Glauche I (2011) Towards a quantitative understanding of stem cell-niche interaction: experiments, models, and technologies. Blood Cells Mol Dis 46(4):308–317
64. Barker N, Clevers H (2007) Tracking down the stem cells of the intestine: strategies to identify adult stem cells. Gastroenterology 133(6):1755–1760
65. Morrison SJ, Scadden DT (2014) The bone marrow niche for haematopoietic stem cells. Nature 505(7483):327–334
66. Hoyt RF Jr, McNelly NA, Sorokin SP (1990) Dynamics of neuroepithelial body (NEB) formation in developing hamster lung: light microscopic autoradiography after 3H-thymidine labeling in vivo. Anat Rec 227(3):340–350
67. Hoyt RF Jr, McNelly NA, McDowell EM, Sorokin SP (1991) Neuroepithelial bodies stimulate proliferation of airway epithelium in fetal hamster lung. Am J Physiol 260(4 Pt 1):L234–L240
68. Reynolds SD, Hong KU, Giangreco A, Mango GW, Guron C, Morimoto Y, Stripp BR (2000) Conditional clara cell ablation reveals a self-renewing progenitor function of pulmonary neuroendocrine cells. Am J Physiol Lung Cell Mol Physiol 278(6):L1256–L1263
69. LaBarge MA, Petersen OW, Bissell MJ (2007) Of microenvironments and mammary stem cells. Stem Cell Rev 3(2):137–146
70. Fuchs E, Tumbar T, Guasch G (2004) Socializing with the neighbors: stem cells and their niche. Cell 116(6):769–778
71. Lee JH, Bhang DH, Beede A, Huang TL, Stripp BR, Bloch KD et al (2014) Lung stem cell differentiation in mice directed by endothelial cells via a BMP4-NFATc1-thrombospondin-1 axis. Cell 156(3):440–455
72. Blank U, Karlsson G, Karlsson S (2008) Signaling pathways governing stem-cell fate. Blood 111(2):492–503
73. Clevers H, Loh KM, Nusse R (2014) Stem cell signaling. An integral program for tissue renewal and regeneration: Wnt signaling and stem cell control. Science 346(6205):1248012
74. Lowry WE, Richter L (2007) Signaling in adult stem cells. Front Biosci 12:3911–3927
75. Volckaert T, De Langhe S (2014) Lung epithelial stem cells and their niches: Fgf10 takes center stage. Fibrogenesis Tissue Repair 7:8
76. Asselin-Labat ML, Filby CE (2012) Adult lung stem cells and their contribution to lung tumourigenesis. Open Biol 2(8):120094
77. Reynolds SD, Zemke AC, Giangreco A, Brockway BL, Teisanu RM, Drake JA et al (2008) Conditional stabilization of beta-catenin expands the pool of lung stem cells. Stem Cells 26 (5):1337–1346
78. Zhang Y, Goss AM, Cohen ED, Kadzik R, Lepore JJ, Muthukumaraswamy K et al (2008) A Gata6-Wnt pathway required for epithelial stem cell development and airway regeneration. Nat Genet 40(7):862–870
79. Guseh JS, Bores SA, Stanger BZ, Zhou Q, Anderson WJ, Melton DA et al (2009) Notch signaling promotes airway mucous metaplasia and inhibits alveolar development. Development 136(10):1751–1759
80. Ito T, Udaka N, Yazawa T, Okudela K, Hayashi H, Sudo T et al (2000) Basic helix-loop-helix transcription factors regulate the neuroendocrine differentiation of fetal mouse pulmonary epithelium. Development 127(18):3913–3921
81. Xing Y, Li A, Borok Z, Li C, Minoo P (2012) NOTCH1 is required for regeneration of Clara cells during repair of airway injury. Stem Cells 30(5):946–955
82. Tsao PN, Wei SC, Wu MF, Huang MT, Lin HY, Lee MC et al (2011) Notch signaling prevents mucous metaplasia in mouse conducting airways during postnatal development. Development 138(16):3533–3543

83. Morimoto M, Liu Z, Cheng HT, Winters N, Bader D, Kopan R (2010) Canonical notch signaling in the developing lung is required for determination of arterial smooth muscle cells and selection of Clara versus ciliated cell fate. J Cell Sci 123(Pt 2):213–224
84. Rock JR, Gao X, Xue Y, Randell SH, Kong YY, Hogan BL (2011) Notch-dependent differentiation of adult airway basal stem cells. Cell Stem Cell 8(6):639–648
85. Park KS, Korfhagen TR, Bruno MD, Kitzmiller JA, Wan H, Wert SE et al (2007) SPDEF regulates goblet cell hyperplasia in the airway epithelium. J Clin Invest 117(4):978–988
86. Ding BS, Nolan DJ, Guo P, Babazadeh AO, Cao Z, Rosenwaks Z et al (2011) Endothelial-derived angiocrine signals induce and sustain regenerative lung alveolarization. Cell 147 (3):539–553
87. Masterson JC, Molloy EL, Gilbert JL, McCormack N, Adams A, O'Dea S (2011) Bone morphogenetic protein signalling in airway epithelial cells during regeneration. Cell Signal 23(2):398–406
88. Ruiz EJ, Oeztuerk-Winder F, Ventura JJ (2014) A paracrine network regulates the cross-talk between human lung stem cells and the stroma. Nat Commun 5:3175
89. Visvader JE (2011) Cells of origin in cancer. Nature 469(7330):314–322
90. Barker N, Bartfeld S, Clevers H (2010) Tissue-resident adult stem cell populations of rapidly self-renewing organs. Cell Stem Cell 7(6):656–670
91. Krivtsov AV, Figueroa ME, Sinha AU, Stubbs MC, Feng Z, Valk PJ et al (2013) Cell of origin determines clinically relevant subtypes of MLL-rearranged AML. Leukemia 27(4):852–860
92. Lapouge G, Youssef KK, Vokaer B, Achouri Y, Michaux C, Sotiropoulou PA, Blanpain C (2011) Identifying the cellular origin of squamous skin tumors. Proc Natl Acad Sci U S A 108 (18):7431–7436
93. Succony L, Janes SM (2014) Airway stem cells and lung cancer. QJM 107(8):607–612
94. Desai TJ, Brownfield DG, Krasnow MA (2014) Alveolar progenitor and stem cells in lung development, renewal and cancer. Nature 507(7491):190–194
95. Xu X, Rock JR, Lu Y, Futtner C, Schwab B, Guinney J et al (2012) Evidence for type II cells as cells of origin of K-Ras-induced distal lung adenocarcinoma. Proc Natl Acad Sci U S A 109 (13):4910–4915
96. Sutherland KD, Song JY, Kwon MC, Proost N, Zevenhoven J, Berns A (2014) Multiple cells-of-origin of mutant K-Ras-induced mouse lung adenocarcinoma. Proc Natl Acad Sci U S A 111(13):4952–4957
97. Mainardi S, Mijimolle N, Francoz S, Vicente-Duenas C, Sanchez-Garcia I, Barbacid M (2014) Identification of cancer initiating cells in K-Ras driven lung adenocarcinoma. Proc Natl Acad Sci U S A 111(1):255–260
98. Dunn GP, Bruce AT, Ikeda H, Old LJ, Schreiber RD (2002) Cancer immunoediting: from immunosurveillance to tumor escape. Nat Immunol 3(11):991–998
99. Harvey RD (2014) Immunologic and clinical effects of targeting PD-1 in lung cancer. Clin Pharmacol Ther 96(2):214–223

Chapter 8
Regenerative Cells in the Ageing Lung

Andrew M. Hoffman

Healthy Ageing Across the Life Span

Healthy ageing is a critical but often overlooked factor in the investigation of fundamental biology and disease progression in the lung. It has a profound effect on the regenerative capacity of the lung thought to be mediated by tissue-resident stem-progenitor cells [1–5]. Both intrinsic and extrinsic factors conspire to diminish the potential of stem-progenitor cells to self-renew and differentiate, participate in homeostasis, and respond to injuries and disease in the ageing lung. While the changes in tissue architecture and pulmonary function that accompany ageing are well known, the relationships between stem-progenitor cell ageing and organ regeneration in the lung are poorly understood. Studies of ageing have been segmented into life stages. These include lung development, a post-development period where lung regeneration can be reinitiated (e.g., by partial pneumonectomy, "PNX"), a period after which reinitiation is no longer feasible, and a "late ageing" period typified by senescence.

During development, the growth of airways and acini follows a pattern of symmorphosis, optimizing ventilation-perfusion matching, and minimizing convective losses, generating the so-called good lung [6]. This is the only period where structure and function of the lung are in perfect balance in healthy mammals.

In the period immediately following the end of lung development (age 2 years in humans [7] and 4 weeks in rodents), somatic growth of the chest continues in the absence of alveolar multiplication. However, the lung remains competent to reinitiate new growth of acini including respiratory bronchioles, alveolar ducts, and alveoli in response to sufficient mechanical stimuli such as partial

A.M. Hoffman (✉)
Regenerative Medicine Laboratory, Department of Clinical Sciences, Cummings School of Veterinary Medicine, Tufts University, Bldg 21, Room 102, 200 Westboro Road, North Grafton, MA 01536, USA
e-mail: andrew.hoffman@tufts.edu

I. Bertoncello (ed.), *Stem Cells in the Lung*, Stem Cell Biology and Regenerative Medicine, DOI 10.1007/978-3-319-21082-7_8

pneumonectomy (PNX) [8, 9]. For example, in immature dogs, as little as 45 % PNX is an effective stimulus, initiating the restoration of alveolar volume, surface area, and gas exchange. Importantly, no other lung injury permits the elucidation of age-dependent regenerative mechanisms in the lung as accurately and comprehensively as the PNX model. In contrast, injuries elicited by bleomycin or naphthalene administration, or influenza infection damage the airway and alveolar epithelia and promote re-epithelialization (normal or aberrant), but without demonstrable neo-alveolarization or neo-angiogenesis.

Ageing of the lung is first evident as discrepancies between post-natal lung growth and the regenerative process that occurs after PNX. Firstly, multiple studies show that alveolar multiplication after PNX in rodents is incomplete, with total lung alveolar numbers reaching approximately 80 % (not 100 %) control values [10, 11]. Secondly, regrowth after PNX is *non-uniform*, with greatest alveolarization in lung lobes (e.g., the cardiac lobe, also called the accessory lobe) [12–16] and lung regions (medial and mediastinal lobes) [17] that experience the greatest mechanical stress. These lobes also display more cellular proliferation [17, 18] and expression of genes encoding growth and differentiation [19]. Thus, it can be quite useful to focus sampling and mechanistic studies on the cardiac (accessory) lobe in this model. Interestingly, non-uniformity of lobar regrowth is present even 1 year after PNX, whether the surgery is performed in the immature (post-natal) or mature (post-development) animal. This implies that reactivation of signals during post-natal lung development does not mitigate or exaggerate structural non-uniformity [14, 20]. Rather this appears to be a pattern established by the topography of mechanical stress placed on the organ after PNX irrespective of age. Growth of alveoli after PNX is accompanied by early bronchiolar lengthening without commensurate bronchial dilation, resulting in dysanaptic growth [9, 21]; therefore, contribution from distal airway epithelium to lung regeneration is important to regeneration per se, but not identical to development. Despite the topographical-functional differences that may exist between alveolarization during late lung development and after PNX, the latter serves as a convenient, minimally injurious model system for the study of reinitiation, in an organ with slow cell and matrix turnover.

Lung regenerative responses are absent or incomplete in the ageing lung, after a transient period whereby reinitiation of alveolarization can be induced by PNX [22–24]. Even 5 years after PNX in mature dogs, there was no evidence of compensatory lung growth [25]. Furthermore, after extensive lung resections (55 %) which evoke substantial mechanical stretch in adult dogs, there was no evidence of new alveolar development [26, 27]. Retinoic acid treatment which mediates early lung bud formation and branching morphogenesis also failed to induce re-alveolarization or improve lung function in pneumonectomized adult dogs [27–30]. In the absence of alveolar multiplication, alveolar septal cells respond to PNX not by hyperplasia of typical stem-progenitor cells, but by local "adaptation" responses, i.e., accommodation of mechanical stresses by transiently intense remodeling of the alveolar septae including hyperplasia of endothelial, interstitial, and epithelial cells and matrix without formation of secondary septae [31]. Precise in vivo identity of the cells that participate in the adaptive response to stress in mature animals has not been determined. Eventually, these adaptations are remodeled to restore the harmonic mean septal barrier thickness, yet, twofold cell numbers persist in those regions. These data demonstrate that ageing to maturity results in a shift in the nature of responses to stress, rather than the absence of responses in the lung. Given that the lung continues to respond to mechanical stress, it is possible that post-PNX lung growth potential has been underestimated in humans, since most studies have not spanned decades or an equivalent period in animal. For example, one report provided compelling evidence for new acinar development which was discovered 15 years after PNX in a 33-year-old woman [32]. Studies of post-resection patients vary considerably with regard to the concise age range over which compensatory lung growth (including alveolar multiplication) occurs. In animals, the alveolar multiplication is observed for several months after sexual maturity is reached [23]. In humans, it is generally thought that efficient compensatory lung growth is arrested much earlier (e.g., 2–3 years of age) [23]. Knowledge arising from longitudinal study of remnant lung in mature living lung donors and lobectomy patients, using advanced noninvasive imaging modalities [33–35] may refine our understanding of regenerative potential in the adult lung.

Advanced ageing in the lung is characterized by irreversible structural and functional decline including severe enlargement of airspaces, diminution of elastic recoil associated with loss of elastin fibers, accumulation of collagen III (col-III) in alveolar septae and col-IV in basement membranes, reduced forced vital capacity, increased residual volume, impaired gas exchange, and increasing vulnerability to infection, fibrosis, and cancer [36, 37]. Forced vital capacity can be less than 50 % of peak values in elderly >80 year of age [38–40]. In rodents, there is similar loss of elastin, accumulation of collagen, and increased mean chord length [41]. The lung of humans (as well as rodents) adopts the appearance of "senile emphysema," characterized by greatly dilated alveolar ducts and alveolar spaces without necessarily a reduction in alveolar numbers [42, 43]. One study posits that in women the decline in estrogen and upregulation of estrogen receptors with ageing promotes this phenotype, given that it is reversible by estrogen replacement therapy in mouse models [44]. Intrinsic and extrinsic mechanisms of ageing contribute to this final transformation, including telomere shortening, oxidative stress, DNA damage, mitochondrial dysfunction, and matrix (and thus niche) disruption, resulting in cellular senescence and stem cell exhaustion [45], acknowledging that these mechanisms are inextricably linked. Knowledge concerning the participation of each of these ageing factors in the attrition of lung stem-progenitor cells and their homeostatic functions at the cellular-molecular level are compelling topics of research with great translational significance, yet have been largely neglected in the lung.

Cellular Participants of Lung Regeneration Vulnerable to Ageing

Alveolarization relies on a specialized population of myofibroblasts derived from PDGFRα^{pos} fibroblasts which are dedicated to secondary septation [46]. The subpopulation of precursor PDGFRα^{pos} fibroblasts and PDGFRα expression itself are under the reciprocal control of PPARγ which drives the elastogenic "structural" phenotype, and FGFR2 which drives α-smooth muscle actin expression and the "contractile" phenotype [47]. The magnitude of PDFGRα expression is also associated with differentiation toward myofibroblasts versus preservation of secretory lipofibroblasts that do not participate in secondary septation [48, 49]. Moreover, decreased Sox9 and increased MRTF-A and α-smooth muscle actin in PDGFRα^{high} alveolar fibroblasts have been associated with differentiation to myofibroblasts from P8 to P12 [49]; interestingly, loss of Sox9 in respiratory epithelium is inconsequential to lung development [50] implying that developmental pathways that persist specifically in interstitial fibroblasts may be crucial for sustaining post-natal development and beyond. Lipofibroblasts identified as CD166^{neg} reserve the potential to differentiate to CD166^{pos} myofibroblasts and support epithelial stem cells through secretion of FGF10, particularly in the absence of TGFβ1 [51]. Attesting to the role for TGFβ1, induction of the post-natal lung exposed to hyperoxia leads to excessive TGFβ1 and bronchopulmonary dysplasia [52] perhaps reducing lipofibroblast frequency or survival. Lipofibroblast differentiation to myofibroblasts is disrupted by an age-related elevation of Wnt4, through a β-catenin-dependent mechanisms [37] thus factors that secure a proper balance between lipofibroblast secretions (e.g., FGF10, SDF1) and differentiation capacity to myofibroblasts are important to the understanding of alveolarization during lung development and beyond.

That interstitial fibroblasts are critical to lung regeneration is further evidenced by the over-expression in lung of genes encoding proteins associated with mesenchymal cell proliferation (Igf1, Cyr61, Tnfrsf12a, Ctgf, Igfbp3, and Ifgbp2) [53, 54]. Further, in isolated lipofibroblasts (Sca-1^{pos}, EpCAM^{neg}, CD45^{neg}, and CD31^{neg}), genes encoding many important growth factors (Midkine, Fgf10, Vegfb, Tgfb3, Ifgbp2/3, and Gdf1/10) and extracellular matrix (ECM) (Col1, Col III, Eln, Fn, and Fbn1) are over-expressed [51, 55–57]. Overall stromal fibroblasts appear to be critically important to lung development and regeneration, but gain and loss of function and genetic lineage-tracing studies are needed to better define the temporal and spatial role for these cells.

Angiogenesis is implicit to lung development and post-PNX neo-alveolarization, effectively matching perfusion with alveolar ventilation. Peak endothelial cell expansion and neo-angiogenesis occurs approximately 7 days after PNX and is concentrated in subpleural regions notably in lobes with greatest mechanical stress and regrowth (i.e., cardiac lobe) [58–61]. VEGF markedly accelerates compensatory lung growth after PNX through stimulation of endothelial (and quite possibly AECII) proliferation, although bFGF which was effective in augmenting liver regeneration in past studies was ineffective in this respect [62]. Endothelial cell production of VEGFA [60], HIF1α [63, 64], $PDGF_{BB}$ [65], nitric oxide synthetase [66], and epoxyeicosanoids [67] are also shown to be important to post-natal alveolarization or post-PNX lung regeneration. Moreover, pulmonary capillary endothelial cells (PCEC) produce MMP14 under regulatory control by VEGFR2 and FGFR1 that cause the shedding of EGFR ligands from matrix and surrounding cells necessary for expansion of neighboring AECII and BASC after PNX [68]. These data expose additional, potentially age-sensitive mechanisms. Senescence is likely to impact several of these mechanisms; for example, nitric oxide signaling is markedly disturbed with ageing as are several protective antioxidant mechanisms (e.g., SOD3) [69], and the HIF1α-adrenomedullin axis is markedly depressed with age in rats [70].

Type II alveolar epithelial cells (AECII) are widely considered to be stem cells of the alveolar compartment, which amplify and differentiate to type I alveolar epithelial cells (AECI) following injury to alveolar epithelial cells [71, 72]. A subpopulation of AECII possesses telomerase expression, further evidence of their long-term self-renewal potential and "stemness" [73]. In the event that AECII are depleted as a consequence of epithelial injury, e.g., injuries due to bleomycin or influenza [74], data assert that non-AECII cells including bronchioalveolar stem cells [75], integrin α6β4 expressing cells [76], c-kit^{pos} multi-lineage potential lung stem cells [77], CD90^{pos} AECII progenitors [78, 79], CD45neg CD31neg EpCAM^{high} integrin α6^{pos} CD104^{pos} CD24^{low} lung cells [80], or low density epithelial precursors [81] from healthy rodent or human lungs may reconstitute AECII. AECII can also arise from distal airway Scgbln1 expressing cells (Club cells) in bleomycin [82, 83] or influenza injury [83, 84] consistent with the idea that damage to AECII promotes cytokine driven autocrine repair and generates an inductive milieu for differentiation of cells to the AECII phenotype [85]. AECII transplantation augments post-PNX compensatory lung growth in rats whose compensatory lung growth was limited suggesting a significant paracrine role for AECII [86]. In this regard, AECII produce paracrine signals that control elastogenesis through FGFR3/4 on interstitial fibroblasts [87] and during development, FGFR3 and FGFR4 double-null mice fail to undergo alveologenesis [88]. Signaling of HGF from endothelial cells to c-Met on the surface of AECII is required for normal lung development; knockdown of c-Met in AECII of adult lungs reduces AECII frequency (consistent with HGF improvement of AECII survival), airspace dilation, oxidative stress and inflammation, as well as pruning of the vascular bed [89], reminiscent of the aged lung. Lung mesenchymal stromal cells ("L-MSC": Sca-1^{pos}, EpCAM^{neg}, CD31^{pos}, CD45^{pos}) supports by co-culture epithelial stem-progenitor cell colony formation and self-renewal, features that can be substituted by L-MSC derived paracrine factors (HGF, FGF10) [80]. Furthermore, Lgr6^{pos} multipotent epithelial stem-progenitor cells [90] under the control of p38α expression secrete SDF-1 which activates stromal fibroblasts in the lung [91]. Hence, epithelial-mesenchymal and epithelial-endothelial crosstalk through PPARγ, FGFR/FGF, HGF/c-Met, and SDF-1/CXCR4 signaling pathways elaborate bidirectional mechanisms that with ageing, are presumed to disrupt alveolar homeostasis and regenerative capacity. Indeed, Paxson [55] was able to show that FGFR1, a cornerstone in lung development and upregulated after PNX, is dramatically downregulated (>600-fold) in isolated lung fibroblasts ($CD45^{neg}$ $CD31^{neg}$ $EpCAM^{neg}$ $Sca\text{-}1^{pos}$) from 12-month-old mice when compared to 3-month-old mice. Significant work needs to be done to refine our understanding of signaling pathways during late development that are most vulnerable to age.

Ageing Transition Between Post-Natal and Post-PNX Lung

The question whether lung regeneration is a recapitulation of development remains critical to our understanding of ageing in the lung [23, 92, 93]. Early studies showed that the post-PNX lung is distinct with regard to several molecular pathways, including erythropoietin receptor processing [94], epidermal growth factor and receptor expressions, and surfactant protein A and C expression [95]. This has led to further genome-wide investigations of transcriptional regulation, comparing lung development in adulthood before or after PNX. Wolff et al. [53] compared whole lung gene expression arrays (G4122A, Agilent) between post-natal (days 1 or 3 after birth) and post-PNX adult (10 week) lungs of C57BL6 mixed gender mice. The post-natal lung (vs. post-PNX) exhibited greater numbers of upregulated growth and differentiation genes, and substantially lower expression of immune dense genes. Genes encoding enzymes and enzyme inhibitors and mitochondrial respiratory chain-related proteins were also markedly downregulated in the post-PNX lung but not in the post-natal lung, suggesting that growth potential depends on mitochondrial energetics, which is repressed in the adult lung even after PNX. Importantly, none of the conflicts between gene expression in the post-natal versus adult or adult post-PNX lung encompassed embryonic gene expression, consistent with the lack of embryonic gene expression in whole gene analyses. Likewise, Paxson [54] found that PNX in adult C57BL6 female (12 weeks) mice influenced many growth-differentiation genes found during post-natal alveolarization, but not those which are activated during embryonic or saccular lung development. Present in both post-natal lung development and post-PNX alveolarization includes Igf signaling, but absent in post-PNX lung was upregulation of Fgf, Pdgf signaling, and HoxA5 [54]. Alpha-smooth muscle actin which is abundantly expressed in the secondary septae of the post-natal lung was conspicuously absent after PNX in young mice [54] but present in older mice after PNX [96]. These data imply that the phenotype and function of myofibroblasts associated with alveolarization at various life stages are diverse and require further exploration.

Kho et al. [19] sought to clarify the developmental context of gene expression after PNX in adult C57BL6 mice within the entire spectrum of gene expression from embryonic (E9.5) through lung development, adulthood, and from 1 to 56 days after PNX, a more prolonged period than previously attempted. The 3–7 days post-PNX transcriptome was concordant with that of early post-natal alveolarization; therefore, the authors contended that PNX induce a sort of "de-differentiation," which resolved into a period of later developmental gene regulation, referred to by the authors as "redevelopment." Accordingly, de-differentiation and re-development are cardinal features of limb regeneration in salamanders [97], considered a classical form of organ regeneration. Thus, the bulk of evidence supports that post-PNX gene expression corroborates events during late, not early lung development. This paradigm is consistent with liver and kidney regeneration, for which gene expression in the adult organ similarly aligns with cellular-molecular events that transpire during late development (hyperplasia and hypertrophy of the functional units, respectively).

One limitation of these whole lung transcriptomic studies is a lack of lineage specificity and thus granularity of the findings. While studies of whole lung gene expression have failed to show that PNX reactivates certain embryonic and pseudoglandular (Ttf1, Rarα/β, Gli2/3, Wnt2/2b), canalicular (FoxA1/A2, Foxfla, HoxA5, Mycn, Ttf1, Tcf2l, Sox2l, Foxjl, Alk3), or saccular stage (Klf2, Nf1B, Gata6, Erm) genes [98], studies of respiratory epithelium provide histologic evidence for over-expressed genes marking definitive endodermal specification and morphogenesis (Sox17, Foxa2, Foxj1, β-catenin in bronchial epithelium; Ttf-1 in alveolar ducts) after PNX [99] implying that reactivation of embryonic signaling pathways is possible, specifically in epithelium [100]. Ttf-1 was found to be expressed in alveolar duct proSP-C^{pos} cells after PNX, and knockdown of Ttf-1 resulted in transient delay (although not a reduction in magnitude) of compensatory lung growth in mice, consistent with reactivation of Ttf-1 and participation of this factor in AECII proliferation or differentiation during post-PNX lung regeneration [101]. Conditional knockdown of other developmental genes or ablation of specific subsets of progenitor cells is warranted to further our understanding of the ageing regenerative capacity of the lung.

Ageing Effects on Competence of the Post-PNX Lung to Regenerate

Following PNX in young mice, progenitor cells including PCEC [62, 102], bronchiolar epithelial cells (Club cells, bronchioalveolar stem cells [17, 18]), AECII [17, 18, 58, 103], interstitial fibroblasts [55, 96], pleural lining cells [58], and macrophages [103, 104] show marked proliferation consistent with their participation in regeneration. L-MSC in particular reach their maximal rate of proliferation early after PNX (days 0–3) while proliferative activity of endothelial and bronchial and alveolar epithelial progenitor cells peak later (days 4–7), implying a key role for MSCs to drive re-alveolarization. The impact of age on L-MSC frequency was more profound than on endothelial or epithelial (bronchial, AECII) progenitor cells [55]. At 9 months (vs. young, or 3 months) of age, there was a dramatic decline in the abundance of L-MSC (prior to PNX) that was not evident for endothelial or epithelial progenitor cells. Shortly after PNX (0–3 days) the initial rise in L-MSC, endothelial cells, or bronchial epithelial progenitor cells is virtually absent in older mice (9 months, 17 months), consistent with arrest of cell division (i.e., early senescence) in these populations. In contrast, baseline abundance and proliferative responses of AECII to PNX were unaffected by age, implying that AECII populations are sustained by unique mechanisms.

The mechanism by which age alters L-MSC was further investigated ex vivo; the CFU and growth rate of L-MSC was inferior in 12-month-old (middle age) mice compared to young (3 months) old mice [55]. Moreover, the phenotype of older L-MSC was significantly more myofibroblastic, implying that attrition of precursors (i.e., lipofibroblasts) is a significant event underlying the restricted regenerative response at the tissue level after PNX. Under-expressed in ageing L-MSC were genes associated with retinoic acid signaling (Aldh1a3, Rbp4), Fgfr1/Wnt signaling (Fgfr1, Sfrp1, Ctnnb1, and Wnt2), and ECM formation (Col1a1, Eln, Fbn1, and Sdc2). In particular, Fgfr1 expression was markedly downregulated (>600-fold) at middle age (12 months) vs. young (3 months) consistent with the abrupt cessation of key regulatory mechanisms essential for alveolarization. Mechanistically, Wnt4 and Wnt5a expressions increase lipofibroblast ($PPAR\gamma^{pos}$) differentiation to myofibroblasts in the aged mouse lung [37]; similarly, activation of Wnt3a causes differentiation of bone marrow MSCs to fibroblasts and myofibroblasts with increased expression of col1, α-smooth muscle actin in vitro, and reduced the ability to control HCl-induced lung fibrosis in vivo [105]. This implies that loss of lipofibroblast by differentiation may limit post-PNX re-alveolarization or lung injury repair, as similar losses impair lung development.

The reason why ageing lung fibroblasts dissipate over time has not been completely elucidated. Multiple lines of evidence exist that differentiation to, as well as senescence of lung myofibroblasts (as well as pulmonary fibrosis) may be driven by activation of canonical Wnt/β-catenin signaling [37, 105]. Vaidya et al. [106] found in young mice that non-homologous end-joining (NHEJ) as a DNA repair mechanism was higher in lung fibroblasts than fibroblasts of other tissues, but declined significantly in lung fibroblasts between 5 and 24 months of age. Age was also associated with larger deletions, and greater reliance on micro-homology mediated end joining (MMEJ) which is more error-prone sub-pathway of NHEJ. This may further contribute to the attrition of lipofibroblasts and/or myofibroblasts in the lung.

Immune effector cells are critical for organ regeneration, for example, liver regeneration after partial hepatectomy [107]. In the post-PNX lung, blood born macrophages (CD11b) migrate into the interstitial and alveolar spaces, and express putative pro-regenerative factors (e.g., Angpt1, Egf, Mmp9, Tgfb2) [104]. Type M2 macrophages are known to be important for post-natal lung development, releasing growth factors (e.g., Igf1) [108] suggesting that M2 macrophages specifically are critical for post-PNX re-alveolarization. In a recent study, it was observed that large numbers of interstitial macrophages were found in close association with AECII in zones of sprouting and intussusceptive angiogenesis and secondary septation after PNX [61], implying that these macrophages directly participate in remodeling associated with neo-angiogenesis. It has not been established whether this role is an essential one for alveolarization after PNX. With respect to ageing, there is an accumulation of macrophages and B cells with age in the lung [37], and these may contribute to "inflammageing."

Ageing and the Extracellular Matrix Control of Lung Regeneration

The ECM is a complex mixture of secreted products originating from resident lung cells that imparts a profound inductive influence on fetal development and stem-progenitor cell capacity for self-renewal and differentiation [109, 110]. These include growth factors, cytokines (i.e., matrikines), cryptic peptides, glycosaminoglycans, and glycoproteins that support cell division, migration, recruitment, quiescence, angiogenesis, antimicrobial functions, preservation of phenotype, self-renewal, and/or differentiation, and protect the lung from intense environmental challenges. As well, ECM delivers mechanical cues that evoke outside-in signaling pathways that facilitate survival of anchorage-dependent cells ("tensegrity") such as endothelial cells and fibroblasts [111]. Indeed, mechanical forces transferred through the ECM are undoubtedly the most important initiation factor in lung regeneration after PNX [112].

Following PNX, desmosine and hydroxyproline are deposited in the lung and resultant elastin and collagens are distributed anatomically in a manner that mirrors development [113]. With ageing, there is significant reduction in elastic recoil and thus FEV1, FVC, and FEV1/FVC [114]. Similarly, lungs in aged mice show an increase in lung compliance and lung volume at pressures equivalent to total lung capacity [115] which is accentuated in certain inbred strains of mice such as DBA/2J [41]. These structural–functional changes in mice coincide with decreases in lung tissue elastin fiber content and Eln gene expression. In humans, elastin life span based on mean carbon residence time is ~74 years, consistent with the absence of new elastin synthesis over the human life span [116]. Despite the apparent longevity of lung elastin, it is evident that in the elderly, elastin in alveolar walls is thin, fragmented, and bundled [36], commensurate with the reduced recoil properties of the organ.

Reduction in elastin content in haploinsufficient mice is also associated with airspace dilation [117]. Following PNX, elastin insufficiency is associated with a suppression of epithelial progenitor cell proliferation and compensatory growth [17], implying an indirect effect of lowering mechanical stress after PNX. Similarly, direct interference with mechanical stress after PNX, using plombage, also reduces epithelial (BASC, AECII) responses and alveolarization in mice [17]. In vitro studies corroborate the importance of mechanical stretch to mechanotransduction [118], secretory function [119], and DNA synthesis in AECII [120]. Insufficient mechanical stress on cells presumably contributes to apoptosis and senescence of stem-progenitor cells (AECII, endothelial cells, and interstitial fibroblasts) in disease processes with disrupted matrix.

In addition to extrinsic alterations produced by age-related modifications of ECM, it has recently been reported that ageing impairs intrinsic mechanotransduction through YAP/TAZ signaling including failure of YAP nuclear translocation in breast epithelial progenitor cells [121]. One might speculate that similar mechanisms fail with age in the lung, thus contributing to failure of PNX to reinitiate growth-related signals that are activated upon mechanostimulation.

The inductive properties of exogenous ECM might be effective to mitigate or stabilize the age phenotype of cells. In the lung, fetal (amniotic) basement membrane sustains AECII morphology and phenotype whereby adult lung-derived basement membrane matrix does not [122]. Fetal lung tissues contain more glycosaminoglycans (chondroitin sulfate 4) that support stem-progenitor cell growth and differentiation [123, 124], compatible with the inductive properties of mesenchyme during branching morphogenesis. Hence, a combination of specific mechanical (extrinsic), mechanotransductive (intrinsic), and inductive signals that are derived from ECM appear to be involved in age-related modifications of stem-progenitor cell behavior in the lung.

Ageing of the Lung with Respect to Telomere Length

Evidence is mounting that telomere shortening is responsible for multiple mechanisms leading to an ageing phenotype including stem cell failure, mitochondrial dysfunction, genotoxic stress, and epigenetic changes [125]. Telomerase is significantly reduced after post-natal day 9 in mice, but can be reactivated by hyperoxia, notably within PCNA positive cells suggesting that telomerase is a stem-progenitor cell marker in the lung [73]. Progressive reduction in telomere length in AECII over multiple generations of Terc null mice is evidence that oxidative damage is sustained by lung cells even under normoxic conditions [126]. In that study, Terc null F4 mice showed reduced abundance of AECII, reduced interstitial collagen, and dilated airspaces similar to the biochemical and morphological hallmarks of ageing in mice [96]. Similar to hyperoxia, PNX induces telomerase in wild type and Terc null AECII, but proliferative responses of AECII and BASC as well as re-alveolarization are markedly reduced in Terc null mice consistent with replicative senescence [127], paving the way for our understanding of how senescence and reinitiation of alveolarization are interrelated.

Disease-Induced Premature Ageing of the Lung Phenotype

Cigarette exposure induces genotoxic senescence in alveolar [128] and bronchial [129] epithelia, and this premature senescence of epithelial cells exacerbates the course of COPD [130]. In human emphysema, similar findings of higher than expected senescence in endothelial and AECII, including higher $p16^{Ink4a}$ and $p21^{Waf1}$ expression, shorter telomeres, and proportional lower pulmonary function are evident [131]. Exposure of epithelial cells or fibroblasts from human lungs to cigarette smoke extract (CSE) results in an increased expression of β-galactosidase, a marker of cellular senescence, and decreased cell proliferation [132]. Moreover, there is evidence that CSE induces mitochondrial fragmentation furthering senescence [121, 133, 134]. As well, cigarette exposure in telomerase null mice induces an emphysema phenotype that is associated with accentuated DNA damage, shorter telomeres, AECII senescence, and limited proliferative capacity [135]. While CSE induced senescence markers, advanced age did not augment CSE-induced senescence, despite shorter telomeres in response to CSE [136]. Thus, the connection between senescence, telomere length, telomerase, and CSE exposure in mice, and the species specificity of these findings requires further clarification. Strong parallels have been made between the senescence associated secretory phenotype (SASP) secondary to DNA damage and damage responses, and COPD phenotype [132] based on similarities in cytokine profiles (e.g., IL-6, IL8). This paradigm has enormous implications for progression of lung disease, in that smoking, genetics, and ageing synergistically act to degrade lung function [137].

Ageing is also associated with a predisposition toward a fibrotic phenotype [138]. Lung fibroblasts with ageing of donor also exhibit more myofibroblastic features including greater α-smooth muscle actin, tenascin, fibronectin, Col I, Col III, lower CFU, larger cell size, and loss of clonogenicity and self-renewal in vitro [96]. Pulmonary fibrosis may thus be a manifestation of telomere shortening syndrome even in the absence of identifiable Terc or TR mutations. The relationship between ageing and fibrotic lung disease has been addressed recently by Thannickal et al. [139], Armanios et al. [125], and Chilosi et al. [140]. Briefly, there are a significant percentage of patients with sporadic IPF that harbor genetic mutations in telomerase components (e.g., reverse transcriptase, RT) [140]. Similarly, patients with sporadic IPF have shorter leukocyte telomeres than age-matched controls [141]. Consistent with this observation, a large fraction of patients with identifiable telomere dysfunction exhibit IPF or related liver cirrhosis [125, 142]. The AECII is thought to be a major target for familial IPF mutations, and AECII are often infected with latent Herpesviridae that induce the unfolded protein response, contributing to DNA damage. In familial IPF (~1–2 % IPF patients) mutations specifically target SP-C and SP-A2 in addition to TERT and TERC, resulting in misfolded proteins which contribute to endothelial reticulum (ER) stress in AECII [143]. As a consequence of ER stress, AECII may contribute to epithelial-mesenchymal transition in the lung of IPF patients. In sporadic IPF, there is also evidence of ER stress within AECII implying that familial and sporadic IPF converge on age-related mechanisms [143]. How these findings implicate AECII in the age-related changes observed in lung regeneration is unclear.

In ageing humans, common cancers express markers associated with regenerative stem-progenitor cells in the lung, thus suggesting cell of origin for squamous cell carcinoma (p63, CK5, Sox2 basal cells) or adenocarcinoma (CK14, TTF1, and SP-C AECII cells) [144]. In Kras mice, both BASC and AECII are considered likely cells of origin for adenocarcinomas [75, 145], thus positing an important role for stem-progenitor cell ageing, genomic instability, and transformation to lung cancer.

Epigenetic Ageing of the Lung

Healthy ageing and longevity are strongly associated with changes in cellular DNA methylation [146]. Epigenome-wide association scans (EWAS) were used to extract a significant relationship between age-related phenotype and FVC [146]. Similarly in mice, Lui et al. [147] compared epigenetic and transcriptomic data at 4 weeks vs. 1 week of age, and found a close relationship between ageing, decreased gene expression of growth regulatory genes, and a decrease in H3K4me3 implying the cessation of post-natal growth in lung and kidney may be epigenetically determined. Epigenetic dysregulation is a common feature of ageing in humans and animals, culminating in significant deterioration of stem, immune, and stromal cells [148], so it can be assumed that cells that are critical to reinitiation of lung regeneration are impacted to varying degrees. For example, human diploid lung fibroblasts undergo senescence concomitant with marked alterations to epigenetic regulators such as SUV39H1 that repress genomic instability, demonstrating the close relationship between mechanisms of ageing and transformation of cells [149].

Conclusion

Very few studies have addressed the cellular-molecular biology of healthy ageing in the lung. Given the similarities in mechanisms that echo from post-natal development through the various life stages, ageing and rejuvenation research in the future may address the entire life span of mammals, taking into account environmental, genetic, and epigenetic factors to answer questions. Strategies for better understanding of ageing of regenerative stem-progenitor cells might include great emphasis on genetic lineage tracing, studies of clonality, conditional ablation, epigenetic mechanisms, ECM inductive properties for, and changes to intrinsic mechanotransduction in stem-progenitor cells. As well, the roles for telomere length, genomic instability, and DNA damage in stem-progenitor cells must be further elucidated to understand the link between ageing and stem cell exhaustion.

Indeed, the explanation for the decline in lung stem-progenitor cell abundance and differentiation and self-renewal properties is another mystery that needs to be solved given the importance of this population to regenerative capacity. Finally, more studies that focus on rejuvenation in the lung using heterochronic parabiosis, in vivo reprogramming, and transplantation of stem-progenitor cells or their derivatives will pave the way for translation.

References

1. Bertoncello I, McQualter JL (2013) Lung stem cells: do they exist? Respirology 18(4): 587–595
2. Thane K, Ingenito EP, Hoffman AM (2014) Lung regeneration and translational implications of the postpneumonectomy model. Transl Res 163(4):363–376
3. Hoffman AM, Ingenito EP (2012) Alveolar epithelial stem and progenitor cells: emerging evidence for their role in lung regeneration. Curr Med Chem 19(35):6003–6008
4. Kotton DN, Morrisey EE (2014) Lung regeneration: mechanisms, applications and emerging stem cell populations. Nat Med 20(8):822–832
5. Beers MF, Morrisey EE (2011) The three R's of lung health and disease: repair, remodeling, and regeneration. J Clin Invest 121(6):2065–2073
6. Weibel ER (2013) It takes more than cells to make a good lung. Am J Respir Crit Care Med 187(4):342–346
7. Thurlbeck WM (1982) Postnatal human lung growth. Thorax 37(8):564–571
8. Takeda S et al (1999) Compensatory alveolar growth normalizes gas-exchange function in immature dogs after pneumonectomy. J Appl Physiol 86(4):1301–1310
9. Hsia CC et al (2000) Regenerative growth of respiratory bronchioles in dogs. Am J Physiol Lung Cell Mol Physiol 279(1):L136–L142
10. Fehrenbach H et al (2008) Neoalveolarization contributes to compensatory lung growth following pneumonectomy in mice. Eur Respir J 31(3):515–522
11. Holmes C, Thurlbeck WM (1979) Normal lung growth and response after pneumonectomy in rats at various ages. Am Rev Respir Dis 120(5):1125–1136
12. Voswinckel R et al (2004) Characterisation of post-pneumonectomy lung growth in adult mice. Eur Respir J 24(4):524–532
13. Yilmaz C et al (2009) Noninvasive quantification of heterogeneous lung growth following extensive lung resection by high-resolution computed tomography. J Appl Physiol 107(5): 1569–1578
14. Ravikumar P et al (2007) Developmental signals do not further accentuate nonuniform postpneumonectomy compensatory lung growth. J Appl Physiol 102(3):1170–1177
15. Ravikumar P et al (2004) Regional lung growth following pneumonectomy assessed by computed tomography. J Appl Physiol 97(4):1567–1574, discussion 1549
16. Gibney BC et al (2012) Detection of murine post-pneumonectomy lung regeneration by 18FDG PET imaging. EJNMMI Res 2(1):48
17. Hoffman AM et al (2010) Matrix modulation of compensatory lung regrowth and progenitor cell proliferation in mice. Am J Physiol Lung Cell Mol Physiol 298(2):L158–L168
18. Nolen-Walston RD et al (2008) Cellular kinetics and modeling of bronchioalveolar stem cell response during lung regeneration. Am J Physiol Lung Cell Mol Physiol 294(6): L1158–L1165
19. Kho AT et al (2013) Identification of dedifferentiation and redevelopment phases during postpneumonectomy lung growth. Am J Physiol Lung Cell Mol Physiol 305(8):L542–L554

20. Eisenhauer P et al (2013) Endogenous distal airway progenitor cells, lung mechanics, and disproportionate lobar growth following long-term postpneumonectomy in mice. Stem Cells 31(7):1330–1339
21. Dane DM, Johnson RL Jr, Hsia CC (2002) Dysanaptic growth of conducting airways after pneumonectomy assessed by CT scan. J Appl Physiol 93(4):1235–1242
22. Buhain WJ, Brody JS (1973) Compensatory growth of the lung following pneumonectomy. J Appl Physiol 35(6):898–902
23. Cagle PT, Thurlbeck WM (1988) Postpneumonectomy compensatory lung growth. Am Rev Respir Dis 138(5):1314–1326
24. Cagle PT, Langston C, Thurlbeck WM (1988) The effect of age on postpneumonectomy growth in rabbits. Pediatr Pulmonol 5(2):92–95
25. Davies P et al (1982) Structural changes in the canine lung and pulmonary arteries after pneumonectomy. J Appl Physiol 53(4):859–864
26. Hsia CC et al (1994) Compensatory lung growth occurs in adult dogs after right pneumonectomy. J Clin Invest 94(1):405–412
27. Yan X et al (2004) Retinoic acid induces nonuniform alveolar septal growth after right pneumonectomy. J Appl Physiol 96(3):1080–1089
28. Malpel S, Mendelsohn C, Cardoso WV (2000) Regulation of retinoic acid signaling during lung morphogenesis. Development 127(14):3057–3067
29. Dane DM et al (2004) Retinoic acid-induced alveolar cellular growth does not improve function after right pneumonectomy. J Appl Physiol 96(3):1090–1096
30. Yan X et al (2005) Lack of response to all-trans retinoic acid supplementation in adult dogs following left pneumonectomy. J Appl Physiol 99(5):1681–1688
31. Hsia CC, Johnson RL Jr (2006) Further examination of alveolar septal adaptation to left pneumonectomy in the adult lung. Respir Physiol Neurobiol 151(2–3):167–177
32. Butler JP et al (2012) Evidence for adult lung growth in humans. N Engl J Med 367(3): 244–247
33. Chen F et al (2012) Short-term outcome in living donors for lung transplantation: the role of preoperative computer tomographic evaluations of fissures and vascular anatomy. Transpl Int 25(7):732–738
34. Chen F et al (2012) Outcomes and pulmonary function in living lobar lung transplant donors. Transpl Int 25(2):153–157
35. Mizobuchi T et al (2013) Radiologic evaluation for volume and weight of remnant lung in living lung donors. J Thorac Cardiovasc Surg 146(5):1253–1258
36. D'Errico A et al (1989) Changes in the alveolar connective tissue of the ageing lung. An immunohistochemical study. Virchows Arch A Pathol Anat Histopathol 415(2):137–144
37. Kovacs T et al (2014) Alteration in the Wnt microenvironment directly regulates molecular events leading to pulmonary senescence. Aging Cell 13(5):838–849
38. Falaschetti E et al (2004) Prediction equations for normal and low lung function from the health survey for England. Eur Respir J 23(3):456–463
39. Hankinson JL, Odencrantz JR, Fedan KB (1999) Spirometric reference values from a sample of the general U.S. population. Am J Respir Crit Care Med 159(1):179–187
40. DeLorey DS, Babb TG (1999) Progressive mechanical ventilatory constraints with aging. Am J Respir Crit Care Med 160(1):169–177
41. Huang K et al (2007) Age-dependent changes of airway and lung parenchyma in C57BL/6J mice. J Appl Physiol 102(1):200–206
42. Verbeken EK et al (1992) The senile lung. Comparison with normal and emphysematous lungs. 1. Structural aspects. Chest 101(3):793–799
43. Lang MR et al (1994) Collagen content of alveolar wall tissue in emphysematous and non-emphysematous lungs. Thorax 49(4):319–326
44. Glassberg MK et al (2014) 17beta-estradiol replacement reverses age-related lung disease in estrogen-deficient C57BL/6J mice. Endocrinology 155(2):441–448
45. Lopez-Otin C et al (2013) The hallmarks of aging. Cell 153(6):1194–1217

46. Perl AK, Gale E (2009) FGF signaling is required for myofibroblast differentiation during alveolar regeneration. Am J Physiol Lung Cell Mol Physiol 297(2):L299–L308
47. Chen L et al (2012) Dynamic regulation of platelet-derived growth factor receptor alpha expression in alveolar fibroblasts during realveolarization. Am J Respir Cell Mol Biol 47(4): 517–527
48. McGowan SE, McCoy DM (2011) Fibroblasts expressing PDGF-receptor-alpha diminish during alveolar septal thinning in mice. Pediatr Res 70(1):44–49
49. McGowan SE, McCoy DM (2014) Regulation of fibroblast lipid storage and myofibroblast phenotypes during alveolar septation in mice. Am J Physiol Lung Cell Mol Physiol 307(8): L618–L631
50. Perl AK et al (2005) Normal lung development and function after Sox9 inactivation in the respiratory epithelium. Genesis 41(1):23–32
51. McQualter JL et al (2013) TGF-beta signaling in stromal cells acts upstream of FGF-10 to regulate epithelial stem cell growth in the adult lung. Stem Cell Res 11(3):1222–1233
52. Lecart C et al (2000) Bioactive transforming growth factor-beta in the lungs of extremely low birthweight neonates predicts the need for home oxygen supplementation. Biol Neonate 77(4):217–223
53. Wolff JC et al (2010) Comparative gene expression profiling of post-natal and post-pneumonectomy lung growth. Eur Respir J 35(3):655–666
54. Paxson JA et al (2009) Global gene expression patterns in the post-pneumonectomy lung of adult mice. Respir Res 10:92
55. Paxson JA et al (2013) Age dependence of lung mesenchymal stromal cell dynamics following pneumonectomy. Stem Cells Dev 22(24):3214–3225
56. Hoffman AM et al (2011) Lung derived mesenchymal stromal cell post-transplantation survival, persistence, paracrine expression, and repair of elastase injured lung. Stem Cells Dev 20(10):1779–1792
57. McQualter JL et al (2009) Endogenous fibroblastic progenitor cells in the adult mouse lung are highly enriched in the sca-1 positive cell fraction. Stem Cells 27(3):623–633
58. Brody JS, Burki R, Kaplan N (1978) Deoxyribonucleic acid synthesis in lung cells during compensatory lung growth after pneumonectomy. Am Rev Respir Dis 117(2):307–316
59. Voswinckel R et al (2003) Circulating vascular progenitor cells do not contribute to compensatory lung growth. Circ Res 93(4):372–379
60. Konerding MA et al (2012) Spatial dependence of alveolar angiogenesis in post-pneumonectomy lung growth. Angiogenesis 15(1):23–32
61. Ackermann M et al (2014) Sprouting and intussusceptive angiogenesis in postpneumonectomy lung growth: mechanisms of alveolar neovascularization. Angiogenesis 17(3): 541–551
62. Sakurai MK et al (2007) Vascular endothelial growth factor accelerates compensatory lung growth after unilateral pneumonectomy. Am J Physiol Lung Cell Mol Physiol 292(3): L742–L747
63. Zhang Q et al (2007) Postpneumonectomy lung expansion elicits hypoxia-inducible factor-1alpha signaling. Am J Physiol Lung Cell Mol Physiol 293(2):L497–L504
64. Zhang Q et al (2006) Regulated expression of hypoxia-inducible factors during postnatal and postpneumonectomy lung growth. Am J Physiol Lung Cell Mol Physiol 290(5):L880–L889
65. Yuan S et al (2002) A role for platelet-derived growth factor-BB in rat postpneumonectomy compensatory lung growth. Pediatr Res 52(1):25–33
66. Leuwerke SM et al (2002) Inhibition of compensatory lung growth in endothelial nitric oxide synthase-deficient mice. Am J Physiol Lung Cell Mol Physiol 282(6):L1272–L1278
67. Panigrahy D et al (2013) Epoxyeicosanoids promote organ and tissue regeneration. Proc Natl Acad Sci U S A 110(33):13528–13533
68. Ding BS et al (2011) Endothelial-derived angiocrine signals induce and sustain regenerative lung alveolarization. Cell 147(3):539–553

69. Jane-Wit D, Chun HJ (2012) Mechanisms of dysfunction in senescent pulmonary endothelium. J Gerontol A Biol Sci Med Sci 67(3):236–241
70. Hwang IS et al (2007) Age-related changes in adrenomedullin expression and hypoxia-inducible factor-1 activity in the rat lung and their responses to hypoxia. J Gerontol A Biol Sci Med Sci 62(1):41–49
71. Evans MJ et al (1973) Renewal of alveolar epithelium in the rat following exposure to NO2. Am J Pathol 70(2):175–198
72. Barkauskas CE et al (2013) Type 2 alveolar cells are stem cells in adult lung. J Clin Invest 123(7):3025–3036
73. Driscoll B et al (2000) Telomerase in alveolar epithelial development and repair. Am J Physiol Lung Cell Mol Physiol 279(6):L1191–L1198
74. Weinheimer VK et al (2012) Influenza a viruses target type II pneumocytes in the human lung. J Infect Dis 206(11):1685–1694
75. Kim CF et al (2005) Identification of bronchioalveolar stem cells in normal lung and lung cancer. Cell 121(6):823–835
76. Chapman HA et al (2011) Integrin alpha6beta4 identifies an adult distal lung epithelial population with regenerative potential in mice. J Clin Invest 121(7):2855–2862
77. Kajstura J et al (2011) Evidence for human lung stem cells. N Engl J Med 364(19):1795–1806
78. Fujino N et al (2012) Analysis of gene expression profiles in alveolar epithelial type II-like cells differentiated from human alveolar epithelial progenitor cells. Respir Invest 50(3): 110–116
79. Fujino N et al (2011) Isolation of alveolar epithelial type II progenitor cells from adult human lungs. Lab Invest 91(3):363–378
80. McQualter JL et al (2010) Evidence of an epithelial stem/progenitor cell hierarchy in the adult mouse lung. Proc Natl Acad Sci U S A 107(4):1414–1419
81. Sen N, Weprin S, Peter Y (2013) Discrimination between lung homeostatic and injury-induced epithelial progenitor subsets by cell-density properties. Stem Cells Dev 22(14): 2036–2046
82. Rock JR et al (2011) Multiple stromal populations contribute to pulmonary fibrosis without evidence for epithelial to mesenchymal transition. Proc Natl Acad Sci U S A 108(52): E1475–E1483
83. Zheng D et al (2013) A cellular pathway involved in Clara cell to alveolar type II cell differentiation after severe lung injury. PLoS One 8(8), e71028
84. Kumar PA et al (2011) Distal airway stem cells yield alveoli in vitro and during lung regeneration following H1N1 influenza infection. Cell 147(3):525–538
85. Buckley S et al (2011) The milieu of damaged alveolar epithelial type 2 cells stimulates alveolar wound repair by endogenous and exogenous progenitors. Am J Respir Cell Mol Biol 45(6):1212–1221
86. Wada H et al (2012) Transplantation of alveolar type II cells stimulates lung regeneration during compensatory lung growth in adult rats. J Thorac Cardiovasc Surg 143(3):711–719, e2
87. Srisuma S et al (2010) Fibroblast growth factor receptors control epithelial-mesenchymal interactions necessary for alveolar elastogenesis. Am J Respir Crit Care Med 181(8):838–850
88. Weinstein M et al (1998) FGFR-3 and FGFR-4 function cooperatively to direct alveogenesis in the murine lung. Development 125(18):3615–3623
89. Calvi C et al (2013) Hepatocyte growth factor, a determinant of airspace homeostasis in the murine lung. PLoS Genet 9(2), e1003228
90. Oeztuerk-Winder F et al (2012) Regulation of human lung alveolar multipotent cells by a novel p38alpha MAPK/miR-17-92 axis. EMBO J 31(16):3431–3441
91. Ruiz EJ, Oeztuerk-Winder F, Ventura JJ (2014) A paracrine network regulates the cross-talk between human lung stem cells and the stroma. Nat Commun 5:3175
92. Warburton D et al (2008) Stem/progenitor cells in lung development, injury repair, and regeneration. Proc Am Thorac Soc 5(6):703–706

93. Warburton D et al (2001) Do lung remodeling, repair, and regeneration recapitulate respiratory ontogeny? Am J Respir Crit Care Med 164(10 Pt 2):S59–S62
94. Foster DJ, Moe OW, Hsia CC (2004) Upregulation of erythropoietin receptor during postnatal and postpneumonectomy lung growth. Am J Physiol Lung Cell Mol Physiol 287(6): L1107–L1115
95. Foster DJ et al (2002) Expression of epidermal growth factor and surfactant proteins during postnatal and compensatory lung growth. Am J Physiol Lung Cell Mol Physiol 283(5): L981–L990
96. Paxson JA et al (2011) Age-dependent decline in mouse lung regeneration with loss of lung fibroblast clonogenicity and increased myofibroblastic differentiation. PLoS One 6(8), e23232
97. Satoh A, Bryant SV, Gardiner DM (2008) Regulation of dermal fibroblast dedifferentiation and redifferentiation during wound healing and limb regeneration in the Axolotl. Dev Growth Differ 50(9):743–754
98. Moodley Y, Thompson P, Warburton D (2013) Stem cells: a recapitulation of development. Respirology 18(8):1167–1176
99. Park KS et al (2006) Transdifferentiation of ciliated cells during repair of the respiratory epithelium. Am J Respir Cell Mol Biol 34(2):151–157
100. Whitsett JA, Haitchi HM, Maeda Y (2011) Intersections between pulmonary development and disease. Am J Respir Crit Care Med 184(4):401–406
101. Takahashi Y et al (2010) Thyroid transcription factor-1 influences the early phase of compensatory lung growth in adult mice. Am J Respir Crit Care Med 181(12):1397–1406
102. Lin M et al (2011) Angiogenesis gene expression in murine endothelial cells during postpneumonectomy lung growth. Respir Res 12:98
103. Chamoto K et al (2013) Alveolar epithelial dynamics in postpneumonectomy lung growth. Anat Rec 296(3):495–503
104. Chamoto K et al (2013) Migration of CD11b+ accessory cells during murine lung regeneration. Stem Cell Res 10(3):267–277
105. Sun Z et al (2014) Inhibition of Wnt/beta-catenin signaling promotes engraftment of mesenchymal stem cells to repair lung injury. J Cell Physiol 229(2):213–224
106. Vaidya A et al (2014) Knock-in reporter mice demonstrate that DNA repair by non-homologous end joining declines with age. PLoS Genet 10(7), e1004511
107. Meijer C et al (2000) Kupffer cell depletion by CI2MDP-liposomes alters hepatic cytokine expression and delays liver regeneration after partial hepatectomy. Liver 20(1):66–77
108. Jones CV et al (2013) M2 macrophage polarisation is associated with alveolar formation during postnatal lung development. Respir Res 14:41
109. Brown BN, Badylak SF (2014) Extracellular matrix as an inductive scaffold for functional tissue reconstruction. Transl Res 163(4):268–285
110. Cortiella J et al (2010) Influence of acellular natural lung matrix on murine embryonic stem cell differentiation and tissue formation. Tissue Eng Part A 16(8):2565–2580
111. Ingber DE (1997) Tensegrity: the architectural basis of cellular mechanotransduction. Annu Rev Physiol 59:575–599
112. Rannels DE, Rannels SR (1988) Compensatory growth of the lung following partial pneumonectomy. Exp Lung Res 14(2):157–182
113. Koh DW et al (1996) Postpneumonectomy lung growth: a model of reinitiation of tropoelastin and type I collagen production in a normal pattern in adult rat lung. Am J Respir Cell Mol Biol 15(5):611–623
114. Griffith KA et al (2001) Predictors of loss of lung function in the elderly: the Cardiovascular Health Study. Am J Respir Crit Care Med 163(1):61–68
115. Ranga V et al (1979) Age-related changes in elastic fibers and elastin of lung. Am Rev Respir Dis 119(3):369–376

116. Shapiro SD et al (1991) Marked longevity of human lung parenchymal elastic fibers deduced from prevalence of D-aspartate and nuclear weapons-related radiocarbon. J Clin Invest 87(5): 1828–1834
117. Shifren A et al (2007) Elastin protein levels are a vital modifier affecting normal lung development and susceptibility to emphysema. Am J Physiol Lung Cell Mol Physiol 292(3):L778–L787
118. Correa-Meyer E et al (2002) Cyclic stretch activates ERK1/2 via G proteins and EGFR in alveolar epithelial cells. Am J Physiol Lung Cell Mol Physiol 282(5):L883–L891
119. Edwards YS et al (1999) Cyclic stretch induces both apoptosis and secretion in rat alveolar type II cells. FEBS Lett 448(1):127–130
120. Pasternack M Jr et al (1997) Regulated stimulation of epithelial cell DNA synthesis by fibroblast-derived mediators. Am J Physiol 272(4 Pt 1):L619–L630
121. Pelissier FA et al (2014) Age-related dysfunction in mechanotransduction impairs differentiation of human mammary epithelial progenitors. Cell Rep 7(6):1926–1939
122. Lwebuga-Mukasa JS, Ingbar DH, Madri JA (1986) Repopulation of a human alveolar matrix by adult rat type II pneumocytes in vitro. A novel system for type II pneumocyte culture. Exp Cell Res 162(2):423–435
123. Horwitz AL, Crystal RC (1975) Content and synthesis of glycosaminoglycans in the developing lung. J Clin Invest 56(5):1312–1318
124. Chen P et al (2005) Formation of lung alveolar-like structures in collagen-glycosaminoglycan scaffolds in vitro. Tissue Eng 11(9–10):1436–1448
125. Armanios M (2013) Telomeres and age-related disease: how telomere biology informs clinical paradigms. J Clin Invest 123(3):996–1002
126. Lee J et al (2009) Lung alveolar integrity is compromised by telomere shortening in telomerase-null mice. Am J Physiol Lung Cell Mol Physiol 296(1):L57–L70
127. Jackson SR et al (2011) Partial pneumonectomy of telomerase null mice carrying shortened telomeres initiates cell growth arrest resulting in a limited compensatory growth response. Am J Physiol Lung Cell Mol Physiol 300(6):L898–L909
128. Tsuji T, Aoshiba K, Nagai A (2004) Cigarette smoke induces senescence in alveolar epithelial cells. Am J Respir Cell Mol Biol 31(6):643–649
129. Walters MS et al (2014) Smoking accelerates aging of the small airway epithelium. Respir Res 15(1):94
130. Tsuji T, Aoshiba K, Nagai A (2010) Alveolar cell senescence exacerbates pulmonary inflammation in patients with chronic obstructive pulmonary disease. Respiration 80(1):59–70
131. Tsuji T, Aoshiba K, Nagai A (2006) Alveolar cell senescence in patients with pulmonary emphysema. Am J Respir Crit Care Med 174(8):886–893
132. Kumar M, Seeger W, Voswinckel R (2014) Senescence-associated secretory phenotype and its possible role in chronic obstructive pulmonary disease. Am J Respir Cell Mol Biol 51(3):323–333
133. Hoffmann RF et al (2013) Prolonged cigarette smoke exposure alters mitochondrial structure and function in airway epithelial cells. Respir Res 14:97
134. Hara H et al (2013) Mitochondrial fragmentation in cigarette smoke-induced bronchial epithelial cell senescence. Am J Physiol Lung Cell Mol Physiol 305(10):L737–L746
135. Alder JK et al (2011) Telomere length is a determinant of emphysema susceptibility. Am J Respir Crit Care Med 184(8):904–912
136. Zhou S et al (2013) Aging does not enhance experimental cigarette smoke-induced COPD in the mouse. PLoS One 8(8), e71410
137. Warburton D et al (2010) Lung organogenesis. Curr Top Dev Biol 90:73–158
138. Sueblinvong V et al (2012) Predisposition for disrepair in the aged lung. Am J Med Sci 344(1):41–51
139. Thannickal VJ (2013) Mechanistic links between aging and lung fibrosis. Biogerontology 14(6):609–615

140. Chilosi M et al (2013) Premature lung aging and cellular senescence in the pathogenesis of idiopathic pulmonary fibrosis and COPD/emphysema. Transl Res 162(3):156–173
141. Alder JK et al (2008) Short telomeres are a risk factor for idiopathic pulmonary fibrosis. Proc Natl Acad Sci U S A 105(35):13051–13056
142. Armanios MY et al (2007) Telomerase mutations in families with idiopathic pulmonary fibrosis. N Engl J Med 356(13):1317–1326
143. Kropski JA et al (2013) Genetic studies provide clues on the pathogenesis of idiopathic pulmonary fibrosis. Dis Model Mech 6(1):9–17
144. Leeman KT, Fillmore CM, Kim CF (2014) Lung stem and progenitor cells in tissue homeostasis and disease. Curr Top Dev Biol 107:207–233
145. Xu X et al (2012) Evidence for type II cells as cells of origin of K-Ras-induced distal lung adenocarcinoma. Proc Natl Acad Sci U S A 109(13):4910–4915
146. Bell JT et al (2012) Epigenome-wide scans identify differentially methylated regions for age and age-related phenotypes in a healthy ageing population. PLoS Genet 8(4), e1002629
147. Lui JC et al (2014) Broad shifts in gene expression during early postnatal life are associated with shifts in histone methylation patterns. PLoS One 9(1), e86957
148. Maegawa S et al (2010) Widespread and tissue specific age-related DNA methylation changes in mice. Genome Res 20(3):332–340
149. Sidler C et al (2014) A role for SUV39H1-mediated H3K9 trimethylation in the control of genome stability and senescence in WI38 human diploid lung fibroblasts. Aging 6(7): 545–563

Part III
Regulation of Lung Regeneration and Repair

Chapter 9
Stromal Regulation of Lung Regeneration and Repair

Jonathan L. McQualter

Introduction

The lung comprises a diverse array of specialized cell types that together form complex intercellular communication networks necessary for its proper development and function. The epithelium and the mesenchyme comprise the two major cellular compartments of the lung, where reciprocal signalling is key to the development and maintenance of the functional architecture of the organ. During development, signals from the mesenchyme drive epithelial proliferation, differentiation, and morphogenetic patterning to generate branching airways and alveoli [1, 2]. These interactions are mediated by a myriad of growth factors that are precisely regulated in their temporal and spatial expression to direct the differentiation of specialized cells and their organization into a complex three-dimensional (3D) structure that facilitates respiration and protection against inhaled pathogens. The relationship between alveolar epithelium and pulmonary capillaries permits for gas-exchange, while the airway epithelium plays a major role in microbial defence, immune regulation, mucociliary clearance of pathogens and toxins; and as a physical barrier to exogenous particles in the air we breathe [3]. Mesenchymal progenitor cells differentiate into various lineages to create cartilage rings, providing structural support in the trachea, and contractile smooth muscle cells, fibroblasts, pericytes, and myofibroblasts throughout the lung [4, 5]. There is also some evidence to suggest that a subset of mesenchymal cells are capable of differentiating into endothelial cells in the lung [5, 6]. The mesenchymal compartment is also responsible for the elaboration of extracellular matrix proteins and cytokines that in addition to epithelial–mesenchymal interactions provide the permissive or restrictive cues during development.

J.L. McQualter (✉)
School of Health Sciences and Health Innovations Research Institute, RMIT University, Bundoora, VIC, Australia
e-mail: jonathan.mcqualter@gmail.com

I. Bertoncello (ed.), *Stem Cells in the Lung*, Stem Cell Biology and Regenerative Medicine, DOI 10.1007/978-3-319-21082-7_9

In the context of regeneration and repair of the adult lung, most of the basic research to date has seen the spotlight focused intensely on understanding the hierarchical organization of epithelial stem and progenitor cells [7, 8]. Consequently, the role of the adult lung mesenchyme in tissue homeostasis has been somewhat neglected and remains poorly understood. Nevertheless, what is known is that epithelial–mesenchymal interactions do play a vital role in adult lung regeneration, just as they do during lung development [1, 2]. The multifarious functions of mesenchymal cells in the lung now challenges us to better understand how these cells communicate with epithelial progenitor cells to regulate regeneration and repair; and more importantly, understand how the breakdown of the communication between mesenchymal and epithelial cell networks leads to disrupted tissue regeneration, remodelling, malfunction, and disease. In this chapter, I will review the current understanding of the organization of mesenchymal lineage cells in the adult lung with special emphasis on their function as the architects of epithelial turnover, maintenance, and repair.

Resident Lung Mesenchymal Stromal Cells

The mesenchymal compartment in the adult lung comprises a multitude of stromal cells including fibroblasts, lipofibroblasts, myofibroblasts, and smooth muscle cells. However, despite the obvious importance of mesenchymal cells in tissue homeostasis, remarkably little is known about their identity and organization in the adult lung. This is compounded by inconsistencies in the criteria for defining mesenchymal cells in adult tissues. Many laboratories have identified mesenchymal lineage cells in the adult lung, which invariably have subtle, and sometimes substantial, differences in their fate potential [9]. To help clarify, we will refer to clonogenic cells with the capacity to differentiate into multiple mesenchymal lineages in vitro (e.g. adipogenic, chondrogenic, osteogenic, and myogenic differentiation) as mesenchymal stromal cells (MSC); and progenitor cells with mesenchymal potential in vivo as tissue-resident mesenchymal progenitor cells.

Numerous studies have reported the isolation of MSC populations from both mouse and human adult lungs that exhibit the capacity for adipogenic, chondrogenic, osteogenic and myogenic differentiation in vitro. Early studies exploited the differential ability of many stem and progenitor cell populations to efflux the Hoechst 33342 dye due to their cell membrane expression of the ATP-binding cassette (ABC) family of transporter proteins [10] to identify, enrich, and characterize adult lung MSC in the side population (SP) fraction of lung cell suspensions [11]. These lung SP cells exhibited a fibroblastic morphology; they expressed the archetypal MSC markers Sca-1, CD106, and CD44; and could be serially passaged in culture and directed to differentiate into smooth muscle, cartilage, bone, and fat [11]. Recent lineage tracing studies using ABCG2-CreERT2 mice have also suggested that $ABCG2^{+}$ cells are a population of pericytes that are precursors of myofibroblasts [12] that exhibit greater colony-forming potential and myofibroblast differentiation potential than NG2$^+$ pericytes [12]. This finding is consistent with other fate-mapping studies using NG2-CreER/Rosa-GFP reporter mice that have shown that NG2+ pericytes proliferated in response to bleomycin injury but did not give rise to myofibroblasts [13]. At least in part, these discrepancies likely reflect mesenchymal heterogeneity in the perivascular space and the challenges of correctly identifying mesenchymal cell subsets. Some perivascular cells defined as pericytes in the literature are probably mesenchymal progenitor cells [14].

Using flow cytometry fractionation in combination with analysis of fibroblastic colony-forming potential, we have demonstrated that lung MSC are enriched in the non-hematopoietic (CD45$^-$) non-endothelial (CD31$^-$) Sca-1$^+$ cell fraction of the dissociated adult mouse lung. This study showed that primary CD45$^-$ CD31$^-$Sca-1$^+$ cells could form fibroblastic colonies and could differentiate into lipofibroblasts and myofibroblasts in vitro [15]. Moreover, we have shown that these cells express Fibroblast Growth Factor (FGF)-10 [16], as do mesenchymal progenitor cells during lung development [17]. Lineage tracing studies using FGF-10-CreER/Rosa-Tomato mice also showed that FGF-10$^+$ cells in the adult mouse lung were capable of giving rise to myofibroblasts and lipofibroblasts in vivo [17]. These studies advocate for the existence of a mesenchymal progenitor cell population in the adult mouse lung that is capable of giving rise to multiple mesenchymal cell lineages in vitro and in vivo.

Substantial evidence also exists for the presence of tissue-resident mesenchymal progenitor cells in the adult human lung. Cells isolated from bronchial biopsies, after trimming off the bronchi and peribronchial vascular structures, express the archetypal MSC markers CD73, CD90, and CD105 after culture and have the capacity to differentiate into adipocytes, chondrocytes, and osteocytes [18]. Another study, which characterized adherent cells isolated from human bronchoalveolar lavage (BAL) samples obtained from lung transplant recipients, has also provided convincing evidence for the existence of resident MSC in the adult human lung [19]. Cells isolated in this study also expressed the archetypal MSC markers and exhibited multilineage differentiation capacity [19]. An important addition to this study was the discovery that almost all (97 %) cultured MSC isolated from BAL sex-mismatched lung transplant recipients up to 11 years after transplant were of donor origin [19]. This suggests that these cells are derived from a renewable, or at least long-lived, population of tissue-resident mesenchymal progenitor cells that reside locally in the adult human lung.

MSC in the Epithelial Stem Cell Niche

Continual exposure to environmental insults such as allergens, pollution, irritants, smoke, and infectious agents damages the cellular architecture of the lung. The lung initiates a wound-healing response to injury involving the co-ordinated proliferation and differentiation of epithelial and mesenchymal progenitor cells. Failure to correctly initiate or resolve this process is thought to contribute to the development and exacerbation of pathologic tissue remodelling, fibrosis, and cancer. Hence, knowledge of the cellular and molecular mechanisms involved in lung regeneration and remodelling has significant implications for developing new therapeutic strategies to maintain lung homeostasis.

Recently, we discovered that epithelial stem/progenitor cells cultured in 3D Matrigel cultures failed to grow unless plated together with lung mesenchymal progenitor cells ($CD45^{-}$ $CD31^{-}$ $EpCAM^{-}$ $Sca\text{-}1^{hi}$ cells) that supported the generation of complex epithelial colonies containing differentiated goblet cells, ciliated cells, and/or alveolar cells [20]. This suggests that the proliferation and differentiation of epithelial stem cells is reliant on trophic support from resident mesenchymal progenitor cells and turned the spotlight back on the importance of the stromal microenvironment for normal lung epithelial regeneration.

To explore epithelial regeneration in vivo, researchers have exploited the cytochrome P450-mediated cytotoxic metabolism of naphthalene to selectively ablate club cells in bronchiolar airways of mice [21] to reveal that a subset of naphthalene-resistant club cell secretory protein $(CCSP)^{+}$ cells are largely responsible for regeneration of the bronchiolar airway epithelium [22–24]. Several studies have shown that sub-epithelial mesenchymal cells are important mediators of epithelial regeneration in vivo. In naphthalene-injured mice, an increase in the incidence of $FGF\text{-}10^{+}$ mesenchymal cells juxtaposed to the club cell-depleted airways has been observed during the primary wound-healing response, suggesting that $FGF\text{-}10^{+}$ mesenchymal cells form a niche for epithelial progenitor cells [25]. Specifically, this study showed that after naphthalene injury, surviving epithelial cells secrete Wnt7b which activates sub-epithelial mesenchymal cells to express FGF-10, which in turn, induces proliferation of epithelial progenitor cells to initiate epithelial repair. The same group also showed that the conditional ablation of c-myc in mesenchymal lineage cells, results in depletion of $FGF\text{-}10^{+}$ cells, thus preventing airway epithelial repair [26]. Interestingly, using our 3D Matrigel co-culture assay, we have also demonstrated that the trophic support provided by primary tissue-resident mesenchymal progenitor cells directly correlates with the level of FGF-10 expression [16]. It remains unclear whether $FGF\text{-}10^{+}$ cells identified in these different studies are one and the same cell subset or phenotypically distinct populations. However, lineage tracing studies in adult FGF-10-CreER/Rosa-Tomato mice have shown that at least a fraction of primary $CD45^{-}$ $CD31^{-}$ $Sca\text{-}1^{+}$ lung MSC are FGF-10-expressing progenitor cells [17]. Irrespective of the cellular diversity, it is clear from these studies that the paracrine release of FGF-10 plays a fundamental role in promoting epithelial regeneration.

A number of studies imply that the reciprocal communication between epithelial stem cells and mesenchymal cells is important in creating an epithelial–mesenchymal trophic unit in the adult lung that is essential for regeneration of the alveolar epithelium. Lipofibroblasts are found juxtaposed to alveolar type II (ATII) cells in the adult lung and have long been thought to contribute to the alveolar stem cell niche [27–29]. This is supported by recent studies showing that the growth and differentiation of alveolar progenitor cells is stimulated by co-culture with primary lung mesenchymal cells, which include lipofibroblasts [20, 28]. There is also evidence of direct mesenchymal–epithelial communication by virtue of the fact that the processes of ABCG2$^+$ MSC in the lung parenchyma form direct contact with epithelial cells with functional gap-junctional communication as defined by the intercellular transfer of calcein dye [12]. Lastly, a recent study has suggested that the recruitment of mesenchymal cells to the Lgr6+ epithelial stem cell niche in the human lung creates a paracrine feedback loop to maintain their own niche. The authors provide evidence that SDF-1 secreted by epithelial stem cells functions to recruit and activate mesenchymal cells (fibroblasts) to produce angiogenic factors and TNFα, which promotes endothelial homeostasis and further activates autocrine production of SDF-1 by epithelial stem cells [30].

The above studies suggest that mesenchymal progenitor cells in the adult lung create an epithelial stem cell niche that is permissive for epithelial regeneration. It is, however, important to consider the influence that the heterogeneity of lung mesenchymal cells has on the nature of the epithelial stem cell niche. Indeed, we have shown that myofibroblast differentiation of mesenchymal progenitor cells results in a loss of epithelial-supportive capacity in these cells. Specifically, we have shown that following ex vivo expansion, primary Sca-1$^+$ lung MSC underwent myofibroblast differentiation, stopped expressing FGF-10 and were no longer able to support the growth of epithelial cells in culture [16]. However, blocking the TGF-β signalling pathway with the TGF-β inhibitor SB431542 could reinstate epithelial support, which suggests that regulating the fate of mesenchymal progenitor cells may be a fundamental determinant in initiating and then limiting epithelial regeneration in the adult lung (Fig. 9.1). Interestingly, studies in older mice have shown that there is an age-dependent decline in mouse lung regeneration, which is associated with loss of clonogenic mesenchymal progenitor cells and increased myofibroblast differentiation [31]. Future studies should seek to better define the

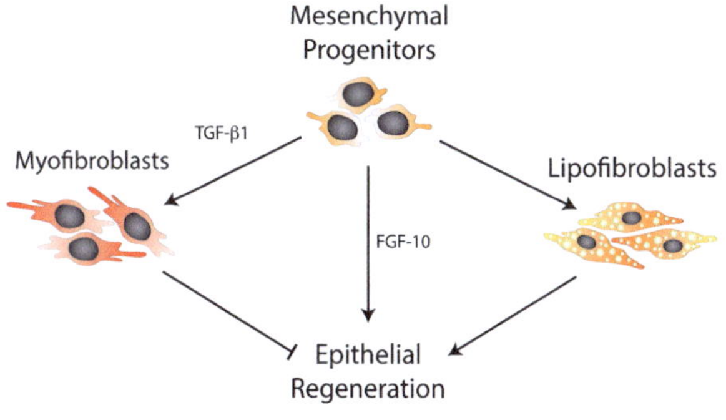

Fig. 9.1 Mesenchymal progenitors support epithelial regeneration via the release of growth factors like FGF-10 which promote the proliferation of epithelial stem/progenitor cells. The differentiation of mesenchymal progenitors cells into myofibroblasts (mediated by TGF-β) results in the down regulation of FGF-10 and other growth factors, and a subsequent loss in epithelial-supportive capacity. In contrast, mesenchymal progenitors that differentiate into lipofibroblasts maintain their epithelial-supportive capacity, although the role of lipofibroblasts on epithelial regeneration is still unclear

heterogeneity of mesenchymal lineage cells in the adult lung and determine how dynamic changes in the stromal architecture of the adult lung contribute to tissue homeostasis, regeneration, and fibrosis.

Endothelial Cells in the Epithelial Stem Cell Niche

Perhaps the most convincing evidence of robust epithelial regeneration in the adult human lung comes from a recent observation of extensive regenerative lung growth in a 33-year-old woman 15 years after a right-sided pneumonectomy for treatment of lung cancer [32]. Comparable studies in mice suggest that pneumonectomy stimulates pulmonary capillary endothelial cells to produce paracrine growth factors that induce proliferation of epithelial progenitor cells [33]. The authors provide evidence that activated endothelial cells support alveologenesis through MMP14-mediated release of EGF-like ligands. These studies clearly demonstrate that endothelial cells are also an important component of the epithelial stem cell niche in vivo.

Cell culture studies also suggest that endothelial cells communicate directly with lung epithelial stem/progenitor cells and support their proliferation and differentiation in vitro. Co-culture of the human bronchial epithelial cell line VA10, which is a $p63^+$ $K5^+$ $K14^+$ basal cell line, with human umbilical vein endothelial cells (HUVECS) results in epithelial branching in 3D culture [34]. Similarly, another study demonstrated that the proliferation of primary human airway basal cells was significantly increased when co-cultured with HUVECs. Mechanistically, they showed that basal cell secretion of VEGFA activated endothelial cells to produce trophic factors, which in turn support basal cell proliferation [35]. Lastly, $CD31^+$ endothelial cells have been shown to be capable of supporting the proliferation and differentiation of multipotent epithelial stem/progenitor cells in vitro and after subcutaneous injection. In this study, the authors showed that BMP4-induced NFATc1-dependent expression of thrombospondin-1 (Tsp1) in lung endothelial cells was involved in driving alveolar lineage specification [36]. Thus, a change in the stromal niche (including endothelial cells) is one mechanism by which the fate and specificity of lung epithelial stem cells might be regulated in response to lung injury in vivo.

Abnormalities in the Epithelial Niche Involved in Tumour Progression

In recent years, it has become increasingly clear that tumours comprise a heterogeneous population of cells that are hierarchically organized with cancer-initiating stem cells at the apex with the greatest tumorigenic potential. In parallel with this idea, it is becoming evident that, similar to normal epithelial stem cells, the tumorigenic and metastatic potential of cancer-initiating stem cells may be regulated by the tumour niche [37, 38]. Based on our growing understanding of the lung epithelial stem cell niche, it is likely that alterations in the composition of MSC in the lung may create a tumorigenic niche that allows for the growth of epithelial tumours. In lung squamous cell carcinoma, the profusion of podoplanin-expressing mesenchymal cells, referred to as cancer-associated fibroblasts (CAFs), in the tumour microenvironment is a negative prognostic indicator for patient survival [39]. CAFs are a population of stromal cells that regulate the fate of epithelial tumours through the secretion of various cytokines, including members of the FGF family, EGF and VEGF, which participate in tumour cell growth, metastasis, and neovascularization as reviewed in [37, 40]. Notably, primary isolated CAFs from non-small cell lung cancer (NSCLC) patients have the capacity to enhance the proliferation of lung tumour cell lines and enrich for cancer-initiating stem cell populations through the insulin-like growth factor-II (IGF-II)/IGF1 receptor (IGF1R) pathway. Moreover, activation of the IGF-II/IGF1R pathway correlates with poor-prognosis for overall survival and relapse-free survival in stage I NSCLC patients [41]. Hence, the preexisting squamous phenotype observed in COPD and the chronic injurious state of this lung microenvironment favour the acquisition of a sufficient number of mutations to facilitate a classical pattern of tumour development progressing from metaplasia, dysplasia, carcinoma in situ, and subsequent malignant transformation.

Conclusion

Although our understanding of the epithelial stem cell niche is incomplete, evidence is accumulating to suggest that resident lung mesenchymal progenitor cells function as the architects of lung regeneration by creating a dynamic and hierarchically organized niche so as to exactly regulate the growth of epithelial stem cells and maintain homeostasis of the lung. I propose a simplified model in which the recruitment and/or activation of mesenchymal progenitor cells creates a permissive niche by providing molecular signals, such as FGF-10, to drive the proliferation of epithelial stem and progenitor cells, while myofibroblast differentiation acts to resolve the regenerative process by creating restrictive niche. With this in mind, I would suggest that regulating the niche is arguably more important than the intrinsic potential of stem cells themselves when it comes to maintaining homeostasis in the lung. Therefore, future studies should include characterization of the stem cell–niche interactions in models of lung disease.

References

1. Shannon JM, Hyatt BA (2004) Epithelial-mesenchymal interactions in the developing lung. Annu Rev Physiol 66:625–645
2. Warburton D, Bellusci S, De Langhe S, Del Moral PM, Fleury V, Mailleux A et al (2005) Molecular mechanisms of early lung specification and branching morphogenesis. Pediatr Res 57(5 Pt 2):26R–37R
3. Knight DA, Holgate ST (2003) The airway epithelium: structural and functional properties in health and disease. Respirology 8(4):432–446
4. Fisher KA, Summer RS (2006) Stem and progenitor cells in the formation of the pulmonary vasculature. Curr Top Dev Biol 74:117–131
5. Peng T, Tian Y, Boogerd CJ, Lu MM, Kadzik RS, Stewart KM et al (2013) Coordination of heart and lung co-development by a multipotent cardiopulmonary progenitor. Nature 500 (7464):589–592
6. Sims-Lucas S, Schaefer C, Bushnell D, Ho J, Logar A, Prochownik E et al (2013) Endothelial progenitors exist within the kidney and lung mesenchyme. PLoS One 8(6):e65993
7. Kotton DN, Morrisey EE (2014) Lung regeneration: mechanisms, applications and emerging stem cell populations. Nat Med 20(8):822–832
8. McQualter JL, Anthony D, Bozinovski S, Prele CM, Laurent GJ (2014) Harnessing the potential of lung stem cells for regenerative medicine. Int J Biochem Cell Biol 56:82–91
9. Sinclair K, Yerkovich ST, Chambers DC (2013) Mesenchymal stem cells and the lung. Respirology 18(3):397–411
10. Golebiewska A, Brons NH, Bjerkvig R, Niclou SP (2011) Critical appraisal of the side population assay in stem cell and cancer stem cell research. Cell Stem Cell 8(2):136–147
11. Summer R, Fitzsimmons K, Dwyer D, Murphy J, Fine A (2007) Isolation of an adult mouse lung mesenchymal progenitor cell population. Am J Respir Cell Mol Biol 37(2):152–159
12. Marriott S, Baskir RS, Gaskill C, Menon S, Carrier EJ, Williams J et al (2014) ABCG2pos lung mesenchymal stem cells are a novel pericyte subpopulation that contributes to fibrotic remodeling. Am J Physiol Cell Physiol 307(8):C684–C698
13. Rock JR, Barkauskas CE, Cronce MJ, Xue Y, Harris JR, Liang J et al (2011) Multiple stromal populations contribute to pulmonary fibrosis without evidence for epithelial to mesenchymal transition. Proc Natl Acad Sci U S A 108(52):E1475–E1483
14. Armulik A, Genove G, Betsholtz C (2011) Pericytes: developmental, physiological, and pathological perspectives, problems, and promises. Dev Cell 21(2):193–215
15. McQualter JL, Brouard N, Williams B, Baird BN, Sims-Lucas S, Yuen K et al (2009) Endogenous fibroblastic progenitor cells in the adult mouse lung are highly enriched in the Sca-1 positive cell fraction. Stem Cells 27:623–633
16. McQualter JL, McCarty RC, Van der Velden J, O'Donoghue RJ, Asselin-Labat ML, Bozinovski S, Bertoncello I (2013) TGF-beta signaling in stromal cells acts upstream of FGF-10 to regulate epithelial stem cell growth in the adult lung. Stem Cell Res 11 (3):1222–1233
17. El Agha E, Herold S, Alam DA, Quantius J, Mackenzie B, Carraro G et al (2014) Fgf10-positive cells represent a progenitor cell population during lung development and postnatally. Development 141(2):296–306
18. Sabatini F, Petecchia L, Tavian M, Jodon de Villeroche V, Rossi GA, Brouty-Boye D (2005) Human bronchial fibroblasts exhibit a mesenchymal stem cell phenotype and multilineage differentiating potentialities. Lab Invest 85(8):962–971
19. Lama VN, Smith L, Badri L, Flint A, Andrei AC, Murray S, Wang Z, Liao H, Toews GB, Krebsbach PH, Peters-Golden M, Pinsky DJ, Martinez FJ, Thannickal VJ (2007) Evidence for tissue-resident mesenchymal stem cells in human adult lung from studies of transplanted allografts. J Clin Invest 117(4):989–996

20. McQualter JL, Yuen K, Williams B, Bertoncello I (2010) Evidence of an epithelial stem/progenitor cell hierarchy in the adult mouse lung. Proc Natl Acad Sci U S A 107(4):1414–1419
21. Chichester CH, Philpot RM, Weir AJ, Buckpitt AR, Plopper CG (1991) Characterization of the cytochrome P-450 monooxygenase system in nonciliated bronchiolar epithelial (Clara) cells isolated from mouse lung. Am J Respir Cell Mol Biol 4(2):179–186
22. Hong KU, Reynolds SD, Giangreco A, Hurley CM, Stripp BR (2001) Clara cell secretory protein-expressing cells of the airway neuroepithelial body microenvironment include a label-retaining subset and are critical for epithelial renewal after progenitor cell depletion. Am J Respir Cell Mol Biol 24(6):671–681
23. Giangreco A, Reynolds SD, Stripp BR (2002) Terminal bronchioles harbor a unique airway stem cell population that localizes to the bronchoalveolar duct junction. Am J Pathol 161 (1):173–182
24. Guha A, Vasconcelos M, Cai Y, Yoneda M, Hinds A, Qian J et al (2012) Neuroepithelial body microenvironment is a niche for a distinct subset of Clara-like precursors in the developing airways. Proc Natl Acad Sci U S A 109(31):12592–12597
25. Volckaert T, Dill E, Campbell A, Tiozzo C, Majka S, Bellusci S, De Langhe SP (2011) Parabronchial smooth muscle constitutes an airway epithelial stem cell niche in the mouse lung after injury. J Clin Invest 121(11):4409–4419
26. Volckaert T, Campbell A, De Langhe S (2013) c-Myc regulates proliferation and Fgf10 expression in airway smooth muscle after airway epithelial injury in mouse. PLoS One 8(8): e71426
27. Rehan VK, Torday JS (2012) PPARgamma signaling mediates the evolution, development, homeostasis, and repair of the lung. PPAR Res 2012:289867
28. Barkauskas CE, Cronce MJ, Rackley CR, Bowie EJ, Keene DR, Stripp BR et al (2013) Type 2 alveolar cells are stem cells in adult lung. J Clin Invest 123(7):3025–3036
29. Chen L, Acciani T, Le Cras T, Lutzko C, Perl AK (2012) Dynamic regulation of platelet-derived growth factor receptor alpha expression in alveolar fibroblasts during realveolarization. Am J Respir Cell Mol Biol 47(4):517–527
30. Ruiz EJ, Oeztuerk-Winder F, Ventura JJ (2014) A paracrine network regulates the cross-talk between human lung stem cells and the stroma. Nat Commun 5:3175
31. Paxson JA, Gruntman A, Parkin CD, Mazan MR, Davis A, Ingenito EP, Hoffman AM (2011) Age-dependent decline in mouse lung regeneration with loss of lung fibroblast clonogenicity and increased myofibroblastic differentiation. PLoS One 6(8):e23232
32. Butler JP, Loring SH, Patz S, Tsuda A, Yablonskiy DA, Mentzer SJ (2012) Evidence for adult lung growth in humans. N Engl J Med 367(3):244–247
33. Ding BS, Nolan DJ, Guo P, Babazadeh AO, Cao Z, Rosenwaks Z et al (2011) Endothelial-derived angiocrine signals induce and sustain regenerative lung alveolarization. Cell 147 (3):539–553
34. Franzdottir SR, Axelsson IT, Arason AJ, Baldursson O, Gudjonsson T, Magnusson MK (2010) Airway branching morphogenesis in three dimensional culture. Respir Res 11:162
35. Curradi G, Walters MS, Ding BS, Rafii S, Hackett NR, Crystal RG (2012) Airway basal cell vascular endothelial growth factor-mediated cross-talk regulates endothelial cell-dependent growth support of human airway basal cells. Cell Mol Life Sci 69(13):2217–2231
36. Lee JH, Bhang DH, Beede A, Huang TL, Stripp BR, Bloch KD et al (2014) Lung stem cell differentiation in mice directed by endothelial cells via a BMP4-NFATc1-thrombospondin-1 axis. Cell 156(3):440–455
37. Kalluri R, Zeisberg M (2006) Fibroblasts in cancer. Nat Rev Cancer 6(5):392–401
38. Ohlund D, Elyada E, Tuveson D (2014) Fibroblast heterogeneity in the cancer wound. J Exp Med 211(8):1503–1523
39. Ono S, Ishii G, Nagai K, Takuwa T, Yoshida J, Nishimura M et al (2013) Podoplanin-positive cancer-associated fibroblasts could have prognostic value independent of cancer cell phenotype in stage I lung squamous cell carcinoma: usefulness of combining analysis of both cancer cell phenotype and cancer-associated fibroblast phenotype. Chest 143(4):963–970

40. Madar S, Goldstein I, Rotter V (2013) 'Cancer associated fibroblasts'—more than meets the eye. Trends Mol Med 19(8):447–453
41. Chen WJ, Ho CC, Chang YL, Chen HY, Lin CA, Ling TY et al (2014) Cancer-associated fibroblasts regulate the plasticity of lung cancer stemness via paracrine signalling. Nat Commun 5:3472

Chapter 10
Immunomodulatory Regulation of Lung Regeneration and Repair

Jonathan L. McQualter, Desiree Anthony, Ross Vlahos, and Steven Bozinovski

Introduction

During respiration, the lung epithelium is continuously exposed to a wide variety of microorganisms, dusts, and pollutants. It is therefore not surprising that the lung is defended by a network of innate immune cells that protect the integrity of the epithelial surface from airborne threats. Innate immune cells at the mucosal surfaces have evolved to mediate complex immune responses to protect the body from bacteria, viruses and parasites [1]. However, an important concept that is gathering support is that specialized innate immune cells have also evolved to play a critical role in the stem cell niche by providing support for effective wound repair and regeneration in the adult lung.

Successful tissue regeneration in the lung requires precise coordination of multiple processes, including eliminating pathogens and scavenging cellular debris, immune modulation, and proliferation and activation of progenitor cells. Macrophages are the most abundant innate immune cells in the lung. They are an incredibly diverse cell population that constantly alter their functional state in response to physiological stimuli or environmental challenges [2]. In addition to their enhanced antimicrobial and antiparasitic functions, macrophages can also be alternatively activated to adopt a wound-healing phenotype and release a vast array of cytokines involved in tissue repair [3]. This certainly holds true for organs like liver, skin, and kidney, where there is considerable evidence to suggest that specialized resident macrophages play a vital role in regulating tissue regeneration

J.L. McQualter (✉) • R. Vlahos • S. Bozinovski
School of Health Sciences and Health Innovations Research Institute, RMIT University, Bundoora, VIC, Australia
e-mail: jonathan.mcqualter@rmit.edu.au

D. Anthony
Lung Health Research Centre, Department of Pharmacology and Therapeutics, University of Melbourne, Melbourne, VIC, Australia

I. Bertoncello (ed.), *Stem Cells in the Lung*, Stem Cell Biology and Regenerative Medicine, DOI 10.1007/978-3-319-21082-7_10

[4–6], and evidence is mounting for a similar role in maintenance of the respiratory epithelium. Innate lymphoid cells (ILCs) are another population of mucosal immune cells that are emerging as key regulators of lung tissue homeostasis. They have also been shown to be critical in regulating epithelial repair and remodeling after inflammation or infection [7].

This review addresses the role of macrophages and ILCs in mediating regeneration of the adult lung to restore normal tissue architecture after lung injury.

Macrophage Heterogeneity in the Adult Lung

Macrophages are a versatile population of immune cells that display remarkable phenotypic and functional diversity. In most tissues, including the lung, macrophages play a role in almost every aspect of tissue homeostasis by acting as sentinels and rapidly responding to inhaled challenges and physiological stresses [8, 9]. The diverse functional states exhibited by macrophages are often loosely classified as classically or alternatively activated. Classically activated macrophages, are induced by IFNγ or toll-like receptor ligands (e.g., LPS), express pro-inflammatory cytokines, (e.g., TNFα, IL-1, IL-12, and IL-6) and have enhanced antimicrobial functions, including upregulation of inducible nitric oxide synthase (iNOS). Alternatively activated macrophages include regulatory macrophages that are induced by IL-10 and produce IL-10 and CCL1; as well as wound-healing macrophages that are induced by IL-4, IL-13, and TNFα and express increased Ym1 (chitinase-like protein), RELMα (resistin-like molecule α) and arginase [2]. However, this is not a rigid classification and there is considerable evidence showing that there is extensive overlap of these functional states. It is likely that macrophages change over time and express temporally restricted phenotypes that instruct other cells through the different phases of repair [10].

Another way of classifying macrophages is based on their ontogeny and tissue residency. Several studies have recently revealed that tissue-resident macrophages ($CD11c^{hi}$ $CD11b^{neg/low}$ $CSF1R^{high}$ $CCR2^{neg/low}$ $CD169^{pos}$) are derived from embryonic yolk sac progenitors and self-renew locally in tissues in which they reside throughout adult life, whereas inflammatory macrophages ($CD11c^{neg/low}$ $CD11b^{hi}$ $CSF1R^{neg/low}$ $CCR2^{high}$ $CD169^{neg}$) are derived from bone marrow-derived hematopoietic progenitors [11–13]. In the context of tissue regeneration, evidence is accumulating to suggest that self-renewing tissue-resident macrophages are assigned the bulk of the wound-healing duties, whereas pro-inflammatory bone marrow-derived macrophages are engaged in host defense [9].

In the adult lung, macrophages display distinct location-specific phenotypes in the airway lumen, alveoli, and interstitial space. Studies in mice have shown that alveolar macrophages (AMs) comprise more than 90 % of macrophages in the adult lung [14] and belong to the tissue-resident lineage of macrophages which are derived from fetal monocytes within the first week of life and later self-renew in the adult tissue [15–17]. Mature AMs phagocytose dead cells and debris, coordinate the activation of innate and adaptive immune responses and contribute to immunosuppression associated with tissue repair. The local proliferation of AMs during homeostatic repletion is dependent on both granulocyte macrophage colony stimulating factor (GM-CSF) and colony stimulating factor-1 (CSF-1) [12].

Macrophages and Defence Against Inhaled Pathogens

In response to acute lung injury, the initial inflammatory response is necessary to eliminate inhaled pathogens before tissue regeneration can succeed. AMs are considered to be major effector cells in first-line innate host defense against inhaled pathogens and irritants by virtue of their phagocytic ability [18]. They mediate antimicrobial defenses through expression of receptors for immunoglobulin (F_cR), complement, β-glucan, mannose, and several types of scavenger receptors which together facilitate phagocytosis [19]. In addition, classically activated pro-inflammatory macrophages, which are recruited to the lung from the circulation, are able to generate reactive nitrogen and oxygen intermediates to defend against microbial infection [20, 21]. These defense mechanisms are boosted by epithelial-derived surfactant protein A, which stimulates macrophage phagocytosis and the production of reactive oxygen–nitrogen intermediates [22]. Recent studies using targeting strategies to selectively ablate different monocyte/macrophage populations have shown that $CCR2^{hi}$ bone marrow-derived pro-inflammatory macrophages are also responsible for neutrophil emigration after acute lung injury [23].

The maintenance of the resident AM population is also important for maintaining the first-line of defense against viral and bacterial pathogens. Studies have shown that viral replication is restricted in AMs. Consequently, viral infection of AMs is nonproductive: no infectious virus is released, thereby reducing viral load [24–26]. Attesting to the central role of macrophages in defense against virus, studies have shown that depletion of resident lung macrophages from mice prior to influenza virus infection leads to increased virus titers in the lung [26–28]. However, the downside of this is that virus infection induces cell death of AMs [29]. Hence, the depletion of AMs following a significant viral infection can facilitate secondary bacterial infection by altering early innate cellular immunity. In a mouse model of secondary bacterial infection, the initial depletion of AMs as a consequence of influenza infection rendered the host susceptible to *Streptococcus pneumonia* colonization and systemic invasion [29]. Importantly, innate host protection to *Streptococcus pneumonia* was not restored until the self-renewing resident AMs population returned to normal levels 2 weeks after influenza infection [29].

Not surprisingly, given their potent antimicrobial function, impaired AM function is thought to be central to high bacterial colonization rates and increased susceptibility to exacerbations in chronic obstructive pulmonary disease (COPD). This may, at least in part, be due to excessive oxidative stress from chronic cigarette smoke exposure [30], which causes a deficiency in phagocytosis of bacteria [31] and efferocytosis of apoptotic cells [32]. Treatment with antioxidants such as procysteine can significantly improve efferocytic function of AMs isolated from experimental models of COPD [33]. Other studies have suggested that the increase in resident and inflammatory macrophage numbers in the lungs of COPD patients plays a central role in mediating collateral tissue damage [34–36]. AMs secrete elastolytic enzymes (proteases), including matrix metalloprotease (MMP)-2, MMP-9, MMP-12, cathepsin K, L, and S in response to irritants and infection, which together may be responsible for destruction of lung parenchyma [37]. Increased concentrations of these enzymes have been observed in patients with emphysema [38, 39].

Macrophages in Resolution of Inflammation and Tissue Regeneration

Although inflammation plays an important role in eliminating pathogens, it can also cause collateral damage and limit epithelial wound healing. To progress from the inflammatory phase to the next phase of tissue repair, inflammation needs to be dampened. It is now becoming apparent that alternative activation of resident AMs plays a key role in mediating the resolution of inflammation after acute lung injury through their ability to engulf apoptotic neutrophils and debris [40, 41]. Efficient clearance of cellular debris prevents the persistence of potentially toxic and immunostimulatory material in the lung and is an important step in resolution of the inflammatory process. The active phagocytosis of dead cells also activates AMs to secrete factors like IL-1β, IL-8, TNFα, and GM-CSF that perform downstream immunomodulatory functions [42, 43]. The inability to efficiently remove exhausted neutrophils has damaging implications in COPD as accumulation of necrotic neutrophils can lead to the indiscriminate release of granule protease pools including neutrophil elastase. Neutrophil elastase localizes to lung elastic fibers in emphysema patients and degrades extracellular matrix components [44].

In addition to their immune functions, alternatively activated macrophages can also adopt a wound-healing phenotype and release a vast array of cytokines involved in tissue repair [3]. Following resolution of inflammatory responses, restoration of the cellular architecture of the lung requires the coordinated proliferation and differentiation of epithelial and mesenchymal progenitor cells. Inferring a role for macrophages in this phase of lung regeneration, several studies have shown that AMs produce cytokines that may influence the fate of adult lung epithelial and mesenchymal cells. AMs are a source of IGF-1, which can prevent the apoptosis of lung mesenchymal stromal cells [45] and stimulate their proliferation and migration [46, 47]. We have shown that resident mesenchymal stromal cells create a niche for epithelial progenitor cells, providing many of the cues necessary for their proliferation and differentiation [48–50]. Other studies have shown that following *Streptococcus pneumonia* infection in mice, AMs that phagocytose apoptotic neutrophils produce hepatocyte growth factor (HGF) [51]; and that the elevated levels of HGF induced by viral stimulation of AMs is associated with alveolar epithelial hyperplasia [52]. Significantly, our studies have shown that HGF is a potent epithelial mitogen that acts in synergy with fibroblast growth factor 10 (FGF-10) released by mesenchymal stromal cells to support the proliferation of lung epithelial progenitor cells [50]. IL-13, produced by alternatively activated macrophages has also been shown to stimulate release of TGF-α from lung epithelial cells, which in turn stimulates epithelial cell proliferation via binding to the epidermal growth factor receptor (EGFR) in an autocrine mechanism [53]. Interestingly, studies have also shown that GM-CSF, which is critical for self-renewal of AMs, is involved in epithelial repair in the lung [54–56]. Alternatively activated AMs also secrete large amounts of transforming growth factor-β (TGF-β), which exhibits immunosuppressive activity while also promoting myofibroblast differentiation and extracellular matrix production [57].

The persistence of alternatively activated AMs may therefore support unrestrained proliferation of mesenchymal stromal niche cells without apoptotic clearance, and thus contribute to pathologic tissue remodeling [58]. Conditional depletion of resident lung macrophages expressing a diphtheria toxin receptor under control of the CD11c promoter (CD11c-DTR) after the establishment of OVA-induced asthma, was sufficient to prevent the advancement of airway remodeling [59]. Increasing evidence also suggests that release of profibrotic factors like IGF-1, IL-13, and TGF-β by wound-healing AMs may play a key role in the development of pulmonary fibrosis, which is thought to involve the dysregulated proliferation and myofibroblastic differentiation of mesenchymal stromal cells that are involved in the regeneration and/or resolution phases of epithelial repair [60]. The fact that corticosteroids are also capable of activating the wound-healing phenotype in macrophages, may also explain why these drugs are not effective against fibrosis and may even be disadvantageous [58]. AMs in COPD also display a unique wound-healing phenotype that is thought to contribute to deleterious remodeling in the lung. Evidence suggests that the relative balance between macrophage polarization states can have a profound impact on disease progression [61]. The induction of CD163 is commonly recognized as a marker for alternatively activated macrophages involved in wound healing [62, 63] and CD163 positive macrophages are highly prominent in the lungs of current and ex-smokers with COPD [64].

A Role for ILCs in Regulating Immune and Tissue Homeostasis

In recent years, ILCs, which are newly identified members of the lymphoid lineage, have emerged as important effectors of epithelial homeostasis [7]. ILCs are primarily defined by the absence of recombination activating gene (RAG)-dependent rearranged antigen receptors, a lack of mature hematopoietic lineage markers and their lymphoid morphology. They can be divided into three subsets based on the cytokines they produce. Group 1 ILCs include conventional NK cells and are defined by the production of IFNγ. Group 2 ILCs require IL-7 for their development and secrete IL-5 and IL-13 in response to stimulation with IL-25 or IL-33. Group 3 ILCs are defined by their production of IL-17A and/or IL-22 [65, 66].

In the context of tissue regeneration, recent studies have highlighted multiple roles for ILC2s in regulating immunity and tissue homeostasis [67]. Notably, ILC2s have been shown to be a major source of IL-13 driving allergen-induced and influenza-induced airway hyperactivity [68, 69]. Moreover, intranasal administration of IL-25 or IL-33, which promotes IL-13 secretion by ILC2s, induced an asthma phenotype in naïve mice that was associated with an increase in the number of ILC2s in the lungs [70]. The release of IL-13 by ILC2s may be one of the first events driving polarization of AMs towards a wound-healing phenotype.

A number of studies have demonstrated a critical role for ILC2s in regulating lung epithelial homeostasis and tissue remodeling after inflammation or infection. For example, depletion of ILC2s after influenza virus infection using an antibody reagent targeting CD90.2 expressed on the surface of ILC2s, resulted in a breakdown in epithelial integrity and a decrease in lung function [71]. Similarly, blockade of IL-33 led to a reduction in ILC2 number and disruption to epithelial repair processes. However, epithelial repair in ILC2-depleted mice could be restored by adoptive transfer of ILC2s. Furthermore, this study also identified amphiregulin, not IL-13, as the key ILC2-derived factor driving epithelial homeostasis after influenza virus infection [71]. Importantly, amphiregulin is a ligand for the EGFR, and has also been shown to be significantly upregulated during naphthalene-induced epithelial regeneration in the lung [72].

It is also interesting to note that a recent study has shown that IL-33 is upregulated in basal cells in the inflamed or viral infected lung [73]. Basal cells are well-characterised as airway epithelial progenitor cells in the upper airways, and this observation suggests that they may serve as a renewable source of IL-33, which can in turn activate ILC2s to initiate the wound healing process and restore epithelial integrity and tissue homeostasis following influenza virus infection. These studies demonstrate a critical role for ILC2s in regulating epithelial homeostasis and tissue remodeling after inflammation or infection.

Conclusion

In this review, we have discussed the emerging evidence that innate immune cells play a critical role in the regulation of lung regeneration and repair. In addition to their role as first-line defenders against inhaled pathogens, innate immune cells function as niche cells in the adult lung by providing direct and indirect support for tissue repair and regeneration. During the resolution phase of wound healing, resident AMs play a crucial role in the phagocytosis of apoptotic neutrophils, and the release of cytokines that enhance the regenerative capacity of the epithelial–mesenchymal trophic unit. There is also emerging evidence that a relatively newly discovered population of mucosal immune cells (ILC2s), may also play a key role in stimulating the wound-healing process after inflammation or infection. Although we still have much to learn about the mechanisms by which innate immune cells interact with regenerative stem and progenitor cells in the lung, it is clear from these limited studies that they are not simply onlookers and have central roles in regulating tissue homeostasis. Moreover, if left unchecked they may be directly responsible for driving pathogenic remodeling in lung diseases.

References

1. Werner JL, Steele C (2014) Innate receptors and cellular defense against pulmonary infections. J Immunol 193(8):3842–3850
2. Gordon S, Martinez FO (2010) Alternative activation of macrophages: mechanism and functions. Immunity 32(5):593–604
3. Brancato SK, Albina JE (2011) Wound macrophages as key regulators of repair: origin, phenotype, and function. Am J Pathol 178(1):19–25
4. Duffield JS, Forbes SJ, Constandinou CM, Clay S, Partolina M, Vuthoori S et al (2005) Selective depletion of macrophages reveals distinct, opposing roles during liver injury and repair. J Clin Invest 115(1):56–65
5. Lee S, Huen S, Nishio H, Nishio S, Lee HK, Choi BS et al (2011) Distinct macrophage phenotypes contribute to kidney injury and repair. J Am Soc Nephrol 22(2):317–326
6. Mahdavian Delavary B, van der Veer WM, van Egmond M, Niessen FB, Beelen RH (2011) Macrophages in skin injury and repair. Immunobiology 216(7):753–762
7. Walker JA, Barlow JL, McKenzie AN (2013) Innate lymphoid cells—how did we miss them? Nat Rev Immunol 13(2):75–87
8. Wynn TA, Chawla A, Pollard JW (2013) Macrophage biology in development, homeostasis and disease. Nature 496(7446):445–455
9. Davies LC, Jenkins SJ, Allen JE, Taylor PR (2013) Tissue-resident macrophages. Nat Immunol 14(10):986–995
10. Gordon S, Pluddemann A, Martinez Estrada F (2014) Macrophage heterogeneity in tissues: phenotypic diversity and functions. Immunol Rev 262(1):36–55
11. Schulz C, Gomez Perdiguero E, Chorro L, Szabo-Rogers H, Cagnard N, Kierdorf K et al (2012) A lineage of myeloid cells independent of Myb and hematopoietic stem cells. Science 336(6077):86–90
12. Hashimoto D, Chow A, Noizat C, Teo P, Beasley MB, Leboeuf M et al (2013) Tissue-resident macrophages self-maintain locally throughout adult life with minimal contribution from circulating monocytes. Immunity 38(4):792–804
13. Yona S, Kim KW, Wolf Y, Mildner A, Varol D, Breker M et al (2013) Fate mapping reveals origins and dynamics of monocytes and tissue macrophages under homeostasis. Immunity 38 (1):79–91
14. van oud Alblas AB, van Furth R (1979) Origin, Kinetics, and characteristics of pulmonary macrophages in the normal steady state. J Exp Med 149(6):1504–1518
15. Guilliams M, De Kleer I, Henri S, Post S, Vanhoutte L, De Prijck S et al (2013) Alveolar macrophages develop from fetal monocytes that differentiate into long-lived cells in the first week of life via GM-CSF. J Exp Med 210(10):1977–1992

16. Maus UA, Janzen S, Wall G, Srivastava M, Blackwell TS, Christman JW et al (2006) Resident alveolar macrophages are replaced by recruited monocytes in response to endotoxin-induced lung inflammation. Am J Respir Cell Mol Biol 35(2):227–235
17. Tarling JD, Lin HS, Hsu S (1987) Self-renewal of pulmonary alveolar macrophages: evidence from radiation chimera studies. J Leukoc Biol 42(5):443–446
18. Franke-Ullmann G, Pfortner C, Walter P, Steinmuller C, Lohmann-Matthes ML, Kobzik L (1996) Characterization of murine lung interstitial macrophages in comparison with alveolar macrophages in vitro. J Immunol 157(7):3097–3104
19. Taylor PR, Martinez-Pomares L, Stacey M, Lin HH, Brown GD, Gordon S (2005) Macrophage receptors and immune recognition. Annu Rev Immunol 23:901–944
20. Nathan CF, Hibbs JB Jr (1991) Role of nitric oxide synthesis in macrophage antimicrobial activity. Curr Opin Immunol 3(1):65–70
21. Fantone JC, Ward PA (1982) Role of oxygen-derived free radicals and metabolites in leukocyte-dependent inflammatory reactions. Am J Pathol 107(3):395–418
22. Hickman-Davis JM, O'Reilly P, Davis IC, Peti-Peterdi J, Davis G, Young KR et al (2002) Killing of Klebsiella pneumoniae by human alveolar macrophages. Am J Physiol Lung Cell Mol Physiol 282:L944–L956
23. Dhaliwal K, Scholefield E, Ferenbach D, Gibbons M, Duffin R, Dorward DA et al (2012) Monocytes control second-phase neutrophil emigration in established lipopolysaccharide-induced murine lung injury. Am J Respir Crit Care Med 186(6):514–524
24. Rodgers B, Mims CA (1981) Interaction of influenza virus with mouse macrophages. Infect Immun 31(2):751–757
25. Cirino NM, Panuska JR, Villani A, Taraf H, Rebert NA, Merolla R et al (1993) Restricted replication of respiratory syncytial virus in human alveolar macrophages. J Gen Virol 74 (Pt 8):1527–1537
26. Tate MD, Pickett DL, van Rooijen N, Brooks AG, Reading PC (2010) Critical role of airway macrophages in modulating disease severity during influenza virus infection of mice. J Virol 84(15):7569–7580
27. Tumpey TM, Garcia-Sastre A, Taubenberger JK, Palese P, Swayne DE, Pantin-Jackwood MJ et al (2005) Pathogenicity of influenza viruses with genes from the 1918 pandemic virus: functional roles of alveolar macrophages and neutrophils in limiting virus replication and mortality in mice. J Virol 79(23):14933–14944
28. Schneider C, Nobs SP, Heer AK, Kurrer M, Klinke G, van Rooijen N et al (2014) Alveolar macrophages are essential for protection from respiratory failure and associated morbidity following influenza virus infection. PLoS Pathog 10(4):e1004053
29. Ghoneim HE, Thomas PG, McCullers JA (2013) Depletion of alveolar macrophages during influenza infection facilitates bacterial superinfections. J Immunol 191(3):1250–1259
30. Rahman I, MacNee W (2000) Oxidative stress and regulation of glutathione in lung inflammation. Eur Respir J 16(3):534–554
31. Marti-Lliteras P, Regueiro V, Morey P, Hood DW, Saus C, Sauleda J, Agusti AG et al (2009) Nontypeable Haemophilus influenzae clearance by alveolar macrophages is impaired by exposure to cigarette smoke. Infect Immun 77(10):4232–4242
32. Hodge S, Hodge G, Ahern J, Jersmann H, Holmes M, Reynolds PN (2007) Smoking alters alveolar macrophage recognition and phagocytic ability: implications in chronic obstructive pulmonary disease. Am J Respir Cell Mol Biol 37(6):748–755
33. Hodge S, Matthews G, Mukaro V, Ahern J, Shivam A, Hodge G et al (2011) Cigarette smoke-induced changes to alveolar macrophage phenotype and function are improved by treatment with procysteine. Am J Respir Cell Mol Biol 44(5):673–681
34. Pesci A, Balbi B, Majori M, Cacciani G, Bertacco S, Alciato P, Donner CF (1998) Inflammatory cells and mediators in bronchial lavage of patients with chronic obstructive pulmonary disease. Eur Respir J 12(2):380–386

35. Keatings VM, Collins PD, Scott DM, Barnes PJ (1996) Differences in interleukin-8 and tumor necrosis factor-alpha in induced sputum from patients with chronic obstructive pulmonary disease or asthma. Am J Respir Crit Care Med 153(2):530–534
36. Di Stefano A, Capelli A, Lusuardi M, Balbo P, Vecchio C, Maestrelli P et al (1998) Severity of airflow limitation is associated with severity of airway inflammation in smokers. Am J Respir Crit Care Med 158(4):1277–1285
37. Barnes PJ, Shapiro SD, Pauwels RA (2003) Chronic obstructive pulmonary disease: molecular and cellular mechanisms. Eur Respir J 22(4):672–688
38. Finlay GA, O'Driscoll LR, Russell KJ, D'Arcy EM, Masterson JB, FitzGerald MX, O'Connor CM (1997) Matrix metalloproteinase expression and production by alveolar macrophages in emphysema. Am J Respir Crit Care Med 156(1):240–247
39. Ohnishi K, Takagi M, Kurokawa Y, Satomi S, Konttinen YT (1998) Matrix metalloproteinase-mediated extracellular matrix protein degradation in human pulmonary emphysema. Lab Invest 78(9):1077–1087
40. Knapp S, Leemans JC, Florquin S, Branger J, Maris NA, Pater J et al (2003) Alveolar macrophages have a protective antiinflammatory role during murine pneumococcal pneumonia. Am J Respir Crit Care Med 167(2):171–179
41. Cox G, Crossley J, Xing Z (1995) Macrophage engulfment of apoptotic neutrophils contributes to the resolution of acute pulmonary inflammation in vivo. Am J Respir Cell Mol Biol 12 (2):232–237
42. Fadok VA, Bratton DL, Konowal A, Freed PW, Westcott JY, Henson PM (1998) Macrophages that have ingested apoptotic cells in vitro inhibit proinflammatory cytokine production through autocrine/paracrine mechanisms involving TGF-beta, PGE2, and PAF. J Clin Invest 4:890–898
43. Voll RE, Herrmann M, Roth EA, Stach C, Kalden JR, Girkontaite I (1997) Immunosuppressive effects of apoptotic cells. Nature 390(6658):350–351
44. Damiano VV, Tsang A, Kucich U, Abrams WR, Rosenbloom J, Kimbel P et al (1986) Immunolocalization of elastase in human emphysematous lungs. J Clin Invest 78(2):482–493
45. Wynes MW, Frankel SK, Riches DW (2004) IL-4-induced macrophage-derived IGF-I protects myofibroblasts from apoptosis following growth factor withdrawal. J Leukoc Biol 76 (5):1019–1027
46. Chetty A, Andersson S, Lassus P, Nielsen HC (2004) Insulin-like growth factor-1 (IGF-1) and IGF-1 receptor (IGF-1R) expression in human lung in RDS and BPD. Pediatr Pulmonol 37 (2):128–136
47. Hung CF, Rohani MG, Lee SS, Chen P, Schnapp LM (2013) Role of IGF-1 pathway in lung fibroblast activation. Respir Res 14:102
48. McQualter JL, Brouard N, Williams B, Baird BN, Sims-Lucas S, Yuen K et al (2009) Endogenous fibroblastic progenitor cells in the adult mouse lung are highly enriched in the sca-1 positive cell fraction. Stem Cells 27(3):623–633
49. McQualter JL, McCarty RC, Van der Velden J, O'Donoghue RJ, Asselin-Labat ML, Bozinovski S, Bertoncello I (2013) TGF-beta signaling in stromal cells acts upstream of FGF-10 to regulate epithelial stem cell growth in the adult lung. Stem Cell Res 11 (3):1222–1233
50. McQualter JL, Yuen K, Williams B, Bertoncello I (2010) Evidence of an epithelial stem/progenitor cell hierarchy in the adult mouse lung. Proc Natl Acad Sci U S A 107(4):1414–1419
51. Morimoto K, Amano H, Sonoda F, Baba M, Senba M, Yoshimine H et al (2001) Alveolar macrophages that phagocytose apoptotic neutrophils produce hepatocyte growth factor during bacterial pneumonia in mice. Am J Respir Cell Mol Biol 24(5):608–615
52. Narasaraju T, Ng HH, Phoon MC, Chow VT (2010) MCP-1 antibody treatment enhances damage and impedes repair of the alveolar epithelium in influenza pneumonitis. Am J Respir Cell Mol Biol 42(6):732–743
53. Booth BW, Adler KB, Bonner JC, Tournier F, Martin LD (2001) Interleukin-13 induces proliferation of human airway epithelial cells in vitro via a mechanism mediated by transforming growth factor-alpha. Am J Respir Cell Mol Biol 25(6):739–743

54. Cakarova L, Marsh LM, Wilhelm J, Mayer K, Grimminger F, Seeger W et al (2009) Macrophage tumor necrosis factor-alpha induces epithelial expression of granulocyte-macrophage colony-stimulating factor: impact on alveolar epithelial repair. Am J Respir Crit Care Med 180 (6):521–532
55. Huffman Reed JA, Rice WR, Zsengeller ZK, Wert SE, Dranoff G, Whitsett JA (1997) GM-CSF enhances lung growth and causes alveolar type II epithelial cell hyperplasia in transgenic mice. Am J Physiol 273(4 Pt 1):L715–L725
56. Paine R III, Wilcoxen SE, Morris SB, Sartori C, Baleeiro CE, Matthay MA, Christensen PJ (2003) Transgenic overexpression of granulocyte macrophage-colony stimulating factor in the lung prevents hyperoxic lung injury. Am J Pathol 163(6):2397–2406
57. Wynn TA, Ramalingam TR (2012) Mechanisms of fibrosis: therapeutic translation for fibrotic disease. Nat Med 18(7):1028–1040
58. Boorsma CE, Draijer C, Melgert BN (2013) Macrophage heterogeneity in respiratory diseases. Mediators Inflamm 2013:769214
59. van Rijt LS, Jung S, Kleinjan A, Vos N, Willart M, Duez C et al (2005) In vivo depletion of lung CD11c+ dendritic cells during allergen challenge abrogates the characteristic features of asthma. J Exp Med 201(6):981–991
60. Selman M, King TE, Pardo A, American Thoracic Society, European Respiratory Society, American College of Chest Physicians (2001) Idiopathic pulmonary fibrosis: prevailing and evolving hypotheses about its pathogenesis and implications for therapy. Ann Intern Med 134 (2):136–151
61. Mills CD, Ley K (2014) M1 and M2 Macrophages: the chicken and the egg of immunity. J Innate Immun 6(6):716–726
62. Mantovani A, Sica A, Sozzani S, Allavena P, Vecchi A, Locati M (2004) The chemokine system in diverse forms of macrophage activation and polarization. Trends Immunol 25 (12):677–686
63. Van Gorp H, Delputte PL, Nauwynck HJ (2010) Scavenger receptor CD163, a Jack-of-all-trades and potential target for cell-directed therapy. Mol Immunol 47(7–8):1650–1660
64. Kunz LI, Lapperre TS, Snoeck-Stroband JB, Budulac SE, Timens W, van Wijngaarden S et al (2011) Smoking status and anti-inflammatory macrophages in bronchoalveolar lavage and induced sputum in COPD. Respir Res 12:34
65. Cortez VS, Robinette ML, Colonna M (2015) Innate lymphoid cells: new insights into function and development. Curr Opin Immunol 32C:71–77
66. Diefenbach A, Colonna M, Koyasu S (2014) Development, differentiation, and diversity of innate lymphoid cells. Immunity 41(3):354–365
67. Spits H, Di Santo JP (2011) The expanding family of innate lymphoid cells: regulators and effectors of immunity and tissue remodeling. Nat Immunol 12(1):21–27
68. Barlow JL, Bellosi A, Hardman CS, Drynan LF, Wong SH, Cruickshank JP, McKenzie AN (2012) Innate IL-13-producing nuocytes arise during allergic lung inflammation and contribute to airways hyperreactivity. J Allergy Clin Immunol 129(1):191–198
69. Bartemes KR, Iijima K, Kobayashi T, Kephart GM, McKenzie AN, Kita H (2012) IL-33-responsive lineage-CD25+ CD44(hi) lymphoid cells mediate innate type 2 immunity and allergic inflammation in the lungs. J Immunol 188(3):1503–1513
70. Klein Wolterink RG, Kleinjan A, van Nimwegen M, Bergen I, de Bruijn M, Levani Y, Hendriks RW (2012) Pulmonary innate lymphoid cells are major producers of IL-5 and IL-13 in murine models of allergic asthma. Eur J Immunol 42(5):1106–1116
71. Monticelli LA, Sonnenberg GF, Abt MC, Alenghat T, Ziegler CG, Doering TA et al (2011) Innate lymphoid cells promote lung-tissue homeostasis after infection with influenza virus. Nat Immunol 12(11):1045–1054
72. Snyder JC, Zemke AC, Stripp BR (2009) Reparative capacity of airway epithelium impacts deposition and remodeling of extracellular matrix. Am J Respir Cell Mol Biol 40(6):633–642
73. Byers DE, Alexander-Brett J, Patel AC, Agapov E, Dang-Vu G, Jin X et al (2013) Long-term IL-33-producing epithelial progenitor cells in chronic obstructive lung disease. J Clin Invest 123(9):3967–3982

Chapter 11
Extracellular Matrix Specification of Regenerative Cells in the Adult Lung

Janette K. Burgess and Gavin Tjin

The Extracellular Matrix of the Lung

The extracellular matrix (ECM), the meshwork of proteins that provides the framework for all tissues and organs, is a complex mixture of more than 300 different components [1]. The functional capabilities of the ECM span from providing a basic structural scaffolding and rigidity on which tissues are organised, to biological and biomechanical cues which regulate cell characteristics including survival, proliferation, migration and differentiation [2–5]. The ECM generates a microenvironment that is readily remodelled and is sensed by the cells within it to modulate their behaviours. Almost all cells produce and interact with ECM proteins, establishing an interdependent relationship within the microenvironment in the tissue.

The organisation of the ECM in the lung is specifically structured to facilitate the primary role of the organ, allowing efficient gas exchange. It provides an extensive framework capable of expanding and contracting that provides elasticity and tensile and compressive strength, which also acts as a protective buffer against the retention of fluid, particularly during oedema [6–9]. It is also recognised that

J.K. Burgess (✉)
Woolcock Institute of Medical Research, The University of Sydney, Sydney, Australia

Discipline of Pharmacology, The University of Sydney, Sydney, Australia

Central Clinical School, The University of Sydney, Sydney, Australia
e-mail: janette.burgess@sydney.edu.au

G. Tjin
Woolcock Institute of Medical Research, The University of Sydney, Sydney, Australia

Central Clinical School, The University of Sydney, Sydney, Australia

I. Bertoncello (ed.), *Stem Cells in the Lung*, Stem Cell Biology and Regenerative Medicine, DOI 10.1007/978-3-319-21082-7_11

the ECM in the lung acts as a reservoir for growth factors and other chemical cellular signals and provides a substrate that cells can adhere to through their surface receptors [9–11].

The organisation of the ECM within the lung is structurally compartmentalised [12]. The walls of the large bronchi and trachea are lined with cartilage. Within the cartilage, the outer regions contain collagen fibres which are essential for opposing tensile forces, whilst within the central regions proteoglycans are present to counter the compressive forces experienced during breathing, thereby ensuring that the airways do not collapse due to changes in intrathoracic pressure [13]. The basement membrane is a specialised area of ECM that acts as an interface between the epithelial or endothelial cells and the surrounding or underlying connective tissue region. Collagen IV makes up a major fraction of this area, with fibronectin, laminin and a number of proteoglycans also playing important roles. The pulmonary interstitium fills the region between the basement membranes below the airway epithelium or around the vascular endothelium and the pleural mesothelium. This region is of central importance as it provides the lung structural rigidity whilst breathing. The ECM in this region contains several types of collagen molecules and elastin [12].

The Importance of the ECM in Directing Fetal Lung Development

The complexity of the ECM structure in the lung is generated during development through branching morphogenesis which is orchestrated by specific programming of distinct pools of progenitor cell populations [14, 15]. The production and deposition of ECM molecules during development is a highly regulated process with distinct spatial and temporal patterning indicating that the ECM is a central part of the programming that ensures that the initial epithelial and mesenchymal cell populations build the highly ordered organ that is the lung. The ECM also acts as a reservoir for growth factors that direct cellular movement and responses within the developing lung [16, 17].

The lung is recognised to inaugurate a pre-programmed provisional matrix during development to provide cues to the infiltrating cells as to their ultimate destination within the developing lung [12]. The occurrence of distinct ECM molecules at specific times and locations during the developmental process provides chemical and mechanical signals to the cells within this microenvironment but also acts as a reservoir for growth factors and other diffusible components that drive the pattern of development.

The role of the ECM in directing fetal lung development has been reviewed elsewhere and will not be discussed in detail in this chapter [14, 18, 19].

The lung also develops a provisional ECM during wound repair. Fibronectin and fibrin are the most common components in these matrices, but thrombospondin, tenascin and secreted protein, acidic and rich in cysteine (SPARC) are often also involved. Once the repair process has been completed, this provisional matrix usually resolves [12]. However, the failure to resolve, or disruption of the usual programming that is needed to turn off the deposition of matrix proteins during the wound repair process, has been suggested to underlie pathologic changes in the ECM in lung diseases.

The ECM Influence on Cellular Homeostasis

The composition of the ECM is important because its particular make-up defines both its biomechanical and biochemical properties [20]. The ECM is now recognised as a biologically active substratum that can regulate cellular behaviours including adhesion, migration, proliferation, differentiation and survival [2, 4]. Individual matrix proteins have been reported to direct cellular behaviours, for example, fibronectin and collagen I enhance cell proliferation, whereas laminin is inhibitory [21–24]. These proteins also provide distinctive cell movement directions [25]. Complex native matrices laid down by cells isolated from people with disease have altered functional effects when cells are exposed to these matrices compared to matrices laid down by cells from people without disease [26, 27]. Interestingly, when the incorporation of individual matrix proteins is blocked in these native matrices, the functional potential of the matrix is altered [28], indicating that apparently small changes to the ECM, such as the reduction in the level of a single ECM protein, can have marked effects on the assembly and the subsequent bioactivity of the ECM [29, 30].

Physical effects of the ECM are of particular importance in the lung. The ECM is responsible for the compressive and tensile strength of the tissue or organ. Increases in ECM stiffness relate to the reduced elastic properties of the lung and increase the work of breathing, but cells also respond to increased matrix stiffness by selectively increasing expression of contractile proteins and decreasing production of fibrosis-inhibiting cytokines [31, 32]. In the context of fibrotic lung disease, it has long been established that the lung becomes stiffer as the disease progresses [33]. However, there have been few studies relating the changes to matrix stiffness with alterations in the ECM profile in patients.

The Role of the ECM as a Stem Cell Niche in the Adult Lung

There are more than 40 different cell types in the adult lung [34–36]. The complexity of the structural organisation, coupled with the number of cell types that constitute this mature organ, decrees that multiple pools of stable stem cells with the capacity for self-regeneration and pluripotency may be required for a response after injury. Within the lung, it is thought that there are potentially multiple specialised niches that house these different progenitor cell populations [37–40].

During development and repair, these cells from multiple lineages must interact in a programmed manner. How the cells are driven to adopt a lung cell phenotype is not currently clear. Some studies provide evidence of cell fusion occurring between the stem cells and the resident lung cells [41–43]. It is becoming clear that the interaction with the stem cell niche (directed by the ECM microenvironment encountered in the lung) is crucial for the specification of the fate and lineage development of the cells [42–45]. The potential for lineage specification is determined by the intrinsic potential of the cells and the extrinsic cues they receive from the microenvironment in which they reside.

It is now clear that stem cells reside in a specialised, highly organised and bioactive microenvironment that is the "niche" that is composed of ECM components and cells in the local environment, a concept first proposed by Raymond Schofield in 1978 [46]. The mechanism through which the ECM-derived niche dictates stem cell fates, which allows the balance of cells in the quiescent, self-perpetuating or differentiating states to be maintained for lung-specific lineages, has been not completely elucidated to date [47]. It is thought that the same mechanisms through which the ECM directs cellular homeostasis also apply within the stem cell niche, with some ECM proteins recently being identified as having defined roles in regulating stem cell activities [48, 49]. Prevention of communication between the cells and the ECM through integrin receptors, either through gene knockout of integrin genes [50] or the transcriptional regulator of integrin expression *Myc* [51], is able to reduce the number of stem cells. In addition, anchorage to the ECM is suggested to maintain stem cell polarity, allow orientation of the mitotic spindle and direct the cells to undergo asymmetric cell division. In this way, the ECM is instrumental in driving the stem cells to their required fate [50, 52, 53]. The biomechanical properties of the ECM niche also contribute to the direction of the stem cell fates [54–56].

Alterations of the ECM in Lung Disease

The ECM in the lung of a patient with chronic lung disease is structurally altered, and the composition changed compared to that in a healthy, age-matched person. Given the importance of the ECM in regulating cellular responses, it is now clear that in a diseased lung the repair capacity and self-renewal of the resident stem cells are impaired. In addition, the continuing injury and repair process that occurs in these lungs is thought to lead to the eventual depletion of the stem cell pool within the affected regions of the lung, which further contributes to the fibrotic remodelling in the absence of effective repair [57]. The remodelling of the ECM may be a key factor in the loss of the specific stem cell niche resulting in disruption to the clearly defined programming of the resident stem cell pool. The disruption of the epithelial stem cell niche in the fibrotic lung has been suggested to be an underlying mechanism connecting idiopathic pulmonary fibrosis (IPF) with ageing and anomalous activation of ATII cells [58, 59].

Asthma

In asthma, structural changes are observed in both the large and small airways but changes are not usually seen in the parenchyma in non-smoking asthmatics [60]. Alterations in the ECM in the airways of cases of fatal asthma were described as early as 1922 [61]. The basement membrane is thickened in asthmatics, and the composition and amount of ECM between the basement membrane and the airway smooth muscle (ASM) bundles is altered [62–64]. Extensive histological studies on post-mortem tissues, explanted lung samples or biopsies from asthmatics have shown increased deposition of collagen (types I, III and V), laminin α2/β2, fibronectin, versican, tenascin, perlecan and hyaluronan, [65–74] whilst collagen IV, elastin and decorin are decreased [71, 75] in comparison to non-diseased airways. The fraction of ECM [64] and its composition [62] is also altered within and surrounding the ASM bundles. The changes observed in different compartments within the airways vary, suggesting that the ECM is differentially regulated in the different regions of the lung. Other changes that are characteristic in severe asthma include an increase in the bulk of the ASM, increased vascularisation and epithelial hyperplasia. The ECM has been suggested to play a role in regulating all of these features of the airways [76].

The mesenchymal cells are recognised as the primary source of ECM proteins in the airways. In vitro ASM cells derived from people with asthma produce a different profile of ECM proteins compared to cells from people without asthma [27]. The proteins produced in vitro reflect the changes seen in the immunohistochemical studies, with perlecan and collagen I being increased and chondroitin sulphate and collagen IV decreased. In addition, fibronectin is increased in ASM cells from people with asthma [26]. The production of ECM proteins from fibroblasts derived from the airways of people with asthma is also altered. The released proteoglycan profile from fibroblasts in vitro has been linked to airway hyperresponsiveness [77]. Up to four times, more proteoglycans are released from fibroblasts from the most hyperresponsive airways compared to fibroblasts from less hyperresponsive or non-responsive airways. A negative correlation exists between the release of small heparin sulphate proteoglycans, biglycan and perlecan and airway responsiveness (measured as PC20 to methacoline) [77].

Chronic Obstructive Pulmonary Disease

Our understanding of the structural changes in the lung in Chronic Obstructive Pulmonary Disease (COPD) is much less advanced [78, 79]. In COPD, the most recognised change is the destruction of the parenchyma, characterised as a destruction of the elastin fibres [80], which leads to the characteristic emphysema. It is now recognised that there are also changes in the airways in COPD, including alveolar wall thickening, fibrosis of the airway wall and remodelling of the pulmonary vasculature [60, 81–87]. Alterations to the ECM underlie many of these changes [88–90]. The proteoglycans, versican (enhanced [91]) and perlecan (reduced [92]), are altered in the large and small airways. Fibroblasts are thought to be the main producers of ECM proteins in the airways in COPD, with differences seen in the ECM protein production from cells isolated from the central and distal airways [93, 94].

Whilst the occurrence of a thickened basement membrane in COPD is controversial [95–98], differences in the ECM composition in this region have been observed in COPD. Total collagen, collagen I and III, but not collagen IV, fibronectin or laminin, are increased in the basement membranes in patients with COPD compared to those without COPD [99]. Greater intensity of staining is reported in asthma than COPD basement membranes for collagen I and laminin. In contrast, collagen IV is suggested to be of greater intensity in COPD than asthma, whilst collagen V does not differ between COPD, asthma or healthy controls [95].

Evidence is also emerging that the structural organisation of the ECM is altered in COPD compared to non-diseased tissues. In COPD lung tissue, the alveolar wall structure is thicker and highly disorganised in the emphysematous regions compared to non-diseased tissues. The collagen and elastin fibres are disrupted and have a web-like appearance extending into the airspaces, compared to the smooth condensed fibres in the non-diseased tissues [81]. These findings are also supported by the report by Tjin et al. [100]. The importance of the structural organisation of the fibres in the airways and surrounding tissues has yet to be fully understood (Fig. 11.1).

Idiopathic Pulmonary Fibrosis

In IPF, remodelling of the lung tissue structure is an important contributor to the pathology [101–106]. These changes include increased deposits of fibrotic ECM proteins in the interstitium and air spaces, thickening of the airway walls and the collapse of air spaces. These tissues have increased staining for hyaluronic acid [11], the foetal form of cellular fibronectin (containing the EIIIA domain) [107] and collagen I and III [80, 108, 109], which are usually located in or around the fibrotic foci of fibroblasts that are characteristic in IPF lung tissues [106, 110, 111].

Re-cellularising the Lung: Where to Next?

The highly complex structure of the ECM and its importance in the regulation of cellular behaviours in vivo have made it a challenging structure to study [112–114]. Early studies using light and electron microscopy showed that the matrices were widespread with distinct structural arrangements that underlay many cells. Exploration of the biology of the development and functionality of the ECM has been,

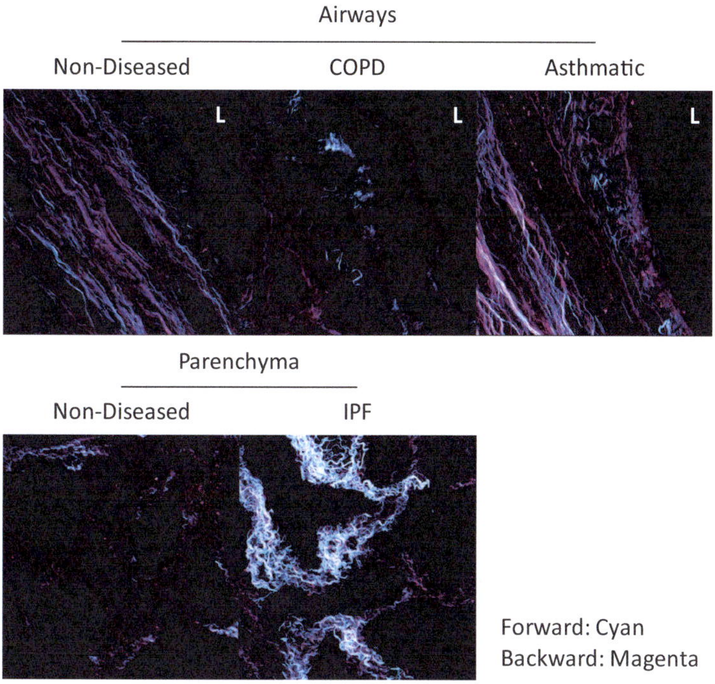

Fig. 11.1 Second Harmonic Generation images of airway and parenchymal tissues from non-diseased and diseased lungs. Second Harmonic Generation signals from fibrillar collagen I propagated in the forward (*Cyan*) (representing organised collagen) and backward (*Magenta*) (representing disorganised collagen) direction relative to the incident light (Dimensions: 246.03 × 246.03 μm). Abbreviations: *COPD* chronic obstructive pulmonary disease, *IPF* idiopathic pulmonary fibrosis, *L* airway lumen

and continues to be, challenging due to the fact that the ECM is, by nature, cross-linked and highly insoluble [1]. The traditional two-dimensional (2D) in vitro cell culture model that is often employed in pursuit of these answers does not sufficiently reflect native biology for cost-effective and efficient translational medical research. Owing to the many limitations of in vitro studies, translating the findings from in vitro studies to in vivo and eventually human studies is an ongoing challenge (Fig. 11.2).

Among the many limitations of the in vitro 2D culture systems is the fact that they do not accurately represent the in vivo 3D structures. Furthermore, most in vitro culture systems utilise a plastic/glass surface—which are both non-biological, in terms of the surface composition presented to the cells, and the mechanical properties are at odds with those in the in vivo system [115]. Whilst advancements have been made in the in vitro systems utilising surfaces coated with ECM-derived materials such as collagen and fibronectin [115] or mixtures of homogenised and dissolved decellularised tissues [116], this is merely a small step towards replicating the true ECM environment in vivo.

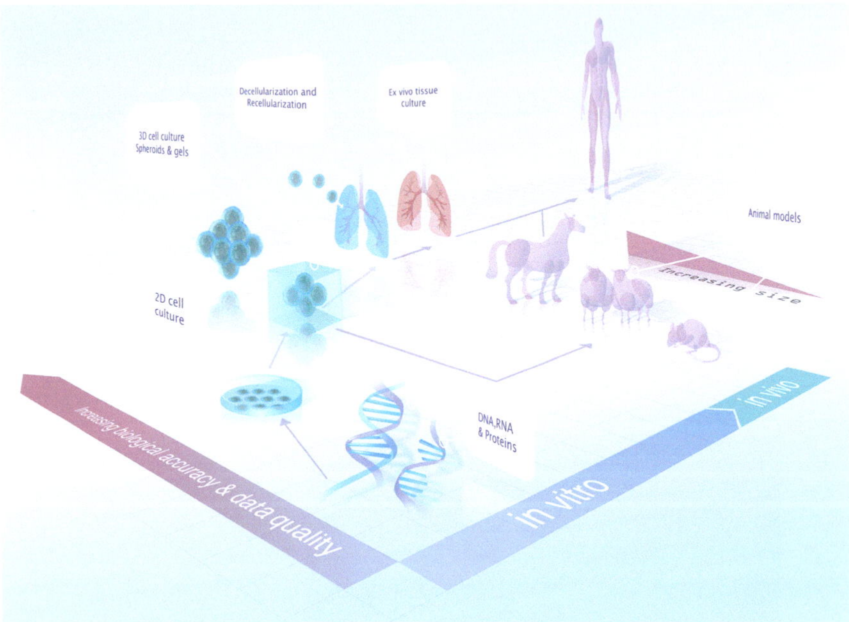

Fig. 11.2 Milestones for translational medical research

Three-dimensional cell culture is a further advancement for the field, bringing the experimental outcome closer to biological significance. However, in vitro 3D cell culture techniques such as spheroid suspensions [117] and 3D gel cultures [115, 118–120] retain some of the limitations described for 2D systems, primarily the question of: "What are the appropriate components and structure of the ECM?" Whilst there are many studies investigating the precise combination of ECM components in various tissues using techniques such as mass spectrometry [121, 122], current in vitro 3D model systems utilise single ECM components such as collagen, fibronectin or fibrin gels [115] or combinations of a few ECM components [123]; which pale in comparison to in vivo ECM that is composed of hundreds of different components [1]. Further complications are added by the interactions between ECM components which ultimately dictate the "net biological properties" of the scaffold. As such, the construction of a biologically accurate artificial/synthetic ECM scaffold remains a major hurdle in translational research.

To counter this highly complex problem, the technique of tissue decellularisation has been developed. Decellularisation is a novel technique that produces scaffolds through the removal of cells from viable tissue (utilising a combination of chemical and mechanical treatments), thus preserving much of the original ECM content and architecture (Fig. 11.3). Scaffolds created using this technique maintain most (but not all) of their ECM components and mechanical properties, making them a suitable platform for translational studies. As with any technique, there are also limitations, the first of which is that the decellularisation process can remove some of the less adhesive components of the ECM and introduce some small

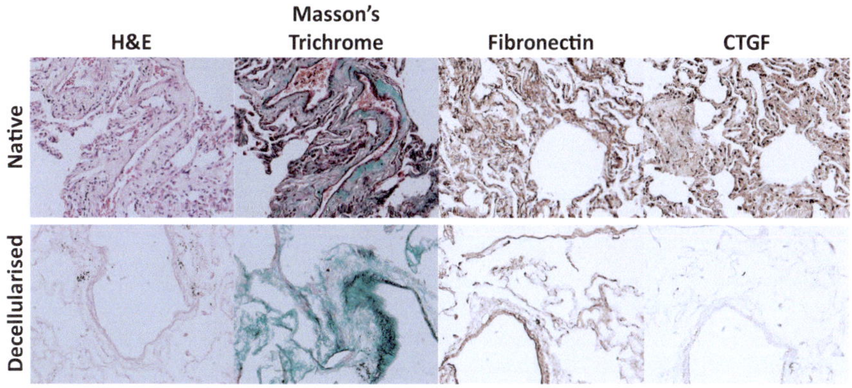

Fig. 11.3 Comparison of lung tissue components pre- and post-decellularisation. Lung tissue sections before and after decellularisation with subsequent washes in 0.1 % Triton X-100, 2 % Sodium Deoxycholate, 1 M NaCl, 30 μg/mL DNAse and sterilised with 0.018 % Peracetic acid in 4.8 % Ethanol with mild agitation were stained and imaged using a BX60 Olympus light microscope and attached DP71 camera to examine Hematoxylin and Eosin stain (blue = cell nuclei; pink = cytoplasm), Masson's Trichrome stain (*black* = cell nuclei; *red* = muscle, cytoplasm, keratin; *green* = collagen), Fibronectin or CTGF using DAB detection of a specific antibody (*brown*). Images were captured at ×20 magnification. Abbreviations: *H&E* hematoxylin and eosin, *CTGF* connective tissue growth factor, *DAB* = 3,3′-diaminobenzidine

structural changes. There are many different decellularisation techniques emerging, each with its own advantages and disadvantages (Table 11.1) [172] but the field is yet to agree on a standardised approach. The ideal approach would be for the protocol used for the production of the decellularised tissue to be tailored towards the type of tissue used and the study in question to ensure preservation of as many of the ECM properties as possible.

Re-cellularisation: The Final Goal

One of the major issues for organ transplantation programmes is the shortage of viable donor organs, followed by the immunological rejection of the implanted organ by the recipient [173–175]. The ultimate goal of tissue decellularisation is to eventually enable the re-cellularisation and repopulation of generated ECM scaffolds with recipient autologous cells to regenerate a viable tissue/organ for implantation.

The ECM scaffold has limited immunogenicity [167, 176] and through re-cellularisation of the scaffold with the recipient's cells it should be possible to create graft tissue material for treatment purposes [141, 151, 157, 176, 177]. Whilst this process has had success in simple organ systems such as skin [172], to date it is yet to be successful in the long term for complex organs such as the lungs [153]. Current research has successfully re-cellularised the lungs from various different species with a variety of cell types (Table 11.2). Whilst re-cellularised

Table 11.1 The advantages and disadvantages of methods of tissue decellularisation

Agent/technique	Mode of action	Effects on ECM	References
Chemical agents			
Acids and bases	Solubilises cytoplasmic components, disrupts nucleic acids and denatures proteins	May cause damage to collagen, GAGs and growth factors	[124–132]
Hypotonic and hypertonic solutions	Cell lysis through osmotic shock with disruption to DNA–protein interaction	Effective for cell lysis but not removal of cellular residues	[129, 131, 133–139]
Non-ionic detergents	Disrupts interactions between DNA–protein, lipid–lipid, lipid–protein and to a lesser extent protein–protein interactions. (e.g. Triton X-100)	Mixed efficacy. Efficacy dependent on tissue, cell removal is more effective on thin tissues and some disruption of the ultrastructure and removal of GAGs may occur	[124, 126–128, 130, 132–135, 138–152]
Ionic detergents	Solubilization of cellular and nucleic membranes. Denatures proteins. (e.g. Sodium Dodecyl Sulfate (SDS), Sodium Deoxycholate and Triton X-200)	Effect varies between detergents. Effectively removes nuclear remnants and cytoplasmic proteins with tendency to disrupt the ultrastructure, removal of GAGs and may damage collagen	[124, 126–128, 130, 131, 134, 135, 139–146, 148–162]
Zwitterionic detergents	Exhibits properties from both ionic and non-ionic detergents	Effective at removal of cells with mild disruption to the ultrastructure of thin tissues	[143, 150, 153, 159, 161]
Solvents	Alcohols and acetone lyse cells through dehydration, also solubilises and removes lipids	Effective at removal of cells from dense tissues and inactivates pyrogens but it causes proteins such as collagen to crosslink and precipitate	[125–128, 136, 137, 155, 163]
	Tributyl Phosphates form stable complexes with metals and disrupts protein–protein interactions	Mixed efficacy results (dependent on tissues) but causes loss of collagen with minimal impact on mechanical properties	

(continued)

Table 11.1 (continued)

Agent/technique	Mode of action	Effects on ECM	References
Biological agents			
Enzymes			
Nucleases	Targets ribonucleotide and deoxy-ribonucleotide chains	Difficult to remove from tissue and can invoke immune a response	[128, 130, 131, 134, 136, 139, 144, 146, 151–153, 157, 160, 164]
Trypsin	Cleaves peptide bonds	Prolonged exposure can disrupt the ECM ultra-structure and removes ECM components (e.g. collagen, laminin, fibronectin, elastin and GAGs—less effective GAG removal compared to detergents)	[124–126, 128, 133–135, 138, 139, 144–147, 149, 154, 165]
Dispase	Cleaves specific peptides (e.g. fibronectin and collagen IV)	Prolonged exposure can disrupt ECM ultrastructure and remove ECM components (e.g. fibronectin and collagen IV)	[125, 154, 166]
Chelating agents	Disrupts cell-ECM adhesion by binding metallic ions. (e.g. EDTA and EGTA)	Ineffective when used alone. Typically used in combination with enzymatic (e.g. trypsin) or other agents	[124–126, 128, 131, 133–135, 138, 139, 144–147, 149, 154, 165, 166]
Physical and miscellaneous agents			
Temperature (freeze thaw)	Intracellular ice crystals develop and disrupt cell membrane	Intracellular ice crystals can potentially disrupt or fracture the ECM	[124, 125, 132, 136, 149, 157, 166, 167]
Direct application of force	Physically damages and bursts cells	Direct application of force can damage the ECM	[129, 134, 137, 149, 166]
Pressure	Pressure can burst cells and help in the removal of cellular material	Pressure can disrupt the ECM	[146, 148, 164]
Electroporation	Pulsed electrical fields can disrupt the cell membrane	Pulsed electrical fields can disrupt the ECM	[168, 169]
Techniques to apply agents			
Perfusion	Increases chemical exposure through the tissue and helps in the removal of cellular material	Excessive pressure during perfusion can disrupt the ECM	[130, 140, 141, 151–153, 156, 157, 161, 162, 169]
Pressure gradient across tissue	Increases chemical exposure through the tissue and helps in the removal of cellular material	Pressure gradient can disrupt the ECM	[125, 131, 163]

(continued)

Table 11.1 (continued)

Agent/technique	Mode of action	Effects on ECM	References
Agitation	Can cause cell lysis and increases chemical exposure through the tissue and helps in the removal of cellular material	Excessive agitation and/or sonication can disrupt the ECM	[126–129, 131–134, 138, 139, 142–145, 147, 149–152, 155, 157–160, 163–167, 170, 171]

Table 11.2 Examples of successful lung tissue decellularisation and re-cellularisation

Source species	Re-cellularised cells	Reference
Murine	Mouse embryonic stem cells	[178, 179]
	Mouse fibroblast	[180]
	Mouse bone marrow-derived mesenchymal stromal cells	[181, 182]
	Mouse C10 lung epithelial cells	[182]
	Mouse pulmonary alveolar epithelial cells	[130]
Rattus	Human small airway epithelial cells	[161]
	Human pulmonary alveolar epithelial cells	[161]
	Human pulmonary basal epithelial cells (A549)	[156]
	Human umbilical vein endothelial cells	[156, 161, 162]
	Mouse embryonic stem cells	[157]
	Rat fetal lung cells	[156, 162]
	Rat microvascular endothelial cells	[153]
	Rat pulmonary alveolar epithelial cells	[153]
Porcine	Human lung fibroblasts	[151, 183]
	Human small airway epithelial cells	[183]
	Human adipose-derived mesenchymal stem cells	[183]
	Human bone marrow-derived mesenchymal stem cells	[151]
	Human vascular endothelial cells (CBF12)	[151]
	Human bronchial epithelial cells	[151]
Non-human primate	Primate adipose-derived mesenchymal stem cells	[184]
	Primate bone marrow-derived mesenchymal stem cells	[184]
Human	Human lung fibroblasts	[151, 152, 183]
	Human small airway epithelial cells	[152, 161, 183]
	Human pulmonary alveolar epithelial cells	[161]
	Human adipose-derived mesenchymal stem cells	[183]
	Human bone marrow-derived mesenchymal stem cells	[151, 152]
	Human umbilical vein endothelial cells	[161]
	Human vascular endothelial cells (CBF12)	[151]
	Human bronchial epithelial cells	[151]
	Human progenitor endothelial cells	[152]

organs can perform organ-specific functions for short periods of time [140], further research is necessary to progress and exploit this new emerging treatment possibility.

Stem cells have been highlighted as ideal cells for re-cellularisation of the ECM scaffold as not only does the ECM scaffold possess the required structure and mechanical properties, but ECM components can also assist in guiding the cell differentiation and migration processes for optimal regeneration of the tissue/organ for implantation [5, 157]. Whether the defined populations of progenitor cells within the lung can be re-established in healthy lung niches is an exciting prospect that needs extensive investigation.

Conclusions

The ECM provides structural support and environmental cues for cells within its microenvirons. This local regulatory system enacts dedicated stem cell niches within the lung which allow for the maintenance of specialised progenitor cell populations. Greater understanding of cell signalling, preferential attachment, migration and differentiation by cells as directed by the ECM microenvironment will provide valuable insights to cell/matrix dynamics. This will lead to advances in knowledge about the roles of various ECM components and their structural interactions with cells. The field is moving towards the ultimate goal of the development of a synthetic matrix scaffold suitable for re-cellularisation [185] and eventual re-construction of a viable lung for transplant purposes.

References

1. Hynes RO, Naba A (2012) Overview of the matrisome—an inventory of extracellular matrix constituents and functions. Cold Spring Harb Perspect Biol 4(1):a004903
2. Roberts CR, Walker DC, Schellenberg RR (2002) Extracellular matrix. Clin Allergy Immunol 16:143–178
3. Frantz C, Stewart KM, Weaver VM (2010) The extracellular matrix at a glance. J Cell Sci 123(Pt 24):4195–4200
4. Tran T, Halayko AJ (2007) Extracellular matrix and airway smooth muscle interactions: a target for modulating airway wall remodelling and hyperresponsiveness? Can J Physiol Pharmacol 85(7):666–671
5. Raghavan S, Bitar KN (2014) The influence of extracellular matrix composition on the differentiation of neuronal subtypes in tissue engineered innervated intestinal smooth muscle sheets. Biomaterials 35(26):7429–7440
6. Pelosi P, Rocco PRM, Negrini D, Passi A (2007) The extracellular matrix of the lung and its role in edema formation. An Acad Bras Cienc 79(2):285–297
7. Marastoni S, Ligresti G, Lorenzon E, Colombatti A, Mongiat M (2008) Extracellular matrix: a matter of life and death. Connect Tissue Res 49(3):203–206
8. Negrini D, Passi A, de Luca G, Miserocchi G (1998) Role of matrix macromolecules in controlling lung fluid balance: transition from a dry tissue to edema. In: Matalon S,

Sznajder JI (eds) Acute respiratory distress syndrome: cellular and molecular mechanisms and clinical management, vol 297. Plenum, New York, pp 65–68

9. Suki B, Ito S, Stamenovic D, Lutchen KR, Ingenito EP (2005) Biomechanics of the lung parenchyma: critical roles of collagen and mechanical forces. J Appl Physiol 98(5): 1892–1899
10. Negrini D, Tenstad O, Passi A, Wiig H (2006) Differential degradation of matrix proteoglycans and edema development in rabbit lung. Am J Physiol Lung Cell Mol Physiol 290(3): L470–L477
11. Papakonstantinou E, Karakiulakis G (2009) The 'sweet' and 'bitter' involvement of glycosaminoglycans in lung diseases: pharmacotherapeutic relevance. Br J Pharmacol 157(7): 1111–1127
12. Dunsmore SE (2008) Treatment of COPD: a matrix perspective. Int J Chron Obstruct Pulmon Dis 3(1):113–122
13. Roberts CR, Rains JK, Pare PD, Walker DC, Wiggs B, Bert JL (1998) Ultrastructure and tensile properties of human tracheal cartilage. J Biomech 31(1):81–86
14. Kim HY, Nelson CM (2012) Extracellular matrix and cytoskeletal dynamics during branching morphogenesis. Organogenesis 8(2):56–64
15. Kumar ME, Bogard PE, Espinoza FH, Menke DB, Kingsley DM, Krasnow MA (2014) Mesenchymal cells. Defining a mesenchymal progenitor niche at single-cell resolution. Science 346(6211):1258810
16. Matsuo I, Kimura-Yoshida C (2014) Extracellular distribution of diffusible growth factors controlled by heparan sulfate proteoglycans during mammalian embryogenesis. Philos Trans R Soc Lond B Biol Sci 369(1657)
17. Hynes RO (2009) The extracellular matrix: not just pretty fibrils. Science 326(5957): 1216–1219
18. McGowan SE (1992) Extracellular matrix and the regulation of lung development and repair. FASEB J 6(11):2895–2904
19. Shannon JM (1994) Induction of alveolar type II cell differentiation in fetal tracheal epithelium by grafted distal lung mesenchyme. Dev Biol 166(2):600–614
20. Antunes MA et al (2009) Different strains of mice present distinct lung tissue mechanics and extracellular matrix composition in a model of chronic allergic asthma. Respir Physiol Neurobiol 165(2–3):202–207
21. Freyer AM, Johnson SR, Hall IP (2001) Effects of growth factors and extracellular matrix on survival of human airway smooth muscle cells. Am J Respir Cell Mol Biol 25(5):569–576
22. Hirst SJ, Twort CH, Lee TH (2000) Differential effects of extracellular matrix proteins on human airway smooth muscle cell proliferation and phenotype. Am J Respir Cell Mol Biol 23(3):335–344
23. Tran T et al (2013) Laminin drives survival signals to promote a contractile smooth muscle phenotype and airway hyperreactivity. FASEB J 27(10):3991–4003
24. Bonacci JV, Harris T, Wilson JW, Stewart AG (2003) Collagen-induced resistance to glucocorticoid anti-mitogenic actions: a potential explanation of smooth muscle hyperplasia in the asthmatic remodelled airway. Br J Pharmacol 138(7):1203–1206
25. Parameswaran K, Radford K, Zuo J, Janssen LJ, O'Byrne PM, Cox PG (2004) Extracellular matrix regulates human airway smooth muscle cell migration. Eur Respir J 24(4):545–551
26. Chan V et al (2006) Extracellular matrix regulates enhanced eotaxin expression in asthmatic airway smooth muscle cells. Am J Respir Crit Care Med 174(4):379–385
27. Johnson PR et al (2004) Extracellular matrix proteins modulate asthmatic airway smooth muscle cell proliferation via an autocrine mechanism. J Allergy Clin Immunol 113(4): 690–696
28. Lau JY et al (2010) Fibulin-1 is increased in asthma–a novel mediator of airway remodeling? PLoS One 5(10)
29. Jarvelainen H, Sainio A, Koulu M, Wight TN, Penttinen R (2009) Extracellular matrix molecules: potential targets in pharmacotherapy. Pharmacol Rev 61(2):198–223

30. Gattazzo F, Urciuolo A, Bonaldo P (2014) Extracellular matrix: a dynamic microenvironment for stem cell niche. Biochim Biophys Acta 1840(8):2506–2519
31. Mih JD, Marinkovic A, Liu F, Sharif AS, Tschumperlin DJ (2012) Matrix stiffness reverses the effect of actomyosin tension on cell proliferation. J Cell Sci 125(Pt 24):5974–5983
32. Liu F et al (2010) Feedback amplification of fibrosis through matrix stiffening and COX-2 suppression. J Cell Biol 190(4):693–706
33. Gibson GJ, Pride NB (1976) Lung distensibility. The static pressure-volume curve of the lungs and its use in clinical assessment. Br J Dis Chest 70(3):143–184
34. Cardoso WV, Whitsett JA (2008) Resident cellular components of the lung: developmental aspects. Proc Am Thorac Soc 5(7):767–771
35. Franks TJ et al (2008) Resident cellular components of the human lung: current knowledge and goals for research on cell phenotyping and function. Proc Am Thorac Soc 5(7):763–766
36. Morrisey EE, Hogan BL (2010) Preparing for the first breath: genetic and cellular mechanisms in lung development. Dev Cell 18(1):8–23
37. Giangreco A, Reynolds SD, Stripp BR (2002) Terminal bronchioles harbor a unique airway stem cell population that localizes to the bronchoalveolar duct junction. Am J Pathol 161(1): 173–182
38. Hong KU, Reynolds SD, Watkins S, Fuchs E, Stripp BR (2004) Basal cells are a multipotent progenitor capable of renewing the bronchial epithelium. Am J Pathol 164(2):577–588
39. Rawlins EL, Clark CP, Xue Y, Hogan BL (2009) The Id2+ distal tip lung epithelium contains individual multipotent embryonic progenitor cells. Development 136(22):3741–3745
40. Rock JR et al (2009) Basal cells as stem cells of the mouse trachea and human airway epithelium. Proc Natl Acad Sci U S A 106(31):12771–12775
41. Spees JL et al (2003) Differentiation, cell fusion, and nuclear fusion during ex vivo repair of epithelium by human adult stem cells from bone marrow stroma. Proc Natl Acad Sci U S A 100(5):2397–2402
42. Camargo FD, Chambers SM, Goodell MA (2004) Stem cell plasticity: from transdifferentiation to macrophage fusion. Cell Prolif 37(1):55–65
43. Wagers AJ, Weissman IL (2004) Plasticity of adult stem cells. Cell 116(5):639–648
44. Hong KU, Reynolds SD, Watkins S, Fuchs E, Stripp BR (2004) In vivo differentiation potential of tracheal basal cells: evidence for multipotent and unipotent subpopulations. Am J Physiol Lung Cell Mol Physiol 286(4):L643–L649
45. Jiang Y et al (2002) Pluripotency of mesenchymal stem cells derived from adult marrow. Nature 418(6893):41–49
46. Schofield R (1978) The relationship between the spleen colony-forming cell and the haemopoietic stem cell. Blood Cells 4(1–2):7–25
47. Volckaert T, De Langhe S (2014) Lung epithelial stem cells and their niches: Fgf10 takes center stage. Fibrogenesis Tissue Repair 7:8
48. Urciuolo A et al (2013) Collagen VI regulates satellite cell self-renewal and muscle regeneration. Nat Commun 4:1964
49. Bentzinger CF, Wang YX, von Maltzahn J, Soleimani VD, Yin H, Rudnicki MA (2013) Fibronectin regulates Wnt7a signaling and satellite cell expansion. Cell Stem Cell 12(1): 75–87
50. Yamashita YM, Fuller MT, Jones DL (2005) Signaling in stem cell niches: lessons from the Drosophila germline. J Cell Sci 118(Pt 4):665–672
51. Frye M, Gardner C, Li ER, Arnold I, Watt FM (2003) Evidence that Myc activation depletes the epidermal stem cell compartment by modulating adhesive interactions with the local microenvironment. Development 130(12):2793–2808
52. Lu P, Takai K, Weaver VM, Werb Z (2011) Extracellular matrix degradation and remodeling in development and disease. Cold Spring Harb Perspect Biol 3(12)
53. Sever M, Mammadov B, Guler MO, Tekinay AB (2014) Tenascin-C mimetic peptide nanofibers direct stem cell differentiation to osteogenic lineage. Biomacromolecules 15(12):4480–4487

54. Engler AJ, Sen S, Sweeney HL, Discher DE (2006) Matrix elasticity directs stem cell lineage specification. Cell 126(4):677–689
55. Gilbert PM et al (2010) Substrate elasticity regulates skeletal muscle stem cell self-renewal in culture. Science 329(5995):1078–1081
56. Lutolf MP, Gilbert PM, Blau HM (2009) Designing materials to direct stem-cell fate. Nature 462(7272):433–441
57. Beers MF, Morrisey EE (2011) The three R's of lung health and disease: repair, remodeling, and regeneration. J Clin Invest 121(6):2065–2073
58. King TE Jr, Pardo A, Selman M (2011) Idiopathic pulmonary fibrosis. Lancet 378(9807): 1949–1961
59. Selman M, Pardo A, Kaminski N (2008) Idiopathic pulmonary fibrosis: aberrant recapitulation of developmental programs? PLoS Med 5(3):e62
60. Jeffery PK (2001) Remodeling in asthma and chronic obstructive lung disease. Am J Respir Crit Care Med 164(10 Pt 2):S28–S38
61. Larsen NP, Paddock R, Alexander HL (1961) Historical document, 1922. Bronchial asthma and allied conditions. Clinical and immunological observations. Ann Allergy 19:771–778
62. Araujo BB et al (2008) Extracellular matrix components and regulators in the airway smooth muscle in asthma. Eur Respir J 32(1):61–69
63. James AL, Maxwell PS, Pearce-Pinto G, Elliot JG, Carroll NG (2002) The relationship of reticular basement membrane thickness to airway wall remodeling in asthma. Am J Respir Crit Care Med 166(12):1590–1595
64. Bai TR, Cooper J, Koelmeyer T, Pare PD, Weir TD (2000) The effect of age and duration of disease on airway structure in fatal asthma. Am J Respir Crit Care Med 162(2 Pt 1):663–669
65. Roberts CR, Burke AK (1998) Remodelling of the extracellular matrix in asthma: proteoglycan synthesis and degradation. Can Respir J 5(1):48–50
66. Amin K, Janson C, Boman G, Venge P (2005) The extracellular deposition of mast cell products is increased in hypertrophic airways smooth muscles in allergic asthma but not in nonallergic asthma. Allergy 60(10):1241–1247
67. Amin K et al (2000) Inflammation and structural changes in the airways of patients with atopic and nonatopic asthma. BHR group. Am J Respir Crit Care Med 162(6):2295–2301
68. Altraja A et al (1996) Expression of laminins in the airways in various types of asthmatic patients: a morphometric study. Am J Respir Cell Mol Biol 15(4):482–488
69. Huang J, Olivenstein R, Taha R, Hamid Q, Ludwig M (1999) Enhanced proteoglycan deposition in the airway wall of atopic asthmatics. Am J Respir Crit Care Med 160(2): 725–729
70. Pini L et al (2007) Differences in proteoglycan deposition in the airways of moderate and severe asthmatics. Eur Respir J 29(1):71–77
71. de Medeiros Matsushita M et al (2005) Airway proteoglycans are differentially altered in fatal asthma. J Pathol 207(1):102–110
72. Vignola AM, Kips J, Bousquet J (2000) Tissue remodeling as a feature of persistent asthma. J Allergy Clin Immunol 105(6 Pt 1):1041–1053
73. Laitinen A, Altraja A, Kampe M, Linden M, Virtanen I, Laitinen LA (1997) Tenascin is increased in airway basement membrane of asthmatics and decreased by an inhaled steroid. Am J Respir Crit Care Med 156(3 Pt 1):951–958
74. Seow CY, Schellenberg RR, Pare PD (1998) Structural and functional changes in the airway smooth muscle of asthmatic subjects. Am J Respir Crit Care Med 158(5 Pt 3):S179–S186
75. Bousquet J et al (1992) Asthma: a disease remodeling the airways. Allergy 47(1):3–11
76. Burgess JK (2009) The role of the extracellular matrix and specific growth factors in the regulation of inflammation and remodelling in asthma. Pharmacol Ther 122(1):19–29
77. Westergren-Thorsson G, Chakir J, Lafreniere-Allard MJ, Boulet LP, Tremblay GM (2002) Correlation between airway responsiveness and proteoglycan production by bronchial fibroblasts from normal and asthmatic subjects. Int J Biochem Cell Biol 34(10):1256–1267

78. Bergeron C, Boulet LP (2006) Structural changes in airway diseases: characteristics, mechanisms, consequences, and pharmacologic modulation. Chest 129(4):1068–1087
79. James AL, Wenzel S (2007) Clinical relevance of airway remodelling in airway diseases. Eur Respir J 30(1):134–155
80. Barnes PJ (2004) Mediators of chronic obstructive pulmonary disease. Pharmacol Rev 56(4): 515–548
81. Abraham T, Hogg J (2010) Extracellular matrix remodeling of lung alveolar walls in three dimensional space identified using second harmonic generation and multiphoton excitation fluorescence. J Struct Biol 171(2):189–196
82. Lang MR, Fiaux GW, Gillooly M, Stewart JA, Hulmes DJ, Lamb D (1994) Collagen content of alveolar wall tissue in emphysematous and non-emphysematous lungs. Thorax 49(4): 319–326
83. Vlahovic G, Russell ML, Mercer RR, Crapo JD (1999) Cellular and connective tissue changes in alveolar septal walls in emphysema. Am J Respir Crit Care Med 160(6): 2086–2092
84. Bosken CH, Wiggs BR, Pare PD, Hogg JC (1990) Small airway dimensions in smokers with obstruction to airflow. Am Rev Respir Dis 142(3):563–570
85. Cosio M et al (1978) The relations between structural changes in small airways and pulmonary-function tests. N Engl J Med 298(23):1277–1281
86. Snider GL (1986) Chronic obstructive pulmonary disease—a continuing challenge. Am Rev Respir Dis 133(5):942–944
87. Snider GL (1986) Experimental studies on emphysema and chronic bronchial injury. Eur J Respir Dis Suppl 146:17–35
88. Saetta M, Turato G, Maestrelli P, Mapp CE, Fabbri LM (2001) Cellular and structural bases of chronic obstructive pulmonary disease. Am J Respir Crit Care Med 163(6):1304–1309
89. Maestrelli P, Saetta M, Mapp CE, Fabbri LM (2001) Remodeling in response to infection and injury. Airway inflammation and hypersecretion of mucus in smoking subjects with chronic obstructive pulmonary disease. Am J Respir Crit Care Med 164(10 Pt 2):S76–S80
90. Saetta M et al (1985) Loss of alveolar attachments in smokers. A morphometric correlate of lung function impairment. Am Rev Respir Dis 132(4):894–900
91. Merrilees MJ, Ching PS, Beaumont B, Hinek A, Wight TN, Black PN (2008) Changes in elastin, elastin binding protein and versican in alveoli in chronic obstructive pulmonary disease. Respir Res 9:41
92. van Straaten JF et al (1999) Proteoglycan changes in the extracellular matrix of lung tissue from patients with pulmonary emphysema. Mod Pathol 12(7):697–705
93. Hallgren O et al (2010) Altered fibroblast proteoglycan production in COPD. Respir Res 11: 55
94. Zandvoort A, Postma DS, Jonker MR, Noordhoek JA, Vos JT, Timens W (2008) Smad gene expression in pulmonary fibroblasts: indications for defective ECM repair in COPD. Respir Res 9:83
95. Liesker JJ, Ten Hacken NH, Zeinstra-Smith M, Rutgers SR, Postma DS, Timens W (2009) Reticular basement membrane in asthma and COPD: similar thickness, yet different composition. Int J Chron Obstruct Pulmon Dis 4:127–135
96. Lacoste JY et al (1993) Eosinophilic and neutrophilic inflammation in asthma, chronic bronchitis, and chronic obstructive pulmonary disease. J Allergy Clin Immunol 92(4): 537–548
97. Chanez P et al (1997) Corticosteroid reversibility in COPD is related to features of asthma. Am J Respir Crit Care Med 155(5):1529–1534
98. Soltani A et al (2010) Basement membrane and vascular remodelling in smokers and chronic obstructive pulmonary disease: a cross-sectional study. Respir Res 11:105
99. Kranenburg AR et al (2006) Enhanced bronchial expression of extracellular matrix proteins in chronic obstructive pulmonary disease. Am J Clin Pathol 126(5):725–735

100. Tjin G, Xu P, Kable SH, Kable EP, Burgess JK (2014) Quantification of collagen I in airway tissues using second harmonic generation. J Biomed Opt 19(3):36005
101. Basset F, Ferrans VJ, Soler P, Takemura T, Fukuda Y, Crystal RG (1986) Intraluminal fibrosis in interstitial lung disorders. Am J Pathol 122(3):443–461
102. Molina-Molina M, Vicens V, Estany S (2010) New aspects of idiopathic pulmonary fibrosis. Clin Pulm Med 17(4):170–176
103. Burkhardt A (1989) Alveolitis and collapse in the pathogenesis of pulmonary fibrosis. Am Rev Respir Dis 140(2):513–524
104. Fukuda Y, Ishizaki M, Masuda Y, Kimura G, Kawanami O, Masugi Y (1987) The role of intraalveolar fibrosis in the process of pulmonary structural remodeling in patients with diffuse alveolar damage. Am J Pathol 126(1):171–182
105. Katzenstein AL (1985) Pathogenesis of "fibrosis" in interstitial pneumonia: an electron microscopic study. Hum Pathol 16(10):1015–1024
106. Kuhn C 3rd, Boldt J, King TE Jr, Crouch E, Vartio T, McDonald JA (1989) An immunohistochemical study of architectural remodeling and connective tissue synthesis in pulmonary fibrosis. Am Rev Respir Dis 140(6):1693–1703
107. Ffrench-Constant C, Van de Water L, Dvorak HF, Hynes RO (1989) Reappearance of an embryonic pattern of fibronectin splicing during wound healing in the adult rat. J Cell Biol 109(2):903–914
108. Scharffetter K et al (1989) Localization of collagen alpha 1(I) gene expression during wound healing by in situ hybridization. J Invest Dermatol 93(3):405–412
109. Welch MP, Odland GF, Clark RA (1990) Temporal relationships of F-actin bundle formation, collagen and fibronectin matrix assembly, and fibronectin receptor expression to wound contraction. J Cell Biol 110(1):133–145
110. Kuhn C, McDonald JA (1991) The roles of the myofibroblast in idiopathic pulmonary fibrosis. Ultrastructural and immunohistochemical features of sites of active extracellular matrix synthesis. Am J Pathol 138(5):1257–1265
111. Ward PA, Hunninghake GW (1998) Lung inflammation and fibrosis. Am J Respir Crit Care Med 157(4 Pt 2):S123–S129
112. Hay ED (ed) (1981) Cell biology of extracellular matrix. Plenum, New York
113. Hay ED (ed) (1991) Cell biology of extracellular matrix, 2nd edn. Plenum, New York
114. Mecham R (2011) The extracellular matrix: an overview. Springer, Berlin
115. Baker BM, Chen CS (2012) Deconstructing the third dimension: how 3D culture microenvironments alter cellular cues. J Cell Sci 125(Pt 13):3015–3024
116. Zhang Y et al (2009) Tissue-specific extracellular matrix coatings for the promotion of cell proliferation and maintenance of cell phenotype. Biomaterials 30(23–24):4021–4028
117. Fennema E, Rivron N, Rouwkema J, van Blitterswijk C, de Boer J (2013) Spheroid culture as a tool for creating 3D complex tissues. Trends Biotechnol 31(2):108–115
118. Hess MW, Pfaller K, Ebner HL, Beer B, Hekl D, Seppi T (2010) 3D versus 2D cell culture implications for electron microscopy. Methods Cell Biol 96:649–670
119. Lee KY, Mooney DJ (2001) Hydrogels for tissue engineering. Chem Rev 101(7):1869–1879
120. Rommerswinkel N, Niggemann B, Keil S, Zanker KS, Dittmar T (2014) Analysis of Cell migration within a three-dimensional collagen matrix. J Vis Exp 92:e51963
121. Xiao Z, Conrads TP, Beck GR, Veenstra TD (2008) Analysis of the extracellular matrix and secreted vesicle proteomes by mass spectrometry. Methods Mol Biol 428:231–244
122. Wang X, Han J, Yang J, Pan J, Borchers CH (2015) Matrix coating assisted by an electric field (MCAEF) for enhanced tissue imaging by MALDI-MS. Chem Sci 6:729–738
123. Gillette BM, Jensen JA, Wang M, Tchao J, Sia SK (2010) Dynamic hydrogels: switching of 3D microenvironments using two-component naturally derived extracellular matrices. Adv Mater 22(6):686–691
124. Brown BN, Valentin JE, Stewart-Akers AM, McCabe GP, Badylak SF (2009) Macrophage phenotype and remodeling outcomes in response to biologic scaffolds with and without a cellular component. Biomaterials 30(8):1482–1491

125. Prasertsung I, Kanokpanont S, Bunaprasert T, Thanakit V, Damrongsakkul S (2008) Development of acellular dermis from porcine skin using periodic pressurized technique. J Biomed Mater Res B Appl Biomater 85(1):210–219
126. Reing JE et al (2010) The effects of processing methods upon mechanical and biologic properties of porcine dermal extracellular matrix scaffolds. Biomaterials 31(33):8626–8633
127. Deeken CR et al (2011) Method of preparing a decellularized porcine tendon using tributyl phosphate. J Biomed Mater Res B Appl Biomater 96(2):199–206
128. Brown BN et al (2011) Comparison of three methods for the derivation of a biologic scaffold composed of adipose tissue extracellular matrix. Tissue Eng Part C Methods 17(4):411–421
129. Freytes DO, Stoner RM, Badylak SF (2008) Uniaxial and biaxial properties of terminally sterilized porcine urinary bladder matrix scaffolds. J Biomed Mater Res B Appl Biomater 84(2):408–414
130. Price AP, England KA, Matson AM, Blazar BR, Panoskaltsis-Mortari A (2010) Development of a decellularized lung bioreactor system for bioengineering the lung: the matrix reloaded. Tissue Eng Part A 16(8):2581–2591
131. Bolland F et al (2007) Development and characterisation of a full-thickness acellular porcine bladder matrix for tissue engineering. Biomaterials 28(6):1061–1070
132. Remlinger NT et al (2010) Hydrated xenogeneic decellularized tracheal matrix as a scaffold for tracheal reconstruction. Biomaterials 31(13):3520–3526
133. Meyer SR, Chiu B, Churchill TA, Zhu L, Lakey JR, Ross DB (2006) Comparison of aortic valve allograft decellularization techniques in the rat. J Biomed Mater Res A 79(2):254–262
134. Yang B et al (2010) Development of a porcine bladder acellular matrix with well-preserved extracellular bioactive factors for tissue engineering. Tissue Eng Part C Methods 16(5): 1201–1211
135. Zhou J et al (2010) Impact of heart valve decellularization on 3-D ultrastructure, immunogenicity and thrombogenicity. Biomaterials 31(9):2549–2554
136. Flynn LE (2010) The use of decellularized adipose tissue to provide an inductive microenvironment for the adipogenic differentiation of human adipose-derived stem cells. Biomaterials 31(17):4715–4724
137. Wicha MS, Lowrie G, Kohn E, Bagavandoss P, Mahn T (1982) Extracellular matrix promotes mammary epithelial growth and differentiation in vitro. Proc Natl Acad Sci U S A 79(10): 3213–3217
138. Stern MM et al (2009) The influence of extracellular matrix derived from skeletal muscle tissue on the proliferation and differentiation of myogenic progenitor cells ex vivo. Biomaterials 30(12):2393–2399
139. Yang M, Chen CZ, Wang XN, Zhu YB, Gu YJ (2009) Favorable effects of the detergent and enzyme extraction method for preparing decellularized bovine pericardium scaffold for tissue engineered heart valves. J Biomed Mater Res B Appl Biomater 91(1):354–361
140. Ott HC et al (2008) Perfusion-decellularized matrix: using nature's platform to engineer a bioartificial heart. Nat Med 14(2):213–221
141. Uygun BE et al (2010) Organ reengineering through development of a transplantable recellularized liver graft using decellularized liver matrix. Nat Med 16(7):814–820
142. Nakayama KH, Batchelder CA, Lee CI, Tarantal AF (2010) Decellularized rhesus monkey kidney as a three-dimensional scaffold for renal tissue engineering. Tissue Eng Part A 16(7): 2207–2216
143. Du L, Wu X, Pang K, Yang Y (2011) Histological evaluation and biomechanical characterisation of an acellular porcine cornea scaffold. Br J Ophthalmol 95(3):410–414
144. Rieder E et al (2004) Decellularization protocols of porcine heart valves differ importantly in efficiency of cell removal and susceptibility of the matrix to recellularization with human vascular cells. J Thorac Cardiovasc Surg 127(2):399–405
145. Kasimir MT et al (2003) Comparison of different decellularization procedures of porcine heart valves. Int J Artif Organs 26(5):421–427

146. Funamoto S et al (2010) The use of high-hydrostatic pressure treatment to decellularize blood vessels. Biomaterials 31(13):3590–3595
147. Grauss RW, Hazekamp MG, Oppenhuizen F, van Munsteren CJ, Gittenberger-de Groot AC, DeRuiter MC (2005) Histological evaluation of decellularised porcine aortic valves: matrix changes due to different decellularisation methods. Eur J Cardiothorac Surg 27(4):566–571
148. Sasaki S et al (2009) In vivo evaluation of a novel scaffold for artificial corneas prepared by using ultrahigh hydrostatic pressure to decellularize porcine corneas. Mol Vis 15:2022–2028
149. Sellaro TL et al (2010) Maintenance of human hepatocyte function in vitro by liver-derived extracellular matrix gels. Tissue Eng Part A 16(3):1075–1082
150. Hudson TW, Liu SY, Schmidt CE (2004) Engineering an improved acellular nerve graft via optimized chemical processing. Tissue Eng 10(9–10):1346–1358
151. Wagner DE et al (2014) Three-dimensional scaffolds of acellular human and porcine lungs for high throughput studies of lung disease and regeneration. Biomaterials 35(9):2664–2679
152. Wagner DE et al (2014) Comparative decellularization and recellularization of normal versus emphysematous human lungs. Biomaterials 35(10):3281–3297
153. Petersen TH et al (2010) Tissue-engineered lungs for in vivo implantation. Science 329(5991):538–541
154. Chen RN, Ho HO, Tsai YT, Sheu MT (2004) Process development of an acellular dermal matrix (ADM) for biomedical applications. Biomaterials 25(13):2679–2686
155. Lumpkins SB, Pierre N, McFetridge PS (2008) A mechanical evaluation of three decellularization methods in the design of a xenogeneic scaffold for tissue engineering the temporomandibular joint disc. Acta Biomater 4(4):808–816
156. Ott HC et al (2010) Regeneration and orthotopic transplantation of a bioartificial lung. Nat Med 16(8):927–933
157. Cortiella J et al (2010) Influence of acellular natural lung matrix on murine embryonic stem cell differentiation and tissue formation. Tissue Eng Part A 16(8):2565–2580
158. Cebotari S et al (2010) Detergent decellularization of heart valves for tissue engineering: toxicological effects of residual detergents on human endothelial cells. Artif Organs 34(3): 206–210
159. Gui L, Chan SA, Breuer CK, Niklason LE (2010) Novel utilization of serum in tissue decellularization. Tissue Eng Part C Methods 16(2):173–184
160. Conconi MT et al (2005) Tracheal matrices, obtained by a detergent-enzymatic method, support in vitro the adhesion of chondrocytes and tracheal epithelial cells. Transpl Int 18(6): 727–734
161. Gilpin SE et al (2014) Perfusion decellularization of human and porcine lungs: bringing the matrix to clinical scale. J Heart Lung Transplant 33(3):298–308
162. Song JJ et al (2011) Enhanced in vivo function of bioartificial lungs in rats. Ann Thorac Surg 92(3):998–1005; discussion 1005–1006
163. Montoya CV, McFetridge PS (2009) Preparation of ex vivo-based biomaterials using convective flow decellularization. Tissue Eng Part C Methods 15(2):191–200
164. Hashimoto Y et al (2010) Preparation and characterization of decellularized cornea using high-hydrostatic pressurization for corneal tissue engineering. Biomaterials 31(14): 3941–3948
165. Schenke-Layland K et al (2003) Impact of decellularization of xenogeneic tissue on extracellular matrix integrity for tissue engineering of heart valves. J Struct Biol 143(3):201–208
166. Hopkinson A et al (2008) Optimization of amniotic membrane (AM) denuding for tissue engineering. Tissue Eng Part C Methods 14(4):371–381
167. Lehr EJ et al (2011) Decellularization reduces immunogenicity of sheep pulmonary artery vascular patches. J Thorac Cardiovasc Surg 141(4):1056–1062
168. Phillips M, Maor E, Rubinsky B (2010) Nonthermal irreversible electroporation for tissue decellularization. J Biomech Eng 132(9):091003

169. Sano MB, Neal RE 2nd, Garcia PA, Gerber D, Robertson J, Davalos RV (2010) Towards the creation of decellularized organ constructs using irreversible electroporation and active mechanical perfusion. Biomed Eng Online 9:83
170. Baiguera S et al (2010) Tissue engineered human tracheas for in vivo implantation. Biomaterials 31(34):8931–8938
171. Macchiarini P et al (2008) Clinical transplantation of a tissue-engineered airway. Lancet 372(9655):2023–2030
172. Crapo PM, Gilbert TW, Badylak SF (2011) An overview of tissue and whole organ decellularization processes. Biomaterials 32(12):3233–3243
173. Martinu T, Chen DF, Palmer SM (2009) Acute rejection and humoral sensitization in lung transplant recipients. Proc Am Thorac Soc 6(1):54–65
174. Sharples LD, McNeil K, Stewart S, Wallwork J (2002) Risk factors for bronchiolitis obliterans: a systematic review of recent publications. J Heart Lung Transplant 21(2): 271–281
175. Orens JB, Garrity ER Jr (2009) General overview of lung transplantation and review of organ allocation. Proc Am Thorac Soc 6(1):13–19
176. Ma R et al (2013) Structural integrity, ECM components and immunogenicity of decellularized laryngeal scaffold with preserved cartilage. Biomaterials 34(7):1790–1798
177. Tsuchiya T, Sivarapatna A, Rocco K, Nanashima A, Nagayasu T, Niklason LE (2014) Future prospects for tissue engineered lung transplantation: decellularization and recellularization-based whole lung regeneration. Organogenesis 10(2):196–207
178. Jensen T et al (2012) A rapid lung de-cellularization protocol supports embryonic stem cell differentiation in vitro and following implantation. Tissue Eng Part C Methods 18(8): 632–646
179. Lecht S et al (2014) Enhanced reseeding of decellularized rodent lungs with mouse embryonic stem cells. Biomaterials 35(10):3252–3262
180. Sun H et al (2014) Fibroblast engraftment in the decellularized mouse lung occurs via a beta1-integrin-dependent, FAK-dependent pathway that is mediated by ERK and opposed by AKT. Am J Physiol Lung Cell Mol Physiol 306(6):L463–L475
181. Daly AB et al (2012) Initial binding and recellularization of decellularized mouse lung scaffolds with bone marrow-derived mesenchymal stromal cells. Tissue Eng Part A 18(1–2):1–16
182. Wallis JM et al (2012) Comparative assessment of detergent-based protocols for mouse lung de-cellularization and re-cellularization. Tissue Eng Part C Methods 18(6):420–432
183. O'Neill JD et al (2013) Decellularization of human and porcine lung tissues for pulmonary tissue engineering. Ann Thorac Surg 96(3):1046–1055; discussion 1055–1046
184. Bonvillain RW et al (2012) A nonhuman primate model of lung regeneration: detergent-mediated decellularization and initial in vitro recellularization with mesenchymal stem cells. Tissue Eng Part A 18(23–24):2437–2452
185. Pati F et al (2014) Printing three-dimensional tissue analogues with decellularized extracellular matrix bioink. Nat Commun 5:3935

Chapter 12
Role of Mechanical Stress in Lung Repair and Regeneration

Connie C.W. Hsia and Priya Ravikumar

Introduction

The mammalian lung and its thoracic container, bounded by the diaphragm and the rib cage and partitioned by the mediastinum, arise from distinct embryonic origins but are mechanically linked as a single functional unit of gas transport. The scaffold is essentially a suspended elastic fiber continuum stabilized by surfactant, subjected to constant and cyclic physical stresses arising from two interdependent pumps (respiratory and cardiac) that bring air and blood into close proximity over a vast interface area. From rest to peak exercise, minute ventilation can increase tenfold while blood flow through lung capillaries, the largest microvascular network in the body that receives the entire right ventricular cardiac output, can increase fivefold in elite athletes. Respiratory mechanical stresses are hardwired for transmission via elastic fibers from central airways to distal alveolar septa and the pleura, then via cell junctions and the cytoskeleton from the cell surface to the nucleus where they transduce a host of molecular and biochemical processes culminating in structural growth and remodeling. As physical tension and compression of tissue constituents is essential for maintaining whole lung integrity (termed tensegrity) [1], growth and function of the lung are mechanically constrained by the size, shape, and compliance of its container, the thorax.

In general, branching morphogenesis is the result of mechanical forces acting at an interface between two material phases (e.g., air–tissue or tissue–fluid). Inhomogeneity in either the applied force or the encountered resistance creates pressure gradients that drive iterative bronchoalveolar and vascular bifurcations until all available intrathoracic spaces are filled [2]. The characteristic fractal or "self-similar" branching networks allow efficient inspired gas delivery by convection

C.C.W. Hsia (✉) • P. Ravikumar
Department of Internal Medicine, University of Texas Southwestern Medical Center, Dallas, TX 75390-9034, USA
e-mail: Connie.Hsia@utsouthwestern.edu

I. Bertoncello (ed.), *Stem Cells in the Lung*, Stem Cell Biology and Regenerative Medicine, DOI 10.1007/978-3-319-21082-7_12

to each of some 30,000 human acini, the basic units of gas exchange, and maximize the diffusion surface area of some 480 million alveolar subdivisions [3]. This anatomical plan fundamentally dictates physiology in that ventilation, perfusion, and diffusion also exhibit fractal behavior at different scales [4]. Through the modulation of intrathoracic pressures, generated passively by the opposing elastic recoil of the chest wall and the lung and actively by respiratory muscle contractions, pathophysiological changes in the lung and the thorax directly influence each other [5]. The simple fact of mechanical thoraco-pulmonary coupling leads to complex consequences on the initiation, perpetuation, and re-initiation of lung growth, remodeling, and functional outcome. In this chapter, we briefly review the nature of thoraco-pulmonary mechanical interactions and their contribution to growth, regrowth, and physiological compensation of the postnatal lung.

Mechanical Stress Modulates Lung Development

Normal respiratory development invokes nearly all of the major homeostatic signaling pathways, leading to coordinated enlargement of the rib cage, diaphragm, and lung. In the embryonic lung bud, the positive pressure exerted by tracheal luminal fluid sustains iterative airway branching. Increasing luminal pressure by tracheal ligation accelerates branching and the increase in lung size. Reducing luminal pressure has the opposite effect. Pressure-accelerated branching morphogenesis proceeds in the absence of fibroblast growth factor-10 (FGF-10), although FGF-10 expression accentuates the response [6, 7]. Respiratory movements are essential for lung maturation in later fetal life [8, 9]. Postnatal lung distention by continuous positive airway pressure [10] or perfluorocarbon instillation [11] increases lung volume and protein and DNA contents. New alveoli form predominantly at the lung periphery [12, 13] where mechanical forces are borne by the septa in the absence of major bronchovascular support. Restricted lung expansion in congenital diaphragmatic hernia causes lung hypoplasia [14]; correcting the diaphragmatic defect allows lung expansion, "*catch-up*" growth, and vascular remodeling, eventually reversing pulmonary arterial hypertension [15] and nearly normalizing lung diffusing capacity in long-term survivors [16, 17]. Temporary tracheal occlusion to increase airway luminal pressure has been used in congenital diaphragmatic hernia to enhance antenatal lung growth as a bridge to surgical repair [18]. In severe childhood kyphoscoliosis chest wall deformity restricts both lung and thorax growth [19–22]. Conversely, inhibition of lung growth may deform the growing rib cage [23].

The balance between the inward lung recoil and the outward rib cage recoil mutually determines their dimensions at a given transpulmonary pressure. During maturation, the enlarging rib cage imposes mechanical traction on lung parenchyma (Fig. 12.1). Stretching lung cells or tissue broadly increases permeability and ion flux [24, 25], signal transduction [26, 27], gene transcription [28], and cytokine release [29, 30], culminating in cell proliferation [27, 31], apoptosis [32], protein

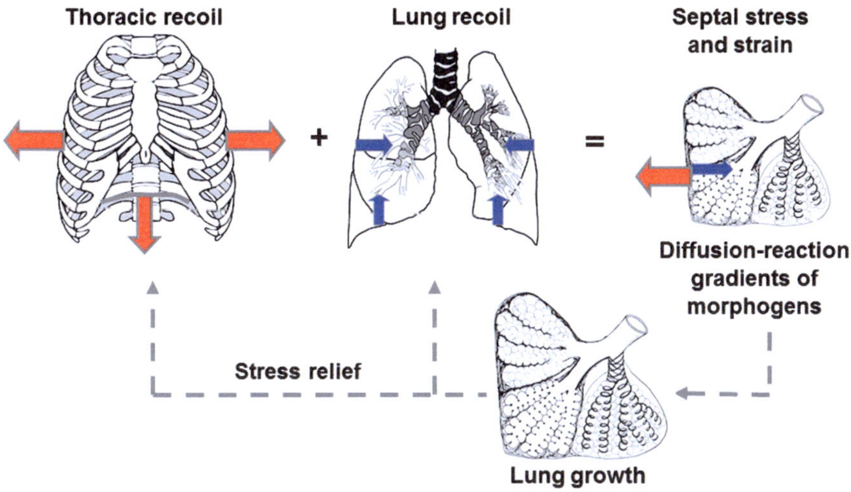

Fig. 12.1 Growth and development of the lung are mechanically coupled to that of the thorax. Adapted from [5]

turnover [24, 33, 34], and other metabolic processes. Tissue growth and remodeling in turn relieve stress and strain. Recoil of the lung exerts reciprocal traction on and modulates the growth of the rib cage. These feedback interactions continue until the closure of bony epiphyses upon reaching somatic maturity. Thereafter, mechanical signals diminish so that both the lung and the thorax stop growing [5]. Subsequent re-initiation of lung growth is possible only if the appropriate thoraco-pulmonary mechanical stress is reimposed and intrathoracic space is available for growth to occur without significant architectural distortion.

Innate Potential for Compensatory Lung Growth

Lower vertebrates such as newts and salamanders are well known for their ability to regrow a completely new and fully functional limb or tail to replace one that has been lost [35]. While mammals cannot regenerate an entire lung, under appropriate stimulation, mature mammalian lungs retain the ability to *add new gas exchange structures* including the intra-acinar airways (respiratory bronchioles, alveolar ducts, and alveolar sacs) and alveolar septa (cells, matrix, fibers, and capillaries). When some lung units are destroyed by disease or surgery, the remaining units expand under the negative intrathoracic pressure leading to unfolding of alveolar surfaces. Simultaneously, pulmonary perfusion to these units increases at a given cardiac output, leading to microvascular recruitment and augmentation of lung diffusing capacity (Fig. 12.2). As loss of lung units continues, physical stress and deformation of tissue and microvasculature exceed a critical threshold, at which point structural growth is stimulated. The newly generated cells, matrix, fibers, and

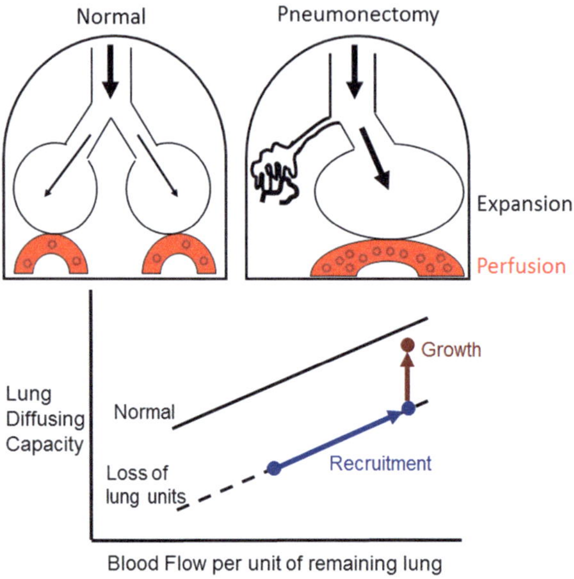

Fig. 12.2 Two types of complex mechanical signals following loss of lung units—expansion of the remaining parenchyma, and increased perfusion to the remaining microvasculature. These signals recruit alveolar-capillary volumes and surface areas, leading to a higher lung diffusing capacity as blood flow per unit of remaining lung increases (*blue line* and *symbols*). When mechanical signals exceed a threshold intensity, structural growth is stimulated with a further increase in lung diffusing capacity at any given pulmonary blood flow above that expected per unit of initially remaining lung

capillaries undergo *remodeling* or architectural adjustment of the alveolar septa, fibroelastic scaffold, and broncho-vasculature, ultimately leading to balanced increases in all major acinar components. Remodeling is a critical step that redistributes mechanical stress, maximizes air–tissue and tissue–blood interface areas, minimizes tissue–blood barrier resistance to diffusion, and optimizes ventilation-to-perfusion and perfusion-to-diffusion matching. The end result is a larger gas exchanger with a higher lung diffusing capacity above that expected from the initially remaining fraction of lung units under a given set of conditions, thus achieving the goal of *functional compensation* (Fig. 12.2). Conversely, generation of structural components that fail to directly or indirectly support the functional goal is not compensatory growth.

Pneumonectomy Model of Compensatory Lung Growth

Owing to the invasive procedures needed to obtain human lung tissue, animal models have been indispensible in the investigation of lung growth. From a mechanistic standpoint, major surgical lung resection (e.g., pneumonectomy), a widely investigated model that consistently demonstrates robust compensatory growth of the remaining lung in all species examined including the mouse, rat, rabbit, ferret, dog [36–45], and indirectly in human subjects with functional loss of one lung [16, 17, 46–49]. Major lung resection mimics the loss of gas exchange units caused by obliterative disease, e.g., pulmonary fibrosis. However, unlike most models of lung injury, pneumonectomy removes a known and highly reproducible fraction of functioning lung units; the resulting signals and responses of the remaining lung units are readily quantifiable. Over more than a century, this model has been extensively characterized at physiological, structural, cellular, and molecular levels and utilized to study the sources and magnitudes of adaptation, the determinants of structural growth and remodeling as well as eventual functional outcome. Cumulative results strongly implicate mechanical stimuli in the re-initiation, modulation, and limitation of lung growth.

Pneumonectomy was first performed in dogs and rabbits in 1881, reviewed by [50]. By the 1920s, it was known that animals function quite well with one expanded remaining lung that fills the entire thoracic cavity [51–53], and surgeons began to perform major lung resection in patients. Studies in the 1950s showed that dogs [54] and patients [55] tolerate staged removal of up to 70 % of lung mass. Subsequently, rodents were used extensively for characterizing the cellular and molecular basis of accelerated post-pneumonectomy lung growth while the canine model remains useful for relating structure to function, defining the sources and limits of adaptation, and evaluating translational interventions.

Mechanical Signal–Response Relationships

In all species, the magnitude of post-pneumonectomy compensatory lung growth correlates inversely with age and maturation stage. Thus, compared to young animals compensatory growth in adults requires a higher threshold for initiation, a longer time course of adaptation [49] with early cell proliferation and progressive scaffold remodeling [56] that only partially normalizes structure–function [57, 58]. Critical pathways that are normally activated during lung development are further upregulated; nondevelopmental pathways may also be recruited [59]. The magnitude and distribution of compensatory growth vary with the fraction of lung units removed. Following canine left pneumonectomy (~42% resection), only the most caudal (infracardiac) remaining lobe exhibits significant growth of alveolar tissue and surface area; compensation is derived predominantly from alveolar-capillary recruitment and parenchyma remodeling. Following right pneumonectomy (58 % resection), alveolar growth intensifies in magnitude and uniformity in all lobes; compensatory gains in structure and function exceed that following 42 % resection [60]. Following 65–70 % lung resection, alveolar growth is still vigorous but with diminishing gains in structure and function [57] (Fig. 12.3). This pattern signifies a *threshold*, an *optimal range*, and an *upper limit* of the mechanical signal–response relationship following increasing loss of lung units.

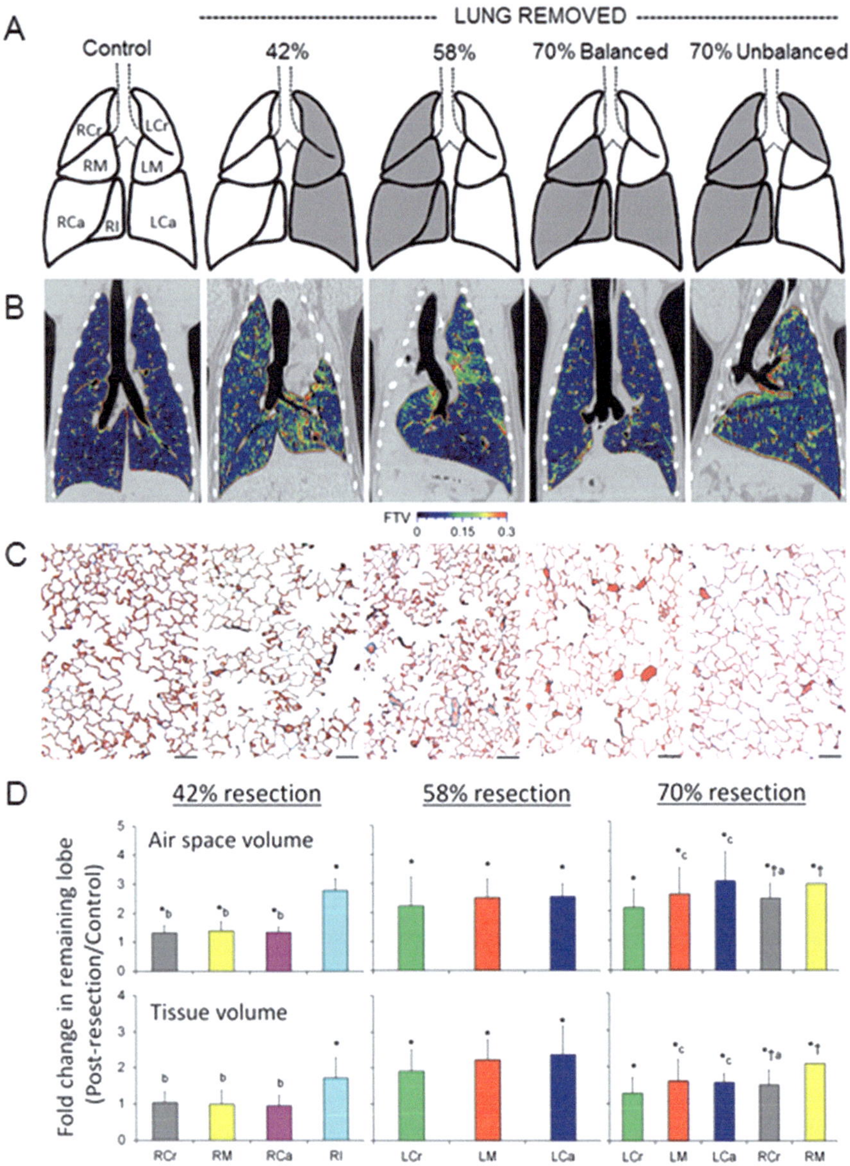

Fig. 12.3 Structural response to increasing loss of lung units by pneumonectomy—42, 58, and 70 % balanced and 70 % unbalanced resection. (**a**) The adult canine lobes removed are shaded in *gray*. (**b**) Coronal HRCT images are shown at the level of the carina. The *color scale* indicates in vivo fractional tissue volume (FTV) of lung parenchyma. (**c**) Representative micrographs of the distal lung. Bar = 200 μm. (**d**) Average fold changes in airspace volume (*upper*) and extravascular alveolar tissue volume (*lower*) in individual remaining lobes following different degrees of lung resection, expressed as ratios with respect to the same lobe in normal control animals. The lobes were fixed by tracheal instillation at a constant airway pressure. *Left*: 42 % resection. *Middle*: 58 % resection. *Right*: 70 % resection (balanced and unbalanced groups combined). *Right lobes*: *RCr* right cranial (*gray*), *RM* right middle (*yellow*), *RCa* right caudal (*magenta*), *RI* right infracardiac

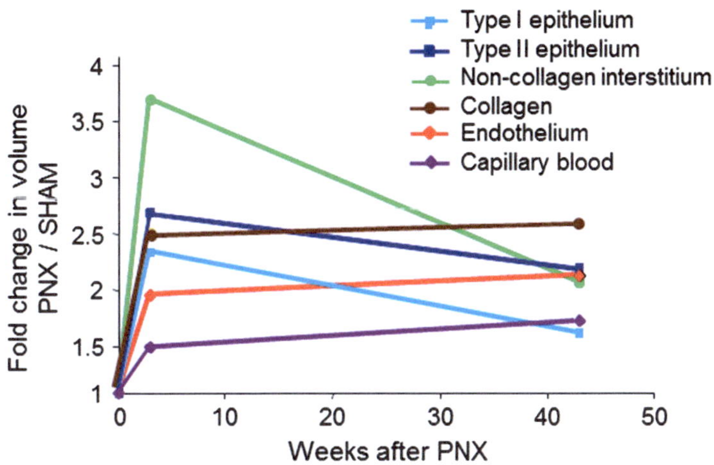

Fig. 12.4 Increases in the compartmental volumes of the alveolar septa in young dogs (2.5 months age) 3 weeks and 1 year following 58 % resection by right pneumonectomy (PNX), expressed as ratios to the corresponding mean values in litter- and gender-matched control animals following sham pneumonectomy. Note the early disproportionate increase in the volume of non-collagen septal interstitium

Following 42 % resection, lobar expansion initially unfolds the remaining alveolar septa with little tissue stress in most lobes except the infracardiac lobe, which lies between the heart and the diaphragm, preferentially expands across the midline, increases nearly twofold increase in tissue volume and partially reconstitutes the cardiac fossa. Type-2 pneumocyte volume increases first, before that of other septal cell types, suggesting its sensitivity to mechanical stress and consistent with its role as a resident progenitor cell [61]. Following 58 % lung resection, expansion and stress in all lobes exceed a critical threshold, leading to uniform growth initiation, stress relief, and greater functional enhancement per unit of initially remaining lung compared to 42 % resection. The initial rapid post-pneumonectomy airspace expansion is followed by further gradual expansion in the subsequent days as intrathoracic air and fluid are resorbed. An increase in tissue permeability recruits fluid, cells (especially monocytes and macrophages), chemokines, and cytokines to the remaining lung leading to a disproportionate (3.6-fold) increase in the volume of septal interstitial cells and matrix (excluding collagen). With time, the gain in interstitial volume wanes, and all major septal components—epithelium, interstitium, endothelium, and capillary blood—increase to about two (1.5–2.5) fold of that in the control lung (sham pneumonectomy) (Fig. 12.4). Signals related to airspace expansion account for approximately half of the observed post-pneumonectomy structure–function compensation (Fig. 12.5);

Fig. 12.3 (continued) (*aqua*). *Left lobes*: *LCr* left cranial (*green*), *LM* left middle (*red*), *LCa* left caudal (*blue*). Mean ± SD. $P \leq 0.05$: *vs. control (1.0); †vs. 42 %; [a]vs. RM, [b]vs. RI, [c]vs. LCr. Adapted from [57]

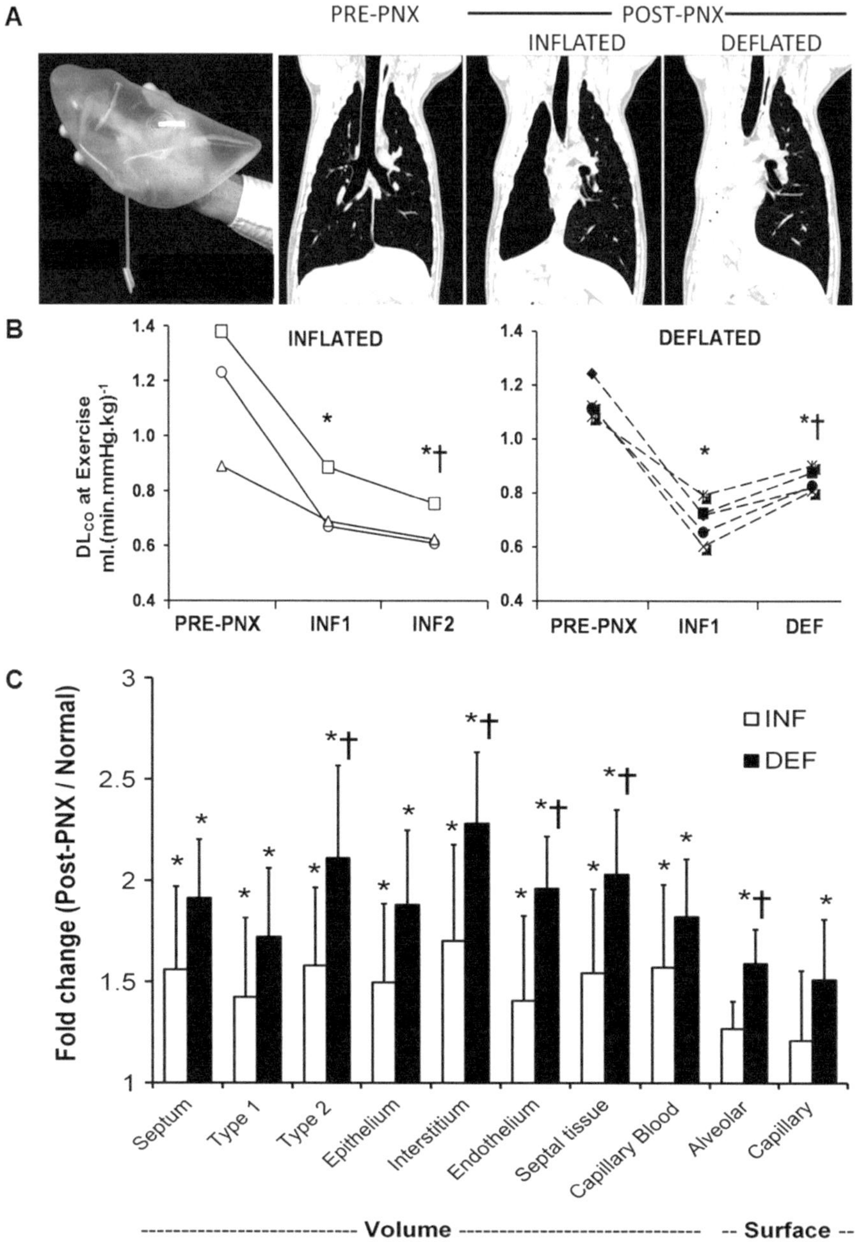

Fig. 12.5 Contribution of lung expansion to post-pneumonectomy (PNX) growth and compensation. *Upper panels*: Adult canine right lung was replaced with custom-shaped inflated silicone prosthesis following right PNX to minimize expansion of the remaining lobes. High-resolution computed tomography was performed at a transpulmonary pressure of 30 cmH_2O pre- and 4 months post-PNX (with inflated prosthesis, INF), and then 4 months following acute deflation of the prosthesis (~8 months after surgery, DEF). Control animals underwent right PNX with continuously inflated prosthesis and also studied at 4 and 8 months post-PNX (INF1 and INF2, respectively). *Middle*: Lung diffusing capacity (DL_{CO}) was measured during exercise and

the remainder is attributable to perfusion-related stimuli although nonmechanical factors may also play a role [62, 63]. As resection increases from 58 to 65–70 %, the diminishing gains in growth and compensation suggest the counter-balancing effects of excessive mechanical stress that may heighten cellular oxidative stress and damage tissue integrity at the expense of growth-related activities. An excessive increase in pulmonary vascular resistance may also impair tissue adaptive response. However, new capillaries (evidenced by an increase in double alveolar-capillary profiles) continue to form with increasing loss of lung units without reaching a plateau up to 70 % resection, suggesting the existence of separate perfusion-related stimuli for microvascular growth [57].

Quantifying Regional Mechanical Stimuli

Using in vivo high-resolution computed tomography, nonrigid image registration and deformation analysis, we visualized regional parenchyma deformation (displacement, strain, and shear) and compliance, and quantified growth of the functional parenchyma, which includes alveolar tissue as well as microvascular blood [56, 64]. Results demonstrate markedly nonuniform regional displacement, deformation, and growth with inflation (Fig. 12.6), with a prolonged course of adaptation marked by initial parenchyma growth, progressive tissue relaxation, stress relief, and gradual functional improvement over many months [56, 58, 64] (Fig. 12.7). A protracted temporal course of adaptation is probably fundamentally important to ensure the coordination and optimization of homeostatic pathways at micro and macro scales and to minimize architectural distortion. Intuitively, one might expect nonuniform spatial stimuli-response to weaken the average or the whole lung compensation at a given stimulus intensity. On the other hand, nonuniformity also prolongs the "*window of susceptibility*" during which mechanical stimuli remain active at least in some parts of the lung, thereby rendering the overall adaptive response amenable to intervention.

Structural Basis of Compensatory Growth and Remodeling

Postnatal compensatory lung growth is thought to be mediated mainly via recruitment and activation of resident progenitor cells in response to local signals while circulating bone marrow-derived cells or vascular progenitor cells play no role or at

Fig. 12.5 (continued) expressed at a constant cardiac output of 400 mL (min kg)$^{-1}$ in individual animals pre-PNX and at the two time-points post-PNX. $P \leq 0.05$: *vs. pre-PNX and †vs. INF1 by repeated measures ANOVA. *Lower*: Fold increase in volumes and surface areas of alveolar septal components measured 1 year following right PNX in animals with inflated or deflated prosthesis, expressed as ratios to the average values in normal controls. Mean ± SD. $P \leq 0.05$: *vs. control (1.0), †vs. INF. From [62, 63]

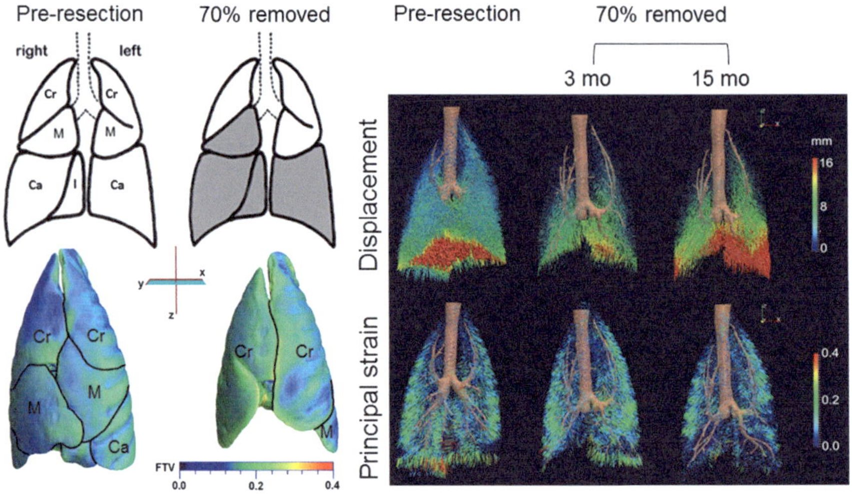

Fig. 12.6 Parenchyma deformation quantified by high-resolution computed tomography. *Left upper panels*: Diagrams of normal adult canine lobes and those remaining following balanced 70 % resection—the removed lobes are shaded in *gray*. *Left lower panels*: Representative three-dimensional reconstruction shows marked expansion of the remaining three lobes (demarcated by *black lines*) mainly in a caudal direction and around the mediastinum. The *color map* shows heterogeneous subpleural fractional tissue volume (FTV). Lobes: *Cr* cranial, *M* middle, *Ca* caudal. *Right panels*: Three-dimensional vector field maps of parenchyma displacement (*upper*) and principal strain (*lower*) during inflation (from 15 to 30 cmH_2O transpulmonary pressure) before and 3 and 15 months after 70 % resection illustrate temporal and spatial mechanical heterogeneity during compensation. At 3 months post-compared to pre-resection, displacement magnitude is reduced and principal strain increases nonuniformly in the enlarged remaining lobes compared to the same lobe pre-resection. At 15 months compared to 3 months post-resection, displacement increases markedly particularly in caudal regions, and regional strain is nonuniformly reduced [88, 89]

best a minor role [65–67]. Alveolar septation is a four-dimensional event involving spatio-temporal coordination of hundreds of genes and thousands of mediators. The bulk of gas exchange occurs through the "*thin side*" of septa, which contains minimal tissue with high barrier conductance. Structural support of septa is concentrated in the "*thick side*," which contains most of the cells, matrix, and fibers. As the elastin fibers coursing throughout the "*thick side*" of alveolar wall are pulled under tension, they may lift and "fold" an existing alveolar capillary, create a tissue pillar that transects the capillary lumen leading to a "double-capillary" profile typical of the developing lung. Tension on elastin fibers may also lift tissue and capillary constituents out of their two-dimensional plane to create a new septum. Subsequently, cell proliferation, matrix deposition, and fiber rearrangement separate the "double-capillary" and remodel it into two single capillaries typical of the mature lung. This process, termed "*intussusception*" [68], could generate a new septum or expand an existing alveolar tissue-capillary sheet depending on the balance of forces (Fig. 12.8).

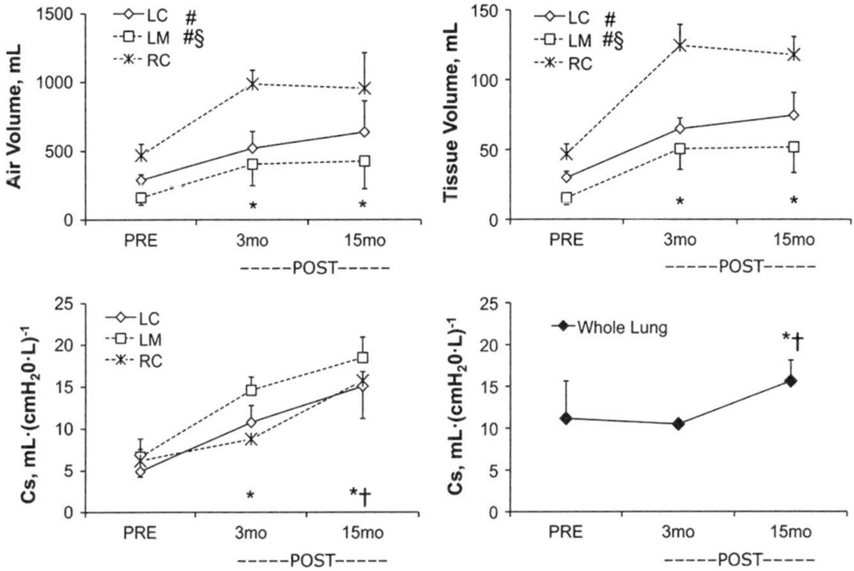

Fig. 12.7 Growth and remodeling phases in the remaining lobes 3 and 15 months following 65–70 % lung resection assessed by high-resolution computed tomography. *Upper panels*: Lobar air and tissue volumes increase significantly from PRE to 3 months POST-resection then remain stable between 3 and 15 months. *Lower left panel*: Specific compliance (Cs) of the remaining three lobes continue to increase from 3 to 15 months POST-resection. *Lower right panel*: Whole lung Cs (representing seven lobes PRE and three lobes POST) did not change from PRE to 3 months POST then increased between 3 and 15 months POST. Mean ± SD. *Lobes*: *RC* right cranial, *LC* left cranial, *LM* left middle. Repeated measures ANOVA with post hoc analysis by Fisher's Protected Least Significant Difference. Comparison with respect to time: $P < 0.05$, *vs. PRE, †vs. 3 months POST in all remaining lobes. Comparison among lobes: $P < 0.05$, #vs. RC, §vs. LC. From [56]

At the level of acinar airways, alveolar ducts contain smooth muscle and other contractile elements in their incomplete walls, which form the entrance rings to alveolar sacs [3]. Following pneumonectomy, alveolar ducts increase in number and volume [69]. Branching of the most distal alveolar ducts could add one more airway generation to double the total alveolar tissue volume and surface areas. This is a likely mechanism because tissue deformation is disproportionally larger in subpleural than central lung regions [56] corresponding to a similar gradient of cell proliferation and growth factor expression [12, 13]. Alternatively, alveolar sacs may bud from the terminal bronchiole that forms the entrance to an acinus, transforming the terminal bronchiole into another generation of respiratory bronchiole. It remains unclear which, or perhaps both, of these mechanisms are operative during compensatory growth.

At the level of the conducting broncho-vasculature, further branching is not possible. A different type of growth occurs post-pneumonectomy, namely traction-related elongation and dilatation, which must also involve the generation of additional tissue components. Elongation along each airway or vascular generation

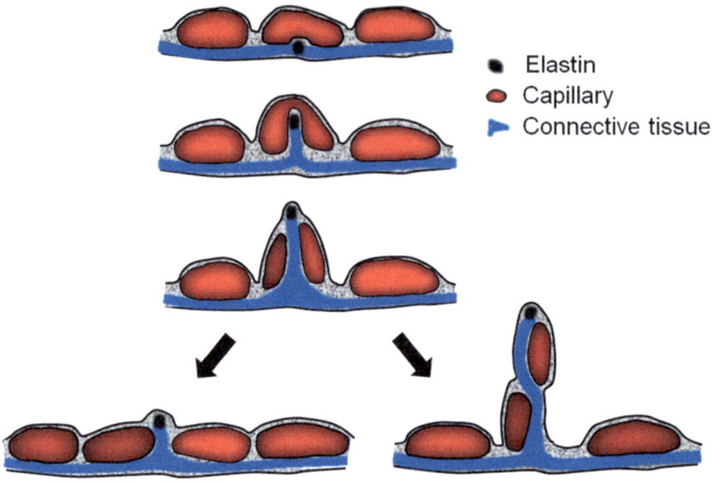

Fig. 12.8 Diagram of intussusceptive capillary growth and formation of a new alveolar septum. See text for explanation

increases, while dilatation mitigates, the increase in flow resistance. As luminal flow resistance is directly proportional to the length and inversely proportional to the fourth power of the radius (Poisuille's Law), only a small increase in radius is needed to offset the effects of lengthening. Bronchovascular adaptation is less vigorous or complete than alveolar adaptation, leading to dissociated compensation or "dysanaptic" lung growth [38, 70], where pulmonary limitation upon exercise is primarily limited by the persistently and disproportionately elevated airway and pulmonary vascular resistances and not by the reduction in lung diffusing capacity [71].

Species Differences in Lung Growth and Compensation

Large mammalian lungs differ from rodent lungs in several aspects of anatomy, development, and maturation that impact adaptation. Bronchovascular stratification is simplified in rodents. Respiratory bronchioles are few and short in the rabbit, guinea pig, hamster, gerbil, rat, and mice [72–74]. In human and canine lungs, acinar airways bifurcate through several generations of respiratory bronchioles and alveolar ducts to end in alveolar sacs. Extensive stratification allows modulation of ventilation–perfusion distributions and the penetration, deposition, and clearance of inhaled particles. Relative to rodent lungs, the highly stratified large lungs need stronger connective tissue and fibers for support, which in turn requires a more rigid rib cage to maintain stability but not too rigid as to restrict truncal flexibility in locomotion. Stratification also creates a longer mean acinar path length, which requires more smooth muscle and contractile elements to fine-tune ventilation–perfusion–diffusion matching among the regions and at different stratified levels. In the vasculature, murine tracheobronchial capacitance vessels may not penetrate into intrapulmonary airways [75, 76], and perturbation of pulmonary blood flow readily stimulates angiogenesis of chest wall and pleural vessels [77]. In large animals, extensive bronchial and pulmonary precapillary anastomoses [78] provide ample collateral circulatory reserves, which may minimize the need for new vessel formation. These structural differences could explain the ease with which the lungs of small animals may be stimulated to regrow while the lungs of large animals rely heavily on nonstructural adaptive mechanisms; structural growth is stimulated only when the nonstructural reserves have been exhausted.

Translational Challenges in Lung Growth Induction

Mechanical signals are the only stimuli known to re-initiate adult lung growth de novo [5, 79]. While the innate growth potential is retained in adult lungs, the incomplete regrowth response is in need of ways of augmentation. This issue is critical because current therapy for chronic lung disease is non-curative except for lung transplantation, which is burdened by problems of donor availability and complications. The key unanswered questions include the following: (1) How to maximally realize the innate ability for regrowth and compensation of functioning lung units in obliterative disease and reduce the need for lung transplantation or replacement? (2) How to re-initiate lung growth in the absence of adequate mechanical stimuli such as in emphysema? (3) How to optimize alveolar cellular repopulation and capillary regrowth in bioengineered lungs?

The pneumonectomy model has proved useful for examining the integrated network interactions elicited by mechanical stimulation and for evaluating the efficacy of therapeutic approaches. Post-pneumonectomy exposure to ambient hypoxia [80, 81] and the administration of exogenous growth promoter molecules, proteins, and DNA [82, 83] have been shown to modestly enhance selective aspects of lung growth in rodents, although none has been shown to improve function. We examined this question in adult canines by administering oral all trans-retinoic acid (RA) for 3 months post-pneumonectomy. In the absence of active endogenous compensatory growth (left pneumonectomy), RA supplementation has no significant effect [83]. In the presence of endogenous growth activities (right pneumonectomy), RA supplementation modestly enhances growth of type-1 pneumocytes, interstitium, endothelium, and capillary blood volume as well as alveolar and capillary surface areas compared to placebo treatment [82, 84]. The expected increase in double-capillary profiles is also accentuated consistent with RA-enhanced capillary formation and remodeling. On the other hand, RA supplementation minimally stimulates the growth of alveolar type-2 pneumocytes, causes thickening of the alveolar-capillary basement membrane and the septa, and fails to significantly enhance long-term whole lung function [82, 84, 85]. These results suggest that pharmacological intervention could augment mechanically induced lung growth but is insufficient for re-initiating growth in the absence of sufficient mechanical signals. What factors prevent the translation of exogenously enhanced structural growth into functional gain? The answer remains speculative although a few possibilities should be considered:

1. Most studies tested a single exogenous growth promoter at a pharmacological dose over a short duration. This approach is conceptually inadequate because endogenous compensatory lung growth involves sustained, low-intensity, balanced amplification of numerous mechano-transduction pathways to gradually enlarge all parts of the gas exchanger, minimize distortion, and optimize outcome. Intense unbalanced stimulation could distort the blood gas barrier and neutralize the intended benefit. Broad stimulation of multiple key pathways at a low dose over a longer duration may be more effective in enhancing overall growth and function.
2. Scaffold remodeling continues long after cessation of accelerated cellular proliferation. As mechanical stress is redistributed and the local mechano-transduction micro-milieu shifts, different molecular pathways may be favored at different time points in the course of response. Selective administration of exogenous growth promoters that sequentially target each specific phase of the natural response may prove more effective.
3. Mechanical stimuli carry inherent risks of oxidative stress and tissue damage. Mildly increased mechanical lung stress (e.g., positive pressure breathing) is beneficial for maintaining airway patency. Moderate mechanical stress transduces growth and remodeling while excessive mechanical stress causes lung injury (e.g., ventilator-induced trauma). Balancing the competing adaptive needs—enlarging the gas exchange units vs. protecting the integrity of existing units—is a crucial part of the mechano-sensitive response spectrum (Fig. 12.9). For a given mechanical stress, the equilibrium interactions between signaling networks of growth/remodeling and those of cytoprotection determine the net gain in structure and function. Consistent with this idea, many of the endogenous growth/remodeling pathways that are upregulated post-pneumonectomy, e.g., hypoxia-inducible factor (HIF), erythropoietin (EPO), vascular endothelial growth factor (VEGF), and hepatocyte growth factor (HGF) [86–92], are multifunctional and simultaneously mediate cytoprotection. Strategies that aim to fortify cytoprotective pathways in addition to growth-promoting pathways may extend the upper limits of mechano-sensitive lung growth and function.

Implications for Chronic Lung Disease and Ex Vivo Lungs

Mechanically induced compensatory lung growth and remodeling enhances long-term gas exchange and mechanical function in adult mammals. Pharmacological manipulation [82, 83] and ambient hypoxia exposure [80, 81] modify already active growth-related pathways. When mechanical stimuli are *below* the growth-initiating threshold, these other attempts have little effect [83]. Thus, ambient hypoxia

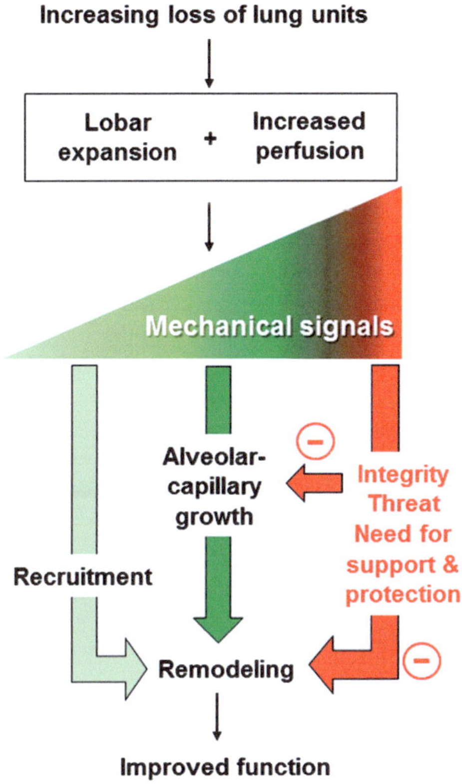

Fig. 12.9 A paradigm for the spectrum of mechanically induced compensatory responses

enhances lung growth and function in developing and not adult animals [93]. Young animals with an enlarging thorax respond more vigorously to pneumonectomy, and compensation is more complete, than older mature animals with fixed thoracic dimensions [94–96]. Exogenous growth factor supplementation enhances lung growth following 58 % and not 42 % resection [82–85]. Where mechanical signals are diminished as in pulmonary emphysema, attempts to pharmacologically stimulate lung growth have been uniformly unsuccessful. Whereas exogenous growth promoter modestly augments some aspects of mechanically induced lung growth, the response may be skewed leading to structural gain without functional enhancement. The fact that independent factors (maturity, hypoxia, parenchyma and vascular sources of mechanical stress) exert additive effects indicates plasticity of the innate regrowth potential and suggests possible approaches for augmentation. On the other hand, the risks of mechanically induced oxidative stress and tissue damage cannot be ignored. In interstitial lung disease, in addition to inflammation the typically patchy contracting fibrotic foci tether and deform the adjacent functioning alveoli, which can be expected to mechanically influence their adaptive response. Do these remaining functioning alveoli undergo compensatory growth/remodeling or suffer secondary physical injury compounding the primary insult? Perhaps both occur to varying degrees in different regions. Strategies that target the protection of mechanically stressed remaining lung units might improve innate compensatory response as well as the response to exogenous interventions. In pulmonary emphysema, the priority will be to find novel ways of reconstituting normal mechano-transduction of the native lung. In ex vivo and bioengineered lungs, it will be imperative to optimize the magnitude and distribution of mechano-transduction response to facilitate cell repopulation and matrix remodeling as well as protect barrier integrity in all parts of the donor lung before and after in vivo implantation. Meeting these challenges require multi-scale integrated understanding of mechano-sensitive networks. There continues to be a need for models that mimic thoraco-pulmonary interactions in the human lung to bridge these knowledge gaps, test high impact interventions to harness the compensatory potential, and avoid translational failure when novel therapeutics are brought to human applications.

Acknowledgements The authors acknowledge the important support by National Institutes of Health grants R01 HL040070 and UO1 HL111146. The contents of this article are solely the responsibility of the authors and do not necessarily represent the official views of the National Heart, Lung, and Blood Institute or of the National Institutes of Health.

References

1. Ingber DE (2003) Tensegrity I. Cell structure and hierarchical systems biology. J Cell Sci 116:1157–1173
2. Weibel ER (2009) What makes a good lung? Swiss Med Wkly 139:375–386
3. Ochs M, Nyengaard JR, Jung A, Knudsen L, Voigt M, Wahlers T, Richter J, Gundersen HJ (2004) The number of alveoli in the human lung. Am J Respir Crit Care Med 169:120–124
4. Robertson HT, Altemeier WA, Glenny RW (2000) Physiological implications of the fractal distribution of ventilation and perfusion in the lung. Ann Biomed Eng 28:1028–1031
5. Hsia CC (2004) Signals and mechanisms of compensatory lung growth. J Appl Physiol 97:1992–1998
6. Unbekandt M, del Moral PM, Sala FG, Bellusci S, Warburton D, Fleury V (2008) Tracheal occlusion increases the rate of epithelial branching of embryonic mouse lung via the FGF10-FGFR2b-Sprouty2 pathway. Mech Dev 125:314–324
7. Papadakis K, Luks FI, De Paepe ME, Piasecki GJ, Wesselhoeft CW Jr (1997) Fetal lung growth after tracheal ligation is not solely a pressure phenomenon. J Pediatr Surg 32:347–351
8. Harding R, Hooper SB, Han VK (1993) Abolition of fetal breathing movements by spinal cord transection leads to reductions in fetal lung liquid volume, lung growth, and IGF-II gene expression. Pediatr Res 34:148–153
9. Tseng BS, Cavin ST, Booth FW, Olson EN, Marin MC, McDonnell TJ, Butler IJ (2000) Pulmonary hypoplasia in the myogenin null mouse embryo. Am J Respir Cell Mol Biol 22:304–315
10. Zhang S, Garbutt V, McBride JT (1996) Strain-induced growth of the immature lung. J Appl Physiol 81:1471–1476
11. Nobuhara KK, Fauza DO, DiFiore JW, Hines MH, Fackler JC, Slavin R, Hirschl R, Wilson JM (1998) Continuous intrapulmonary distension with perfluorocarbon accelerates neonatal (but not adult) lung growth. J Pediatr Surg 33:292–298
12. Massaro GD, Massaro D (1993) Postnatal lung growth: evidence that the gas-exchange region grows fastest at the periphery. Am J Physiol 265:L319–L322

13. Foster DJ, Yan X, Bellotto DJ, Moe OW, Hagler HK, Estrera AS, Hsia CCW (2002) Expression of epidermal growth factor and surfactant proteins during postnatal and compensatory lung growth. Am J Physiol Lung Cell Mol Physiol 283:L981–L990
14. Nagaya M, Akatsuka H, Kato J, Niimi N, Ishiguro Y (1996) Development in lung function of the affected side after repair of congenital diaphragmatic hernia. J Pediatr Surg 31:349–356
15. Beals DA, Schloo BL, Vacanti JP, Reid LM, Wilson JM (1992) Pulmonary growth and remodeling in infants with high-risk congenital diaphragmatic hernia. J Pediatr Surg 27:997–1001; discussion 1001–1002
16. Ijsselstijn H, Tibboel D, Hop WJ, Molenaar JC, de Jongste JC (1997) Long-term pulmonary sequelae in children with congenital diaphragmatic hernia. Am J Respir Crit Care Med 155:174–180
17. Marven SS, Smith CM, Claxton D, Chapman J, Davies HA, Primhak RA, Powell CV (1998) Pulmonary function, exercise performance, and growth in survivors of congenital diaphragmatic hernia. Arch Dis Child 78:137–142
18. McHugh K, Afaq A, Broderick N, Gabra HO, Roebuck DJ, Elliott MJ (2010) Tracheomegaly: a complication of fetal endoscopic tracheal occlusion in the treatment of congenital diaphragmatic hernia. Pediatr Radiol 40:674–680
19. Berend N, Marlin GE (1979) Arrest of alveolar multiplication in kyphoscoliosis. Pathology 11:485–491
20. Davies G, Reid L (1971) Effect of scoliosis on growth of alveoli and pulmonary arteries and on right ventricle. Arch Dis Child 46:623–632
21. Olgiati R, Levine D, Smith JP, Briscoe WA, King TK (1982) Diffusing capacity in idiopathic scoliosis and its interpretation regarding alveolar development. Am Rev Respir Dis 126:229–234
22. Boffa P, Stovin P, Shneerson J (1984) Lung developmental abnormalities in severe scoliosis. Thorax 39:681–682
23. Lund DP, Mitchell J, Kharasch V, Quigley S, Kuehn M, Wilson JM (1994) Congenital diaphragmatic hernia: the hidden morbidity. J Pediatr Surg 29:258–262; discussion 262–254
24. Xu J, Liu M, Liu J, Caniggia I, Post M (1996) Mechanical strain induces constitutive and regulated secretion of glycosaminoglycans and proteoglycans in fetal lung cells. J Cell Sci 109:1605–1613
25. Liu M, Xu J, Tanswell AK, Post M (1994) Inhibition of mechanical strain-induced fetal rat lung cell proliferation by gadolinium, a stretch-activated channel blocker. J Cell Physiol 161:501–507
26. Liu M, Qin Y, Liu J, Tanswell AK, Post M (1996) Mechanical strain induces pp60src activation and translocation to cytoskeleton in fetal rat lung cells. J Biol Chem 271:7066–7071
27. Chess PR, Toia L, Finkelstein JN (2000) Mechanical strain-induced proliferation and signaling in pulmonary epithelial H441 cells. Am J Physiol Lung Cell Mol Physiol 279:L43–L51
28. Sanchez-Esteban J, Tsai SW, Sang J, Qin J, Torday JS, Rubin LP (1998) Effects of mechanical forces on lung-specific gene expression. Am J Med Sci 316:200–204
29. Liu M, Liu J, Buch S, Tanswell AK, Post M (1995) Antisense oligonucleotides for PDGF-B and its receptor inhibit mechanical strain-induced fetal lung cell growth. Am J Physiol 269: L178–L184
30. Waters CM, Chang JY, Glucksberg MR, DePaola N, Grotberg JB (1997) Mechanical forces alter growth factor release by pleural mesothelial cells. Am J Physiol 272:L552–L557
31. Liu M, Skinner SJ, Xu J, Han RN, Tanswell AK, Post M (1992) Stimulation of fetal rat lung cell proliferation in vitro by mechanical stretch. Am J Physiol 263:L376–L383
32. Sanchez-Esteban J, Wang Y, Cicchiello LA, Rubin LP (2002) Cyclic mechanical stretch inhibits cell proliferation and induces apoptosis in fetal rat lung fibroblasts. Am J Physiol Lung Cell Mol Physiol 282:L448–L456
33. Breen EC (2000) Mechanical strain increases type I collagen expression in pulmonary fibroblasts in vitro. J Appl Physiol 88:203–209

34. Smith PG, Moreno R, Ikebe M (1997) Strain increases airway smooth muscle contractile and cytoskeletal proteins in vitro. Am J Physiol 272:L20–L27
35. Simon A, Tanaka EM (2013) Limb regeneration. Wiley interdisciplinary reviews. Dev Biol 2:291–300
36. Cagle PT, Thurlbeck WM (1988) Postpneumonectomy compensatory lung growth. Am Rev Respir Dis 138:1314–1326
37. Hsia CCW, Berberich MA, Driscoll B, Laubach VE, Lillehei CW, Massaro DJ et al (2004) Mechanisms and limits of induced postnatal lung growth. Am J Respir Crit Care Med 170:319–343
38. Hsia CC (2004) Lessons from a canine model of compensatory lung growth. Curr Top Dev Biol 64:17–32
39. Burri PH, Sehovic S (1979) The adaptive response of the rat lung after bilobectomy. Am Rev Respir Dis 119:769–777
40. Langston C, Sachdeva P, Cowan MJ, Haines J, Crystal RG, Thurlbeck WM (1977) Alveolar multiplication in the contralateral lung after unilateral pneumonectomy in the rabbit. Am Rev Respir Dis 115:7–13
41. Sekhon HS, Thurlbeck WM (1992) A comparative study of postpneumonectomy compensatory lung response in growing male and female rats. J Appl Physiol 73:446–451
42. Thurlbeck WM, Langston C (1989) Long-term consequences of compensatory lung growth. J Appl Physiol 66:2891–2894
43. Davies P, McBride J, Murray GF, Wilcox BR, Shallal JA, Reid L (1982) Structural changes in the canine lung and pulmonary arteries after pneumonectomy. J Appl Physiol 53:859–864
44. Kirchner KK, McBride JT (1990) Changes in airway length after unilateral pneumonectomy in weanling ferrets. J Appl Physiol 68:187–192
45. McBride JT (1985) Postpneumonectomy airway growth in the ferret. J Appl Physiol 58:1010–1014
46. Laros CD, Westermann CJ (1987) Dilatation, compensatory growth, or both after pneumonectomy during childhood and adolescence. A thirty-year follow-up study. J Thorac Cardiovasc Surg 93:570–576
47. Ciric I, Meyers MP, Mayba J, Anthonisen NR (2003) Autopneumonectomy with compensatory lung growth. Can Respir J 10:271–273
48. McBride JT, Wohl ME, Strieder DJ, Jackson AC, Morton JR, Zwerdling RG et al (1980) Lung growth and airway function after lobectomy in infancy for congenital lobar emphysema. J Clin Invest 66:962–970
49. Butler JP, Loring SH, Patz S, Tsuda A, Yablonskiy DA, Mentzer SJ (2012) Evidence for adult lung growth in humans. N Engl J Med 367:244–247
50. Schilling JA (1965) Pulmonary resection and sequelae of thoracic surgery. In: Fenn WO, Rahn H (eds) Handbook of physiology section 3: respiration. American Physiological Society, Washington, DC, pp 1531–1563
51. Heuer GJ, Dunn GR (1920) Experimental pneumonectomy. Bull Johns Hopkins Hosp 31:31–42
52. Heuer GJ, Andrus WDW (1922) Alveolar and blood gas changes following pneumonectomy. Bull Johns Hopkins Hosp 33:130–134
53. Andrus WDW (1923) Observations on the total lung volume and blood flow following pneumonectomy. Bull Johns Hopkins Hosp 34:119–121
54. Schilling JA, Harvey RB, Balke B, Rattunde HF (1956) Extensive pulmonary resection in dogs: altitude tolerance, work capacity, and pathologic-physiologic changes. Ann Surg 144:635–646
55. DeGraff AC Jr, Taylor HF, Ord JW, Chuang TH, Johnson RL Jr (1965) Exercise limitation following extensive pulmonary resection. J Clin Invest 44:1514–1522
56. Yilmaz C, Tustison NJ, Dane DM, Ravikumar P, Takahashi M, Gee JC, Hsia CC (2011) Progressive adaptation in regional parenchyma mechanics following extensive lung resection assessed by functional computed tomography. J Appl Physiol 111:1150–1158

57. Ravikumar P, Yilmaz C, Dane DM, Bellotto DJ, Estrera AS, Hsia CC (2014) Defining a stimuli-response relationship in compensatory lung growth following major resection. J Appl Physiol 116(7):816–824
58. Hsia CC, Herazo LF, Ramanathan M, Johnson RL Jr, Wagner PD (1993) Cardiopulmonary adaptations to pneumonectomy in dogs. II. VA/Q relationships and microvascular recruitment. J Appl Physiol 74:1299–1309
59. Zhang Q, Zhang J, Moe OW, Hsia CC (2008) Synergistic upregulation of erythropoietin receptor (EPO-R) expression by sense and antisense EPO-R transcripts in the canine lung. Proc Natl Acad Sci U S A 105:7612–7617
60. Hsia CCW, Herazo LF, Ramanathan M, Johnson RL Jr (1994) Cardiopulmonary adaptations to pneumonectomy in dogs. IV. Membrane diffusing capacity and capillary blood volume. J Appl Physiol 77:998–1005
61. Hsia CC, Johnson RL Jr (2006) Further examination of alveolar septal adaptation to left pneumonectomy in the adult lung. Respir Physiol Neurobiol 151:167–177
62. Dane DM, Yilmaz C, Estrera AS, Hsia CC (2013) Separating in vivo mechanical stimuli for postpneumonectomy compensation: physiological assessment. J Appl Physiol 114:99–106
63. Ravikumar P, Yilmaz C, Bellotto DJ, Dane DM, Estrera AS, Hsia CC (2013) Separating in vivo mechanical stimuli for postpneumonectomy compensation: imaging and ultrastructural assessment. J Appl Physiol 114:961–970
64. Yilmaz C, Ravikumar P, Dane DM, Bellotto DJ, Johnson RL Jr, Hsia CC (2009) Noninvasive quantification of heterogeneous lung growth following extensive lung resection by high-resolution computed tomography. J Appl Physiol 107:1569–1578
65. Voswinckel R, Ziegelhoeffer T, Heil M, Kostin S, Breier G, Mehling T et al (2003) Circulating vascular progenitor cells do not contribute to compensatory lung growth. Circ Res 93:372–379
66. Suga A, Ueda K, Takemoto Y, Nishimoto A, Hosoyama T, Li TS, Hamano K (2013) Significant role of bone marrow-derived cells in compensatory regenerative lung growth. J Surg Res 183:84–90
67. Eisenhauer P, Earle B, Loi R, Sueblinvong V, Goodwin M, Allen GB, Lundblad L, Mazan MR, Hoffman AM, Weiss DJ (2013) Endogenous distal airway progenitor cells, lung mechanics, and disproportionate lobar growth following long-term post- pneumonectomy in mice. Stem Cells 31:1330–1339
68. Burri PH (2006) Structural aspects of postnatal lung development—alveolar formation and growth. Biol Neonate 89:313–322
69. Hsia CCW, Zhou XS, Bellotto DJ, Hagler HK (2000) Regenerative growth of respiratory bronchioles in dogs. Am J Physiol Lung Cell Mol Physiol 279:L136–L142
70. Dane DM, Johnson RL Jr, Hsia CCW (2002) Dysanaptic growth of conducting airways after pneumonectomy assessed by CT scan. J Appl Physiol 93:1235–1242
71. Hsia CC, Dane DM, Estrera AS, Wagner HE, Wagner PD, Johnson RL Jr (2008) Shifting sources of functional limitation following extensive (70%) lung resection. J Appl Physiol 104:1069–1079
72. Juhos LT, Green DP, Furiosi NJ, Freeman G (1980) A quantitative study of stenosis in the respiratory bronchiole of the rat in NO2-induced emphysema. Am Rev Respir Dis 121:541–549
73. Bal HS, Ghoshal NG (1988) Morphology of the terminal bronchiolar region of common laboratory mammals. Lab Anim 22:76–82
74. Irvin CG, Bates JH (2003) Measuring the lung function in the mouse: the challenge of size. Respir Res 4:4
75. Widdicombe J (1996) The tracheobronchial vasculature: a possible role in asthma. Microcirculation 3:129–141
76. Mitzner W, Lee W, Georgakopoulos D, Wagner E (2000) Angiogenesis in the mouse lung. Am J Pathol 157:93–101
77. Weibel ER (1960) Early stages in the development of collateral circulation to the lung in the rat. Circ Res 8:353–376

78. Ochs M, Weibel ER (2008) Functional design of the human lung for gas exchange. In: Fishman AP, Elias JA, Fishman JA, Grippi MA, Senior RM, Pack AI (eds) Fishman's pulmonary diseases and disorders, 4th edn. McGraw Hill, New York, pp 23–69
79. Wirtz HR, Dobbs LG (2000) The effects of mechanical forces on lung functions. Respir Physiol 119:1–17
80. Hsia CCW, Polo Carbayo JJ, Yan X, Bellotto DJ (2005) Enhanced alveolar growth and remodeling in guinea pigs raised at high altitude. Respir Physiol Neurobiol 147:105–115
81. Burri PH, Weibel ER (1971) Morphometric estimation of pulmonary diffusion capacity. II. Effect of PO2 on the growing lung, adaption of the growing rat lung to hypoxia and hyperoxia. Respir Physiol 11:247–264
82. Ravikumar P, Dane DM, McDonough P, Yilmaz C, Estrera AS, Hsia CC (2011) Long-term post-pneumonectomy pulmonary adaptation following all-trans-retinoic acid supplementation. J Appl Physiol 110:764–773
83. Yan X, Bellotto DJ, Dane DM, Elmore RG, Johnson RL Jr, Estrera AS, Hsia CC (2005) Lack of response to all-trans retinoic acid supplementation in adult dogs following left pneumonectomy. J Appl Physiol 99:1681–1688
84. Yan X, Bellotto DJ, Foster DJ, Johnson RL Jr, Hagler HH, Estrera AS, Hsia CC (2004) Retinoic acid induces nonuniform alveolar septal growth after right pneumonectomy. J Appl Physiol 96:1080–1089
85. Dane DM, Yan X, Tamhane RM, Johnson RL Jr, Estrera AS, Hogg DC, Hogg RT, Hsia CCW (2004) Retinoic acid-induced alveolar cellular growth does not improve function after right pneumonectomy. J Appl Physiol 96:1090–1096
86. Toffoli S, Roegiers A, Feron O, Van Steenbrugge M, Ninane N, Raes M, Michiels C (2009) Intermittent hypoxia is an angiogenic inducer for endothelial cells: role of HIF-1. Angiogenesis 12:47–67
87. Zhang Q, Bellotto DJ, Ravikumar P, Moe OW, Hogg RT, Hogg DC et al (2007) Postpneumonectomy lung expansion elicits hypoxia-inducible factor-1alpha signaling. Am J Physiol Lung Cell Mol Physiol 293:L497–L504
88. Zhang Q, Moe OW, Garcia JA, Hsia CC (2006) Regulated expression of hypoxia-inducible factors during postnatal and postpneumonectomy lung growth. Am J Physiol Lung Cell Mol Physiol 290:L880–L889
89. Foster DJ, Moe OW, Hsia CC (2004) Upregulation of erythropoietin receptor during postnatal and postpneumonectomy lung growth. Am J Physiol Lung Cell Mol Physiol 287:L1107–L1115
90. Sakamaki Y, Matsumoto K, Mizuno S, Miyoshi S, Matsuda H, Nakamura T (2002) Hepatocyte growth factor stimulates proliferation of respiratory epithelial cells during postpneumonectomy compensatory lung growth in mice. Am J Respir Cell Mol Biol 26:525–533
91. Nakamura T, Sakai K, Nakamura T, Matsumoto K (2011) Hepatocyte growth factor twenty years on: much more than a growth factor. J Gastroenterol Hepatol 26:188–202
92. Sakurai MK, Lee S, Arsenault DA, Nose V, Wilson JM, Heymach JV, Puder M (2007) Vascular endothelial growth factor accelerates compensatory lung growth after unilateral pneumonectomy. Am J Physiol Lung Cell Mol Physiol 292:L742–L747
93. Johnson RL Jr, Cassidy SS, Grover RF, Schutte JE, Epstein RH (1985) Functional capacities of lungs and thorax in beagles after prolonged residence at 3,100 m. J Appl Physiol 59:1773–1782
94. Takeda S, Hsia CCW, Wagner E, Ramanathan M, Estrera AS, Weibel ER (1999) Compensatory alveolar growth normalizes gas exchange function in immature dogs after pneumonectomy. J Appl Physiol 86:1301–1310
95. Takeda S, Ramanathan M, Wu EY, Estrera AS, Hsia CCW (1996) Temporal course of gas exchange and mechanical compensation after right pneumonectomy in immature dogs [published corrigenda in J. Appl. Physiol. 80(6): after table of contents, 1996.]. J Appl Physiol 80:1304–1312
96. Takeda S, Wu EY, Epstein RH, Estrera AS, Hsia CCW (1997) In vivo assessment of changes in air and tissue volumes after pneumonectomy. J Appl Physiol 82:1340–1348

Part IV
Cellular Therapies for Lung Disease

Chapter 13
Cellular Origins of Fibrotic Lung Diseases

Vibha N. Lama

Introduction to Fibrotic Lung Diseases

The term fibrotic lung disease can be utilized for diseases marked by mesenchymal cell infiltration and collagen deposition of the functional units of the lung. There is a large spectrum of fibrotic lung diseases with varied anatomic involvement and clinical presentations. Massive infiltration of the alveolar airspaces by mesenchymal cells marks diseases presenting more acutely such as cryptogenic organizing pneumonia [1]. Fibrosis is also the dreaded complication in patients with Acute Respiratory Distress Syndrome (ARDS) and contributes significantly to irreversibility of respiratory failure [2, 3]. Chronic fibrotic lung diseases such as idiopathic pulmonary fibrosis (IPF) have a more insidious presentation and contribute significantly to the disease burden of chronic lung diseases [4]. IPF and fibrotic lung diseases associated with connective tissue diseases are characterized by interstitial fibrosis with loss of alveolar units [5]. In contrast, other fibrotic lung diseases such as bronchiolitis obliterans (BO) predominantly target small airways and lead to chronic respiratory failure marked by airflow obstruction [6]. BO is a common presentation after lung or bone marrow transplantation, conditions characterized by immune mismatch. Furthermore, autoimmune diseases such as rheumatoid arthritis and environmental exposures can also be associated with BO. However, although not characterized as fibrotic lung diseases, it is important to note that fibrosis as a pathological process marks almost all chronic lung diseases including chronic obstructive pulmonary disease (COPD) [7].

V.N. Lama (✉)
Division of Pulmonary and Critical Care Medicine, Department of Internal Medicine, University of Michigan Medical School, Ann Arbor, MI 48109, USA
e-mail: vlama@umich.edu

I. Bertoncello (ed.), *Stem Cells in the Lung*, Stem Cell Biology and Regenerative Medicine, DOI 10.1007/978-3-319-21082-7_13

Fibrosis and Epithelial Progenitors

Epithelial cell injury is a common histologic feature across various fibrotic diseases [8–10]. In IPF, the delicate air exchange units of alveoli are replaced by honeycomb cysts which are lined by cuboidal epithelium [11]. Dysmorphic epithelium overlies mesenchymal cells in fibrotic foci, the pathognomonic lesions of IPF. Similarly, airway fibrosis is marked by loss of normal epithelium with decreased numbers of club cell specific factor (CCSP) positive cells [12]. Fibrosis can hence be characterized as failure of epithelial repair focusing attention on epithelial progenitors in the lung [13]. Animal studies demonstrate that isolated loss of epithelial progenitors can in fact induce fibrosis. It has been shown that isolated targeted injury to type II alveolar epithelial cells (AEC-II), the putative stem cell that regenerates type 1 AECs (AEC-I) [14], results in pulmonary fibrosis [15]; and that exogenous transplantation of AEC-II has protective effects against fibrotic insult induced by bleomycin [16].

In the airways, loss of basal cells precedes bronchiolitis obliterans-like pathological changes in a murine model of chlorine gas inhalation [17], and conditional deletion of the progenitor club cells induces peribronchiolar fibrosis characteristic of many airway fibrotic lung diseases [18]. That loss of endogenous stem cells can skew a lung towards fibrosis after an insult has also been recently suggested by studies of fulminant viral injury where selective ablation of distal airway epithelial cells, characterized as expressing Trp63 (p63) and keratin 5, prevented normal regeneration and promoted fibrotic responses [19]. These murine studies solidify the hypothesis that epithelial progenitors have a significant role in fibrotic diseases, and their inability to maintain epithelial integrity is an important pathogenic step in ensuing fibroproliferation.

Fibrosis and Mesenchymal Progenitors

The mesenchymal cell is the primary effector cell of fibrosis and hence its origin is of key significance in developing anti-fibrotic strategies targeting these cells [20]. In a fibrotic lung, mesenchymal cells demonstrate evidence of proliferation and differentiation into myofibroblasts, identified by α-smooth muscle actin (SMA) expression and increased matrix secretory function [21]. In this section, we review the various mesenchymal progenitors that can potentially participate in fibroproliferative responses and the present knowledge on their contribution to fibrotic diseases of human lungs.

Lung-resident mesenchymal progenitors: The term mesenchymal stem cell (MSC) is used to denote mesenchymal progenitor cells which are identified in vitro after plastic adherence and demonstrate an ability to differentiate into multiple mesenchymal lineages [22]. Multipotent human MSCs were first characterized from the bone marrow (BM) [23], and it was believed that BM is a source of these precursor cells for other non-hematopoietic tissues and organs such as the lung [24, 25]. However, human studies have conclusively demonstrated that MSCs in solid organs are resident cells and not derived from the bone marrow. These studies have utilized the ability to differentiate local versus hematopoietic origin of the cells on the basis of donor versus recipient status in transplanted organs. Single-cell cytogenetic studies of MSCs isolated from human lung grafts in gender mismatched transplantations demonstrated donor origin of lung-resident MSC at even greater than 10 years post-transplantation [26]. Similar findings were replicated in the studies of transplanted human heart and kidneys with MSCs derived from these organs demonstrating donor origin [27, 28].

Lung-resident MSCs demonstrate a gene expression profile which suggests that they are remnants of embryonic lung mesenchyme. Expression of Forkhead box gene FOXF1 in the splanchnic mesoderm during organogenesis [29] is essential for lung development [30, 31], and lung-derived MSCs demonstrate 30,000 higher fold expression of FOXF1 than bone marrow-resident MSCs [32]. Similarly, the HOX gene expression in lung-derived MSCs [32] mirrors that noted in embryonic mesenchyme of the developing lung [33, 34]. Regionally restricted spatial expression of HOX genes has been shown to be critical in region-specific patterning [35]. While this work establishes the presence of a lung-specific endogenous mesenchymal cell population in the human lung, it is important to recognize that there are likely going to be further subsets of mesenchymal progenitor cells within the adult lung with specialized structural functions. In fact, an ABCG2 positive MSC population associated with the alveolar capillary network has been described in human lungs which likely represent a pericyte precursor [36]. A recent investigation of murine lung mesenchyme during embryogenesis using advanced cellular labeling techniques to label mesenchymal progenitors at a single-cell resolution supports this notion and demonstrates that each differentiated mesenchymal cell type has a distinct mode of progenitor recruitment and regulated boundaries [37]. Such investigation in adult lungs and in context of injury will likely shed more light on lungs endogenous MSC populations.

The mesenchymal epithelial unit—implications for fibrosis: Fibrosis develops in context of tissue response to injury, and resident mesenchymal progenitor cells are likely to be important cellular players in this failed/overactive reparative process. However, investigation of this role requires knowledge of functions of mesenchymal progenitors within the lung microenvironment during normal homeostasis. In the bone marrow, MSCs are an important component of the hematopoietic stem cell niche and regulate hematopoiesis by their interaction with other resident cells via secretory molecules as well as by direct contact [38]. The requisite role of embryonic mesenchyme in defining epithelial cell fate during lung morphogenesis [39, 40], and similar unique lung-specific gene expression in adult lung-derived MSCs [26, 32], also suggests a tissue-specific function. In situ hybridization studies of human lung tissue demonstrate that FOXF1 expressing mesenchymal cells lie in close contact with AEC-II in the alveolar corners and communicate with AECs via gap junctions [41] suggesting that these mesenchymal progenitors potentially play a significant role in regulating the AEC-II cellular niche. The ABCG2 positive MSC population lies in contact with microvascular epithelial cells and AEC-I and also demonstrates communication with these cell types via gap junctions [42]. Mesenchymal epithelial interaction is also important in the airways, where murine investigations reiterate the role of parabronchial smooth muscle cells in supporting bronchial epithelial stem cells [43]. These mesenchymal progenitors are in turn also regulated by their niche as suggested by recent work demonstrating that changes in control signals can disrupt the differentiation and patterning of mesenchymal progenitor cells in a developing murine lung [37]. Hence, investigating the dynamics of crosstalk between the lung's endogenous resident mesenchymal and epithelial progenitors is crucial in deciphering the evolution of fibrosis.

Endogenous lung-resident mesenchymal progenitors in lung fibrosis: Mobilization/recruitment and differentiation of mesenchymal cells are well recognized events in fibrosis. Recent investigations in human lung transplant recipients have demonstrated that lung-resident MSCs are mobilized in lung during injury and participate in fibrogenesis in the allograft [44]. MSC numbers, quantitated as the number of mesenchymal colony-forming units (CFUs) in bronchoalveolar lavage, are high early post-lung transplant, a time of intense ischemia–reperfusion and immunological insults. The number of MSCs then decreases and remains low in quiescent lungs free of chronic rejection. An increase in the number of MSCs later precedes development of chronic rejection or BO. Bioactive lipid lysophosphatidic acid (LPA) plays a principal role in the migration of human lung-resident MSCs through a signaling pathway involving LPA1-induced β-catenin activation [45]. Lung-resident MSCs can be induced into myofibroblast differentiation by pro-fibrotic cytokines, and in situ hybridization demonstrates FOXF1 expression in myofibroblasts in human fibrotic lesions suggesting that they are a lung-resident mesenchymal cell population [32]. FOXF1 is mesenchyme specific and is lacking in lung epithelial cells [29]. Furthermore, it has been shown that FOXF1 expression is not induced by in vitro epithelial mesenchymal transformation [32], underscoring the utility of FOXF1 in delineating local mesenchymal origin of cells.

Another methodology which can be utilized in humans to discriminate between pulmonary versus extrapulmonary derivation of cells is the investigation of donor versus recipient status in tissues in gender mismatched lung transplants. Such studies were employed initially to study chimerism in the epithelium, and although not the focus of that study, fibrotic tissue was suggested to be of donor origin [46]. Two subsequent studies have since addressed this in fibrotic BO lesions in transplanted lungs. While the initial study focused on delineating recipient contribution and suggested the presence of some cells with recipient karyotype in fibrotic lesions [47], a recent report utilizing whole explanted lungs, and stringent criteria to exclude inflammatory cells, has demonstrated that myofibroblasts in fibroproliferative lesions of BO are largely of donor origin [48]. Thus, these recent human investigations demonstrating donor origin of MSCs isolated from transplanted human lungs [26] as well as donor origin of myofibroblasts in situ [48] provide evidence for a crucial role for resident mesenchymal progenitors during lung fibrogenesis.

Animal models have also been utilized to address this important question of the origin of myofibroblasts and effector mesenchymal cells in fibrotic lungs. However, the majority of these studies have addressed the possibility of epithelial mesenchymal transformation, or the potential contribution of bone marrow-derived progenitors to generate mesenchymal cells in the fibrotic lung [49–54]. The lack of studies directly investigating the lung-resident mesenchymal cells can be attributed to lack of specific markers for these cells. The bleomycin lung injury model is most commonly utilized in these studies. This model has several shortcomings including a rather mixed inflammatory fibrotic response, and the reversibility of fibrosis over time [55]. Unlike human IPF where fibrotic foci with a distinct collection of myofibroblasts are present, fibrotic areas in bleomycin treated lungs, although recognizable by trichrome collagen staining, contain scattered myofibroblasts with inflammatory cells, thus making analysis of individual cells harder. In complex fibrotic tissue, staining for myofibroblasts markers such as α-SMA can give the appearance of co-localization in apposing cells. However, several recent investigations utilizing confocal imaging have now clarified the role of local mesenchymal cell populations in fibrotic responses by demonstrating no evidence of epithelial–mesenchymal transformation [56]. In the lung, lineage tracing of surfactant protein C positive AEC-II has shown no significant contribution of these cells to myofibroblasts in a bleomycin lung [57]. Furthermore, lineage tracing of ABCG2 positive mesenchymal progenitor cells in a murine bleomycin fibrosis model has revealed their contribution to vascular remodeling by differentiating into NG2 positive, α-SMA expressing cells [58].

FOXD1 is another marker which has been used to analyze the origin of myofibroblasts in fibrosis. FOXD1 is essential for renal morphogenesis, and FOXD1 lineage tracing has previously been utilized to demonstrate the local mesenchymal origin of myofibroblasts in murine models of renal fibrosis [56]. FOXD1 positive cells thought to represent pericytes in the lung were also recently investigated in a murine bleomycin lung injury model. FOXD1 progenitor-derived pericytes were shown to expand after bleomycin lung injury, and activate expression of collagen-I(α)1 and αSMA in fibrotic foci [59].

In summary, recent human and animal investigations have demonstrated an essential and primary role of lung-resident mesenchymal cell populations in lung fibrosis, underscoring the need to delineate subsets of mesenchymal progenitors and the modulation of their differentiation.

Extrapulmonary Mesenchymal Progenitors and Lung Fibrosis

It has been suggested that bone marrow can be a source of progenitor cells for solid organs [60]. The extrapulmonary origin of fibroblasts in a fibrotic lung has also been an area of active investigations [50, 61]. An important question is whether bone marrow-derived populations, fibrocytes or MSCs, differentiate to myofibroblasts and contribute to collagen deposition during lung fibrosis.

Fibrocytes and pulmonary fibrosis: The term fibrocyte was coined to describe a circulating leukocyte subpopulation which was found to express some fibroblast markers, especially type I collagen. It was proposed that fibrocytes, recruited to sites of tissue injury, contribute to pathologic fibrotic responses [62]. Fibrocytes have been investigated in animal models of pulmonary fibrosis [52, 53] and have also been detected in the blood of patients with pulmonary fibrosis [63]. However, no significant functional contribution of hematopoietic cell-derived type I collagen to lung fibrogenesis has been demonstrated in recent investigations utilizing a double-transgenic mouse model in which the type I collagen gene was specifically and permanently deleted in hematopoietic cells by crossing mice carrying the Vav-Cre transgene with col1a1$^{fl/fl}$ mice [64]. These investigations suggest that the intracellular collagen detected in fibrocytes is predominantly due to the uptake of collagen from neighboring collagen-secreting cells [64]. Thus, while fibrocytes likely play an important role in fibrosis by their interactions with resident mesenchymal cells, their potential role in collagen turnover [65], and their paracrine actions [66], they are not the precursor cell to myofibroblasts in the lung or a significant source of collagen during fibrosis.

Bone marrow MSCs and pulmonary fibrosis: Bone marrow MSCs (BM-MSCs) exhibit myofibroblast differentiation potential and hence can potentially contribute to fibrosis. BM-MSCs also demonstrate immunoregulatory and anti-inflammatory functions in vitro and in vivo after exogenous administration and have been shown to ameliorate fibrosis in murine models [67, 68]. However, there is an important distinction between endogenous and exogenous BM-MSCs in the context of a fibrotic lung. At present, there is no substantial evidence to support the hypothesis that endogenous MSCs are recruited from the bone marrow to the lung during injury or fibrosis, and that they contribute to pathogenesis of fibrotic lung diseases. As mesenchymal progenitors in the organs demonstrate tissue specificity, the role of BM-MSC in the human body during hemostasis and repair is likely local and involves the regulation of hematopoietic functions [69]. However, several characteristics of BM-MSCs make them suitable candidates for exogenous cellular therapy. These include their ability to be easily cultured and expanded; their immunoprivileged status enabling them to be utilized in HLA mismatched recipients; and most importantly, their ability to modulate the inflammatory milieu [70]. This has led to exploration of the ability of these cells to modulate fibrosis when administered exogenously in murine models. Interestingly, unlike lung-resident MSCs which demonstrate specific homing and long-term engraftment after intratracheal administration even in uninjured lungs [41], exogenous BM-MSC does not show any significant tissue engraftment after intravenous or intratracheal injections [71, 72]. Thus, MSCs derived from the bone marrow have not yet been shown to be present in normal or diseased human lungs and fail to reside in the lung even after exogenous administration. Hence, current data does not support a role for endogenous BM-MSCs as a direct cellular player in the lung milieu or as a source of mesenchymal progenitors in the lung.

Conclusion

Fibrotic diseases of the lung are a major cause of mortality and morbidity. Fibrosis evolves as normal homeostatic/reparative mechanisms fail. Hence, knowledge of the endogenous stem/progenitor cell niche(s) within the adult lung is essential in further unraveling the mechanism(s) of fibrosis. Recent studies have consolidated the key role of lung-resident epithelial and mesenchymal progenitor cells in fibrosis. Better understanding of mesenchymal progenitors and their niche, and their functions in an adult lung will help devise methodologies to restrain aberrant fibroproliferative responses and discover novel therapeutic options for fibrotic lung diseases.

References

1. Cordier JF (2006) Cryptogenic organising pneumonia. Eur Respir J 28:422–446
2. Burnham EL, Janssen WJ, Riches DW, Moss M, Downey GP (2014) The fibroproliferative response in acute respiratory distress syndrome: mechanisms and clinical significance. Eur Respir J 43:276–285
3. Marshall RP, Bellingan G, Webb S, Puddicombe A, Goldsack N, McAnulty RJ, Laurent GJ (2000) Fibroproliferation occurs early in the acute respiratory distress syndrome and impacts on outcome. Am J Respir Crit Care Med 162:1783–1788
4. King TE Jr, Pardo A, Selman M (2011) Idiopathic pulmonary fibrosis. Lancet 378:1949–1961
5. Visscher DW, Myers JL (2006) Histologic spectrum of idiopathic interstitial pneumonias. Proc Am Thorac Soc 3:322–329
6. Barker AF, Bergeron A, Rom WN, Hertz MI (2014) Obliterative bronchiolitis. N Engl J Med 370:1820–1828
7. Hogg JC, Chu F, Utokaparch S, Woods R, Elliott WM, Buzatu L, Cherniack RM, Rogers RM, Sciurba FC, Coxson HO, Pare PD (2004) The nature of small-airway obstruction in chronic obstructive pulmonary disease. N Engl J Med 350:2645–2653
8. Kasper M, Haroske G (1996) Alterations in the alveolar epithelium after injury leading to pulmonary fibrosis. Histol Histopathol 11:463–483
9. Kawanami O, Ferrans VJ, Crystal RG (1982) Structure of alveolar epithelial cells in patients with fibrotic lung disorders. Lab Invest 46:39–53
10. Myers JL, Katzenstein AL (1988) Ultrastructural evidence of alveolar epithelial injury in idiopathic bronchiolitis obliterans-organizing pneumonia. Am J Pathol 132:102–109
11. Heppleston AG (1956) The pathology of honeycomb lung. Thorax 11:77–93
12. Kelly FL, Kennedy VE, Jain R, Sindhwani NS, Finlen Copeland CA, Snyder LD, Eu JP, Meltzer EB, Brockway BL, Pavlisko E, Stripp BR, Palmer SM (2012) Epithelial clara cell injury occurs in bronchiolitis obliterans syndrome after human lung transplantation. Am J Transplant 12:3076–3084
13. Zoz DF, Lawson WE, Blackwell TS (2011) Idiopathic pulmonary fibrosis: a disorder of epithelial cell dysfunction. Am J Med Sci 341:435–438
14. Barkauskas CE, Cronce MJ, Rackley CR, Bowie EJ, Keene DR, Stripp BR, Randell SH, Noble PW, Hogan BL (2013) Type 2 alveolar cells are stem cells in adult lung. J Clin Invest 123:3025–3036
15. Sisson TH, Mendez M, Choi K, Subbotina N, Courey A, Cunningham A, Dave A, Engelhardt JF, Liu X, White ES, Thannickal VJ, Moore BB, Christensen PJ, Simon RH (2010) Targeted

injury of type II alveolar epithelial cells induces pulmonary fibrosis. Am J Respir Crit Care Med 181:254–263
16. Serrano-Mollar A, Nacher M, Gay-Jordi G, Closa D, Xaubet A, Bulbena O (2007) Intratracheal transplantation of alveolar type II cells reverses bleomycin-induced lung fibrosis. Am J Respir Crit Care Med 176:1261–1268
17. O'Koren EG, Hogan BL, Gunn MD (2013) Loss of basal cells precedes bronchiolitis obliterans-like pathological changes in a murine model of chlorine gas inhalation. Am J Respir Cell Mol Biol 49:788–797
18. Perl AK, Riethmacher D, Whitsett JA (2011) Conditional depletion of airway progenitor cells induces peribronchiolar fibrosis. Am J Respir Crit Care Med 183:511–521
19. Zuo W, Zhang T, Wu DZ, Guan SP, Liew AA, Yamamoto Y, Wang X, Lim SJ, Vincent M, Lessard M, Crum CP, Xian W, McKeon F (2015) p63(+)Krt5(+) distal airway stem cells are essential for lung regeneration. Nature 517:616–620
20. Sivakumar P, Ntolios P, Jenkins G, Laurent G (2012) Into the matrix: targeting fibroblasts in pulmonary fibrosis. Curr Opin Pulm Med 18:462–469
21. Zhang K, Rekhter MD, Gordon D, Phan SH (1994) Myofibroblasts and their role in lung collagen gene expression during pulmonary fibrosis. A combined immunohistochemical and in situ hybridization study. Am J Pathol 145:114–125
22. Dominici M, Le Blanc K, Mueller I, Slaper-Cortenbach I, Marini F, Krause D, Deans R, Keating A, Prockop D, Horwitz E (2006) Minimal criteria for defining multipotent mesenchymal stromal cells. The International Society for Cellular Therapy position statement. Cytotherapy 8:315–317
23. Pittenger MF, Mackay AM, Beck SC, Jaiswal RK, Douglas R, Mosca JD, Moorman MA, Simonetti DW, Craig S, Marshak DR (1999) Multilineage potential of adult human mesenchymal stem cells. Science 284:143–147
24. Pereira RF, Halford KW, O'Hara MD, Leeper DB, Sokolov BP, Pollard MD, Bagasra O, Prockop DJ (1995) Cultured adherent cells from marrow can serve as long-lasting precursor cells for bone, cartilage, and lung in irradiated mice. Proc Natl Acad Sci U S A 92:4857–4861
25. Prockop DJ (1997) Marrow stromal cells as stem cells for nonhematopoietic tissues. Science 276:71–74
26. Lama VN, Smith L, Badri L, Flint A, Andrei AC, Murray S, Wang Z, Liao H, Toews GB, Krebsbach PH, Peters-Golden M, Pinsky DJ, Martinez FJ, Thannickal VJ (2007) Evidence for tissue-resident mesenchymal stem cells in human adult lung from studies of transplanted allografts. J Clin Invest 117:989–996
27. Bruno S, Bussolati B, Grange C, Collino F, di Cantogno LV, Herrera MB, Biancone L, Tetta C, Segoloni G, Camussi G (2009) Isolation and characterization of resident mesenchymal stem cells in human glomeruli. Stem Cells Dev 18:867–880
28. Hoogduijn MJ, Crop MJ, Peeters AM, Korevaar SS, Eijken M, Drabbels JJ, Roelen DL, Maat AP, Balk AH, Weimar W, Baan CC (2009) Donor-derived mesenchymal stem cells remain present and functional in the transplanted human heart. Am J Transplant 9:222–230
29. Peterson RS, Lim L, Ye H, Zhou H, Overdier DG, Costa RH (1997) The winged helix transcriptional activator HFH-8 is expressed in the mesoderm of the primitive streak stage of mouse embryos and its cellular derivatives. Mech Dev 69:53–69
30. Kalinichenko VV, Lim L, Stolz DB, Shin B, Rausa FM, Clark J, Whitsett JA, Watkins SC, Costa RH (2001) Defects in pulmonary vasculature and perinatal lung hemorrhage in mice heterozygous null for the Forkhead Box f1 transcription factor. Dev Biol 235:489–506
31. Mahlapuu M, Enerback S, Carlsson P (2001) Haploinsufficiency of the forkhead gene Foxf1, a target for sonic hedgehog signaling, causes lung and foregut malformations. Development 128:2397–2406
32. Walker N, Badri L, Wettlaufer S, Flint A, Sajjan U, Krebsbach PH, Keshamouni VG, Peters-Golden M, Lama VN (2011) Resident tissue-specific mesenchymal progenitor cells contribute to fibrogenesis in human lung allografts. Am J Pathol 178:2461–2469

33. Mandeville I, Aubin J, LeBlanc M, Lalancette-Hebert M, Janelle MF, Tremblay GM, Jeannotte L (2006) Impact of the loss of Hoxa5 function on lung alveogenesis. Am J Pathol 169:1312–1327
34. Volpe MV, Pham L, Lessin M, Ralston SJ, Bhan I, Cutz E, Nielsen HC (2003) Expression of Hoxb-5 during human lung development and in congenital lung malformations. Birth Defects Res A Clin Mol Teratol 67:550–556
35. Mallo M, Wellik DM, Deschamps J (2010) Hox genes and regional patterning of the vertebrate body plan. Dev Biol 344:7–15
36. Jun D, Garat C, West J, Thorn N, Chow K, Cleaver T, Sullivan T, Torchia EC, Childs C, Shade T, Tadjali M, Lara A, Nozik-Grayck E, Malkoski S, Sorrentino B, Meyrick B, Klemm D, Rojas M, Wagner DH Jr, Majka SM (2011) The pathology of bleomycin-induced fibrosis is associated with loss of resident lung mesenchymal stem cells that regulate effector T-cell proliferation. Stem Cells 29:725–735
37. Kumar ME, Bogard PE, Espinoza FH, Menke DB, Kingsley DM, Krasnow MA (2014) Mesenchymal cells. Defining a mesenchymal progenitor niche at single-cell resolution. Science 346:1258810
38. Ehninger A, Trumpp A (2011) The bone marrow stem cell niche grows up: mesenchymal stem cells and macrophages move in. J Exp Med 208:421–428
39. Costa RH, Kalinichenko VV, Lim L (2001) Transcription factors in mouse lung development and function. Am J Physiol Lung Cell Mol Physiol 280:L823–L838
40. Maeda Y, Dave V, Whitsett JA (2007) Transcriptional control of lung morphogenesis. Physiol Rev 87:219–244
41. Badri L, Walker NM, Ohtsuka T, Wang Z, Delmar M, Flint A, Peters-Golden M, Toews GB, Pinsky DJ, Krebsbach PH, Lama VN (2011) Epithelial interactions and local engraftment of lung-resident mesenchymal stem cells. Am J Respir Cell Mol Biol 45:809–816
42. Marriott S, Baskir RS, Gaskill C, Menon S, Carrier EJ, Williams J, Talati M, Helm K, Alford CE, Kropski JA, Loyd J, Wheeler L, Johnson J, Austin E, Nozik-Grayck E, Meyrick B, West JD, Klemm DJ, Majka SM (2014) ABCG2pos lung mesenchymal stem cells are a novel pericyte subpopulation that contributes to fibrotic remodeling. Am J Physiol Cell Physiol 307:C684–C698
43. Volckaert T, Dill E, Campbell A, Tiozzo C, Majka S, Bellusci S, De Langhe SP (2011) Parabronchial smooth muscle constitutes an airway epithelial stem cell niche in the mouse lung after injury. J Clin Invest 121:4409–4419
44. Badri L, Murray S, Liu LX, Walker NM, Flint A, Wadhwa A, Chan KM, Toews GB, Pinsky DJ, Martinez FJ, Lama VN (2011) Mesenchymal stromal cells in bronchoalveolar lavage as predictors of bronchiolitis obliterans syndrome. Am J Respir Crit Care Med 183:1062–1070
45. Badri L, Lama VN (2012) Lysophosphatidic acid induces migration of human lung-resident mesenchymal stem cells through the beta-catenin pathway. Stem Cells 30:2010–2019
46. Suratt BT, Cool CD, Serls AE, Chen L, Varella-Garcia M, Shpall EJ, Brown KK, Worthen GS (2003) Human pulmonary chimerism after hematopoietic stem cell transplantation. Am J Respir Crit Care Med 168:318–322
47. Brocker V, Langer F, Fellous TG, Mengel M, Brittan M, Bredt M, Milde S, Welte T, Eder M, Haverich A, Alison MR, Kreipe H, Lehmann U (2006) Fibroblasts of recipient origin contribute to bronchiolitis obliterans in human lung transplants. Am J Respir Crit Care Med 173:1276–1282
48. Yousem SA, Sherer C, Fuhrer K, Cieply K (2013) Myofibroblasts of recipient origin are not the predominant mesenchymal cell in bronchiolitis obliterans in lung allografts. J Heart Lung Transplant 32:266–268
49. Dolgachev VA, Ullenbruch MR, Lukacs NW, Phan SH (2009) Role of stem cell factor and bone marrow-derived fibroblasts in airway remodeling. Am J Pathol 174:390–400
50. Hashimoto N, Jin H, Liu T, Chensue SW, Phan SH (2004) Bone marrow-derived progenitor cells in pulmonary fibrosis. J Clin Invest 113:243–252

51. Kim KK, Kugler MC, Wolters PJ, Robillard L, Galvez MG, Brumwell AN, Sheppard D, Chapman HA (2006) Alveolar epithelial cell mesenchymal transition develops in vivo during pulmonary fibrosis and is regulated by the extracellular matrix. Proc Natl Acad Sci U S A 103:13180–13185
52. Moore BB, Kolodsick JE, Thannickal VJ, Cooke K, Moore TA, Hogaboam C, Wilke CA, Toews GB (2005) CCR2-mediated recruitment of fibrocytes to the alveolar space after fibrotic injury. Am J Pathol 166:675–684
53. Phillips RJ, Burdick MD, Hong K, Lutz MA, Murray LA, Xue YY, Belperio JA, Keane MP, Strieter RM (2004) Circulating fibrocytes traffic to the lungs in response to CXCL12 and mediate fibrosis. J Clin Invest 114:438–446
54. Tanjore H, Xu XC, Polosukhin VV, Degryse AL, Li B, Han W, Sherrill TP, Plieth D, Neilson EG, Blackwell TS, Lawson WE (2009) Contribution of epithelial-derived fibroblasts to bleomycin-induced lung fibrosis. Am J Respir Crit Care Med 180:657–665
55. Moore B, Lawson WE, Oury TD, Sisson TH, Raghavendran K, Hogaboam CM (2013) Animal models of fibrotic lung disease. Am J Respir Cell Mol Biol 49:167–179
56. Humphreys BD, Lin SL, Kobayashi A, Hudson TE, Nowlin BT, Bonventre JV, Valerius MT, McMahon AP, Duffield JS (2010) Fate tracing reveals the pericyte and not epithelial origin of myofibroblasts in kidney fibrosis. Am J Pathol 176:85–97
57. Rock JR, Barkauskas CE, Cronce MJ, Xue Y, Harris JR, Liang J, Noble PW, Hogan BL (2011) Multiple stromal populations contribute to pulmonary fibrosis without evidence for epithelial to mesenchymal transition. Proc Natl Acad Sci U S A 108:E1475–E1483
58. Matsubara H, Sakakibara K, Kunimitsu T, Matsuoka H, Kato K, Oyachi N, Dobashi Y, Matsumoto M (2012) Non-small cell lung carcinoma therapy using mTOR-siRNA. Int J Clin Exp Pathol 5:119–125
59. Hung C, Linn G, Chow YH, Kobayashi A, Mittelsteadt K, Altemeier WA, Gharib SA, Schnapp LM, Duffield JS (2013) Role of lung pericytes and resident fibroblasts in the pathogenesis of pulmonary fibrosis. Am J Respir Crit Care Med 188:820–830
60. Korbling M, Estrov Z (2003) Adult stem cells for tissue repair—a new therapeutic concept? N Engl J Med 349:570–582
61. Lama VN, Phan SH (2006) The extrapulmonary origin of fibroblasts: stem/progenitor cells and beyond. Proc Am Thorac Soc 3:373–376
62. Bucala R, Spiegel LA, Chesney J, Hogan M, Cerami A (1994) Circulating fibrocytes define a new leukocyte subpopulation that mediates tissue repair. Mol Med 1:71–81
63. Moeller A, Gilpin SE, Ask K, Cox G, Cook D, Gauldie J, Margetts PJ, Farkas L, Dobranowski J, Boylan C, O'Byrne PM, Strieter RM, Kolb M (2009) Circulating fibrocytes are an indicator of poor prognosis in idiopathic pulmonary fibrosis. Am J Respir Crit Care Med 179:588–594
64. Kleaveland KR, Velikoff M, Yang J, Agarwal M, Rippe RA, Moore BB, Kim KK (2014) Fibrocytes are not an essential source of type I collagen during lung fibrosis. J Immunol 193:5229–5239
65. Madsen DH, Ingvarsen S, Jurgensen HJ, Melander MC, Kjoller L, Moyer A, Honore C, Madsen CA, Garred P, Burgdorf S, Bugge TH, Behrendt N, Engelholm LH (2011) The non-phagocytic route of collagen uptake: a distinct degradation pathway. J Biol Chem 286:26996–27010
66. Kleaveland KR, Moore BB, Kim KK (2014) Paracrine functions of fibrocytes to promote lung fibrosis. Expert Rev Respir Med 8:163–172
67. Ortiz LA, Gambelli F, McBride C, Gaupp D, Baddoo M, Kaminski N, Phinney DG (2003) Mesenchymal stem cell engraftment in lung is enhanced in response to bleomycin exposure and ameliorates its fibrotic effects. Proc Natl Acad Sci U S A 100:8407–8411
68. Rojas M, Xu J, Woods CR, Mora AL, Spears W, Roman J, Brigham KL (2005) Bone marrow-derived mesenchymal stem cells in repair of the injured lung. Am J Respir Cell Mol Biol 33:145–152

69. Kfoury Y, Scadden DT (2015) Mesenchymal cell contributions to the stem cell niche. Cell Stem Cell 16:239–253
70. Uccelli A, Moretta L, Pistoia V (2006) Immunoregulatory function of mesenchymal stem cells. Eur J Immunol 36:2566–2573
71. Bianco P, Cao X, Frenette PS, Mao JJ, Robey PG, Simmons PJ, Wang CY (2013) The meaning, the sense and the significance: translating the science of mesenchymal stem cells into medicine. Nat Med 19:35–42
72. McNulty K, Janes SM (2012) Stem cells and pulmonary fibrosis: cause or cure? Proc Am Thorac Soc 9:164–171

Chapter 14
Mesenchymal Stromal Cell-Based Therapies for Lung Disease

Daniel Chambers

Introduction

Although our current understanding of the identity of, and interactions between, lung stem/progenitor cells in the adult lung remains a work in progress, particularly in humans, great strides are being made. What is very clear is that understanding how the adult lung achieves homeostasis and repairs injury will be of fundamental importance if we are to completely understand the pathogenesis of lung diseases, particularly the increasingly common chronic, and mostly treatment refractory, degenerative lung diseases. As occurred for bone-marrow disorders three or four decades ago, it is highly likely that cellular therapies will play a role in the future management of these diseases [1]. This chapter will focus on mesenchymal stromal cell (MSC) therapy since this cell type, or a closely related cell, is likely to be the first of these new therapeutic modalities to enter the clinic and will highlight potential pulmonary disease targets, whilst also emphasising the limitations of the tools we currently use to isolate and characterise lung stromal cell populations. MSCs have a broad functional repertoire. Their immunosuppressive, antibacterial and antifibrotic activity; their ability to elaborate growth factors; and their proliferative potential and ability to differentiate into multiple cell lineages has generated great interest in their role in disease pathology, and as architects of organ repair and regeneration. Special emphasis will be given to idiopathic pulmonary fibrosis (IPF) as a potential target for regenerative medicine approaches utilising MSCs, since the incidence of this quintessential degenerative lung disease is rapidly increasing as populations age, and since evidence is accumulating that lung progenitor cell depletion or dysfunction may lie at the heart of its pathogenesis.

D. Chambers, MBBS, MRCP, FRACP, MD. (✉)
School of Medicine, The University of Queensland, Brisbane, QLD, Australia

Queensland Lung Transplant Service, The Prince Charles Hospital, Brisbane, QLD, Australia
e-mail: Daniel_Chambers@Health.qld.gov.au

I. Bertoncello (ed.), *Stem Cells in the Lung*, Stem Cell Biology and Regenerative Medicine, DOI 10.1007/978-3-319-21082-7_14

Therapeutic Potential of MSCs

MSCs are a specialised stromal cell type originally identified in the 1960s by heterotypic transplantation and liquid culture of rodent and guinea pig bone marrow and spleen cell suspensions [2, 3]. Originally characterised by their ability to form fibroblast-like, plastic adherent colonies from single cells when plated at clonogenic levels, and their multipotent (bone, cartilage, and adipose tissue) differentiation capacity when propagated in defined media, the minimal criteria endorsed by the International Society for Cellular Therapy (ISCT) for defining multipotent MSC now includes a $CD105^{+}CD90^{+}CD73^{+}CD45^{-}$ immunophenotypic signature profile [4]. MSCs or MSC-like cells have now been identified in many organs and tissues, including the lung [5]. Sabatini et al. first identified a population of cells consistent with the currently accepted ISCT definition of an MSC in human lung digests [6]. Similar cells have since been isolated from bronchoalveolar lavage of sex-mismatched lung transplant patients using plastic adherence [4]. Clonogenic cells derived from these patients retain the sex of the donor even many years after sex-mismatched lung transplantation, suggesting that they are very long-lived and retain an inherent capacity for self-renewal [4].

However, it is important to note that a similar phenotype is shared by many stromal cells, including the humble dermal fibroblast [7, 8], so many studies using the term ‘MSC’ have likely described the properties of a heterogeneous population of cells [9–11]. The ‘stemness’ (capacity for self-renewal assessed with serial transplantation) of putative MSCs differentiates them from other stromal cell types, but is rarely assessed in published studies [12]. Hence the term ‘mesenchymal stromal cell’ rather than ‘mesenchymal stem cell’ is favoured. It is possible, indeed even probable, that enrichment of heterogeneous MSC populations for stemness will improve therapeutic efficacy [10], Consequently, the development of assays to quantitate potency is an active area for research [13].

Notwithstanding this lack of precision with respect to definition, MSCs perform a remarkable, perhaps even surprising, array of functions and hence carry considerable therapeutic potential. This broad functionality may of course be related to cellular heterogeneity rather than a sweeping repertoire of cell-specific skills. However the currently available literature almost exclusively uses the ISCT definition of an MSC, with its inherent limitations, so this possibility remains largely unexplored. Despite this caveat, a large body of literature, some of which now includes well-controlled randomised trials, suggests that MSCs are highly likely to provide new therapeutic options for a broad range of diseases, and particularly to provide regenerative options for degenerative disease [14].

Relative immune privilege: Aside from this broad functionality, MSCs have the added advantage of being easy to propagate and relatively immune-privileged. Since MSCs can escape lysis by cytotoxic T-cells and natural killer cells, they may be transplantable between HLA-mismatched individuals without the need for immunosuppression. This relative immunoprivilege has been confirmed in multiple studies using xenogeneic and major histocompatibility mismatched models. For instance, when human bone-marrow-derived MSCs were transplanted into lambs they were able to engraft and persist for up to 13 months [15]. While human MSCs do not express HLA class II antigen, they do weakly express HLA class I, however in co-culture experiments human MSCs fail to induce proliferation of allogeneic lymphocytes [16]. More recently, the degree of immunological privilege accorded MSCs has been questioned. Studies in small animal cardiac models suggest that the immunogenicity of MSCs increases with the differentiation state of the cells—so that multipotent cells remain immune-privileged, but major histocompatibility complex expression is upregulated in their terminally differentiated progeny (e.g., myocytes and endothelial cells) leading to destruction by complement- and/or cell-mediated lysis [17, 18], re-enforcing the objective to enrich MSC populations for stemness when devising optimal therapies.

On the other hand, even multipotent cells have been found to be more immunogenic than previously anticipated. In a primate model, multiple administrations of high-dose allogeneic MSC resulted in the production of alloantibodies in two of six animals [19]. MSCs have also been reported to induce memory T-cell responses in a murine model [20] and furthermore, MSCs express the activating NK cell-receptor ligands NKG2D and UL16 [21] limiting their ability to avoid lysis by NK cells [22]. The practical implication is that preclinical work in major histocompatibility complex-matched and/or immunosuppressed animals needs to be cautiously interpreted in the planning of human Phase I studies which are likely to involve HLA-mismatching, particularly if treatment with relatively well differentiated cells is proposed.

Immunosuppression and tolerance: One of the most consistently observed properties of MSCs has been their immunosuppressive function. They are able to abrogate T-cell responses by production of paracrine factors such as PGE_2 [23, 24], and the induction of a regulatory phenotype in $CD4^+$ lymphocytes [25–27]. This provides a strong rationale for preclinical and clinical studies utilising MSC in transplantation, where tolerance remains the Holy Grail. The possibility that MSC-treatment may foster operational tolerance as part of an immunosuppression-minimisation protocol holds great promise [27] with the potential for substantially improved post-transplant outcomes, and has already been confirmed in a randomised controlled trial in renal transplantation [28]. In the lung, MSCs show promise as an adjunct therapy in patients undergoing lung transplantation with encouraging efficacy data from animal models [29–31], and safety data from a Phase 1 study of human transplant patients with chronic lung rejection [32].

Lung homeostasis and epithelial repair: Complex organisms possess a remarkable capacity for extensive and sustained tissue renewal throughout a lifetime. This regenerative capacity is maintained by reservoirs of self-renewing somatic tissue stem cells which are responsible for organ homeostasis and repair following injury [33]. Analysis of the organisation and regulation of the archetypal hematopoietic stem cell (HSC) hierarchy has revealed that bone-marrow MSCs are a critical element of the HSC niche responsible for maintaining stem cell potential, facilitating hematopoiesis [34], as well as playing a key role in the mobilization of HSC into the circulation, and their homing and lodgement in the marrow following transplantation [12].

Since the stroma provides critical cues to support respiratory epithelial progenitors during lung development, homeostasis and repair, it follows that an analogous niche may also exist in adult lung. The critical importance of stromal inputs to lung regeneration and repair has been apparent for many years, but has been brought into sharper focus by the development of assays for identifying and characterising candidate lung epithelial stem cells which have provided powerful tools for analysis of the niche interactions between stem and stromal cells [35–37]. While the cell types providing these cues and the cues themselves remain enigmatic in humans [11], the stem cell function of type 2 pneumocytes during alveolar repair was recently proven in mice and is dependent on cross-talk from a population of stromal cells including alveolar fibroblasts and lipofibroblasts within a niche [38]. Furthermore, ablation of MSCs in the murine lung has been associated with experimental bleomycin-induced fibrosis [39]. MSCs were more recently shown to increase the proliferative potential of a key epithelial progenitor cell—the bronchoalveolar stem cell [40] and, remarkably, to restore bioenergetics in lung epithelium and induce repair programmes through donation of mitochondria [41, 42]. Fully understanding the components and relationships within the niche, or, as is much more likely, niches, is likely to be crucial to understanding the pathogenesis of degenerative lung disease. It is also for this reason that the exogenous delivery of stromal cells is an attractive idea for the treatment of degenerative lung disease.

But could MSCs not just orchestrate, but actually participate in epithelial repair by respecification and transdifferentiation of MSC to generate epithelial cell lineages? As is the case for MSC-like cells derived from other organs, MSC-like cells derived from human lung have been shown to differentiate into non-mesodermal cell lineages, including epithelium [43]. Others have also shown that MSCs derived from lung tissue differentiate into cells expressing club cell secretory protein and aquaporin-5, markers of small airway and alveolar epithelial cells respectively when cultured in suitable media [44]. However it is now broadly accepted that any such trans-differentiation occurs rarely, if at all, in vivo, and is not extensive or robust enough to contribute meaningfully to epithelial repair [45].

Despite this caveat, there remains intense interest in delivering MSCs to the lung to treat inflammatory diseases such as chronic obstructive pulmonary disease (COPD) and asthma or post-transplant rejection; and to manage acute lung injury (ALI) and chronic degenerative diseases such as IPF.

Delivering Cell Therapy to the Lung

There are two potential routes of pulmonary delivery of a candidate cell product–endobronchial and intravenous. A third possible route of delivery, during ex vivo perfusion of whole organs, is feasible and may prove important in lung bioengineering [46] but will not be further discussed here. Whilst the endobronchial route has been used [47] and is readily accessible via bronchoscopy, delivery of large numbers of cells to the distal lung is problematic and unpredictable. In contrast the intravenous route is highly attractive because of the so-called 'first-pass effect', whereby cells delivered intravenously are required to transit the lung so that there is extensive and homogeneous, although admittedly temporary, retention of MSCs as they pass through the pulmonary circulation [48]. Hence, only a small proportion of infused cells pass through into the systemic circulation [49]. This effect is particularly pertinent to MSC-based cell therapy due to the large physical size of MSCs.

While the first-pass effect has impeded the development of regenerative therapy approaches such as MSC therapy for non-pulmonary target organs including the heart [50], the ability to deliver cellular therapy to the lung via a simple intravenous approach is a major advantage and affords the opportunity for large-scale retention of reparative cells. Even more attractively, the apparent preferential retention of infused cells at sites of lung injury [51], provides a distinct advantage for designers of lung cell therapies.

A downside of intravenous delivery of cell therapy is the risk that embolization of MSC will lead to adverse hemodynamic events, but this problem has been largely discounted in early phase human studies [52, 53]. The need for infused cells to bind to and then transit the endothelial layer, in order to reach the site of injury and provide any conceivable therapeutic effect is also a potential impediment to the delivery of an effective cellular therapy. However, since evidence points increasingly to a perivascular location of lung MSCs [10], and hence the epithelial stem cell niches, this problem may not be insurmountable. Despite these advances, large evidence gaps remain, not the least of which are the elucidation of the mechanisms of MSC homing and engraftment to targeted tissue microenvironments. In summary, whilst it must be kept in mind that mere transit to the lung microvasculature does not equal functional engraftment; nevertheless the ease of pulmonary delivery of cellular therapy via the intravenous route is attractive. Future studies will need to focus on the chemokine signals which encourage MSC homing and on the ligand/receptor interactions at the endothelial surface which encourage migration, margination, extravasation and engraftment.

MSC source: At present most MSCs used in human trials are derived from unrelated bone-marrow donors, although placenta, umbilical cord and adipose tissue are other, potentially more convenient sources. The literature presently sheds little light on the similarities and differences between these cell types, with all meeting the (admittedly broad) current ISCT-endorsed definition of an MSC. Further complicating the field, MSCs prepared for clinical use, regardless of source, are heterogeneous in therapeutic efficacy despite attempts to standardise ex vivo expansion protocols. Determining the most accessible and potent source of MSC for therapeutic product development will of course be highly dependent on the target disease, and in the case of degenerative disease, will further depend upon accurately defining the extent of and mode of delivery of stromal cell support to the pulmonary epithelium and enriching heterogeneous cell populations for the relevant activity.

Whole cells, microvesicles or conditioned medium? Whilst it is now clear that MSCs are unable, at least in any significant number, to transdifferentiate to aid in epithelial repair, it remains to be determined whether the support signals orchestrating epithelial repair by MSCs are soluble and delivered in a paracrine manner or cell contact dependent. However, recent exciting studies in the lung provide evidence that cell–cell contact and the transfer of cytoplasmic contents, including mitochondria, from the donor MSC to epithelial progenitors may be critical in providing the cues regulating stem cell proliferation, differentiation and repair [41].

Mitochondria were donated to eukaryotes approximately 1.5 billion years ago by an ancient prokaryote, facilitating aerobic respiration and the subsequent evolution of the eukaryotic cell, and later, complex multicellular organisms. Mitochondrial dysfunction has recently been identified as a key trigger for cellular senescence and apoptosis, and may act as a break on tissue stem cell proliferation in organs with a generally slow turnover of stem cell pools, such as the lung [54]. The potential for MSCs (and other cell types) to have retained the ancient ability to transmit and donate mitochondria and hence restore mitochondrial function to tissue stem cells to effectively slow tissue ageing is of considerable interest. While the literature in this field remains scant in the lung, there is emerging evidence that delivery of cell components (microvesicles or exosomes) carrying organelles and microRNAs, but not the cells themselves, can provide substantial therapeutic benefit [55]. This exciting possibility opens the way to the development of novel therapeutic approaches whereby MSCs may become the 'postmen' of the regenerative medicine world, delivering their designer therapeutic packages to specific target cells.

But what are the likely disease targets? Whilst the answer to this question must of course be founded on a sound understanding of MSC biology, it would be a mistake to design early human studies without considering what the path to successful translation might look like.

Ensuring That the Therapeutic Potential of MSCs Is Not Lost in Translation

The term 'translational research' is used, somewhat confusingly, to refer to two distinct phases of therapeutic product development and application. To add to the confusion, very often, the phase being referred to is not specified. The first translational 'step', or as is sometimes the case, 'leap' is that taken when moving from preclinical studies to first-in-human trials or from the 'bench' to the 'bedside'. The decision to move to a first-in-human study is difficult, but can be considered within the framework outlined in Fig. 14.1. The case for conducting a first-in-human study is strongest when there is an important unmet clinical need for a new therapy; when the therapy has been successfully tested in appropriate preclinical studies which

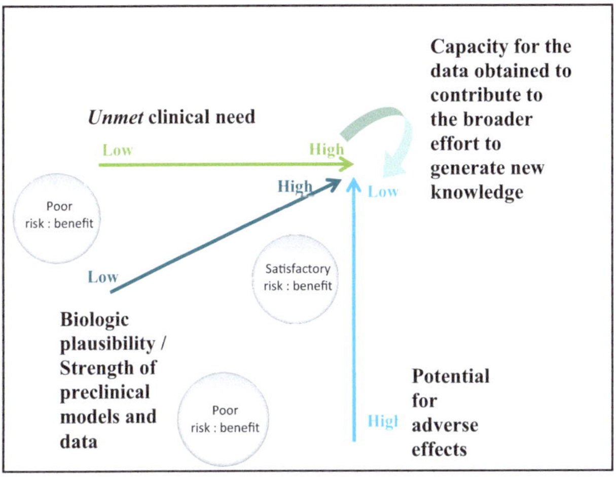

Fig. 14.1 Turning the translational 'leap' into a step. The decision to move to a first-in-human study is difficult, but can be considered within the framework outlined in the figure. The case for conducting a first-in-human study is strongest when there is a sound biologic rationale for a new treatment; when the therapy has been successfully tested in appropriate preclinical studies which reliably reflect the human disease; when the potential for adverse events is low; when the probability that the data generated during early phase human studies will shed new light on the pathogenesis of a serious human disease is high; and when there is an important unmet clinical need. It is important to make the distinction between 'unmet' and 'met' clinical need. In the latter case, while there may be a large burden of disease, effective therapeutic strategies are available and the burden of remaining disease largely relates to ineffective implementation of these strategies (the second translational step). Naturally it is common for not all of these pieces of the puzzle to fall neatly into place, but careful consideration of the weight of evidence along each axis will improve the risk: benefit of a first-in-human study

reliably reflect the human disease; and when the potential for adverse events is low. Of course the proposed new therapy also needs to make biologic sense. The probability that the data generated during an early phase human study will shed important light on a serious human disease further enhances the attractiveness of a first-in-human study. Naturally it is common for not all of these pieces of the puzzle to fall neatly into place. The second translational step refers to the adoption and dissemination of a practise or product which is already known from well-designed clinical trials to be beneficial to human health. The implementation of evidence-based treatment regimens and management algorithms for asthma is a good example of this second step.

The burden of lung disease in the twenty-first century—where is the unmet clinical need going to be? Lung disease remains the third most common cause of death globally, behind ischemic heart disease and cerebrovascular disease [56], with most of these respiratory deaths being related to pneumonia. In developing countries tuberculosis remains a very serious problem and smoking-related lung disease (overwhelmingly lung cancer and COPD) causes significant morbidity and mortality worldwide. However a major change is underway in developed nations, with declining cigarette smoking rates, effective antibiotic therapy and population ageing meaning that non-communicable, chronic degenerative diseases will become, beside lung cancer, the major cause of respiratory death in the coming decades [56].

Currently, the most common chronic respiratory diseases in developed nations remain asthma and COPD. In Australia the prevalence of asthma is approximately 10 % and COPD 3 %. Highly effective and safe therapies are already available for asthma so that mortality is now rare, with the major challenges being in achieving a timely and accurate diagnosis, ensuring adherence to treatment to minimise morbidity and mortality, and in identifying the underlying causes to reduce prevalence. COPD on the other hand remains difficult to treat as the structural nature of the disease means that pharmacologic agents are only partially effective, but COPD is completely preventable if cigarette smoking is avoided. Hence COPD-related mortality is declining in developed nations. In contrast, lung cancer remains a leading cause of death, even with declining smoking rates, since a minority of cases are unrelated to cigarette smoking.

If we look ahead to the latter half of the twenty-first century then, the prevalence of COPD and lung cancer will be decreasing but these diseases will remain significant sources of morbidity and mortality. In contrast the prevalence of degenerative and age-related lung diseases will be increasing rapidly without effective treatments to meet this increasing burden. Indeed this change is already underway, with the incidence of IPF, the most common of these diseases, increasing to somewhere (depending on case definition) between 30 and 93/100,000 [57–59] and the annual cumulative prevalence rising rapidly from 202 cases per 100,000 people in 2001 to 494 in 2011 [59]. IPF is a lethal disease with the median survival from diagnosis being 3.5–4 years, even in the modern era [57–59].

Acute lung injury (ALI) is a clinical syndrome of diverse aetiology characterised by widespread pulmonary infiltration and rapidly developing respiratory failure. It is also prominent among lung diseases for its high mortality, lack of effective treatments, and increasing incidence with population ageing. This incidence is now 60–80/100,000 [60], and the mortality rate is 40 % and static [60]. It is likely that ALI will become an increasing problem in developed nations in the coming years. Several pieces of evidence point to a role for MSC treatment in ALI [41, 61–64], and early phase human studies are now underway.

Although therapeutic options for both IPF and ALI remain limited, there has been significant recent progress for patients with IPF, for the first time, confirmation of efficacy of two small molecules in large randomised controlled trials [65, 66]. However, due to a lack of truly effective preventative or therapeutic measures and a rising incidence, in developed nations this century the clinical need will be most pressingly unmet for lung cancer, ALI and IPF. The therapeutic potential of MSCs comes into clear focus when considered from this perspective. It is unlikely that regenerative strategies will play a role in the management of cancer, however they may well go some way to expanding the therapeutic options available for patients with ALI and IPF.

IPF—a degenerative disease in need of a regenerative solution. It is against this backdrop that focus is shifting toward understanding and developing effective therapeutic strategies to manage degenerative lung disease. These non-communicable, non-malignant lung diseases are characterised by failed or ineffective organ repair after injury, with the clinical phenotype (for instance ALI vs. IPF) in large part depending on the nature, severity and acuity of the lung injury as well as host factors such as the effectiveness of repair. In fact IPF is characterised by sudden deteriorations, called exacerbations, which are a form of ALI [67]. In order to appreciate the potential place of future regenerative strategies for human lung disease, it is instructive to take a step back and review recent advances in our understanding of IPF pathogenesis.

Although the moniker ‘idiopathic’ remains appropriate, compelling insight into IPF pathogenesis comes from the relatively rare germline mutations in the human TERT and TERC genes [68]. Together, TERT and TERC form the specialised enzyme telomerase. In stem/progenitor cells the telomerase complex functions to synthesise telomeric DNA and so protect the chromosome ends during cell division. TERT provides reverse transcriptase activity to the complex, and uses TERC, the RNA component of telomerase, as a template. Hence together the TERT and TERC genes maintain telomere length and prevent cellular senescence during recurrent cycles of cellular replication. Telomerase activity is thus a commonly used measure of ‘stemness’, and overexpression of TERT prevents replicative senescence in MSCs and other cells [69]. The most common phenotype in humans with a germline loss-of-function TERT/TERC mutation is a form of pulmonary fibrosis indistinguishable from IPF [70]. Short telomere length, independent of TERT/TERC mutations, is also a strong risk factor for IPF itself, with a clear dose–response relationship between telomere length and survival [71]. These pieces of evidence, along with the epidemiology of IPF (it is overwhelmingly a disease of ageing [58]) indicate that cellular processes which rely on the maintenance of telomere length are fundamental to IPF pathogenesis. A provocative but tantalising conclusion is that IPF results from dysfunction and/or depletion of lung stem/progenitor cell pools over a lifetime. Candidate cells include epithelial progenitors [72], the stromal progenitors which orchestrate epithelial repair, and potentially others, although recent evidence points to depletion of lung resident MSCs in both animal [39] and human [10] lung fibrosis. Repletion of these pools through the delivery of exogenous MSCs or epithelial progenitors [14], or pharmacologic, cell or exosome treatment to improve native adult lung progenitor cell function thus holds promise for the management of IPF [52, 73], ALI [62] and other diseases.

Given this background, it is clear that understanding lung homeostasis and regeneration, and in turn defining the cells which complete these functions, is highly likely to provide new regenerative treatment options. However, currently the ultimate and, at this point, only, regenerative strategy for lung disease is whole organ transplantation, which has now been a viable and evidence-based treatment for selected patients with IPF and other end-stage lung diseases for over three decades. However, lung transplantation necessarily involves the allogeneic replacement of all lung cells, even those with normal function, via a highly invasive operative approach and utilising a very precious and scarce, but also highly immunogenic, resource. More targeted, less invasive and potentially even non- or 'hypo'-allogeneic approaches to lung regenerative medicine should be possible.

Given the fundamental defects in epithelial repair which underlie these conditions, delivering a cell product like an MSC with the aim of improving repair makes biologic sense, especially since depletion of the MSC pool has now been demonstrated in human IPF [10]. Furthermore, and notwithstanding the limitation of these models, multiple preclinical studies provide robust support for the therapeutic efficacy of MSCs (reviewed by Sinclair et al. [11]) in animal models of lung fibrosis resulting from exposure to bleomycin. Table 14.1 demonstrates the diversity in these studies with respect to cellular source, the use of immunodeficient or immunocompetent animals, and the timing and route of MSC delivery. Most of these studies have utilised allogeneic MSCs which have been isolated from bone marrow using plastic adherence and delivered intravenously or endobronchially. If we return to Fig. 14.1, there appears to be strong evidence for efficacy of MSC treatment in preclinical models of these diseases, but the ability of these models to accurately reflect the corresponding human disease is questionable. This is particularly so for the bleomycin model of pulmonary fibrosis where over the last few decades many compounds have been apparently effective, only to prove ineffective in human trials.

Although there appears to be a consistent effect of MSCs if delivered soon after the administration of bleomycin, the therapeutic effect diminishes considerably if treatment is delayed until 7 days after administration [51, 74, 75]. This effect highlights a well-known deficiency of the bleomycin model, and is particularly important to recognise since the timing of MSC delivery appears to determine the fate of the engrafting cell, with later delivery favouring the differentiation of MSCs into cells which are pro-fibrotic [51]. The latter possibility remains an ongoing concern for investigators contemplating human IPF trials and will be a key safety outcome. Thus the potential efficacy of MSC-based cell therapy in IPF and ALI remains controversial [80, 81], with the potential for profibrosis being the main drawback. Since MSCs can be driven down a myofibroblastic differentiation pathway given the correct context, these paradoxical findings again highlight the importance of the fundamental work being carried out to discover the secrets of the lung stem cell niche(s), and potentially provide a rationale for delivering therapy earlier in the disease course when epithelial disrepair is at its height, but when extensive fibrosis has not yet ensued. In summary, there is biologic plausibility around MSC treatment for IPF and ALI, there is a large body of preclinical data (admittedly in models which have a questionable relationship to the human disease), there is substantial evidence for safety from human studies for other indications, and there is a large unmet clinical need. From Fig. 14.1, the pieces of the puzzle are in place to proceed to human studies of MSC treatment for IPF and ALI [73], with perhaps a combination of these two conditions, the acute and often lethal, exacerbation of IPF being a prime target.

Aside from lung fibrosis (Table 14.1) and ALI [61, 62], animal studies have provided support for therapeutic efficacy in other lung diseases where the need is

Table 14.1 Preclinical studies of MSCs or MSC-like cells in pulmonary fibrosis

Author	Intervention	Model	Outcome	Engraftment?
Ortiz et al. [74]	Allogeneic 5×10^5 BM-MSC @ 0, 7 days via jugular vein	Mouse bleomycin	↓ Hydroxyproline—not significant with day 7 infusion	Yes, increased in fibrotic areas
Cui et al. [75]	BM-MSC @ 1, 7 days via tail vein	Rat bleomycin	↓ Hydroxyproline and lung fibrotic score—more pronounced with day 1 infusion	Yes
Zhao et al. [76]	5×10^6 BM-MSC @ 12 h via tail vein	Rat bleomycin	↓ Hydroxyproline and pro-fibrotic cytokines	Yes
Moodley et al. [77]	Xenogeneic umbilical cord-derived MSC 1×10^6 @ 1 day	Mouse bleomycin	↓ Hydroxyproline, collagen and pro-fibrotic cytokines	Yes, only in fibrotic areas
Bitencourt et al. [78]	Autologous MSC engraftment encouraged by hyaluronidase	Mouse bleomycin	↓ Collagen content and fibrotic score	Yes
Choi et al. [79]	Xenogeneic BM-MSC 2×10^5 IV or microvesicles @ 12 and 14 weeks	Mouse silica	↓ Collagen content and fibrotic score, more pronounced with MSC	Yes + ATII differentiation
Yan et al. [51]	Isogeneic BM-MSC 2×10^5 IV @ 0, 60, 120 days	Mouse radiation	↑ Fibrosis with late delivery	Yes
Jun et al. [39]	Isogeneic Lung MSC (Hoechst) 2.5×10^5 IV day 0	Mouse bleomycin	Bleomycin causes lung MSC depletion with repletion attenuating fibrosis	No—?rescue of lung resident MSC

currently unmet. This evidence base is well summarised in Sinclair et al. [11] and so will not be repeated here, but is significant for the increasingly common (as survival following extreme preterm birth continues to improve) bronchopulmonary dysplasia [40] and also for pulmonary hypertension [82–84]. With respect to bronchopulmonary dysplasia it is noteworthy that these babies are often left with obstructive lung disease which manifests in adulthood as emphysema [85]. Amongst the preclinical studies in pulmonary hypertension, of particular note are the studies which demonstrated that genetic engineering of MSCs, for instance to overexpress heme oxygenase [83] or prostacyclin [84], conferred enhanced efficacy, perhaps providing a glimpse of the future of the field.

Human Studies of MSC Therapy in Lung Disease

The multifaceted activity of MSC has translated into a large body of clinical trial activity outside the lung, most notably in the treatment of steroid refractory graft versus host disease following allogeneic bone-marrow transplant, but also in other immune-mediated diseases like Crohn's disease, multiple sclerosis, lupus and in the renal transplant setting [28]. The tissue repair capability of MSCs is being investigated in clinical trials for cardiac repair, bone disorders (osteogenesis imperfecta), bone fracture and following meniscectomy. As a result, many thousands of human subjects have received intravenous MSC therapy with very few adverse effects [86], providing key safety data for moving to clinical trials in lung disease. While relatively few human trials are underway for patients with lung disease, and while even fewer have been published, the diseases which have been targeted reflect MSC biology, the strength of the preclinical data, and the seriousness and lack of availability of alternate treatments for the target disease.

Chronic obstructive pulmonary disease: Weiss et al. conducted what remains the largest trial in humans with lung disease using allogeneic bone-marrow-derived MSCs [53]. They randomised 62 patients with COPD to receive 4 monthly intravenous infusions of either MSCs (100×10^6 cells) or vehicle control in a double-blind manner. Patients were followed for 2 years. There were no infusional toxicities and no treatment-related deaths or serious adverse events. Although MSC treatment was not associated with any improvement in the efficacy measures (lung function, walk distance, or dyspnea score), this study does provide excellent safety data in humans with moderate-to-severe lung disease [53]. Another study targeting emphysema is currently listed as recruiting (NCT01849159, www.clinicaltrials.gov, accessed 20th Feb 2015).

Obliterative bronchiolitis: Recruitment to a phase I trial of MSC therapy for post-transplant obliterative bronchiolitis has recently been completed (http://clinicaltrials.gov/ct2/show/NCT01175655). In this study ten patients with moderate or severe chronic lung allograft dysfunction received allogeneic bone-marrow-derived MSCs (2×10^6 cells/kg twice weekly for 2 weeks) and will be followed for 1 year. MSC therapy was feasible and appeared well tolerated in the short-term, with long-term results awaited [32]. A similar Phase 1 study has recently commenced recruitment (NCT02181712) in the United States. A Phase 2 study is now being planned in Australia.

IPF: The short-term safety of MSC treatment in IPF was recently confirmed in a Phase 1 study of intravenous, allogeneic, placenta-derived MSC in moderate-severely affected patients [52]. In this study, eight patients were treated with 1–2×10^6 MSCs/kg and were followed for 6 months. There was no evidence of acute hemodynamic or gas exchange compromise and no evidence of worsening fibrosis [52]. In the only other published human study, 14 patients with IPF received 0.5×10^6/kg autologous adipose-derived stromal cells endobronchially. No adverse effects of this treatment were seen. Two other clinical trials are ongoing in IPF. A US trial with very similar design to the study by Chambers et al. but involving intravenous delivery of allogeneic bone-marrow-derived MSCs, has almost completed recruitment (NCT02013700, M. Glassberg personal communication) while the other trial (NCT01919827) involves the non-randomised endobronchial delivery of allogeneic adipose-derived MSCs.

ALI: No studies have been published for MSC treatment of ALI, but there are currently two randomised clinical trials and one non-randomised trial underway.

The first (NCT01775774) is a dose-escalation (1×10^6, 5×10^6, and 10×10^6 cells/kg) study delivering allogeneic bone-marrow-derived MSCs to three cohorts of patients ($n = 3$/cohort). The other randomised, double-blind, placebo-controlled trial (NCT01902082) delivers allogeneic adipose-derived MSCs. The inclusion criteria are similar for these two studies. In the third non-randomised study, patients with viral infection-induced and extra-corporeal membrane oxygenation-dependent ALI will receive open-label allogeneic bone-marrow-derived MSCs (NCT02215811). In all three studies MSCs are delivered intravenously.

Other disease targets: The only other clinical trials of MSC therapy for lung disease listed as active at www.clinicaltrials.gov are in bronchopulmonary dysplasia, where a Phase 1 study has recently reported promising results [87]; in pulmonary hypertension (NCT01795950) where a Phase 1 study involves three dosing cohorts of $0.5–2 \times 10^6$ placenta-derived MSCs/kg delivered intravenously; and in radiation-induced lung fibrosis (NCT02277145).

Conclusions

Together, this preclinical and clinical trial activity showcases the diversity and potential of MSCs in regenerative medicine. MSCs have been shown to be much more than a cellular population with immune-suppressive activity and multipotent capacity. MSCs can also promote tissue repair by regulating the functionality of tissue stem cell pools and rescuing aged, senescent or damaged tissue. While a number of lung diseases, most notably asthma, are now able to be relatively safely and effectively treated due to significant improvements in available pharmacologics, substantial therapeutic gaps remain, particularly for degenerative and fibrotic lung diseases. It is pleasing to think that a deeper understanding of the role of tissue-resident stem cells in the maintenance of lung health, and hence the development of pharmacologic and/or cell therapies aimed at restoring that health, may one day fill these gaps. However in order for this promise to be achieved safely, and in order to avoid a repetition of the problematic headlong introduction of gene therapies in large-scale clinical trials [88], a deeper understanding of basic MSC biology is needed, alongside the careful conduct of early phase human trials. The elucidation of the interrelationships between epithelial and MSC hierarchies; the role of MSCs in the construction of lung stem cell niches; and the dynamics of the cellular interactions within these niches are key to achieving this objective.

References

1. Prockop DJ, Prockop SE, Bertoncello I (2014) Are clinical trials with mesenchymal stem/progenitor cells too far ahead of the science? Lessons from experimental hematology. Stem Cells 32(12):3055–3061

2. Friedenstein AJ, Chailakhjan RK, Lalykina KS (1970) The development of fibroblast colonies in monolayer cultures of guinea-pig bone marrow and spleen cells. Cell Tissue Kinet 3 (4):393–403
3. Friedenstein AJ, Piatetzky-Shapiro II, Petrakova KV (1966) Osteogenesis in transplants of bone marrow cells. J Embryol Exp Morphol 16(3):381–390
4. Lama VN, Smith L, Badri L, Flint A, Andrei AC, Murray S et al (2007) Evidence for tissue-resident mesenchymal stem cells in human adult lung from studies of transplanted allografts. J Clin Invest 117(4):989–996
5. Keating A (2012) Mesenchymal stromal cells: new directions. Cell Stem Cell 10(6):709–716
6. Sabatini F, Petecchia L, Tavian M, de Villeroche VJ, Rossi GA, Brouty-Boye D (2005) Human bronchial fibroblasts exhibit a mesenchymal stem cell phenotype and multilineage differentiating potentialities. Lab Invest 85(8):962–971
7. Alt E, Yan Y, Gehmert S, Song YH, Altman A, Vykoukal D, Bai X (2011) Fibroblasts share mesenchymal phenotypes with stem cells, but lack their differentiation and colony-forming potential. Biol Cell 103(4):197–208
8. Blasi A, Martino C, Balducci L, Saldarelli M, Soleti A, Navone SE et al (2011) Dermal fibroblasts display similar phenotypic and differentiation capacity to fat-derived mesenchymal stem cells, but differ in anti-inflammatory and angiogenic potential. Vasc Cell 3(1):5
9. Ho AD, Wagner W, Franke W (2008) Heterogeneity of mesenchymal stromal cell preparations. Cytotherapy 10(4):320–330
10. Marriott S, Baskir RS, Gaskill C, Menon S, Carrier EJ, Williams J et al (2014) ABCG2pos lung mesenchymal stem cells are a novel pericyte subpopulation that contributes to fibrotic remodeling. Am J Physiol Cell Physiol 307(8):C684–C698
11. Sinclair K, Yerkovich ST, Chambers DC (2013) Mesenchymal stem cells and the lung. Respirology 18(3):397–411
12. Frenette PS, Pinho S, Lucas D, Scheiermann C (2013) Mesenchymal stem cell: keystone of the hematopoietic stem cell niche and a stepping-stone for regenerative medicine. Annu Rev Immunol 31:285–316
13. Lv FJ, Tuan RS, Cheung KM, Leung VY (2014) Concise review: the surface markers and identity of human mesenchymal stem cells. Stem Cells 32(6):1408–1419
14. Chambers DC, Hopkins PM (2013) Idiopathic pulmonary fibrosis: a degenerative disease requiring a regenerative approach. Am J Respir Crit Care Med 188(2):252–253
15. Mackenzie TC, Flake AW (2001) Human mesenchymal stem cells persist, demonstrate site-specific multipotential differentiation, and are present in sites of wound healing and tissue regeneration after transplantation into fetal sheep. Blood Cells Mol Dis 27(3):601–604
16. Le Blanc K, Tammik C, Rosendahl K, Zetterberg E, Ringden O (2003) HLA expression and immunologic properties of differentiated and undifferentiated mesenchymal stem cells. Exp Hematol 31(10):890–896
17. Huang XP, Sun Z, Miyagi Y, McDonald Kinkaid H, Zhang L, Weisel RD, Li RK (2010) Differentiation of allogeneic mesenchymal stem cells induces immunogenicity and limits their long-term benefits for myocardial repair. Circulation 122(23):2419–2429
18. Poncelet AJ, Vercruysse J, Saliez A, Gianello P (2007) Although pig allogeneic mesenchymal stem cells are not immunogenic in vitro, intracardiac injection elicits an immune response in vivo. Transplantation 83(6):783–790
19. Beggs KJ, Lyubimov A, Borneman JN, Bartholomew A, Moseley A, Dodds R et al (2006) Immunologic consequences of multiple, high-dose administration of allogeneic mesenchymal stem cells to baboons. Cell Transplant 15(8–9):711–721
20. Nauta AJ, Westerhuis G, Kruisselbrink AB, Lurvink EG, Willemze R, Fibbe WE (2006) Donor-derived mesenchymal stem cells are immunogenic in an allogeneic host and stimulate donor graft rejection in a nonmyeloablative setting. Blood 108(6):2114–2120
21. Poggi A, Prevosto C, Zancolli M, Canevali P, Musso A, Zocchi MR (2007) NKG2D and natural cytotoxicity receptors are involved in natural killer cell interaction with self-antigen presenting cells and stromal cells. Ann N Y Acad Sci 1109:47–57

22. Sotiropoulou PA, Perez SA, Gritzapis AD, Baxevanis CN, Papamichail M (2006) Interactions between human mesenchymal stem cells and natural killer cells. Stem Cells 24(1):74–85
23. Cutler AJ, Limbani V, Girdlestone J, Navarrete CV (2010) Umbilical cord-derived mesenchymal stromal cells modulate monocyte function to suppress T cell proliferation. J Immunol 185(11):6617–6623
24. Oh DY, Cui P, Hosseini H, Mosse J, Toh BH, Chan J (2012) Potently immunosuppressive 5-fluorouracil-resistant mesenchymal stromal cells completely remit an experimental autoimmune disease. J Immunol 188(5):2207–2217
25. Chao YH, Wu HP, Wu KH, Tsai YG, Peng CT, Lin KC et al (2014) An increase in CD3 + CD4 + CD25+ regulatory T cells after administration of umbilical cord-derived mesenchymal stem cells during sepsis. PLoS One 9(10), e110338
26. Kota DJ, Wiggins LL, Yoon N, Lee RH (2013) TSG-6 produced by hMSCs delays the onset of autoimmune diabetes by suppressing Th1 development and enhancing tolerogenicity. Diabetes 62(6):2048–2058
27. Obermajer N, Popp FC, Soeder Y, Haarer J, Geissler EK, Schlitt HJ, Dahlke MH (2014) Conversion of Th17 into IL-17Aneg regulatory T cells: a novel mechanism in prolonged allograft survival promoted by mesenchymal stem cell-supported minimized immunosuppressive therapy. J Immunol 193(10):4988–4999
28. Tan J, Wu W, Xu X, Liao L, Zheng F, Messinger S et al (2012) Induction therapy with autologous mesenchymal stem cells in living-related kidney transplants: a randomized controlled trial. JAMA 307(11):1169–1177
29. Grove DA, Xu J, Joodi R, Torres-Gonzales E, Neujahr D, Mora AL, Rojas M (2011) Attenuation of early airway obstruction by mesenchymal stem cells in a murine model of heterotopic tracheal transplantation. J Heart Lung Transplant 30(3):341–350
30. Guo Z, Zhou X, Li J, Meng Q, Cao H, Kang L, Ni Y, Fan H, Liu Z (2013) Mesenchymal stem cells reprogram host macrophages to attenuate obliterative bronchiolitis in murine orthotopic tracheal transplantation. Int Immunopharmacol 15(4):726–734
31. Raza K, Price AP, Matson A, Tolar J, Hertz MI, Panoskaltsis-Mortari A (2011) MSC therapy for obliterative bronchiolitis (OB). Am J Respir Crit Care Med 183:A5289
32. Chambers DC, Hopkins PM, Sturm M, Lawrence S, Enever D, Sparks L et al (2012) Mesenchymal stromal cell therapy for bronchiolitis obliterans syndrome—preliminary data in humans. J Heart Lung Transplant 31(4):S67
33. Sharpless NE, DePinho RA (2007) How stem cells age and why this makes us grow old. Nat Rev Mol Cell Biol 8(9):703–713
34. Mendez-Ferrer S, Michurina TV, Ferraro F, Mazloom AR, Macarthur BD, Lira SA et al (2010) Mesenchymal and haematopoietic stem cells form a unique bone marrow niche. Nature 466 (7308):829–834
35. McQualter JL, Yuen K, Williams B, Bertoncello I (2010) Evidence of an epithelial stem/progenitor cell hierarchy in the adult mouse lung. Proc Natl Acad Sci U S A 107(4):1414–1419
36. Kumar ME, Bogard PE, Espinoza FH, Menke DB, Kingsley DM, Krasnow MA (2014) Mesenchymal cells. Defining a mesenchymal progenitor niche at single-cell resolution. Science 346(6211):1258810
37. Lee JH, Kim CF (2014) Developmental biology. Mesenchymal progenitor panoply. Science 346(6211):810–811
38. Barkauskas CE, Cronce MJ, Rackley CR, Bowie EJ, Keene DR, Stripp BR et al (2013) Type 2 alveolar cells are stem cells in adult lung. J Clin Invest 123(7):3025–3036
39. Jun D, Garat C, West J, Thorn N, Chow K, Cleaver T et al (2011) The pathology of bleomycin-induced fibrosis is associated with loss of resident lung mesenchymal stem cells that regulate effector T-cell proliferation. Stem Cells 29(4):725–735
40. Tropea KA, Leder E, Aslam M, Lau AN, Raiser DM, Lee JH et al (2012) Bronchioalveolar stem cells increase after mesenchymal stromal cell treatment in a mouse model of bronchopulmonary dysplasia. Am J Physiol Lung Cell Mol Physiol 302(9):L829–L837

41. Islam MN, Das SR, Emin MT, Wei M, Sun L, Westphalen K et al (2012) Mitochondrial transfer from bone-marrow-derived stromal cells to pulmonary alveoli protects against acute lung injury. Nat Med 18(5):759–765
42. Spees JL, Olson SD, Whitney MJ, Prockop DJ (2006) Mitochondrial transfer between cells can rescue aerobic respiration. Proc Natl Acad Sci U S A 103(5):1283–1288
43. Sueblinvong V, Loi R, Eisenhauer PL, Bernstein IM, Suratt BT, Spees JL, Weiss DJ (2008) Derivation of lung epithelium from human cord blood-derived mesenchymal stem cells. Am J Respir Crit Care Med 177(7):701–711
44. Karoubi G, Cortes-Dericks L, Breyer I, Schmid RA, Dutly AE (2009) Identification of mesenchymal stromal cells in human lung parenchyma capable of differentiating into aquaporin 5-expressing cells. Lab Invest 89(10):1100–1114
45. Prockop DJ (2007) "Stemness" does not explain the repair of many tissues by mesenchymal stem/multipotent stromal cells (MSCs). Clin Pharmacol Ther 82(3):241–243
46. Song JJ, Ott HC (2011) Bioartificial lung engineering. Am J Transplant 12(2):283–288
47. Tzouvelekis A, Paspaliaris V, Koliakos G, Ntolios P, Bouros E, Oikonomou A (2013) A prospective, non-randomized, no placebo-controlled, phase Ib clinical trial to study the safety of the adipose derived stromal cells-stromal vascular fraction in idiopathic pulmonary fibrosis. J Transl Med 11:171
48. Brooke G, Tong H, Levesque JP, Atkinson K (2008) Molecular trafficking mechanisms of multipotent mesenchymal stem cells derived from human bone marrow and placenta. Stem Cells Dev 17(5):929–940
49. Fischer UM, Harting MT, Jimenez F, Monzon-Posadas WO, Xue H, Savitz SI et al (2009) Pulmonary passage is a major obstacle for intravenous stem cell delivery: the pulmonary first-pass effect. Stem Cells Dev 18(5):683–692
50. Schrepfer S, Deuse T, Reichenspurner H, Fischbein MP, Robbins RC, Pelletier MP (2007) Stem cell transplantation: the lung barrier. Transplant Proc 39(2):573–576
51. Yan X, Liu Y, Han Q, Jia M, Liao L, Qi M, Zhao RC (2007) Injured microenvironment directly guides the differentiation of engrafted Flk-1(+) mesenchymal stem cell in lung. Exp Hematol 35(9):1466–1475
52. Chambers DC, Enever D, Ilic N, Sparks L, Whitelaw K, Ayres J et al (2014) A phase 1b study of placenta-derived mesenchymal stromal cells in patients with idiopathic pulmonary fibrosis. Respirology 19(7):1013–1018
53. Weiss DJ, Casaburi R, Flannery R, LeRoux-Williams M, Tashkin DP (2013) A placebo-controlled, randomized trial of mesenchymal stem cells in COPD. Chest 143(6):1590–1598
54. Sahin E, Depinho RA (2010) Linking functional decline of telomeres, mitochondria and stem cells during ageing. Nature 464(7288):520–528
55. Sdrimas K, Kourembanas S (2014) MSC microvesicles for the treatment of lung disease: a new paradigm for cell-free therapy. Antioxid Redox Signal 21(13):1905–1915
56. WHO (2014) World health statistics 2014. WHO, Geneva
57. Navaratnam V, Fleming KM, West J, Smith CJ, Jenkins RG, Fogarty A, Hubbard RB (2011) The rising incidence of idiopathic pulmonary fibrosis in the U.K. Thorax 66(6):462–467
58. Raghu G, Weycker D, Edelsberg J, Bradford WZ, Oster G (2006) Incidence and prevalence of idiopathic pulmonary fibrosis. Am J Respir Crit Care Med 174(7):810–816
59. Raghu G, Chen SY, Yeh WS, Maroni B, Li Q, Lee YC, Collard HR (2014) Idiopathic pulmonary fibrosis in US Medicare beneficiaries aged 65 years and older: incidence, prevalence, and survival, 2001–11. Lancet Respir Med 2(7):566–572
60. Johnson ER, Matthay MA (2010) Acute lung injury: epidemiology, pathogenesis, and treatment. J Aerosol Med Pulm Drug Deliv 23(4):243–252
61. Zhu YG, Feng XM, Abbott J, Fang XH, Hao Q, Monsel A et al (2014) Human mesenchymal stem cell microvesicles for treatment of E. coli endotoxin-induced acute lung injury in mice. Stem Cells 32(1):116–125

62. Matthay MA, Thompson BT, Read EJ, McKenna DH Jr, Liu KD, Calfee CS, Lee JW (2010) Therapeutic potential of mesenchymal stem cells for severe acute lung injury. Chest 138 (4):965–972
63. Lee JW, Krasnodembskaya A, McKenna DH, Song Y, Abbott J, Matthay MA (2013) Therapeutic effects of human mesenchymal stem cells in ex vivo human lungs injured with live bacteria. Am J Respir Crit Care Med 187(7):751–760
64. Lee JW, Fang X, Krasnodembskaya A, Howard JP, Matthay MA (2011) Concise review: mesenchymal stem cells for acute lung injury: role of paracrine soluble factors. Stem Cells 29 (6):913–919
65. King TE Jr, Bradford WZ, Castro-Bernardini S, Fagan EA, Glaspole I, Glassberg MK et al (2014) A phase 3 trial of pirfenidone in patients with idiopathic pulmonary fibrosis. N Engl J Med 370(22):2083–2092
66. Richeldi L, du Bois RM, Raghu G, Azuma A, Brown KK, Costabel U et al (2014) Efficacy and safety of nintedanib in idiopathic pulmonary fibrosis. N Engl J Med 370(22):2071–2082
67. Mukhopadhyay S, Parambil JG (2012) Acute interstitial pneumonia (AIP): relationship to Hamman-Rich syndrome, diffuse alveolar damage (DAD), and acute respiratory distress syndrome (ARDS). Semin Respir Crit Care Med 33(5):476–485
68. Chambers DC, Clarke BE, McGaughran J, Garcia CK (2012) Lung fibrosis, premature graying, and macrocytosis. Am J Respir Crit Care Med 186(5):e8–e9
69. Tsai CC, Chen CL, Liu HC, Lee YT, Wang HW, Hou LT, Hung SC (2010) Overexpression of hTERT increases stem-like properties and decreases spontaneous differentiation in human mesenchymal stem cell lines. J Biomed Sci 17:64
70. Armanios M, Blackburn EH (2012) The telomere syndromes. Nat Rev Genet 13(10):693–704
71. Stuart BD, Lee JS, Kozlitina J, Noth I, Devine MS, Glazer CS et al (2014) Effect of telomere length on survival in patients with idiopathic pulmonary fibrosis: an observational cohort study with independent validation. Lancet Respir Med 2(7):557–565
72. Chilosi M, Doglioni C, Murer B, Poletti V (2010) Epithelial stem cell exhaustion in the pathogenesis of idiopathic pulmonary fibrosis. Sarcoidosis Vasc Diffuse Lung Dis 27(1):7–18
73. Toonkel RL, Hare JM, Matthay MA, Glassberg MK (2013) Mesenchymal stem cells and idiopathic pulmonary fibrosis. Potential for clinical testing. Am J Respir Crit Care Med 188 (2):133–140
74. Ortiz LA, Gambelli F, McBride C, Gaupp D, Baddoo M, Kaminski N, Phinney DG (2003) Mesenchymal stem cell engraftment in lung is enhanced in response to bleomycin exposure and ameliorates its fibrotic effects. Proc Natl Acad Sci U S A 100(14):8407–8411
75. Cui A, Dai HP, Dai JW, Pang BS, Niu SJ, Lu YP, Wang C (2007) Effects of bone marrow mesenchymal stem cells on bleomycin induced pulmonary fibrosis in rats. Zhonghua Jie He He Hu Xi Za Zhi 30(9):677–682
76. Zhao F, Zhang YF, Liu YG, Zhou JJ, Li ZK, Wu CG, Qi HW (2008) Therapeutic effects of bone marrow-derived mesenchymal stem cells engraftment on bleomycin-induced lung injury in rats. Transplant Proc 40(5):1700–1705
77. Moodley Y, Atienza D, Manuelpillai U, Samuel CS, Tchongue J, Ilancheran S, Boyd R, Trounson A (2009) Human umbilical cord mesenchymal stem cells reduce fibrosis of bleomycin-induced lung injury. Am J Pathol 175(1):303–313
78. Bitencourt CS, Pereira PA, Ramos SG, Sampaio SV, Arantes EC, Aronoff DM, Faccioli LH (2011) Hyaluronidase recruits mesenchymal-like cells to the lung and ameliorates fibrosis. Fibrogenesis Tissue Repair 4(1):3
79. Choi M, Ban T, Rhim T (2014) Therapeutic use of stem cell transplantation for cell replacement or cytoprotective effect of microvesicle released from mesenchymal stem cell. Mol Cells 37(2):133–139
80. Ntolios P, Janes SM (2013) Mesenchymal stem cell therapy for lung diseases: oasis or mirage? Respiration 85(4):279–280
81. Weiss DJ, Ortiz LA (2013) Cell therapy trials for lung diseases: progress and cautions. Am J Respir Crit Care Med 188(2):123–125

82. Baber SR, Deng W, Master RG, Bunnell BA, Taylor BK, Murthy SN et al (2007) Intratracheal mesenchymal stem cell administration attenuates monocrotaline-induced pulmonary hypertension and endothelial dysfunction. Am J Physiol Heart Circ Physiol 292(2):H1120–H1128
83. Liang OD, Mitsialis SA, Chang MS, Vergadi E, Lee C, Aslam M et al (2011) Mesenchymal stromal cells expressing heme oxygenase-1 reverse pulmonary hypertension. Stem Cells 29 (1):99–107
84. Takemiya K, Kai H, Yasukawa H, Tahara N, Kato S, Imaizumi T (2010) Mesenchymal stem cell-based prostacyclin synthase gene therapy for pulmonary hypertension rats. Basic Res Cardiol 105(3):409–417
85. Wong PM, Lees AN, Louw J, Lee FY, French N, Gain K et al (2008) Emphysema in young adult survivors of moderate-to-severe bronchopulmonary dysplasia. Eur Respir J 32 (2):321–328
86. Lalu MM, McIntyre L, Pugliese C, Fergusson D, Winston BW, Marshall JC et al (2012) Safety of cell therapy with mesenchymal stromal cells (SafeCell): a systematic review and meta-analysis of clinical trials. PLoS One 7(10), e47559
87. Chang YS, Ahn SY, Yoo HS, Sung SI, Choi SJ, Oh WI, Park WS (2014) Mesenchymal stem cells for bronchopulmonary dysplasia: phase 1 dose-escalation clinical trial. J Pediatr 164 (5):966–972, e966
88. Couzin J, Kaiser J (2005) Gene therapy. As Gelsinger case ends, gene therapy suffers another blow. Science 307(5712):1028

Chapter 15
Lung Vascular Regeneration and Repair

Mervin C. Yoder and Bernard Thébaud

Overview of Proposed Mechanisms for Replacing or Regenerating Vascular Endothelium

Given the astounding number of publications retrieved with the search term "angiogenesis" (new vessel growth from preexisting vessels) in the Pubmed database (74,557 papers identified on November 1, 2014), it is shocking how little is known about certain fundamental properties of vascular endothelium; we do not know the baseline level of endothelial cell turnover required for blood vessel homeostasis or during vessel repair in most vertebrate organisms. A great deal is known about somatic stem cell contributions to tissue maintenance for some cell lineages such as blood, gut, and skin [1]. For example, in the adult human body, billions of circulating blood cells are replaced on an hourly basis in such an exquisite fashion that the circulating blood cell counts do not change. All of the circulating blood cells are derived from hematopoietic stem cells and downstream hematopoietic progenitor cells residing in the bone marrow compartment [2]. The adult mouse small intestinal mucosal lining is completely replaced every 3–5 days via a hierarchy comprised of intestinal stem cells and transient amplifying cells [3]. This cycle of intestinal cell turnover occurs several hundred times during the 2-year lifespan of an average laboratory mouse. Similarly, in the adult human, an estimated 100 billion intestinal epithelial cells are lost and replaced daily [4]. Finally, the entire epidermal layer of human skin is estimated to be replaced

M.C. Yoder (✉)
Department of Pediatrics, Wells Center for Pediatric Research, Indiana University School of Medicine, 1044 West Walnut Street, R4-W125, Indianapolis, IN 46202, USA
e-mail: myoder@iu.edu

B. Thébaud
Division of Neonatology, Department of Pediatrics, Sinclair Centre for Regenerative Medicine, Ottawa Hospital Research Institute, Children's Hospital of Eastern Ontario, University of Ottawa, 501 Smyth Road, Ottawa, ON, Canada, K1H 8L6

I. Bertoncello (ed.), *Stem Cells in the Lung*, Stem Cell Biology and Regenerative Medicine, DOI 10.1007/978-3-319-21082-7_15

every 28 days [5]. These few examples, highlight the dynamic nature of cellular turnover that constitutes normal tissue homeostasis. One would naturally anticipate that the endothelial lining of the vast circulatory system feeding all of these dynamic tissues and organs must also harbor some replacement mechanism, especially since it is well recognized that circulating apoptotic and necrotic endothelial cells are present in human subjects. It is apparent that no one can yet answer some of the simple questions: (1) Do all endothelial cells possess proliferative potential or is this a heterogeneous property only displayed by some cells (as is the case for many epithelial tissues and organs)? (2) How is proliferative potential maintained in endothelial cells when mature endothelial cells comprising the vascular intimal lining must be constrained by tight inter-endothelial junctions to control barrier properties and homogenously participate in other tissue specific functions? (3) Are there certain blood vessels that require more endothelial repair and regeneration than other vessels in the body and do they, therefore, contain endothelial cells with greater proliferative potential than those vessels requiring less repair? Compare these questions to those raised by Dr. Rudolph Altschul six decades ago; "As already mentioned, cell division of the endothelium is a difficult problem. There are several important questions which are still not definitively answered and which are ignored by many investigators. It is difficult for me to understand how the pathology and pathogenesis of arteriosclerosis can be successfully investigated as long as we do not know: (a) when the mitotic division of endothelium occurring in fetal life comes to an end, if at all; (b) how much desquamation or molting of endothelial cells occurs during life; and (c) whether or not it is correct to assume that desquamated endothelial cells are being replaced by means of direct division (of adjacent cells) [6]."

Whether endothelial cells in vessels proliferated at all was a matter of controversy prior to 1950, with some authors concluding that endothelial cells were highly proliferative in the fetal period, but then ceased to divide soon after birth and were maintained within the vessel lifelong [6]. By the 1960s, it had become well accepted that vascular endothelial cells multiply by mitosis from adjacent endothelial cells as a mechanism of repair [7–11]. In the 1970s, it became clear that a variety of physiologic stressors could influence the rate of endothelial proliferation within the aorta (age of the host, various metabolic disorders, and infection or exposure to endotoxin) [12–15]. Some evidence also emerged that not all endothelial cells lining the aorta possessed the same proliferative potential. Particularly in young rodents, some aortic endothelial cells displayed little proliferation while other areas displayed clustered foci of high replication rates approaching 60 % [16]. The question that remains elusive is to understand what factors induce the endothelial cells to begin replicating within the vessels?

One stimulus for inducing local endothelial cell proliferation was to experimentally inflict an endothelial denudation injury to a vessel. For example, careful insertion of a stainless steel wire into the femoral or carotid artery of a rodent can literally scrape and remove a narrow strip of endothelial cells and to allow assessment of the cellular events of the reparative process [13, 17, 18]. It is apparent that small injuries can be repaired simply by surviving neighboring endothelial cell migration and spreading to cover the defect, but larger injuries require proliferation of nearby endothelial cells [19]. Use of tritiated thymidine infusion studies into rodents suggested endothelial turnover in normal tissues to be highly variable and range from a few months [20] to 47–10,000 days [21]. Several groups note that <1 % of endothelial cells are labeled with a brief pulse of tritiated thymidine [5, 22]. However, tools to assess the origin of the proliferating cells were limited in the late 1970s and 1980s and some questions remained about whether some of the repair of the damaged vessel was contributed by circulating blood-borne cells rather than the resident endothelial cells [23, 24]. Indeed, formation of endothelial cells from circulating blood lymphocytes at sites of thrombosis had been postulated as early as 1950 [25].

In 1997, Asahara and colleagues [26] reported that a subset of human circulating blood cells, some expressing CD34 (15.7 % of peripheral blood cells) and/or vascular endothelial growth factor receptor 2 (VEGFR2) (also called kinase insert domain receptor or KDR) (20 % of peripheral blood cells) displayed the ability to form clusters of round cells on fibronectin-coated dishes within 12 h of plating. Spindle-shaped adherent cells emerged from the base of the clusters and these cells possessed the capacity form cellular networks and tube-like structures. These spindle-shaped cells displayed limited expression of hematopoietic antigens CD45 (27.2 %), but increasing amounts of "endothelial" antigens such as CD34 (32.8 %), CD31 (71.5 %), Tie2 (54.6 %), and general uptake of the lectin Ulex Europaeus Agglutinin 1 (UEA1) upon in vitro culture in the presence of vascular endothelial growth factor (VEGF) and serum. Injection of a fraction of human blood cells (isolated by magnetic bead isolation of $CD34^+$ cells) into athymic nude mice that had undergone unilateral hind limb femoral artery excision revealed localization of the human cells into the capillary bed of the muscle of the ischemic limb in areas of neovascularization. The authors termed these blood cells progenitor endothelial cells and proposed a number of potential uses for these cells as vehicles for treating human disease states [26]. With great speed, the field of "endothelial progenitor cell" biology grew from a few hundred papers in 1997 to a recent total of 15,568 cited works (keyword: endothelial progenitor cell, Pubmed, November 8, 2014). A new paradigm arose that bone-marrow-derived circulating endothelial progenitor cells were recruited to sites of vascular injury or repair and these cells differentiated into replacement endothelial cells that formed a portion of the neovasculature via a process of postnatal vasculogenesis (reviewed in [27–29]).

A critical reevaluation of the methods to identify and classify the circulating endothelial progenitor cells has revealed that the vast majority of these EPCs represent various subsets of proangiogenic hematopoietic stem and progenitor cells (HSPC) that are not direct precursors of any endothelial cells [30–32]. The HSPC are recruited to the site of vessel injury and repair to provide paracrine support to initiate or augment local resident endothelial cell migration and proliferation to provide for the vascular repair (Fig. 15.1). While HSPC may adhere to the exposed vascular basement membrane (or to adherent platelets), the HSPC do not reprogram into endothelial cells (change their fate from blood to endothelium) that provide long-term repair of the endothelial intima [31]. Proof of the role of the local resident endothelial cells in providing the endothelial reparative function has been provided by the use of a number of transgenic murine model systems

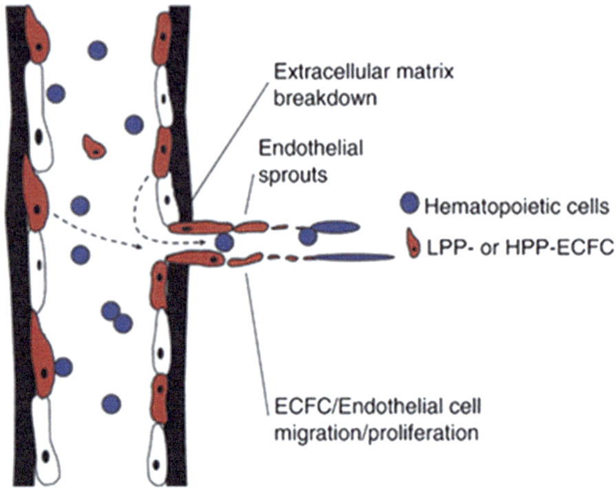

Fig. 15.1 Circulating and resident cells involved in neoangiogenesis. This figure depicts the rare circulating low proliferative potential (LPP) and high proliferative potential endothelial colony forming cells (HPP-ECFC) that may become the new vessels at neoangiogenesis sites. More abundant circulating hematopoietic cell subsets modulate the initiation, recruitment, and formation of the new vessels via stimulation of the circulating and resident ECFC

[33–37]. For example, Purhonen et al. [38] used chimeric mice reconstituted with fluorescent-labeled bone marrow cells to demonstrate that no bone-marrow-derived endothelial or endothelial progenitor cell contributes to repair or incorporation into the site of tumor neovascularization upon implantation of cancer cells. However, evidence was presented for robust recruitment of bone marrow labeled cells to the perivascular site of endothelial repair and some cells differentiate into periendothelial macrophages (an expected hematopoietic derivative). One recent review [39] highlights the many approaches used to present evidence for and against the direct conversion of marrow-derived endothelial progenitor cells into endothelial cells at sites of vascular injury or disease. The preponderance of evidence does not support the contribution of circulating bone marrow-derived endothelial progenitor cells into the endothelium regenerated in injured arterial vessels, rather, arterial endothelial injuries appear to be repaired by migration and proliferation of local resident endothelial cells. For example, transgenic mice prone to develop atherosclerosis were transplanted with bone marrow cells expressing a fluorescence reporter and no bone-marrow-derived circulating cells were shown to contribute to the endothelium covering the site of plaque development in the aorta of affected mice [40] (Fig. 15.2). Similarly, re-endothelialization of a denuded segment of the carotid artery (induced by wire insertion) did not occur through recruitment of circulating bone-marrow-derived cells expressing a fluorescence reporter [37], but rather through migration and possible proliferation of resident endothelial cells from adjacent non-injured endothelial intima [33–35]. This recent review concludes that the paradigm of the 1970s stating that injury to arterial vascular endothelium is repaired by migrating and proliferating local resident endothelium remains valid and without contribution of circulating cells [39].

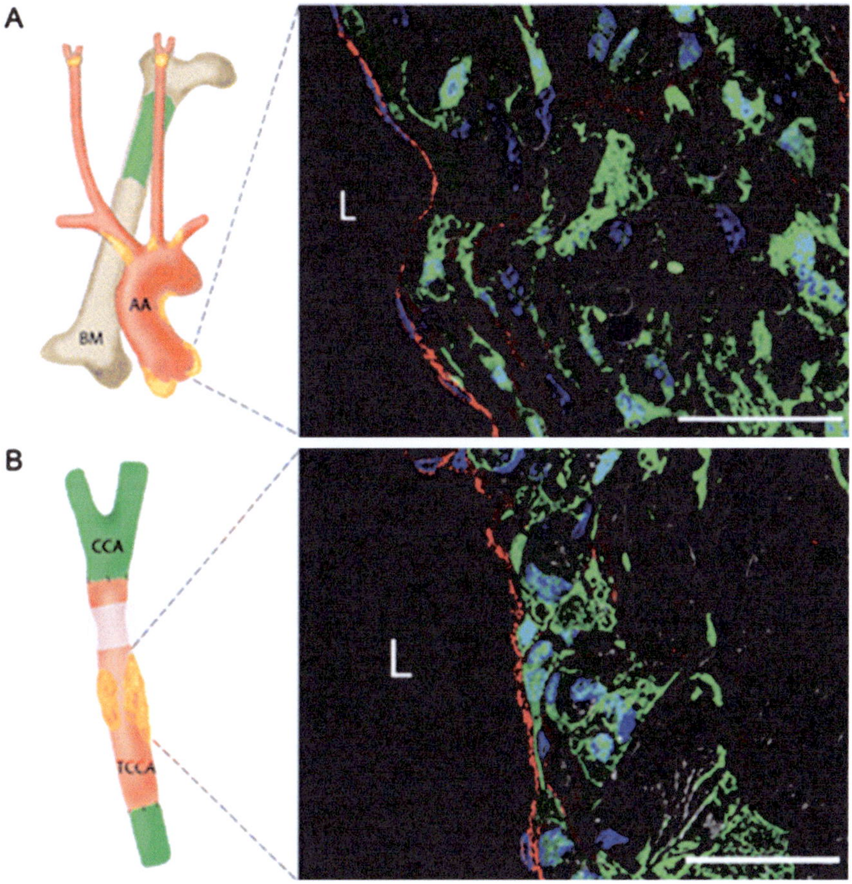

Fig. 15.2 Plaque ECs were derived from the local arterial wall. (**a**) Experiment to investigate whether or not bone marrow (BM)-derived endothelial progenitor cells (EPCs) contribute to plaque ECs during atherogenesis. An aortic root plaque from an $apoE^{-/-}$ mouse transplanted with BM from an $eGFP^{+}apoE^{-/-}$ mouse. No $eGFP^{+}vWF^{+}$ double-positive cells are present. (**b**) An experiment to investigate whether or not any types of blood-borne EPCs contribute to plaque ECs during atherogenesis. A common carotid artery (CCA) segment from an $apoE^{-/-}$ mouse was orthotopically transplanted into $eGFP^{+}apoE^{-/-}$ mice (isotransplantation except for the eGFP transgene). None of the vWF^{+} cells is $eGFP^{+}$. *Green* indicates eGFP; *red* vWF; *blue* nuclei; *gray* DIC. *L* lumen; *BM* bone marrow; *AA* aortic arch; *CCA* common carotid artery; *TCCA* transplanted common carotid artery. Scale bars = 50 μm. Reproduced with permission from Hagensen et al. [39]

In contrast, it has been known for more than 50 years that rare viable circulating endothelial cells (CECs) are present in the blood stream of goats, baboons, dogs, pigs, and human subjects and that these CECs can function to form an intraluminal endothelial lining covering implanted Dacron materials or intravascular devices [41]. Of interest, the circulating blood cells in young pigs also contain fibroblast and smooth muscle cells that form an adventitial subendothelial network on Dacron materials suspended in the flowing blood in the thoracic aorta. A variety of factors appear to play important roles in determining whether a large implanted device will develop an intact coating of endothelial cells and these include, (1) host species, (2) age of host, (3) device surface composition, and (4) dimensions of the device (reviewed in [42]).

The CEC present in adult peripheral and umbilical cord blood display heterogeneity in clonal proliferative potential [43, 44]. Given the ability of the CEC to generate clonal colonies of endothelial cells, these cells have been called endothelial colony forming cells (ECFC) (Fig. 15.3). Cord blood and adult blood ECFC display similar cell surface antigen expression of proteins common to all endothelial cells, fail to express blood markers such as CD45 or CD14, and form human blood vessels that integrate with the host circulation upon transplantation into immunodeficient mice [30, 45–49]. Differences in expression of telomerase and maintenance of telomeres exists between cord blood and adult peripheral blood ECFC, consistent with significantly greater population doublings achieved by cultured cord blood ECFC before reaching replicative senescence. No unique identifying marker has yet been found to identify the ECFC within the human blood cells to permit prospective isolation, though all of the viable ECFC present in cord blood are present within a subset of cells with the phenotype $CD34^{+}CD31^{+}CD146^{+}CD45^{-}CD133^{-}$ [50]. It is unclear what role the circulating ECFC play within the human vasculature, however, recent studies implicate a role for the in vivo selectin-mediated adhesion of circulating human ECFC at the site of thrombus stabilization via recruited neutrophils. This neutrophil and ECFC interaction activates the proangiogenic functions of the ECFC and may be involved in vessel repair [51]. The number of circulating ECFC has been shown to highly correlate with the severity of coronary artery stenosis in symptomatic patients after adjusting for age, gender, cardiac risk factors, left ventricular ejection fraction, angiotensin-converting enzyme inhibitor or statin use [52]. The presence of higher

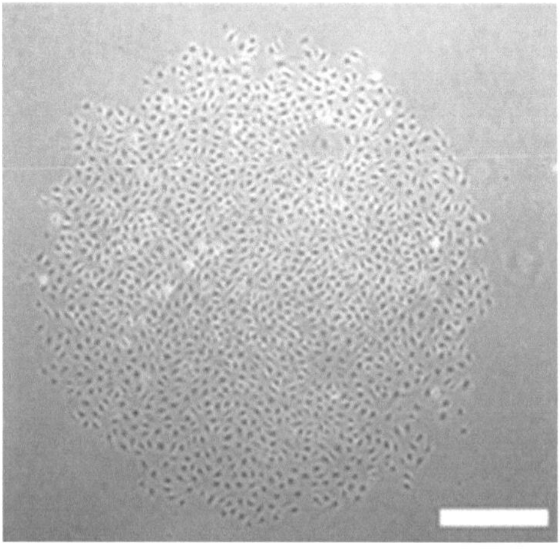

Fig. 15.3 Representative photomicrograph of endothelial colony. Colony formed 7 days after adult and cord blood-derived endothelial cells were plated at low cell density (×50 magnification). Scale bar represents 100 μm

concentrations of circulating ECFC is associated with reduced microvascular obstruction after acute myocardial infarction, leading to reduced infarct size and less left ventricular remodeling in patients compared to those patients in whom circulating ECFC could not be isolated [53]. The circulating concentration of ECFC has also been documented to increase tenfold following an acute myocardial infarction in human subjects [54], but a comparison of patient outcomes based upon the extent of the acute mobilization of these cells during myocardial infarction has not yet been conducted. These data all suggest a relationship between the presence of myocardial ischemic protection and circulating ECFC concentrations in human subjects. Of interest, the source of the circulating ECFC remains unknown.

Whereas circulating apoptotic and necrotic endothelial cells have long been thought to be sloughed vascular endothelial cells that increase in most cardiovascular diseases and cancer [55, 56], the specific site and mechanisms for generating viable ECFC remain obscure [42]. Duong et al. [57] have reported that directly plating of pulmonary artery endothelial cells (PAEC) isolated from vessels from patients with pulmonary hypertension revealed significantly greater number of clonogenic ECFC from the patient samples than from control subject lung vessels. They concluded that patient-derived PAEC contain significantly more ECFC that display greater proliferative potential than control subject PAEC. We have also identified ECFC resident in the endothelium isolated from umbilical cord and aortic endothelium of human subjects [58] that display high proliferative clonogenic potential. In human subjects with peripheral arterial disease, circulating ECFC display diminished proliferative potential that is matched by diminished ECFC numbers and proliferative potential by resident arterial samples [59]. Lin et al. [44] identified the most proliferative circulating ECFC to arise from the donor bone marrow tissue in patients that had undergone a sex-mismatched bone marrow transplant. Whether the ECFC were derived from vessels contained in the marrow tissue or from the hematopoietic cells that reconstituted the host subjects has not been clarified.

In sum, the strength of evidence supports an important role for resident local vascular endothelial cells to participate in repair of experimentally and genetic-induced endothelial injury or disruption in rodent models and that process is facilitated via paracrine molecules provided by circulating proangiogenic hematopoietic cells (previously called endothelial progenitor cells) recruited from the bone marrow. There is limited evidence for circulating endothelial cell contribution to vascular injury models in mice or rats, however, in larger mammalian species, substantial evidence implicates the presence of viable circulating endothelial cells that can cover intravascular implanted devices and circulating ECFC play a role in protection of the cardiac microcirculation following an acute infarction that promotes better patient outcomes.

Vascular Regeneration in Post-pneumonectomy Lung Growth

It has been recognized for more than a century that surgical removal of lung tissue in numerous animal species is associated with compensatory lung growth by the remaining lung tissue (reviewed in [60]). While compensatory growth was successfully demonstrated in dogs and rabbits in 1881, attempts to successfully remove lung tissue in human subjects failed to lead to successful outcomes until the 1930s (reviewed in [61]). Whether or not the human lung responds to lung resection with compensatory growth appears to be related to many factors including age. Several reports have documented long-term increases in lung volume (without an increase in residual volume) indicative of new alveolar growth following pneumonectomy [62–64]. The debate over whether the human adult lung can undergo compensatory growth following pneumonectomy remains controversial (reviewed in [60]). Continued advances in studying the mechanisms leading to compensatory growth in animal model systems have provided important understanding for how the host responds to loss of lung tissue with changes in cardiac and respiratory physiology, while advances in respiratory support and numerous surgical tools and approaches have steadily improved outcomes in this challenging operation.

Many physiologic factors are known to regulate the compensatory growth of the lung to pneumonectomy in animal model systems. These variables include the species, sex, and age of the animal tested [65, 66]. The impact of stretch on the remaining lung is known to be a major factor in the extent of compensatory lung growth observed post-pneumonectomy [67, 68]. Since the entire cardiac output is diverted into the remaining lung post-pneumonectomy, significant changes in pulmonary blood flow also impact the compensatory growth response [69, 70]. The specific roles played by the numerous cell types that comprise the lung in the realveolarization process that exemplifies compensatory lung growth have only recently begun to be elucidated.

Since sprouting angiogenesis is a prominent feature of the pulmonary capillary bed during normal lung development [71–73] and there is a parallel increase in the number of endothelial cells and the nearly 30 % increase in alveoli post-pneumonectomy [74], numerous authors have considered that angiogenesis may play a prominent role in post-pneumonectomy lung growth [75–77]. It is well known that a host of angiogenic growth factors are elevated in the lungs of animals post-pneumonectomy [78–82]. In fact, there are several waves of changes in gene expression that occur in the capillary bed post-pneumonectomy that may drive the angiogenic response [83].

Ding et al. [84] have provided recent analysis of some of the cellular and molecular mechanisms involved in compensatory lung growth following pneumonectomy in mice. Within 2 weeks of removal of the left lung, a 1.5-fold increase in right lung weight and 1.8-fold increase in right lung volume was observed. A significant increase in proliferation was documented in Clara cell secreted protein (CCSP) expressing bronchoalveolar stem cells (BASC) as early as day 3 post-pneumonectomy prior to any change in proliferation of type II alveolar epithelial cells (AECIIs) or pulmonary capillary endothelial cells (PCECs). However, proliferation in the AECIIs and PCECs is significantly increased by day 7, such that the proliferating PCECs constitute 7 % of the total lung mononuclear cells. By day 15 post-pneumonectomy, both AECIIs, which are known to be precursors of the type 1 AEC, and PCECs are increased threefold over sham control populations. Since growth activated endothelial cells are known to secrete a variety of angiocrine factors that can promote tissue repair and regeneration [85, 86], Ding et al. examined whether PCEC displayed evidence of VEGFR2 or fibroblast growth factor receptor 1 (FGFR1) activation. Though the overall level of VEGFR2 was not increased in PCEC, VEGFR2 phosphorylation was significantly increased on days 1–9 post-pneumonectomy, while FGFR1 was increased on days 3–11 post-pneumonectomy. These results implicated activation of these growth factor receptors in the PCEC as a potential key element in the PCEC expansion that occurred during lung regrowth. When VEGFR2 and/or FGFR1 were subsequently deleted specifically in endothelial cells in transgenic mice, left lung pneumonectomy failed to increase proliferation in the BASC population of the right lung at day 3 post-pneumonectomy, completely abrogated proliferation in the AECII and PCEC in the remaining lung at day 7 and beyond, and severely diminished restoration of lung weight and total lung volume in the right lung post-pneumonectomy. These results suggested that the growth factor-stimulated PCEC may be playing a significant role in the realveolarization of the lung post-pneumonectomy. In a series of additional studies the authors provided evidence that the PCEC of the regenerating right lung secreted high concentrations of matrix metalloproteinase 14 (MMP14), that PCEC MMP14 enhanced proliferation of AECII cells and BASC in a cell–cell contact interaction, and that inhibition of PCEC MMP14 with a blocking antibody in vivo blocked alveolarization in the remaining lung post-pneumonectomy. The specific mechanism through which MMP14 stimulated alveolarization appeared to be via unmasking of cryptic epidermal growth factor (EGF)-like ligands (from cleaved laminin matrix molecules) that stimulated the epithelial regrowth via the EGF receptor. Indeed, intravenous infusion of recombinant EGF restored alveologenesis post-pneumonectomy even in transgenic mice lacking both VEGFR2 alleles and one FGFR1 allele in PCEC. Of interest, intravenous infusion of lung PCEC, but not liver sinusoidal endothelial cells, from wild-type littermate controls also rescued alveologenesis and lung regrowth post-pneumonectomy in transgenic mice lacking PCEC expression of VEGFR2 and FGFR1. Ding et al. concluded that pneumonectomy induces a pulmonary specific activation of PCEC in the remaining lung that results in release of specific angiocrine factors that promote alveologenesis from BASC and AECII precursors and this collective endothelial–epithelial interaction leads to compensatory lung growth in adult mice (Fig. 15.4).

Other investigators have focused on understanding the different patterns of neovascularization observed in the remaining lung following pneumonectomy. New blood vessel growth can occur via sprouting angiogenesis from preexisting vessels or by non-sprouting intussusceptive angiogenesis [87]. During intussusceptive vessel remodeling, the opposing walls of the vessel are bridged together and

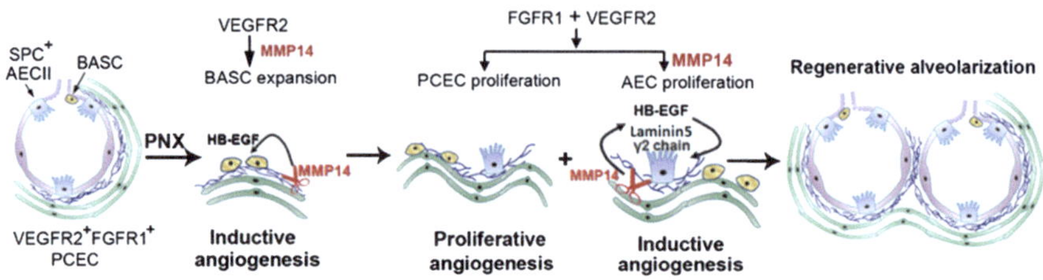

Fig. 15.4 Proposed model illustrating the inductive role of VEGFR2 and FGFR1 primed PCECs in lung regenerative alveolarization. Upon left pneumonectomy, activation of VEGFR2 in PCECs leads to MMP14 production and HB-EGF release to stimulate the expansion of epithelial progenitor cells (BASCs and AECIIs). Subsequent activation of FGFR1 along with VEGFR2 stimulates proliferation of PCECs maintaining MMP14 expression. MMP14 unmasks cryptic EGFR ligands through shedding of HB-EGF and cleaving laminin5 γ2 chain, which by activating EGFR induce proliferation of SPC + E-cadherin + AECs. After pneumonectomy, sequential propagation of epithelial cells induced by PCEC-derived MMP14 and increase in bioavailability of EGFR ligands, culminates in full reconstitution of physiologically functional alveolar-capillary sacs. Proliferation of the PCECs mediated through VEGFR2 and FGFR1, vascularizes the regenerating lung tissue to restore the blood supply and gas exchange function. *Abbreviations*: *VEGFR2* vascular endothelial growth factor receptor 2; *FGFR1* fibroblast growth factor receptor 1; *PCEC* pulmonary capillary endothelial cells; *MMP14* matrix metalloproteinase 14; *HB-EGF* heparin binding epidermal growth factor; *BASCs* basal alveolar stem cells; *AECIIs* alveolar epithelial type II cells; *EGFR* epidermal growth factor receptor; *SPC* surfactant protein C; *E-cadherin* epithelial-cadherin. Reproduced with permission from Ding et al. [84]

become sealed as a pillar that now separates the vessel into two vessels and as the new vessels mature, small holes 1–5 μm in diameter appear in between the vessels where the pillars originally formed [88–90] (Fig. 15.5). A detailed analysis of post-pneumonectomy lung growth by scanning electron microscopy, microCT scans, synchrotron radiation tomographic microscopy, and light and transmission electron microscopy has revealed that there is asymmetric growth of the right lung when the left lung is removed in rodents with the right cardiac lobe displaying the greatest expansion into the left pleural space [75, 91]. Surprisingly, many of the features of normal lung development are observed in the regenerating lung with upfolding of new alveolar septa and a duplicated capillary bed. This duplication is caused by massive intussusceptive angiogenesis and appears to be critical for the septal alveolarization in the regenerating lung just as it is in normal lung development [91] (Fig. 15.6). These data strongly implicate the resident capillary endothelial cells as the primary agents driving the vascular reparative response. However, the capillary endothelial cells do not function in a vacuum, as there are robust increases in recruited CD11b^{+} myeloid cells, alveolar macrophages, and AECII in close apposition to the pulmonary capillary endothelial cells and these recruited cells display an increase in angiogenic growth factor secretion during this phase of angiogenesis; disruption of the recruitment of these accessory cells diminishes the angiogenic response and the extent of lung regeneration indicating an important role for these diverse cell types [91–93].

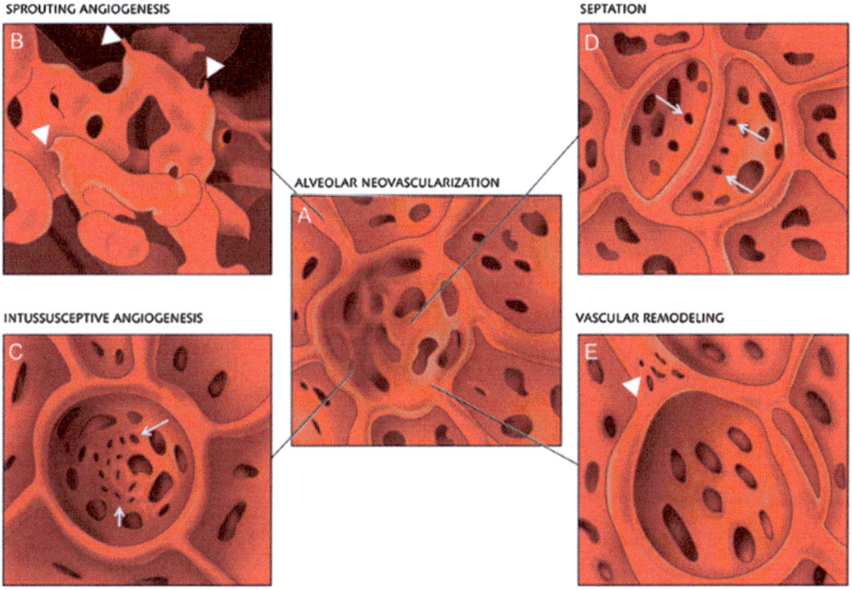

Fig. 15.5 Schematic illustration of alveolar neovascularization (**a**) as seen in microvascular corrosion cast replicas with sprouting angiogenesis (**b**), which is frequently seen subpleurally and in compact growth zones. Sprouts are evident as blind ends or protrusions (*arrowheads*). (**c**) Alveolar intussusceptive angiogenesis, recognizable by the presence of numerous small caliber holes with diameters between 1 and 5 μm as hallmarks of pillar formation (*arrows*). (**d**) The formation of new alveolar septa is accompanied by the occurrence of parallely orientated intussusceptive pillars (*arrows*)—visible in the cast as holes—ensuring a rapid expansion of the alveolar microvascular network. Vascular remodeling also occurs on the AER vessels (*arrowheads*). Intussusceptive pillars may merge and split up the primary vessel. Reproduced with permission from Ackermann et al. [91]

Chamoto et al. [92, 94] have recently proposed that blood-borne $CD34^+$ endothelial progenitor cells displaying enhanced proliferative status are recruited into the pulmonary circulation post-pneumonectomy and become resident endothelial cells during the period of compensatory lung growth. Whereas only 1 % of lung endothelial cells expressed CD34 at homeostasis, a dramatic 12-fold increase in $CD34^+$ lung endothelial cells was detected at day 7 post-pneumonectomy and the frequency of $CD34^+$ endothelial cells returned to nearly normal levels by day 21 following lung resection. At day 7 post-pneumonectomy, cell division was fourfold greater in the $CD34^+$ lung endothelial cells than in the $CD34^-$ lung endothelial cells. In an attempt to interrogate the contribution of blood-borne $CD34^+$ cells to the $CD34^+$ lung endothelial pool post-pneumonectomy, the authors surgically connected a wild-type mouse with a transgenic constitutive green fluorescence protein (GFP) expressing mouse via parabiosis [92, 95]. Following a 28-day recovery period, the wild-type parabiont underwent left pneumonectomy and 7 days later the regenerating right lung was analyzed for evidence of GFP^+

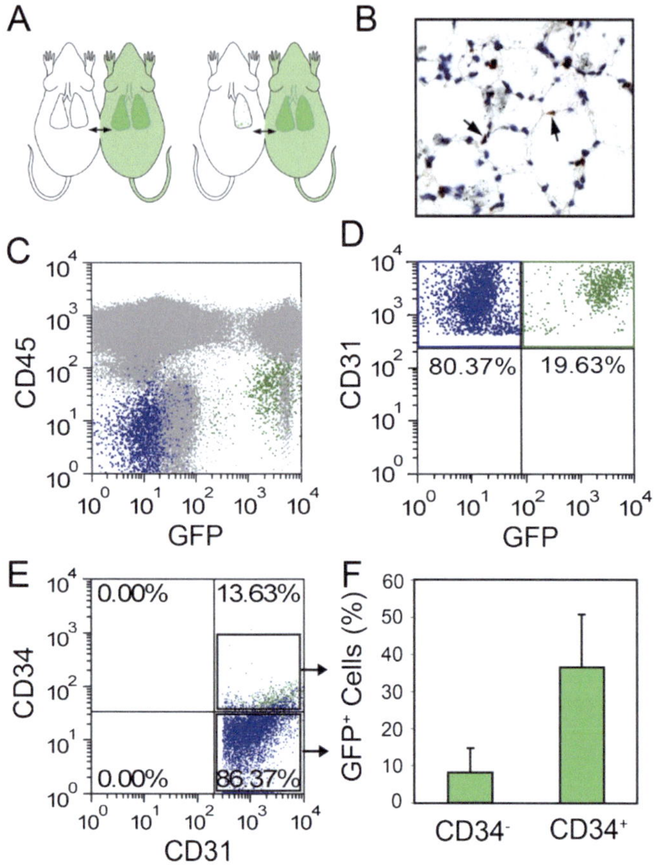

Fig. 15.6 Parabiotic demonstration of blood-borne CD34$^+$ endothelial cells on day 7 after pneumonectomy. (**a**) Schematic of experimental design involving a left pneumonectomy in the wild-type parabiont of wild-type/GFP$^+$ parabiotic twins (C57/B6). Parabiosis was established for 28 days prior to left pneumonectomy. The remaining lung was studied 7 days after pneumonectomy. (**b**) Anti-GFP avidin–biotin complex (ABC) immunohistochemical staining demonstrated GFP$^+$ cells within the alveolar septae. Flow cytometry of the lung digests demonstrated GFP$^+$ cells that were CD45$^-$ (**c**) and CD31$^+$ (**d**). (**e, f**) Analysis of CD31$^+$ lung endothelial cells demonstrated that most of the GFP expression occurred in the CD34$^+$ endothelial cell population. Reproduced with permission from Chamoto et al. [92]

blood-borne cells (Fig. 15.6). Though poorly visualized by fluorescence microscopy, GFP$^+$ cells were detectable by immunohistochemistry in the alveolar septa. Approximately 11 % (range 4–20 %) of the CD31$^+$ lung endothelial cells were GFP$^+$ and more than 1/3 of the CD34$^+$ endothelial cells in the lung were GFP$^+$ (Fig. 15.6). Intravenous infusion of several different fluorochrome labeled lectin molecules prior to sacrifice of the animals and digestion of the lung tissue, demonstrated uniform labeling of the CD34$^+$ and the CD34$^-$ lung endothelial cells, suggesting integration of the CD34$^+$ cells into the endothelial intima. These and other data led the authors to conclude that blood-borne bone marrow-derived $CD34^+$ EPCs were recruited and integrated into the lung circulation during the post-pneumonectomy lung growth. While the use of flow cytometry enhanced the ability of the authors to detect the presence of the dim GFP^+ cells within the lung endothelium, this tool lacks the ability to provide anatomic localization of the cells within the vasculature. Ohle et al. [96] have previously reported that confocal microscopy analysis of putative bone marrow-derived GFP^+ EPC contributions to injured lung has revealed that in essentially all events examined, the flat GFP^+ cells near a vessel lumen were located in a periendothelial location and were not *bona fide* GFP^+ endothelial cells. Since Chamoto et al. [92] showed that $CD34^+$ expression increases in the lung post-pneumonectomy independent of recruited GFP^+ cells (Fig. 15.6e), it is possible that some of the recruited $CD34^+GFP^+$ cells from the GFP^+ parabiont may have been derived as mobilized lung endothelial cells. A direct analysis of the cells circulating in the blood of the parabionts at rest and following pneumonectomy would be required to test this hypothesis. At present the majority of published work (above) suggests that the primary endothelial cells that participate in the angiogenic response that drives neoalveolarization and lung regeneration post-pneumonectomy arise from lung resident endothelial cells.

Vascular Repair and Regeneration in Lung Injury

Abounding evidence suggest the supportive role of circulating proangiogenic hematopoietic cells in the repair of acute and chronic lung injury. In the classic model of murine acute inflammatory lung injury induced by lipopolysaccharide (LPS), there was a rapid release of putative EPCs into the circulation that contribute in concert with other bone marrow-derived progenitor cells to lung repair [97]. Transplantation of autologous putative EPCs as a therapeutic strategy attenuated endotoxin-induced lung injury in rabbits [98]. Huang and Zhao [99] elegantly demonstrated the importance of the resident local vascular endothelial cells and their interaction with circulating bone-marrow-derived progenitor cells (BMPC) to insure lung repair by using a mouse model with endothelial cell-restricted disruption of FoxM1 (FoxM1 CKO) [100]. In wild-type mice, BMPC treatment improved LPS-induced lung inflammation and survival and lead to a rapid induction of FoxM1 expression. Conversely, these effects were abrogated in FoxM1 CKO mice and BMPC treatment failed to induce lung EC proliferation suggesting that endothelial expression of the reparative transcriptional factor FoxM1 is required for the protective effects of BMPCs. Human studies also suggest a supportive reparative role of circulating putative EPCs in acute and chronic lung diseases. Patients with acute lung injury have twofold higher numbers of circulating putative EPCs than healthy control subjects, suggesting some biological role for the mobilization of these cells during lung disease [101]. Interestingly and similar to the prognostic role of EPCs in ischemic vascular diseases, improved patient survival in acute lung injury correlates with increased circulating putative EPCs [102] and severity of illness [101]. Likewise, the number of circulating putative EPCs is significantly increased in patients with pneumonia and patients with low EPC counts tend to have persistent fibrotic changes in their lungs even after recovery from pneumonia. Likewise, the number of circulating putative EPCs is significantly increased in patients with pneumonia and patients with low EPC counts tend to have persistent fibrotic changes in their lungs even after recovery from pneumonia [103]. Decreased numbers and impaired mobilization of circulating putative EPCs have also been reported in patients with pulmonary fibrosis and chronic obstructive pulmonary disease (COPD), but remain controversial [103–105]. More interesting is the study of the function of these cells. Blood outgrowth endothelial cells (BOECs; also called ECFC) from smokers and COPD patients showed increased DNA damage and senescence as well as impaired angiogenic ability compared to nonsmokers [106]. BOECs senescence could be reversed by activating the protein deacetylase SIRT1. In summary, these data largely confirm the supportive role of circulating proangiogenic hematopoietic cells in the pathophysiology and repair of acute and chronic lung injury (though in many cases it is unclear exactly which circulating blood cell subset was comprising the EPC population evaluated).

Additional evidence for the important role of lung resident endothelial cells in lung repair comes from the literature in the developing lung. Bronchopulmonary dysplasia (BPD), the chronic lung disease that follows ventilator and oxygen therapy for acute respiratory failure after premature birth, now predominantly occurs in infants born <28 weeks gestation, is characterized by an arrest in alveolar development and decreased lung vascular growth. Borghesi et al. [107] studied ECFC in the cord blood of 98 preterm infants (gestational age <32 weeks). ECFCs in cord blood were lower in infants who later developed BPD and even though ECFCs decreased with decreasing gestational age, extremely low gestational age infants (gestational age <28 weeks) with higher numbers of cord blood ECFCs were protected from BPD. The endothelial and hematopoietic cell subsets studied by flow cytometry were comparable in infants with and without BPD and rapidly decreased after birth. Conversely, the same group showed that the percentages of circulating angiogenic cells ($CD34^+VEGFR\text{-}2^+$; $CD34^+CD133^+VEGFR\text{-}2^+$; and $CD45^-CD34^+CD133^+VEGFR\text{-}2^+$ cells) in the peripheral blood at birth, day 7 and day 28 of life in 142 preterm neonates (gestational age <32 weeks) were not able to predict the development of BPD [107]. Baker et al. quantified ECFCs and the angiogenic circulating progenitor cells (CPC)/nonangiogenic-CPC ratio (CPC/non-CPC) in cord blood samples from 62 preterm infants (24–36 weeks gestational age) [108]. ECFC number and CPC/non-CPC ratio were significantly decreased in cord blood of preterm infants who subsequently developed moderate or severe BPD. These data suggest that decreased circulating ECFCs in the fetal circulation impairs postnatal vasculogenesis thereby contributing to the severity of chronic lung disease in preterm infants.

Histological abnormalities of BPD are recapitulated in neonatal rodents exposed to chronic hyperoxia [109]. This model exhibits decreased circulating, lung and bone-marrow-derived putative EPCs [110]. Interestingly, hyperoxic adult mice did not display alveolar damage and had increased circulating putative EPCs, implying that decreased EPCs may contribute to the arrested lung growth seen in the neonatal animals. Infusion of these bone marrow-derived angiogenic cells restored alveolar and lung vascular growth in neonatal mice exposed to hyperoxia [111]. The recent discovery in the adult rat lung of resident microvascular endothelial progenitor cells that share features of human cord blood and lung ECFCs provided an opportunity to explore the role of resident ECFCs in experimental BPD models. Alphonse et al. [112] were able to show that the developing human fetal and neonatal rat lungs contained ECFCs with robust proliferative potential, secondary colony formation, and de novo blood vessel formation in vivo when transplanted into a matrigel plug under the skin of immunodeficient mice [112]. In contrast, human fetal lung ECFCs exposed to hyperoxia in vitro and neonatal rat ECFCs isolated from hyperoxic alveolar growth-arrested rat lungs mimicking BPD proliferated less, showed decreased clonogenic capacity, and formed fewer capillary-like networks. This suggested that impaired ECFC function may contribute to explain the incapacity of lung repair and the persistent alveolar growth in BPD. It also provided the rationale to test the therapeutic potential of exogenous supplementation of human cord blood-derived ECFCs. Because of the increased toxicity of hyperoxia on ECFCs [113, 114], intrajugular administration of human cord blood-derived ECFCs was performed after established oxygen-induced arrested alveolar growth. ECFCs restored lung function, alveolar and lung vascular growth, and attenuated pulmonary hypertension in immune-deficient rats and mice. Lung ECFC colony- and capillary-like network-forming capabilities were also restored. The therapeutic benefit persisted at 10 months of age with no adverse effects and persistent improvement in lung structure, exercise capacity, and pulmonary hypertension. However, lung engraftment, similar to what has been observed with other types of cell therapies (mesenchymal stromal cells for example) was low, suggesting that the effect of these exogenous ECFCs was through a paracrine effect. Accordingly, administration of cell-free ECFC-derived conditioned media exerted similar therapeutic benefits in the hyperoxia- and bleomycin-induced models of BPD in rodents [112, 115]. Overall, these findings further support the crucial role of lung resident endothelial cells in lung repair and offers promising new therapeutic strategies to prevent/restore lung damage.

References

1. Fuchs E (2009) The tortoise and the hair: slow-cycling cells in the stem cell race. Cell 137(5): 811–819
2. Orkin SH, Zon LI (2008) Hematopoiesis: an evolving paradigm for stem cell biology. Cell 132(4):631–644
3. Barker N, van de Wetering M, Clevers H (2008) The intestinal stem cell. Genes Dev 22(14): 1856–1864
4. Barker N (2014) Adult intestinal stem cells: critical drivers of epithelial homeostasis and regeneration. Nat Rev Mol Cell Biol 15(1):19–33

5. Hoath SB, Leahy DG (2003) The organization of human epidermis: functional epidermal units and phi proportionality. J Invest Dermatol 121(6):1440–1446
6. Altschul R (1954) Endothelium: its development, morphology, function, and pathology. Macmillan, New York
7. Florentin RA, Nam SC, Lee KT, Thomas WA (1969) Increased 3H-thymidine incorporation into endothelial cells of swine fed cholesterol for 3 days. Exp Mol Pathol 10(3):250–255
8. Poole JC, Sanders AG, Florey HW (1958) The regeneration of aortic endothelium. J Pathol Bacteriol 75(1):133–143
9. Robertson HR, Moore JR, Mersereau WA (1959) Observations on thrombosis and endothelial repair following application of external pressure to a vein. Can J Surg 3:5–16
10. Stehbens WE (1965) Endothelial cell mitosis and permeability. Q J Exp Physiol Cogn Med Sci 50:90–92
11. Wright HP (1968) Endothelial mitosis around aortic branches in normal guinea pigs. Nature 220(5162):78–79
12. Caplan BA, Schwartz CJ (1973) Increased endothelial cell turnover in areas of in vivo evans blue uptake in the pig aorta. Atherosclerosis 17(3):401–417
13. Prescott MF, Muller KR (1983) Endothelial regeneration in hypertensive and genetically hypercholesterolemic rats. Arteriosclerosis 3(3):206–214
14. Schwartz SM, Benditt EP (1977) Aortic endothelial cell replication. I. Effects of age and hypertension in the rat. Circ Res 41(2):248–255
15. Schwartz SM, Gajdusek CM, Reidy MA, Selden SC 3rd, Haudenschild CC (1980) Maintenance of integrity in aortic endothelium. Fed Proc 39(9):2618–2625
16. Schwartz SM, Benditt EP (1976) Clustering of replicating cells in aortic endothelium. Proc Natl Acad Sci U S A 73(2):651–653
17. Li F, Downing BD, Smiley LC, Mund JA, Distasi MR, Bessler WK et al (2014) Neurofibromin-deficient myeloid cells are critical mediators of aneurysm formation in vivo. Circulation 129(11):1213–1224
18. Wara AK, Manica A, Marchini JF, Sun X, Icli B, Tesmenitsky Y et al (2013) Bone marrow-derived kruppel-like factor 10 controls reendothelialization in response to arterial injury. Arterioscler Thromb Vasc Biol 33(7):1552–1560
19. Schwartz SM, Gajdusek CM, Selden SC 3rd (1981) Vascular wall growth control: the role of the endothelium. Arteriosclerosis 1(2):107–126
20. Tannock IF, Hayashi S (1972) The proliferation of capillary endothelial cells. Cancer Res 32(1):77–82
21. Hobson B, Denekamp J (1984) Endothelial proliferation in tumours and normal tissues: continuous labelling studies. Br J Cancer 49(4):405–413
22. Sholley MM, Cavallo T, Cotran RS (1977) Endothelial proliferation in inflammation. I. Autoradiographic studies following thermal injury to the skin of normal rats. Am J Pathol 89(2):277–296
23. Haudenschild C, Studer A (1971) Early interactions between blood cells and severely damaged rabbit aorta. Eur J Clin Invest 2(1):1–7
24. Schwartz SM, Stemerman MB, Benditt EP (1975) The aortic intima: II. Repair of the aortic lining after mechanical denudation. Am J Pathol 81(1):15–42
25. Moschcowitz E (1950) Hyperplastic arteriosclerosis versus atherosclerosis. J Am Med Assoc 143(10):861–865
26. Asahara T, Murohara T, Sullivan A, Silver M, van der Zee R, Li T et al (1997) Isolation of putative progenitor endothelial cells for angiogenesis. Science 275(5302):964–967
27. Asahara T, Kawamoto A, Masuda H (2011) Concise review: circulating endothelial progenitor cells for vascular medicine. Stem Cells 29(11):1650–1655
28. Fadini GP, Losordo D, Dimmeler S (2012) Critical reevaluation of endothelial progenitor cell phenotypes for therapeutic and diagnostic use. Circ Res 110(4):624–637
29. Yoder MC (2012) Human endothelial progenitor cells. Cold Spring Harb Perspect Med 2(7): a006692

30. Medina RJ, O'Neill CL, Sweeney M, Guduric-Fuchs J, Gardiner TA, Simpson DA, Stitt AW (2010) Molecular analysis of endothelial progenitor cell (EPC) subtypes reveals two distinct cell populations with different identities. BMC Med Genomics 3:18
31. Ohtani K, Vlachojannis GJ, Koyanagi M, Boeckel JN, Urbich C, Farcas R, Bonig H, Marquez VE, Zeiher AM, Dimmeler S (2011) Epigenetic regulation of endothelial lineage committed genes in pro-angiogenic hematopoietic and endothelial progenitor cells. Circ Res 109(11):1219–1229
32. Rose JA, Erzurum S, Asosingh K (2015) Biology and flow cytometry of proangiogenic hematopoietic progenitors cells. Cytometry A 87(1):5–19
33. Hagensen MK, Raarup MK, Mortensen MB, Thim T, Nyengaard JR, Falk E, Bentzon JF (2012) Circulating endothelial progenitor cells do not contribute to regeneration of endothelium after murine arterial injury. Cardiovasc Res 93(2):223–231
34. Itoh Y, Toriumi H, Yamada S, Hoshino H, Suzuki N (2010) Resident endothelial cells surrounding damaged arterial endothelium reendothelialize the lesion. Arterioscler Thromb Vasc Biol 30(9):1725–1732
35. Nishimura R, Kawasaki T, Sekine A, Suda R, Urushibara T, Suzuki T et al (2015) Hypoxia-induced proliferation of tissue-resident endothelial progenitor cells in the lung. Am J Physiol Lung Cell Mol Physiol 308(8):L746–L758. doi:10.1152/ajplung.00243.2014
36. Rinkevich Y, Lindau P, Ueno H, Longaker MT, Weissman IL (2011) Germ-layer and lineage-restricted stem/progenitors regenerate the mouse digit tip. Nature 476(7361): 409–413
37. Tsuzuki M (2009) Bone marrow-derived cells are not involved in reendothelialized endothelium as endothelial cells after simple endothelial denudation in mice. Basic Res Cardiol 104(5):601–611
38. Purhonen S, Palm J, Rossi D, Kaskenpaa N, Rajantie I, Yla-Herttuala S et al (2008) Bone marrow-derived circulating endothelial precursors do not contribute to vascular endothelium and are not needed for tumor growth. Proc Natl Acad Sci U S A 105(18):6620–6625
39. Hagensen MK, Vanhoutte PM, Bentzon JF (2012) Arterial endothelial cells: still the craftsmen of regenerated endothelium. Cardiovasc Res 95(3):281–289
40. Hagensen MK, Shim J, Thim T, Falk E, Bentzon JF (2010) Circulating endothelial progenitor cells do not contribute to plaque endothelium in murine atherosclerosis. Circulation 121(7): 898–905
41. Stump MM, Jordan GL Jr, Debakey ME, Halpert B (1963) Endothelium grown from circulating blood on isolated intravascular dacron hub. Am J Pathol 43:361–367
42. Yoder MC (2010) Is endothelium the origin of endothelial progenitor cells? Arterioscler Thromb Vasc Biol 30(6):1094–1103
43. Ingram DA, Mead LE, Tanaka H, Meade V, Fenoglio A, Mortell K et al (2000) Identification of a novel hierarchy of endothelial progenitor cells using human peripheral and umbilical cord blood. Blood 104(9):2752–2760
44. Lin Y, Weisdorf DJ, Solovey A, Hebbel RP (2000) Origins of circulating endothelial cells and endothelial outgrowth from blood. J Clin Invest 105(1):71–77
45. Au P, Daheron LM, Duda DG, Cohen KS, Tyrrell JA, Lanning RM et al (2008) Differential in vivo potential of endothelial progenitor cells from human umbilical cord blood and adult peripheral blood to form functional long-lasting vessels. Blood 111(3):1302–1305
46. Medina RJ, O'Neill CL, Humphreys MW, Gardiner TA, Stitt AW (2010) Outgrowth endothelial cells: characterization and their potential for reversing ischemic retinopathy. Invest Ophthalmol Vis Sci 51(11):5906–5913
47. Melero-Martin JM, De Obaldia ME, Kang SY, Khan ZA, Yuan L, Oettgen P, Bischoff J (2008) Engineering robust and functional vascular networks in vivo with human adult and cord blood-derived progenitor cells. Circ Res 103(2):194–202
48. Melero-Martin JM, Khan ZA, Picard A, Wu X, Paruchuri S, Bischoff J (2007) In vivo vasculogenic potential of human blood-derived endothelial progenitor cells. Blood 109(11): 4761–4768

49. Yoder MC, Mead LE, Prater D, Krier TR, Mroueh KN, Li F et al (2007) Redefining endothelial progenitor cells via clonal analysis and hematopoietic stem/progenitor cell principals. Blood 109(5):1801–1809
50. Mund JA, Estes ML, Yoder MC, Ingram DA Jr, Case J (2012) Flow cytometric identification and functional characterization of immature and mature circulating endothelial cells. Arterioscler Thromb Vasc Biol 32(4):1045–1053
51. Hubert L, Darbousset R, Panicot-Dubois L, Robert S, Sabatier F, Fallague K et al (2014) Neutrophils recruit and activate human endothelial colony-forming cells at the site of vessel injury via P-selectin glycoprotein ligand-1 and L-selectin. J Thromb Haemost 12(7): 1170–1181
52. Guven H, Shepherd RM, Bach RG, Capoccia BJ, Link DC (2006) The number of endothelial progenitor cell colonies in the blood is increased in patients with angiographically significant coronary artery disease. J Am Coll Cardiol 48(8):1579–1587
53. Meneveau N, Deschaseaux F, Seronde MF, Chopard R, Schiele F, Jehl J et al (2011) Presence of endothelial colony-forming cells is associated with reduced microvascular obstruction limiting infarct size and left ventricular remodelling in patients with acute myocardial infarction. Basic Res Cardiol 106(6):1397–1410
54. Massa M, Campanelli R, Bonetti E, Ferrario M, Marinoni B, Rosti V (2009) Rapid and large increase of the frequency of circulating endothelial colony-forming cells (ECFCs) generating late outgrowth endothelial cells in patients with acute myocardial infarction. Exp Hematol 37(1):8–9
55. Blann AD, Woywodt A, Bertolini F, Bull TM, Buyon JP, Clancy RM et al (2005) Circulating endothelial cells. Biomarker of vascular disease. Thromb Haemost 93(2):228–235
56. Woywodt A, Blann AD, Kirsch T, Erdbruegger U, Banzet N, Haubitz M, Dignat-George F (2006) Isolation and enumeration of circulating endothelial cells by immunomagnetic isolation: proposal of a definition and a consensus protocol. J Thromb Haemost 4(3):671–677
57. Duong HT, Comhair SA, Aldred MA, Mavrakis L, Savasky BM, Erzurum SC, Asosingh K (2011) Pulmonary artery endothelium resident endothelial colony-forming cells in pulmonary arterial hypertension. Pulm Circ 1(4):475–486
58. Ingram DA, Mead LE, Moore DB, Woodard W, Fenoglio A, Yoder MC (2005) Vessel wall-derived endothelial cells rapidly proliferate because they contain a complete hierarchy of endothelial progenitor cells. Blood 105(7):2783–2786
59. Prasain N, Lee MR, Vemula S, Meador JL, Yoshimoto M, Ferkowicz MJ et al (2014) Differentiation of human pluripotent stem cells to cells similar to cord-blood endothelial colony-forming cells. Nat Biotechnol 32(11):1151–1157
60. Paisley D, Bevan L, Choy KJ, Gross C (2014) The pneumonectomy model of compensatory lung growth: insights into lung regeneration. Pharmacol Ther 142(2):196–205
61. Fuentes PA (2003) Pneumonectomy: historical perspective and prospective insight. Eur J Cardiothorac Surg 23(4):439–445
62. Laros CD, Westermann CJ (1987) Dilatation, compensatory growth, or both after pneumonectomy during childhood and adolescence. A thirty-year follow-up study. J Thorac Cardiovasc Surg 93(4):570–576
63. Stiles QR, Meyer BW, Lindesmith GG, Jones JC (1969) The effects of pneumonectomy in children. J Thorac Cardiovasc Surg 58(3):394–400
64. Werner HA, Pirie GE, Nadel HR, Fleisher AG, LeBlanc JG (1993) Lung volumes, mechanics, and perfusion after pulmonary resection in infancy. J Thorac Cardiovasc Surg 105(4): 737–742
65. Khadempour MH, Ofulue AF, Sekhon HS, Cherukupalli KM, Thurlbeck WM (1992) Changes of growth hormone, somatomedin c, and bombesin following pneumonectomy. Exp Lung Res 18(3):421–432
66. Sekhon HS, Thurlbeck WM (1992) A comparative study of postpneumonectomy compensatory lung response in growing male and female rats. J Appl Physiol 73(2):446–451

67. Brown LM, Rannels SR, Rannels DE (2001) Implications of post-pneumonectomy compensatory lung growth in pulmonary physiology and disease. Respir Res 2(6):340–347
68. Fisher JM, Simnett JD (1973) Morphogenetic and proliferative changes in the regenerating lung of the rat. Anat Rec 176(4):389–395
69. Dane DM, Yilmaz C, Estrera AS, Hsia CC (2013) Separating in vivo mechanical stimuli for postpneumonectomy compensation: physiological assessment. J Appl Physiol 114(1):99–106
70. McBride JT, Kirchner KK, Russ G, Finkelstein J (1992) Role of pulmonary blood flow in postpneumonectomy lung growth. J Appl Physiol 73(6):2448–2451
71. Metzger RJ, Klein OD, Martin GR, Krasnow MA (2008) The branching programme of mouse lung development. Nature 453(7196):745–750
72. White AC, Lavine KJ, Ornitz DM (2007) FGF9 and SHH regulate mesenchymal Vegfa expression and development of the pulmonary capillary network. Development 134(20): 3743–3752
73. Yamamoto H, Yun EJ, Gerber HP, Ferrara N, Whitsett JA, Vu TH (2007) Epithelial-vascular cross talk mediated by VEGF-A and HGF signaling directs primary septae formation during distal lung morphogenesis. Dev Biol 308(1):44–53
74. Fehrenbach H, Voswinckel R, Michl V, Mehling T, Fehrenbach A, Seeger W, Nyengaard JR (2008) Neoalveolarisation contributes to compensatory lung growth following pneumonectomy in mice. Eur Respir J 31(3):515–522
75. Konerding MA, Gibney BC, Houdek JP, Chamoto K, Ackermann M, Lee GS et al (2012) Spatial dependence of alveolar angiogenesis in post-pneumonectomy lung growth. Angiogenesis 15(1):23–32
76. Takeda SI, Ramanathan M, Estrera AS, Hsia CC (1999) Postpneumonectomy alveolar growth does not normalize hemodynamic and mechanical function. J Appl Physiol 87(2):491–497
77. Yan X, Bellotto DJ, Foster DJ, Johnson RL Jr, Hagler HK, Estrera AS, Hsia CC (2004) Retinoic acid induces nonuniform alveolar septal growth after right pneumonectomy. J Appl Physiol 96(3):1080–1089
78. Jancelewicz T, Grethel EJ, Chapin CJ, Clifton MS, Nobuhara KK (2010) Vascular endothelial growth factor isoform and receptor expression during compensatory lung growth. J Surg Res 160(1):107–113
79. Landesberg LJ, Ramalingam R, Lee K, Rosengart TK, Crystal RG (2001) Upregulation of transcription factors in lung in the early phase of postpneumonectomy lung growth. Am J Physiol Lung Cell Mol Physiol 281(5):L1138–L1149
80. Sakamaki Y, Matsumoto K, Mizuno S, Miyoshi S, Matsuda H, Nakamura T (2002) Hepatocyte growth factor stimulates proliferation of respiratory epithelial cells during postpneumonectomy compensatory lung growth in mice. Am J Respir Cell Mol Biol 26(5): 525–533
81. Yuan S, Hannam V, Belcastro R, Cartel N, Cabacungan J, Wang J et al (2002) A role for platelet-derived growth factor-bb in rat postpneumonectomy compensatory lung growth. Pediatr Res 52(1):25–33
82. Zhang Q, Bellotto DJ, Ravikumar P, Moe OW, Hogg RT, Hogg DC et al (2007) Postpneumonectomy lung expansion elicits hypoxia-inducible factor-1alpha signaling. Am J Physiol Lung Cell Mol Physiol 293(2):L497–L504
83. Lin M, Chamoto K, Gibney BC, Lee GS, Collings-Simpson D, Houdek J et al (2011) Angiogenesis gene expression in murine endothelial cells during post-pneumonectomy lung growth. Respir Res 12:98
84. Ding BS, Nolan DJ, Guo P, Babazadeh AO, Cao Z, Rosenwaks Z et al (2011) Endothelial-derived angiocrine signals induce and sustain regenerative lung alveolarization. Cell 47(3): 539–553
85. Ding BS, Nolan DJ, Butler JM, James D, Babazadeh AO, Rosenwaks Z et al (2010) Inductive angiocrine signals from sinusoidal endothelium are required for liver regeneration. Nature 468(7321):310–315

86. Hooper AT, Butler JM, Nolan DJ, Kranz A, Iida K, Kobayashi M et al (2009) Engraftment and reconstitution of hematopoiesis is dependent on VEGFR2-mediated regeneration of sinusoidal endothelial cells. Cell Stem Cell 4(3):263–274
87. Carmeliet P, Jain RK (2011) Molecular mechanisms and clinical applications of angiogenesis. Nature 473(7347):298–307
88. Burri PH, Tarek MR (1990) A novel mechanism of capillary growth in the rat pulmonary microcirculation. Anat Rec 228(1):35–45
89. Caduff JH, Fischer LC, Burri PH (1986) Scanning electron microscope study of the developing microvasculature in the postnatal rat lung. Anat Rec 216(2):154–164
90. Djonov VG, Kurz H, Burri PH (2002) Optimality in the developing vascular system: branching remodeling by means of intussusception as an efficient adaptation mechanism. Dev Dyn 224(4):391–402
91. Ackermann M, Houdek JP, Gibney BC, Ysasi A, Wagner W, Belle J (2014) Sprouting and intussusceptive angiogenesis in postpneumonectomy lung growth: mechanisms of alveolar neovascularization. Angiogenesis 17(3):541–551
92. Chamoto K, Gibney BC, Ackermann M, Lee GS, Lin M, Konerding MA et al (2012) Alveolar macrophage dynamics in murine lung regeneration. J Cell Physiol 227(9): 3208–3215
93. Chamoto K, Gibney BC, Lee GS, Ackermann M, Konerding MA, Tsuda A, Mentzer SJ (2013) Migration of CD11b+ accessory cells during murine lung regeneration. Stem Cell Res 10(3):267–277
94. Chamoto K, Gibney BC, Lee GS, Lin M, Collings-Simpson D, Voswinckel R et al (2012) CD34+ progenitor to endothelial cell transition in post-pneumonectomy angiogenesis. Am J Respir Cell Mol Biol 46(3):283–289
95. Gibney BC, Chamoto K, Lee GS, Simpson DC, Miele LF, Tsuda A et al (2012) Cross-circulation and cell distribution kinetics in parabiotic mice. J Cell Physiol 227(2):821–828
96. Ohle SJ, Anandaiah A, Fabian AJ, Fine A, Kotton DN (2012) Maintenance and repair of the lung endothelium does not involve contributions from marrow-derived endothelial precursor cells. Am J Respir Cell Mol Biol 47(1):11–19
97. Yamada M, Kubo H, Kobayashi S, Ishizawa K, Numasaki M, Ueda S et al (2004) Bone marrow-derived progenitor cells are important for lung repair after lipopolysaccharide-induced lung injury. J Immunol 172(2):1266–1272
98. Lam CF, Liu YC, Hsu JK, Yeh PA, Su TY, Huang CC et al (2008) Autologous transplantation of endothelial progenitor cells attenuates acute lung injury in rabbits. Anesthesiology 108(3):392–401
99. Huang X, Zhao YY (2012) Transgenic expression of foxm1 promotes endothelial repair following lung injury induced by polymicrobial sepsis in mice. PLoS One 7(11), e50094
100. Zhao YD, Huang X, Yi F, Dai Z, Qian Z, Tiruppathi C et al (2014) Endothelial foxm1 mediates bone marrow progenitor cell-induced vascular repair and resolution of inflammation following inflammatory lung injury. Stem Cells 32(7):1855–1864
101. Burnham EL, Taylor WR, Quyyumi AA, Rojas M, Brigham KL, Moss M (2005) Increased circulating endothelial progenitor cells are associated with survival in acute lung injury. Am J Respir Crit Care Med 172(7):854–860
102. Yamada M, Kubo H, Ishizawa K, Kobayashi S, Shinkawa M, Sasaki H (2005) Increased circulating endothelial progenitor cells in patients with bacterial pneumonia: evidence that bone marrow derived cells contribute to lung repair. Thorax 60(5):410–413
103. Malli F, Koutsokera A, Paraskeva E, Zakynthinos E, Papagianni M, Makris D et al (2013) Endothelial progenitor cells in the pathogenesis of idiopathic pulmonary fibrosis: an evolving concept. PLoS One 8(1), e53658
104. Smadja DM, Mauge L, Nunes H, d'Audigier C, Juvin K, Borie R et al (2013) Imbalance of circulating endothelial cells and progenitors in idiopathic pulmonary fibrosis. Angiogenesis 16(1):147–157

105. Brittan M, Hoogenboom MM, Padfield GJ, Tura O, Fujisawa T, Maclay JD et al (2013) Endothelial progenitor cells in patients with chronic obstructive pulmonary disease. Am J Physiol Lung Cell Mol Physiol 305(12):L964–L969
106. Paschalaki KE, Starke RD, Hu Y, Mercado N, Margariti A, Gorgoulis VG et al (2013) Dysfunction of endothelial progenitor cells from smokers and chronic obstructive pulmonary disease patients due to increased DNA damage and senescence. Stem Cells 31(12): 2813–2826
107. Borghesi A, Massa M, Campanelli R, Garofoli F, Longo S, Cabano R et al (2013) Different subsets of circulating angiogenic cells do not predict bronchopulmonary dysplasia or other diseases of prematurity in preterm infants. Int J Immunopathol Pharmacol 26(3):809–816
108. Baker CD, Balasubramaniam V, Mourani PM, Sontag MK, Black CP, Ryan SL et al (2012) Cord blood angiogenic progenitor cells are decreased in bronchopulmonary dysplasia. Eur Respir J 40(6):1516–1522
109. O'Reilly MA (2012) Angiotensin II: tapping the cell cycle machinery to kill endothelial cells. Am J Physiol Lung Cell Mol Physiol 303(7):L575–L576
110. Balasubramaniam V, Mervis CF, Maxey AM, Markham NE, Abman SH (2007) Hyperoxia reduces bone marrow, circulating, and lung endothelial progenitor cells in the developing lung: implications for the pathogenesis of bronchopulmonary dysplasia. Am J Physiol Lung Cell Mol Physiol 292(5):L1073–L1084
111. Balasubramaniam V, Ryan SL, Seedorf GJ, Roth EV, Heumann TR, Yoder MC et al (2010) Bone marrow-derived angiogenic cells restore lung alveolar and vascular structure after neonatal hyperoxia in infant mice. Am J Physiol Lung Cell Mol Physiol 298(3):L315–L323
112. Alphonse RS, Vadivel A, Fung M, Shelley WC, Critser PJ, Ionescu L et al (2014) Existence, functional impairment, and lung repair potential of endothelial colony-forming cells in oxygen-induced arrested alveolar growth. Circulation 129(21):2144–2157
113. Baker CD, Ryan SL, Ingram DA, Seedorf GJ, Abman SH, Balasubramaniam V (2009) Endothelial colony-forming cells from preterm infants are increased and more susceptible to hyperoxia. Am J Respir Crit Care Med 180(5):454–461
114. Fujinaga H, Baker CD, Ryan SL, Markham NE, Seedorf GJ, Balasubramaniam V, Abman SH (2009) Hyperoxia disrupts vascular endothelial growth factor-nitric oxide signaling and decreases growth of endothelial colony-forming cells from preterm infants. Am J Physiol Lung Cell Mol Physiol 297(6):L1160–L1169
115. Baker CD, Seedorf GJ, Wisniewski BL, Black CP, Ryan SL, Balasubramaniam V, Abman SH (2013) Endothelial colony-forming cell conditioned media promote angiogenesis in vitro and prevent pulmonary hypertension in experimental bronchopulmonary dysplasia. Am J Physiol Lung Cell Mol Physiol 305(1):L73–L81

Chapter 16
Directed Differentiation of Human Pluripotent Stem Cells into Lung and Airway Epithelial Cells

Hans-Willem Snoeck

Introduction

The capability to generate, expand, and differentiate human pluripotent stem cells (hPSCs) is changing the face of biomedical research. Derived from the inner cell mass of the blastocyst of mammals such as mouse and human, embryonic stem cells (ESCs) can be maintained in a pluripotent state in vitro and give rise to every cell type in the organism [1]. Induced pluripotent stem cells (iPSCs) are generated by reprogramming of somatic cells to a pluripotent state [2–7]. The ability to generate tissues and organs from human induced pluripotent state (iPS) cells for tissue or organ replacement therapy is a distant but very important goal [1, 8]. In the shorter term, hPSC-derived differentiated cells are used for disease modeling and drug discovery [9, 10].

Rationale for Generation of Lung and Airway Epithelial Cells from PSCs

Lung regeneration: Nonmalignant lung disease kills between 100,000 and 200,000 people in the USA every year [11]. Lung transplantation, currently the only therapeutic option in eligible patients with end-stage lung disease, is accompanied by a high incidence of medical and immunological complications and hampered by severe shortage of donor organs [12]. Engraftment with adult stem cells is envisaged. However, their identity, functional characteristics, and physical location in the lung, in particular in humans, are unclear. Furthermore, stem cells thus far

H.-W. Snoeck, M.D., Ph.D. (✉)
Columbia Center for Translational Immunology, Columbia University Medical Center, Black Building, 630 W. 168th Street, Room 1501E, New York, NY 10032, USA
e-mail: hs2680@columbia.edu

I. Bertoncello (ed.), *Stem Cells in the Lung*, Stem Cell Biology and Regenerative Medicine, DOI 10.1007/978-3-319-21082-7_16

cannot be prospectively isolated with high purity, expanded in vitro, and engrafted in injured lungs [13–15].

The field is currently attempting to achieve replacement of diseased lung and airways using hPSC-derived cells. One potential approach is to reconstruct autologous donor lung by seeding iPS-derived cells on decellularized lung matrices. Indeed, some proof-of-principle has been published in the rat model using fetal lung and airway epithelial cells and endothelial cells [16, 17]. Major technical hurdles clearly remain before this type approach can be translated to the clinic. Sufficient numbers of cells with the appropriate maturity, variety, and ratio of epithelial cells normally found in the lung have to be generated from patient-derived iPS cells. These have to home to their correct niches along the bronchial tree and within the alveoli. Vascular leakage is a major problem. Furthermore, current decellularization approaches appear to disproportionally destroy elastin, a component of the lung extracellular matrix that is essential to its biomechanical properties [18]. To effect evacuation of debris, cilia of multiciliated cells have to beat synchronously and coordinately. It is not known however how planar polarity in the airway is established. Finally, the regionally distinct postnatal stem and progenitor cells must be included to endow the graft with endogenous regenerative capacity [13].

Some clinical successes have been achieved with engineered trachea and large airways, though none involved hPSC-derived cells. Approaches included a decellularized human trachea scaffold seeded with autologous respiratory epithelial cells and mesenchymal stem cell-derived chondrocytes [19] or autologous respiratory cells and bone marrow mononuclear cells [20, 21], a bioartificial nanocomposite seeded with autologous bone marrow mononuclear cells [22], or more complex approaches involving heterotopic transplantation of an allogeneic graft in the forearm, followed by orthotopic placement of the trachea [23]. Overall, the results of long-term follow-up are unclear and controversy exists on the optimal approach [21, 24].

Disease modeling (Table 16.1): A major application of hPSCs-derived lung and airway epithelial cells is and will be modeling of congenital and acquired human lung disease [10].

Cystic fibrosis (CF), the most common lethal genetic disease among Caucasians (1/2000 births), is caused by mutations in the CFTR chloride ion channel resulting in altered ion flux in exocrine organs and hyperproduction of viscous mucus. Impairment of mucociliary function in the conducting airways leads to colonization of pathogenic bacteria and intractable, infectious complications. Path clamp studies on iPS-derived airway epithelial cells have indeed shown altered chloride conductance [25, 26]. These cells could therefore be used for drug screening. More complex three-dimensional (3D) models might begin to explain how altered chloride conductance leads to the observed severe impairment of mucociliary function. Such models do not exist currently, however.

A subset of almost universally lethal pediatric lung diseases are due to surfactant deficiency, caused by mutations in genes encoding surfactant proteins (SP-C, SP-B), or factors required for surfactant trafficking (ABCA3, CSFR2A)

Table 16.1 Human lung diseases amenable to modeling using iPSCs

Tracheoesophageal fistula/esophageal atresia
Congenital acinar hypoplasia
Congenital alveolar hypoplasia
Surfactant deficiency
Bronchopulmonary dysplasia
Cystic fibrosis
Neuroendocrine hyperplasia of infancy
Chronic obstructive pulmonary disease
Asthma
Idiopathic pulmonary fibrosis
Infectious disease

[27]. The clinical and pathological features of these diseases vary and are, among others, dependent on the exact mutations. Examining how a mutation affects the function of surfactant producing type II alveolar epithelial (ATII) cells and screening for drugs that might act as potential "correctors" requires modeling using hPSC-derived ATII cells. Several severe congenital pediatric lung diseases are not associated with surfactant deficiency, and their etiology is unknown. Pulmonary acinar dysplasia, for example, is characterized by developmental arrest prior to the canalicular or saccular stage of lung development [28, 29]. In two other diseases, congenital alveolar dysplasia [30] and alveolar growth abnormalities [31], developmental arrest or delay appears to occur later in lung development. Neuroendocrine hyperplasia of infancy is another severe pediatric lung disease of which the etiology and the likely developmental origin are unknown [31, 32]. The availability of iPS-based models of human lung development would shed light on the pathogenesis of pediatric congenital lung disease and might lead to novel therapeutic approaches. Equally importantly, such studies would provide unique insights into normal human lung development and into novel approaches to treat bronchopulmonary dysplasia of prematurely born infants, characterized by decreased alveolarization and impaired secondary septation of alveoli and one of the most threatening problems in prematurely born infants [33].

Directed differentiation of hPSCs into respiratory epithelium would also provide a novel and biologically relevant in vitro model to study infectious disease. Although mouse adapted strains of influenza exist [34], influenza infects humans, ferrets, and birds. Other viruses, such as respiratory syncytial virus, show different cellular tropism in mice and humans [35].

Modeling using hiPSCs might provide important insight into chronic obstructive pulmonary disease (COPD), one of the most prevalent lung diseases. COPD is characterized by, among others, mucous cell metaplasia and hyperplasia. Notch signaling enhances the formation of secretory cell types during development [36], directs differentiation of basal cells, the stem cells of the large airway, towards a secretory fate (club and goblet) [37], and induces mucous metaplasia in human tracheal explants [38]. hiPSC-derived airway cells could be used to examine to what extent this pathway also regulates the overabundance of goblet cells in COPD.

Furthermore, hiPS-derived airway cells would allow molecular studies into genetically determined susceptibility to noxious insults [39] and into the role of inflammatory stimuli [39].

Idiopathic pulmonary fibrosis is an intractable, fibrotic lung disease [40, 41] with a survival of 3–4 years. Although recent trials suggest that two drugs (pirfenidone and nintedanib) [42, 43] slow disease progression to some extent, the only definitive treatment is lung transplantation. IPF is characterized by fibroblastic foci and obliteration of alveoli. Genetic predisposition and environmental exposure, such as smoking and metal dust, likely determine disease penetrance [40, 41]. While most cases are sporadic, several genetic polymorphisms and mutations predispose to IPF [44]. The nature of some of these mutations suggests a critical role for ATII cells, the alveolar cells that secrete surfactant. These can replace damaged ATI cells to restore alveolar integrity after injury [45, 46]. Mutation in SP-C and SP-A2 [47, 48], which is selectively expressed in ATII cells, predisposes to IPF, likely by causing aberrant processing of the protein, resulting in an unfolded protein response. It is not known however how the fibrotic process is initiated and sustained. Other mutations or polymorphisms affect genes that are not specifically expressed in ATII cells, such as telomerase [49]. Impaired proliferative potential or function of ATII is therefore likely central to IPF pathogenesis. Another predisposing allele involves the regulatory region of MUC5B, which is not known to be expressed in ATII cells however, and its role in pathogenesis is entirely unknown [50].

ATII cells can currently not be isolated ex vivo and maintained in vitro. ATII cells isolated from patients after diagnosis may not be informative for disease pathogenesis and predisposition, as many observed changes in expression patterns may be secondary to the terminal fibrotic disease process. The mouse furthermore does not model the human disease well. Administration of bleomycin to mice causes patchy fibrosis that resolves over time, and its validity as a model for IPF is debatable [51, 52]. The best opportunity to gain insight into the pathogenesis of IPF is attempting to model this disease using iPSCs.

Finally, one of the most frequent developmental abnormalities of the airway and gastrointestinal tract is tracheoesophageal fistula/esophageal atresia (TEF/EA). TEF/EA can occur in isolation or as part of a variety of syndromes and associations, such as CHARGE, VACTERL, and Opitz syndrome [53]. Although several genetic mouse knockout models show features of TEF/EA, very few of those genes are mutated in humans with this congenital anomaly, and the etiology of most cases of TEF/EA in humans is not known [54].

Lung development *and the case for* human *PSCs*: hPSC-derived cells are a discovery tool that complements genetic approaches in the mouse. The availability of adequate numbers of human cells at well-defined stages of development allows genome-wide expression and epigenetic studies as well as RNAi and CRISPR/Cas9 screens for factors involved in lineage specification that are currently not feasible in mouse embryos. A better understanding of human lung development will in turn improve directed differentiation of hPSCs, the use of carefully timed signals that recapitulate development in the mouse [55–62] to induce specification and maturation of hPSCs into a specific lineage [1, 56]. The generation of functional, mature cells is still a major challenge, however. This may in part be due to the fact that in many organs, human and mouse organogenesis diverge more as development proceeds, likely because of a 3000-fold difference in body mass, profoundly affecting metabolism, organ size, architecture, and function and because of their very different habitats and positions in the food web.

The developmental dynamics and architecture of the respiratory system differ in mouse and human [33, 63]. The development of three-dimensional models could shed light on the underlying mechanisms. A pseudostratified epithelium containing basal cells, the stem cells of the airway, extends to terminal bronchioles in the human but is limited to trachea and large airways in mice, where it is replaced more distally by a columnar epithelium. Secretory club cells and chemosensory "tuft" or "brush" cells are more frequent in rodents than in humans, while the opposite is true for mucus-producing goblet cells [13, 64, 65]. Only the proximal airways contain submucosal glands in mice, while these are also found distally in the main bronchi in humans [66]. The architecture of the distal lung, consisting respiratory bronchioles, alveolar ducts, and associated alveoli, and collectively called the acinus, is more complex in humans than in rodents. The timing of milestones in lung development is also different in mice and humans [33, 63]. The saccular stage, where alveolar sacs containing distinguishable ATI and ATII cells form at the end of the stalks generated during the preceding process of branching morphogenesis [64, 67, 68], begins in late gestation, at E17.5, in the mouse, but relatively much earlier, at week 24, in humans [33]. In humans, further exponential postnatal alveolar growth occurs in the first 2 years, and a slower increase in alveolar number continues into puberty. In rodents, virtually all alveolar growth occurs postnatally in the first 20 days of life and stops more abruptly [69].

Finally, modeling human lung development using hPSCs may also shed light on the biology of human adult stem cells. As their identity and the mechanisms underlying their specification are largely unclear, hPSCs may provide a "forward" approach to gain insight into adult stem cells and into mechanisms involved in regeneration.

Lung Development

As recapitulating lung development in vitro is the scientific underpinning of directed differentiation of PSCs, we will first briefly review the critical stages of the development of the respiratory system (Fig. 16.1).

Lung and airway are derived from the definitive endoderm (DE), one of the germ layers that arises during gastrulation (E7 in mouse and in the third week of gestation in humans). Specification into mesoderm vs. endoderm depends on the strength of nodal signaling: high nodal signaling drives a DE fate while cells exposed to lower levels of nodal signaling are fated towards mesoderm [61, 62, 70, 71]. The DE

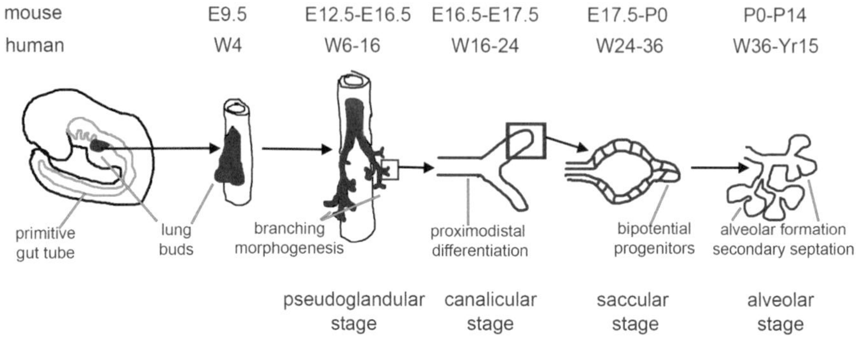

Fig. 16.1 Schematic representation of lung development (from [158])

develops into a tube where organ domains are progressively defined through transcriptional and epigenetic mechanisms along the anterior–posterior axis.

Proximodistal specification is indicated by differential expression of the transcription factors *Sox2*, *Hhex*, and *Foxa2* in the anterior half, while the posterior half is marked by *Cdx2* [72]. Retinoic acid, Wnt, and FGF4 are important for posterior endoderm specification [59, 73–75]. Recent studies in hPSCs have shown that blocking TGF-β and BMP signaling promotes an anterior fate [76]. Furthermore, timing of the exit from the primitive streak appears to determine anteroposterior identity [77], suggesting that the initial patterning may at least in part be established during gastrulation.

Organ domain specification is driven by permissive and instructive signals from surrounding tissues [71, 78]. Organ domains may also be prepatterned through epigenetic mechanisms during the early development of DE and of the gut tube, as has been shown for pancreas and liver [79]. The lung field arises on the ventral AFE and is characterized by the expression of *Nkx2.1*, which is also expressed in the forebrain and the thyroid, which is additionally marked by *Pax8* [71]. Canonical WNT, BMP, and FGF10 signaling from the ventral mesoderm to the AFE are involved in the specification of the lung field [80–85] (Fig. 16.2). RA is required for lung bud formation and in the mouse appears to act indirectly through regulation of Wnt, TGF-β, and FGF10 signaling [86–89]. A dorsoventral BMP signaling gradient, established by expression of BMP4 in the ventral mesenchyme and its antagonist, Noggin, dorsally and in the notochord, regulates patterning of trachea and esophagus [64, 90] and establishes a reciprocal dorsoventral *Sox2*/*Nkx2.1* gradient [91] that differentiates the gut endoderm into the dorsal esophagus (marked by *Dlx3* and high expression of *Sox2*) and a ventral trachea (marked by *Nkx2.1* and *Nkx2.5* and lower expression of *Sox2*) [64, 92, 93] (Fig. 16.2). *Sox2* is subsequently more highly expressed in the proximal airway and plays a critical role in tracheal development and in the maintenance of tracheal and proximal airway epithelium [94]. *Foxf1*, expressed in mesoderm, required for the correct separation of trachea and esophagus and is induced by endodermally derived Sonic Hedgehog [78, 95].

Fig. 16.2 Schematic representation of dorsoventral patterning of the AFE

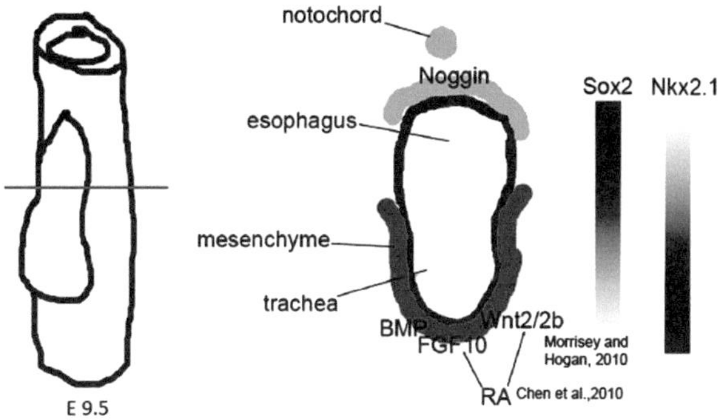

Subsequently, the lung buds undergo a complex branching process, driven by proliferating progenitors at the tips of the buds [96] (pseudoglandular stage, Fig. 16.1). In the stalks, cell cycle activity ceases, and differentiation of the airway epithelium occurs, with the emergence of basal, goblet, club, ciliated, and other cell types [96] (canalicular stage, Fig. 16.1) SOX2 is expressed in the developing airways, while SOX9 is expressed in the distal tips [97–99]. Cessation of proliferation coincides with loss of SOX9 expression in the mouse [96]. The regulation of branching morphogenesis and subsequent specification and terminal differentiation of specific cell types is extremely complex and involves interactions of multiple signals between the pulmonary endoderm and the surrounding mesenchyme. Among the critical factors secreted by the mesenchyme are FGF10 and BMP4, while Hedgehog signals from the endoderm to the mesenchyme [64]. Canonical and non-canonical Wnt signaling as well as BMP4 are involved in promoting a distal fate [85, 100, 101], while RA and Notch signaling promote a more proximal fate [102, 103]. Notch also enhances differentiation of airway basal cells into secretory and goblet cells, but not into ciliated cells or neuroendocrine cells [36, 37]. Based on single cell RNAseq and marker expression studies, bipotent progenitors were identified in the mouse at the tips that express select early ATI (such as Podoplanin, PDPN) and ATII (such as Surfactant Proteins B and C, SP-B, and SP-C) markers. These are presumed to give rise to cells expressing only ATI or ATII markers just proximal to these progenitors in developing alveolar sacs (saccular stage of alveolar development, Fig. 16.1) [46, 104]. Recent studies using differentiation of hPSCs did identify cells with a phenotype compatible with bipotential alveolar progenitors, however. The pulmonary mesenchyme thins, accompanied by the development of an extensive capillary network surrounding the forming alveoli [64, 67, 68]. Subsequently, secondary septation and alveolar growth increase the surface area available to gas exchange (alveolar stage, Fig. 16.1) [64, 92]. In human most, and in rodents almost all, alveolar growth takes place postnatally [69]. During early postnatal development, lung and tracheobronchial stem cells that provide extensive regenerative capacity are laid down as well [13].

Directed Differentiation of PSCs in the Lung and Airway

Studies reporting lung differentiation from stochastically differentiating PSCs failed to show the efficiency or document depletion of other lineages [105]. Based on the paradigms established in studies on lung development in the mouse, directed differentiation of hPSCs into lung and airway cells should be accomplished by inducing definitive endoderm (DE) and subsequently specifying anterior foregut endoderm (AFE), ventral AFE, and lung field, which is then further differentiated into proximal and distal lineages.

Differentiation of PSCs into definitive endoderm. The most critical factor for DE specification in vitro is Activin A, which mimics nodal signaling in the embryo [57, 61, 62, 106, 107]. Removal of the cells from renewal conditions leads to spontaneous differentiation. In non-adherent culture conditions, the cells will form the so-called embryoid bodies, spheres of cells where spontaneous gastrulation takes place [1]. Primitive streak formation is induced by agonizing BMP4 and canonical Wnt signaling for the first 24 h. Although Wnt signaling is required for primitive streak induction [70, 106], it is often not included in differentiation protocols [76, 108, 109]. Endogenous Wnt signaling in the cultures may be sufficient for DE induction in the presence of high concentrations of Activin A. Use of an inhibitor (Y-27632) of the Rho kinase, ROCK, increases the yield of DE cells during the first day of differentiation by maintaining viability of dissociated hPSCs [110, 111]. Subsequently, the cells are exposed to high concentrations of Activin A, inducing the DE surface markers EPCAM [72], c-KIT [57], and CXCR4 [61] and the transcription factors *Sox17* and *Foxa2* [56, 112]. Low concentrations of BMP4 and FGF2 have also been shown to enhance DE formation in EBs [108]. However, high BMP signaling in the absence of FGF signaling promotes primitive ectoderm and trophectoderm formation [61], while BMP exposure after endoderm induction induces a hepatic fate [57]. The appropriate concentrations and duration of exposure to these factors varies between laboratories [76, 108, 109, 111] and needs to be determined empirically. Small molecules, such as two putative HDAC inhibitors, Ide1 and Ide2, may be a potential alternative to Activin A [113], although this approach has not been confirmed. The strategy used to specify DE may determine its ultimate differentiation potential. For example, we observed that lung potential arises relatively early in DE induction (d4-4.5), and is lost later on, despite continued expression of DE markers [114]. Up to 95 % pure DE can be achieved in hPSCs, although significant variability exists among lines in their capacity to generate DE. Yields appear invariably lower in mPSCs.

Anteroposterior patterning and induction of anterior foregut endoderm from PSCs. Proper specification of DE to an AFE fate is critical to generate lung field progenitors. DE generated using the strategy described above will adopt a more posterior fate [76, 115], likely explaining why the generation of lung and airway epithelial cells has lagged behind the success in the generation of pancreatic, hepatic, and intestinal cells [1, 57, 106, 116, 117]. A morphogen screen showed that exposure of human DE to a combination of BMP and TGF-β inhibition (BMPi/

TGFi) led to expression of the foregut marker *SOX2*, suppression of the posterior marker *CDX2*, and maintenance of the endoderm marker *FOXA2* [76], indicative of AFE specification. Timing proved to be essential, however. Application of BMPi/TGFi for longer than 48 h made the cells unresponsive to the subsequent induction of lung progenitors [76]. These findings illustrate how during development cells are appropriately receptive to signaling inputs during narrow windows of time.

It is possible that the requirement for BMPi/TGFi is explained by the dynamics of early DE development. The AFE is derived from cells that move through the primitive streak first, and then leave the node [71]. Hence, this part of the DE is the farthest removed from the nodal signaling area of the epiblast for the longest time [118]. Furthermore, these cells are exposed to the nodal inhibitor, Lefty, expressed in the visceral endoderm that surrounds the epiblast [119, 120]. During its anterior migration, the cells fated to become AFE also pass through a zone where the BMP4 inhibitor Noggin is expressed [121]. This may explain why blocking BMP and TGF-β signaling after a period of exposure to Activin A is required to specify this part of the DE. This strategy to generate AFE has been reproduced in several reports [26, 122–124]. A variation on this protocol was reported by Firth et al., where after initial BMPi/TGFi for 2 days, TGFi in the presence of BMP4 was applied for an additional 3 days [26]. Mou et al. [115] suggested that in monolayer differentiation of mouse ESCs, only inhibition of TGF-β was required for anterior patterning. The efficiency of subsequent differentiation into lung progenitors was very low, however.

Further work showed that sequential inhibition of BMP and TGF-β followed by inhibition of TGF-β and Wnt signaling over the course of 48 h improved the competence of the cells to subsequently generate lung field progenitors [114]. Indeed, in the mouse embryo, DE cells fated to become AFE pass through a zone where they are exposed to the Wnt inhibitor, Dkk1 [121], after being exposed to area where Nodal/Activin inhibitor Lefty and the BMP4 inhibitor Noggin are expressed [125, 126].

Wong et al. [25] generated AFE from hiPSCs by exposing DE to SHH and FGF2. Huang et al. however found that AFE induction using SHH and FGF2 was less efficient than using sequential BMPi/TGFi and WNTi/TGFi [114]. Clearly, the two protocols provide alternative strategies to induce AFE, suggesting the existence of complex and hierarchical relationships between TGF-β, BMP, Wnt, SHH, and FGF2 signaling in the specification of AFE from DE that warrant further study.

Specification of PSC-derived anterior foregut into ventral anterior foregut and lung field. The next step is the induction of ventral AFE and lung field. Green et al. [76] and Huang et al. [114] accomplished this in hPSCs by adding factors that are essential for lung field specification in the mouse: Wnt (or the small molecule CHIR), BMP4, FGF7, FGF10, and RA [80, 89–91, 98, 127, 128]. These manipulations resulted in cultures where, depending on the hPSCs line, up to 90 % of the cells express $FOXA2^+$ and $NKX2.1^+$ between days 15 and 25 [114]. The efficiency of this differentiation protocol varied across hPSC lines, a reflection of the well-documented variability in the lineage-specific differentiation potential of ES [129] and iPS [130, 131] lines. Furthermore, morphogen concentration and timing of culture stages may require optimization for each individual line. This notion is supported by the observation that ES lines are heterogeneous with respect to endogenous Wnt signaling [132]. Blocking RA, Wnt, or BMP signaling entirely abolished the generation of NKX2.1$^+$ cells, while FGF signaling appeared dispensable in this model [114]. It is interesting to note that in vivo in the mouse model, the effect of RA is explained by inhibition of Dkk1, a Wnt inhibitor, and alleviation of TGF-β-mediated inhibition of FGF10 expression, suggesting a primarily indirect effect [89]. As in hPSCs, RA appears required even in the presence of BMP4, Wnt agonism, and FGF10; these findings suggest a direct effect of RA on lung field specification in humans [114]. These cells could be transplanted under the kidney capsule of immunodeficient mice where they gave rise to macroscopic growths, which contained cystic and tubular structures lined by a uniformly FOXA2$^+$NKX2.1$^+$ epithelium that ranged from pseudostratified containing cells consistent with basal, ciliated, club, and goblet cells to a monolayer consisting of flatter cells expressing markers of ATI and ATII cells. Glandular structures resembling submucosal glands were also present. No other endodermal elements were observed. However, the growths also contained smooth muscle, cartilage, and areas containing looser connective tissue, which were of human origin.

Using a ventralization strategy similar to Huang et al. [114], Gotoh et al. [123] generated human lung progenitors with similar efficiency and identified a surface marker specific for NKX2.1$^+$ lung cells, carboxypeptidase M (CPM), which allows further flow cytometric purification. In human fetal lung, CPM was found in cells expressing ATII and ATI markers.

Mou et al. [115] induced lung field endoderm after generation of human AFE using only TGF-β inhibition using BMP4, FGF2, and Wnt. This modification produced NKX2.1$^+$ pulmonary epithelial cells with an efficiency of 15 % in human and 10 % in mouse pluripotent stem cells, a minority of which could be induced to express proximal (*p63*, *Sox2*) markers through BMP7 agonism. Similar maturation studies in the CF iPSCs failed to show any mature cell types in vitro or in vivo. The low efficiency of lung progenitor differentiation is likely explained by suboptimal generation of AFE in the presence of only TGFi and the absence of RA in the ventralization cocktail.

Wong et al. [25], who used SHH and FGF2 to induce AFE, matured these cells towards putative lung and airway progenitors by sequential exposure to FGF7, FGF10, and BMP4 followed by FGF7, FGF10, and FGF18. The efficiency of this approach is unclear, as no immunofluorescence documentation of NKX2.1 and FOXA2 expression was provided [25]. Analysis of mRNA expression suggested some bias towards cells expressing markers of proximal airway, likely because of the presence of FGF18, which has been reported to enhance proximal and inhibit distal differentiation programs [133]. Finally, Ghaedi et al. [124] employed a variety of differentiation strategies throughout that require serum supplementation [124]. Defined conditions in the absence of serum as differentiation inducer are currently the standard of the field. This protocol [124] furthermore lacks addition of BMP4 and RA (which may be present in the serum, however), while the Wong et al. protocol lacks Wnt agonism during lung field induction [25], which we have shown to be indispensable [114], a finding that is entirely consistent with studies in mouse lung specification [64, 90, 91].

In the mouse, Longmire et al. [122] applied the protocol of Green et al. [76] with slight modifications to ESCs with a genetic reporter for pulmonary epithelium (Nkx2.1:GFP) [122]. In contrast to the human system, ventralization appeared to require FGF2, a factor that was also used by Wong et al. in human iPSCs to induce human AFE [25]. As the efficiency (as evidenced by expression of Nkx2.1:GFP) was only approximately 20 %, cell sorting was required to isolate the cells for further studies. In contrast to data in human PSCs, some of the Nkx2.1:GFP$^+$ cells showed evidence of thyroid specification, indicated by expression of *Pax8* [122]. Evident thyroid development was never observed in studies using hPSCs.

Differentiation of PSC-derived lung *progenitor into more mature cell types.* Terminal differentiation into mature lung and airway typically involves further culture of the cells in conditions that favor either proximal or distal differentiation.

Green et al. showed that removing BMP-4 from the ventralization cocktail (Wnt, BMP4, FGF10, FG7, RA) induced expression of SPC, a marker of ATII cells, and of early lung progenitors at E11 in the mouse (Fig. 16.2) [76]. This finding was somewhat surprising, as BMP4 has been shown to promote a more distal fate in the developing lung [85]. Consistent with these findings, however, Wong et al. showed that low concentrations of BMP4 induced proximal markers [25]. Similarly, Firth et al. found that low concentrations of BMP4 induced proximal markers, although distal markers were not assessed [26]. Finally, Huang et al. [114], also found that BMP4 addition during the terminal stages reduced expression of distal markers such as surfactant proteins and ATI markers PDPN and AQP5. They furthermore observed that continued exposure to RA also reduced distal gene expression, a finding in accordance with data from mouse genetic models. Indeed, after initiation of branching morphogenesis in mouse embryos, RA signaling inhibits distal lung and favors proximal airway development [102, 127].

Generation of proximal airway epithelium is typically accomplished by culture of lung and airway progenitors in air–liquid interphase conditions. The efficiency of the published protocols is unclear, and it is possible that this culture system is selective for those cells with proximal airway potential. Wong et al. applied ALI conditions to AFE cells further differentiated in the presence of FGF7, 10 and 18, which yielded CFTR expressing cells. Applying this strategy to iPSCs from CF patients with the F508del mutation, they could show that CFTR did not localize to the plasma membrane, as expected. After addition of a small "corrector" molecule analogue of a drug that is currently clinically used, some restoration of surface membrane localization was achieved [25]. Somewhat surprisingly, they observed expression of NKX2.1 in these presumed proximal airway epithelia. In mature lungs, however, expression of NKX2.1 is limited to ATII and club cells. Of interest is that Firth et al. [26] showed that differentiation can be biased towards ciliated cells by inhibiting Notch signaling using a γ-secretase inhibitor. Notch favors a secretory cell fate in the proximal airway epithelium by silencing the ciliated cell genetic program [36–38]. Notch inhibition, however, has also been reported to favor a distal over a proximal fate in the mouse lung in vivo [103]. The role of manipulation of Notch signaling in the differentiation of PSCs into lung and airway therefore merits further study.

Mouse ESC-derived Nkx2.1:GFP$^+$ cells could be further differentiated in the presence of FGF2 and FGF10 and expressed a variety of airway and alveolar markers [122]. Furthermore, addition of maturation components consisting of dexamethasone, butyrylcAMP, and isobutylmethylxanthine (DCI) [134] further upregulated the expression of SP-C, SP-B, and CC-10 [122]. Nkx2.1:GFP$^+$ cells could home to decellularized lung matrix and in some cases generate cells with marker expression and morphology consistent with ATI cells. While this work clearly showed that it is possible to differentiate PSCs into several lineages of lung and airway epithelial cells, the fact that a Nkx2.1:GFP reporter has to be used to purify cells specified to a lung field is a limitation.

Reliable distal differentiation was achieved by Huang et al. [114, 135] By continued culture in the presence of FGF7, FGF10, and Wnt agonism in addition to DCI, they achieved cultures where >50 % of the cells expressed the ATII marker, SP-B. Furthermore, after seeding onto human decellularized lung matrix, ample expression of SP-C was observed. Functionality of the cells was verified by their capacity to take up and release fluorescent SP-B. Further examination of the cultures in the long term (>50 days) revealed areas of proliferation, where cells express the ATII marker, SP-B or co-express SP-B and the ATI marker, podoplanin (PDPN) [135], suggestive of bipotential alveolar progenitors [46].

Alternative Methods

Direct reprogramming by transduction with transcription factors that drive differentiation of specific lineages is a potential alternative to directed differentiation. Direct reprogramming of ESCs has been reported for thyroid in the mouse model [136]. More often, this approach is applied to somatic cells and has met with some success for neurons [137–140], hepatocytes [141–143], cardiac myocytes [144, 145], hematopoietic stem cells [146, 147], and thymus [148] among others, but may not be useful in the lung, given the plethora of cell types required to build a functional lung and airway epithelium. Furthermore, detailed genomic analysis has revealed that cells resulting from direct reprogramming differ significantly from their normal counterparts [149]. Finally, for several organs, such as heart, reprogramming using human cells was less efficient and incomplete [150], while in others such as thymus and hematopoietic cells, no data on human cells are available yet.

A third approach is the creation of ES lines stably expressing a selectable marker driven by regulatory regions of a gene specifically expressed in the mature cell of choice. In the lung arena, the neomycin resistance gene was expressed under the control of the promoter of the ATII marker, *Sftpc* (SP-C), followed by selection for neomycin-resistant cells from randomly differentiating ES cells [151]. However, genetic modification and selection carries the risk of mutations in the resulting cells.

Furthermore, this approach only yields one cell type, of which the functional characteristics are uncertain, with low quantitative efficiency.

Conclusion

Directed differentiation of PSCs into lung and airway epithelial cells requires discrete steps that recapitulate development as much as possible. Among published papers, there is a broad consensus on the role of Activin A in DE specification, on the use of BMPi/TGFi for anterior foregut specification. For the efficient induction of cells equivalent to the lung field, current evidence suggests that agonism of Wnt, BMP4, and RA signaling is essential. In terms of terminal differentiation, it appears that proximal and distal fates are to some extent mutually exclusive and strongly dependent on the presence of appropriate drivers of proximodistal differentiation, culture conditions, and, for distal differentiation, on the presence of physiological maturation components such as glucocorticoids.

The generation of 3D lung organoids would be a major advance. This has not been achieved yet, however. Development of 3D broncholalveolar organoids would be a first step towards this goal. 3D organoid cultures are revolutionizing the study of human development. The lung, which has lagged behind other organs such as pituitary [152], liver [153], kidney [154, 155], brain [156], eye [157], and intestine [59] in this respect [9], offers a major opportunity, as the lung consists of multiple, repeated broncho-alveolar units. Availability of such technology would also allow addressing the role of extracellular matrix in lung development and in directed differentiation of PSCs into lung and airway.

References

1. Murry CE, Keller G (2008) Differentiation of embryonic stem cells to clinically relevant populations: lessons from embryonic development. Cell 132(4):661–680
2. Hanna JH, Saha K, Jaenisch R (2010) Pluripotency and cellular reprogramming: facts, hypotheses, unresolved issues. Cell 143(4):508–525
3. Takahashi K, Yamanaka S (2006) Induction of pluripotent stem cells from mouse embryonic and adult fibroblast cultures by defined factors. Cell 126(4):663–676
4. Yamanaka S (2009) A fresh look at iPS cells. Cell 137(1):13–17
5. Okita K, Yamanaka S (2011) Induced pluripotent stem cells: opportunities and challenges. Philos Trans R Soc Lond B Biol Sci 366(1575):2198–2207
6. Park IH, Zhao R, West JA, Yabuuchi A, Huo H, Ince TA et al (2008) Reprogramming of human somatic cells to pluripotency with defined factors. Nature 451(7175):141–146
7. Yu J, Vodyanik MA, Smuga-Otto K, Antosiewicz-Bourget J, Frane JL, Tian S et al (2007) Induced pluripotent stem cell lines derived from human somatic cells. Science 318 (5858):1917–1920

8. Fox IJ, Daley GQ, Goldman SA, Huard J, Kamp TJ, Trucco M (2014) Stem cell therapy. Use of differentiated pluripotent stem cells as replacement therapy for treating disease. Science 345(6199):1247391
9. Lancaster MA, Knoblich JA (2014) Organogenesis in a dish: modeling development and disease using organoid technologies. Science 345(6194):1247125
10. Robinton DA, Daley GQ (2012) The promise of induced pluripotent stem cells in research and therapy. Nature 481(7381):295–305
11. Lewis DR, Clegg LX, Johnson NJ (2009) Lung disease mortality in the United States: the national longitudinal mortality study. Int J Tuberc Lung Dis 13(8):1008–1014
12. McCurry KR, Shearon TH, Edwards LB, Chan KM, Sweet SC, Valapour M et al (2009) Lung transplantation in the United States, 1998–2007. Am J Transplant 9(4 Pt 2):942–958
13. Rock JR, Hogan BL (2011) Epithelial progenitor cells in lung development, maintenance, repair, and disease. Annu Rev Cell Dev Biol 27:493–512
14. Hogan BL, Barkauskas CE, Chapman HA, Epstein JA, Jain R, Hsia CC et al (2014) Repair and regeneration of the respiratory system: complexity, plasticity, and mechanisms of lung stem cell function. Cell Stem Cell 15(2):123–138
15. Kotton DN, Morrisey EE (2014) Lung regeneration: mechanisms, applications and emerging stem cell populations. Nat Med 20(8):822–832
16. Petersen TH, Calle EA, Zhao L, Lee EJ, Gui L, Raredon MB et al (2010) Tissue-engineered lungs for in vivo implantation. Science 329(5991):538–541
17. Ott HC, Clippinger B, Conrad C, Schuetz C, Pomerantseva I, Ikonomou L et al (2010) Regeneration and orthotopic transplantation of a bioartificial lung. Nat Med 16(8):927–933
18. O'Neill JD, Anfang R, Anandappa A, Costa J, Javidfar J, Wobma HM et al (2013) Decellularization of human and porcine lung tissues for pulmonary tissue engineering. Ann Thorac Surg 96(3):1046–1055; discussion 1055–1056
19. Macchiarini P, Jungebluth P, Go T, Asnaghi MA, Rees LE, Cogan TA et al (2008) Clinical transplantation of a tissue-engineered airway. Lancet 372(9655):2023–2030
20. Laurance J (2010) British boy receives trachea transplant built with his own stem cells. BMJ 340:c1633
21. Badylak SF, Weiss DJ, Caplan A, Macchiarini P (2012) Engineered whole organs and complex tissues. Lancet 379(9819):943–952
22. Jungebluth P, Alici E, Baiguera S, Le Blanc K, Blomberg P, Bozoky B et al (2011) Tracheobronchial transplantation with a stem-cell-seeded bioartificial nanocomposite: a proof-of-concept study. Lancet 378(9808):1997–2004
23. Delaere PR, Vranckx JJ, Meulemans J, Vander Poorten V, Segers K, Van Raemdonck D et al (2012) Learning curve in tracheal allotransplantation. Am J Transplant 12(9):2538–2545
24. Delaere PR (2012) Tracheal transplantation. Curr Opin Pulm Med 18(4):313–320
25. Wong AP, Bear CE, Chin S, Pasceri P, Thompson TO, Huan LJ et al (2012) Directed differentiation of human pluripotent stem cells into mature airway epithelia expressing functional CFTRTR protein. Nat Biotechnol 30(9):876–882
26. Firth AL, Dargitz CT, Qualls SJ, Menon T, Wright R, Singer O et al (2014) Generation of multiciliated cells in functional airway epithelia from human induced pluripotent stem cells. Proc Natl Acad Sci U S A 111(17):E1723–E1730
27. Whitsett JA, Wert SE, Weaver TE (2010) Alveolar surfactant homeostasis and the pathogenesis of pulmonary disease. Annu Rev Med 61:105–119
28. Gillespie LM, Fenton AC, Wright C (2004) Acinar dysplasia: a rare cause of neonatal respiratory failure. Acta Paediatr 93(5):712–713
29. Rutledge JC, Jensen P (1986) Acinar dysplasia: a new form of pulmonary maldevelopment. Hum Pathol 17(12):1290–1293
30. Mac MH (1948) Congenital alveolar dysplasia: a developmental anomaly involving pulmonary alveoli. Pediatrics 2(1):43–57
31. Dishop MK (2011) Paediatric interstitial lung disease: classification and definitions. Paediatr Respir Rev 12(4):230–237

32. Deterding RR, Pye C, Fan LL, Langston C (2005) Persistent tachypnea of infancy is associated with neuroendocrine cell hyperplasia. Pediatr Pulmonol 40(2):157–165
33. Smith LJ, McKay KO, van Asperen PP, Selvadurai H, Fitzgerald DA (2010) Normal development of the lung and premature birth. Paediatr Respir Rev 11(3):135–142
34. Barnard DL (2009) Animal models for the study of influenza pathogenesis and therapy. Antiviral Res 82(2):A110–A122
35. Bem RA, Domachowske JB, Rosenberg HF (2011) Animal models of human respiratory syncytial virus disease. Am J Physiol Lung Cell Mol Physiol 301(2):L148–L156
36. Tsao PN, Vasconcelos M, Izvolsky KI, Qian J, Lu J, Cardoso WV (2009) Notch signaling controls the balance of ciliated and secretory cell fates in developing airways. Development 136(13):2297–2307
37. Rock JR, Gao X, Xue Y, Randell SH, Kong YY, Hogan BL (2011) Notch-dependent differentiation of adult airway basal stem cells. Cell Stem Cell 8(6):639–648
38. Guseh JS, Bores SA, Stanger BZ, Zhou Q, Anderson WJ, Melton DA et al (2009) Notch signaling promotes airway mucous metaplasia and inhibits alveolar development. Development 136(10):1751–1759
39. Holtzman MJ, Byers DE, Alexander-Brett J, Wang X (2014) The role of airway epithelial cells and innate immune cells in chronic respiratory disease. Nat Rev Immunol 14 (10):686–698
40. Noble PW, Barkauskas CE, Jiang D (2012) Pulmonary fibrosis: patterns and perpetrators. J Clin Invest 122(8):2756–2762
41. Ryu JH, Moua T, Daniels CE, Hartman TE, Yi ES, Utz JP et al (2014) Idiopathic pulmonary fibrosis: evolving concepts. Mayo Clin Proc 89(8):1130–1142
42. King TE Jr, Bradford WZ, Castro-Bernardini S, Fagan EA, Glaspole I, Glassberg MK et al (2014) A phase 3 trial of pirfenidone in patients with idiopathic pulmonary fibrosis. N Engl J Med 370(22):2083–2092
43. Richeldi L, du Bois RM, Raghu G, Azuma A, Brown KK, Costabel U et al (2014) Efficacy and safety of nintedanib in idiopathic pulmonary fibrosis. N Engl J Med 370(22):2071–2082
44. Fingerlin TE, Murphy E, Zhang W, Peljto AL, Brown KK, Steele MP et al (2013) Genome-wide association study identifies multiple susceptibility loci for pulmonary fibrosis. Nat Genet 45(6):613–620
45. Barkauskas CE, Cronce MJ, Rackley CR, Bowie EJ, Keene DR, Stripp BR et al (2013) Type 2 alveolar cells are stem cells in adult lung. J Clin Invest 123:3025–3036
46. Desai TJ, Brownfield DG, Krasnow MA (2014) Alveolar progenitor and stem cells in lung development, renewal and cancer. Nature 507(7491):190–194
47. Nogee LM, Dunbar AE 3rd, Wert SE, Askin F, Hamvas A, Whitsett JA (2001) A mutation in the surfactant protein c gene associated with familial interstitial lung disease. N Engl J Med 344(8):573–579
48. Wang Y, Kuan PJ, Xing C, Cronkhite JT, Torres F, Rosenblatt RL et al (2009) Genetic defects in surfactant protein A2 are associated with pulmonary fibrosis and lung cancer. Am J Hum Genet 84(1):52–59
49. Armanios MY, Chen JJ, Cogan JD, Alder JK, Ingersoll RG, Markin C et al (2007) Telomerase mutations in families with idiopathic pulmonary fibrosis. N Engl J Med 356 (13):1317–1326
50. Seibold MA, Wise AL, Speer MC, Steele MP, Brown KK, Loyd JE et al (2011) A common MUC5B promoter polymorphism and pulmonary fibrosis. N Engl J Med 364(16):1503–1512
51. Moeller A, Ask K, Warburton D, Gauldie J, Kolb M (2008) The bleomycin animal model: a useful tool to investigate treatment options for idiopathic pulmonary fibrosis? Int J Biochem Cell Biol 40(3):362–382
52. Matute-Bello G, Frevert CW, Martin TR (2008) Animal models of acute lung injury. Am J Physiol Lung Cell Mol Physiol 295(3):L379–L399
53. El-Gohary Y, Gittes GK, Tovar JA (2010) Congenital anomalies of the esophagus. Semin Pediatr Surg 19(3):186–193

54. Bednarczyk D, Sasiadek MM, Smigiel R (2013) Chromosome aberrations and gene mutations in patients with esophageal atresia. J Pediatr Gastroenterol Nutr 57(6):688–693
55. Yang L, Soonpaa MH, Adler ED, Roepke TK, Kattman SJ, Kennedy M et al (2008) Human cardiovascular progenitor cells develop from a KDR+ embryonic-stem-cell-derived population. Nature 453(7194):524–528
56. Nostro MC, Keller G (2012) Generation of beta cells from human pluripotent stem cells: potential for regenerative medicine. Semin Cell Dev Biol 23:701–710
57. Gouon-Evans V, Boussemart L, Gadue P, Nierhoff D, Koehler CI, Kubo A et al (2006) BMP-4 is required for hepatic specification of mouse embryonic stem cell-derived definitive endoderm. Nat Biotechnol 24(11):1402–1411
58. Chambers SM, Fasano CA, Papapetrou EP, Tomishima M, Sadelain M, Studer L (2009) Highly efficient neural conversion of human es and iPS cells by dual inhibition of SMAD signaling. Nat Biotechnol 27(3):275–280
59. Spence JR, Mayhew CN, Rankin SA, Kuhar MF, Vallance JE, Tolle K et al (2011) Directed differentiation of human pluripotent stem cells into intestinal tissue in vitro. Nature 470 (7332):105–109
60. Pagliuca FW, Millman JR, Gurtler M, Segel M, Van Dervort A, Ryu JH et al (2014) Generation of functional human pancreatic beta cells in vitro. Cell 159(2):428–439
61. D'Amour KA, Agulnick AD, Eliazer S, Kelly OG, Kroon E, Baetge EE (2005) Efficient differentiation of human embryonic stem cells to definitive endoderm. Nat Biotechnol 23 (12):1534–1541
62. Kubo A, Shinozaki K, Shannon JM, Kouskoff V, Kennedy M, Woo S et al (2004) Development of definitive endoderm from embryonic stem cells in culture. Development 131 (7):1651–1662
63. Plopper CG, Hyde DM (2008) The non-human primate as a model for studying COPD and asthma. Pulm Pharmacol Ther 21(5):755–766
64. Morrisey EE, Hogan BL (2010) Preparing for the first breath: genetic and cellular mechanisms in lung development. Dev Cell 18(1):8–23
65. Krasteva G, Kummer W (2012) "Tasting" the airway lining fluid. Histochem Cell Biol 138:365–383
66. Rawlins EL, Hogan BL (2005) Intercellular growth factor signaling and the development of mouse tracheal submucosal glands. Dev Dyn 233(4):1378–1385
67. Herriges M, Morrisey EE (2014) Lung development: orchestrating the generation and regeneration of a complex organ. Development 141(3):502–513
68. Warburton D, El-Hashash A, Carraro G, Tiozzo C, Sala F, Rogers O et al (2010) Lung organogenesis. Curr Top Dev Biol 90:73–158
69. Herring MJ, Putney LF, Wyatt G, Finkbeiner WE, Hyde DM (2014) Growth of alveoli during postnatal development in humans based on stereological estimation. Am J Physiol Lung Cell Mol Physiol 307(4):L338–L344
70. Tam PP, Loebel DA (2007) Gene function in mouse embryogenesis: get set for gastrulation. Nat Rev Genet 8(5):368–381
71. Zorn AM, Wells JM (2009) Vertebrate endoderm development and organ formation. Annu Rev Cell Dev Biol 25:221–251
72. Sherwood RI, Chen TY, Melton DA (2009) Transcriptional dynamics of endodermal organ formation. Dev Dyn 238(1):29–42
73. Dessimoz J, Opoka R, Kordich JJ, Grapin-Botton A, Wells JM (2006) FGF signaling is necessary for establishing gut tube domains along the anterior-posterior axis in vivo. Mech Dev 123(1):42–55
74. McLin VA, Rankin SA, Zorn AM (2007) Repression of Wnt/beta-catenin signaling in the anterior endoderm is essential for liver and pancreas development. Development 134 (12):2207–2217
75. Wells JM, Melton DA (2000) Early mouse endoderm is patterned by soluble factors from adjacent germ layers. Development 127(8):1563–1572

76. Green MD, Chen A, Nostro MC, d'Souza SL, Schaniel C, Lemischka IR et al (2011) Generation of anterior foregut endoderm from human embryonic and induced pluripotent stem cells. Nat Biotechnol 29(3):267–272
77. Lawson KA, Pedersen RA (1987) Cell fate, morphogenetic movement and population kinetics of embryonic endoderm at the time of germ layer formation in the mouse. Development 101(3):627–652
78. van den Brink GR (2007) Hedgehog signaling in development and homeostasis of the gastrointestinal tract. Physiol Rev 87(4):1343–1375
79. Xu CR, Cole PA, Meyers DJ, Kormish J, Dent S, Zaret KS (2011) Chromatin "prepattern" and histone modifiers in a fate choice for liver and pancreas. Science 332(6032):963–966
80. Goss AM, Tian Y, Tsukiyama T, Cohen ED, Zhou D, Lu MM et al (2009) Wnt2/2b and beta-catenin signaling are necessary and sufficient to specify lung progenitors in the foregut. Dev Cell 17(2):290–298
81. Bellusci S, Grindley J, Emoto H, Itoh N, Hogan BL (1997) Fibroblast growth factor 10 (FGF10) and branching morphogenesis in the embryonic mouse lung. Development 124 (23):4867–4878
82. Min H, Danilenko DM, Scully SA, Bolon B, Ring BD, Tarpley JE et al (1998) Fgf-10 is required for both limb and lung development and exhibits striking functional similarity to drosophila branchless. Genes Dev 12(20):3156–3161
83. Serls AE, Doherty S, Parvatiyar P, Wells JM, Deutsch GH (2005) Different thresholds of fibroblast growth factors pattern the ventral foregut into liver and lung. Development 132 (1):35–47
84. Sekine K, Ohuchi H, Fujiwara M, Yamasaki M, Yoshizawa T, Sato T et al (1999) Fgf10 is essential for limb and lung formation. Nat Genet 21(1):138–141
85. Weaver M, Yingling JM, Dunn NR, Bellusci S, Hogan BL (1999) Bmp signaling regulates proximal-distal differentiation of endoderm in mouse lung development. Development 126 (18):4005–4015
86. Rossant J, Zirngibl R, Cado D, Shago M, Giguere V (1991) Expression of a retinoic acid response element-hsplacZ transgene defines specific domains of transcriptional activity during mouse embryogenesis. Genes Dev 5(8):1333–1344
87. Desai TJ, Chen F, Lu J, Qian J, Niederreither K, Dolle P et al (2006) Distinct roles for retinoic acid receptors alpha and beta in early lung morphogenesis. Dev Biol 291(1):12–24
88. Desai TJ, Malpel S, Flentke GR, Smith SM, Cardoso WV (2004) Retinoic acid selectively regulates Fgf10 expression and maintains cell identity in the prospective lung field of the developing foregut. Dev Biol 273(2):402–415
89. Chen F, Cao Y, Qian J, Shao F, Niederreither K, Cardoso WV (2010) A retinoic acid-dependent network in the foregut controls formation of the mouse lung primordium. J Clin Invest 120(6):2040–2048
90. Li Y, Gordon J, Manley NR, Litingtung Y, Chiang C (2008) Bmp4 is required for tracheal formation: a novel mouse model for tracheal agenesis. Dev Biol 322(1):145–155
91. Domyan ET, Ferretti E, Throckmorton K, Mishina Y, Nicolis SK, Sun X (2011) Signaling through bmp receptors promotes respiratory identity in the foregut via repression of Sox2. Development 138(5):971–981
92. Maeda Y, Dave V, Whitsett JA (2007) Transcriptional control of lung morphogenesis. Physiol Rev 87(1):219–244
93. Kasahara H, Bartunkova S, Schinke M, Tanaka M, Izumo S (1998) Cardiac and extracardiac expression of Csx/Nkx2.5 homeodomain protein. Circ Res 82(9):936–946
94. Que J, Luo X, Schwartz RJ, Hogan BL (2009) Multiple roles for Sox2 in the developing and adult mouse trachea. Development 136(11):1899–1907
95. Mahlapuu M, Enerback S, Carlsson P (2001) Haploinsufficiency of the forkhead gene Foxf1, a target for sonic hedgehog signaling, causes lung and foregut malformations. Development 128(12):2397–2406

96. Rawlins EL, Clark CP, Xue Y, Hogan BL (2009) The Id2+ distal tip lung epithelium contains individual multipotent embryonic progenitor cells. Development 136(22):3741–3745
97. Que J, Choi M, Ziel JW, Klingensmith J, Hogan BL (2006) Morphogenesis of the trachea and esophagus: current players and new roles for noggin and Bmps. Differentiation 74 (7):422–437
98. Que J, Okubo T, Goldenring JR, Nam KT, Kurotani R, Morrisey EE et al (2007) Multiple dose-dependent roles for Sox2 in the patterning and differentiation of anterior foregut endoderm. Development 134(13):2521–2531
99. Liu Y, Hogan BL (2002) Differential gene expression in the distal tip endoderm of the embryonic mouse lung. Gene Expr Patterns 2(3–4):229–233
100. Shu W, Guttentag S, Wang Z, Andl T, Ballard P, Lu MM et al (2005) Wnt/beta-catenin signaling acts upstream of N-myc, BMP4, and FGF signaling to regulate proximal-distal patterning in the lung. Dev Biol 283(1):226–239
101. Li C, Xiao J, Hormi K, Borok Z, Minoo P (2002) Wnt5a participates in distal lung morphogenesis. Dev Biol 248(1):68–81
102. Wongtrakool C, Malpel S, Gorenstein J, Sedita J, Ramirez MI, Underhill TM et al (2003) Down-regulation of retinoic acid receptor alpha signaling is required for sacculation and type I cell formation in the developing lung. J Biol Chem 278(47):46911–46918
103. Tsao PN, Chen F, Izvolsky KI, Walker J, Kukuruzinska MA, Lu J et al (2008) Gamma-secretase activation of notch signaling regulates the balance of proximal and distal fates in progenitor cells of the developing lung. J Biol Chem 283(43):29532–29544
104. Treutlein B, Brownfield DG, Wu AR, Neff NF, Mantalas GL, Espinoza FH et al (2014) Reconstructing lineage hierarchies of the distal lung epithelium using single-cell RNA-seq. Nature 509(7500):371–375
105. Winkler ME, Mauritz C, Groos S, Kispert A, Menke S, Hoffmann A et al (2008) Serum-free differentiation of murine embryonic stem cells into alveolar type II epithelial cells. Cloning Stem Cells 10(1):49–64
106. Gadue P, Huber TL, Paddison PJ, Keller GM (2006) Wnt and TGF-beta signaling are required for the induction of an in vitro model of primitive streak formation using embryonic stem cells. Proc Natl Acad Sci U S A 103(45):16806–16811
107. Yasunaga M, Tada S, Torikai-Nishikawa S, Nakano Y, Okada M, Jakt LM et al (2005) Induction and monitoring of definitive and visceral endoderm differentiation of mouse es cells. Nat Biotechnol 23(12):1542–1550
108. Xu X, Browning VL, Odorico JS (2011) Activin, BMP and FGF pathways cooperate to promote endoderm and pancreatic lineage cell differentiation from human embryonic stem cells. Mech Dev 128(7–10):412–427
109. Nostro MC, Sarangi F, Ogawa S, Holtzinger A, Corneo B, Li X et al (2011) Stage-specific signaling through TGFbeta family members and WNT regulates patterning and pancreatic specification of human pluripotent stem cells. Development 138(5):861–871
110. Watanabe K, Ueno M, Kamiya D, Nishiyama A, Matsumura M, Wataya T et al (2007) A rock inhibitor permits survival of dissociated human embryonic stem cells. Nat Biotechnol 25 (6):681–686
111. Ungrin MD, Clarke G, Yin T, Niebrugge S, Nostro MC, Sarangi F et al (2012) Rational bioprocess design for human pluripotent stem cell expansion and endoderm differentiation based on cellular dynamics. Biotechnol Bioeng 109(4):853–866
112. Van Hoof D, D'Amour KA, German MS (2009) Derivation of insulin-producing cells from human embryonic stem cells. Stem Cell Res 3(2–3):73–87
113. Borowiak M, Maehr R, Chen S, Chen AE, Tang W, Fox JL et al (2009) Small molecules efficiently direct endodermal differentiation of mouse and human embryonic stem cells. Cell Stem Cell 4(4):348–358
114. Huang SX, Islam MN, O'Neill J, Hu Z, Yang YG, Chen YW et al (2014) Efficient generation of lung and airway epithelial cells from human pluripotent stem cells. Nat Biotechnol 32 (1):84–91

115. Mou H, Zhao R, Sherwood R, Ahfeldt T, Lapey A, Wain J et al (2012) Generation of multipotent lung and airway progenitors from mouse ESCs and patient-specific cystic fibrosis iPSCs. Cell Stem Cell 10(4):385–397
116. D'Amour KA, Bang AG, Eliazer S, Kelly OG, Agulnick AD, Smart NG et al (2006) Production of pancreatic hormone-expressing endocrine cells from human embryonic stem cells. Nat Biotechnol 24(11):1392–1401
117. Cai J, Zhao Y, Liu Y, Ye F, Song Z, Qin H et al (2007) Directed differentiation of human embryonic stem cells into functional hepatic cells. Hepatology 45(5):1229–1239
118. Tam PP, Khoo PL, Lewis SL, Bildsoe H, Wong N, Tsang TE et al (2007) Sequential allocation and global pattern of movement of the definitive endoderm in the mouse embryo during gastrulation. Development 134(2):251–260
119. Rossant J, Tam PP (2004) Emerging asymmetry and embryonic patterning in early mouse development. Dev Cell 7(2):155–164
120. Arnold SJ, Robertson EJ (2009) Making a commitment: cell lineage allocation and axis patterning in the early mouse embryo. Nat Rev Mol Cell Biol 10(2):91–103
121. del Barco BI, Davidson G, Grone HJ, Westphal H, Niehrs C (2003) Dkk1 and noggin cooperate in mammalian head induction. Genes Dev 17(18):2239–2244
122. Longmire TA, Ikonomou L, Hawkins F, Christodoulou C, Cao Y, Jean JC et al (2012) Efficient derivation of purified lung and thyroid progenitors from embryonic stem cells. Cell Stem Cell 10(4):398–411
123. Gotoh S, Ito I, Nagasaki T, Yamamoto Y, Konishi S, Korogi Y et al (2014) Generation of alveolar epithelial spheroids via isolated progenitor cells from human pluripotent stem cells. Stem Cell Rep 3(3):394–403
124. Ghaedi M, Calle EA, Mendez JJ, Gard AL, Balestrini J, Booth A et al (2013) Human iPS cell-derived alveolar epithelium repopulates lung extracellular matrix. J Clin Invest 123 (11):4950–4962
125. Yamamoto M, Saijoh Y, Perea-Gomez A, Shawlot W, Behringer RR, Ang SL et al (2004) Nodal antagonists regulate formation of the anteroposterior axis of the mouse embryo. Nature 428(6981):387–392
126. Perea-Gomez A, Vella FD, Shawlot W, Oulad-Abdelghani M, Chazaud C, Meno C et al (2002) Nodal antagonists in the anterior visceral endoderm prevent the formation of multiple primitive streaks. Dev Cell 3(5):745–756
127. Malpel S, Mendelsohn C, Cardoso WV (2000) Regulation of retinoic acid signaling during lung morphogenesis. Development 127(14):3057–3067
128. Weaver M, Dunn NR, Hogan BL (2000) Bmp4 and Fgf10 play opposing roles during lung bud morphogenesis. Development 127(12):2695–2704
129. Osafune K, Caron L, Borowiak M, Martinez RJ, Fitz-Gerald CS, Sato Y et al (2008) Marked differences in differentiation propensity among human embryonic stem cell lines. Nat Biotechnol 26(3):313–315
130. Bock C, Kiskinis E, Verstappen G, Gu H, Boulting G, Smith ZD et al (2011) Reference maps of human ES and iPS cell variation enable high-throughput characterization of pluripotent cell lines. Cell 144(3):439–452
131. Boulting GL, Kiskinis E, Croft GF, Amoroso MW, Oakley DH, Wainger BJ et al (2011) A functionally characterized test set of human induced pluripotent stem cells. Nat Biotechnol 29 (3):279–286
132. Blauwkamp TA, Nigam S, Ardehali R, Weissman IL, Nusse R (2012) Endogenous Wnt signalling in human embryonic stem cells generates an equilibrium of distinct lineage-specified progenitors. Nat Commun 3:1070
133. Whitsett JA, Clark JC, Picard L, Tichelaar JW, Wert SE, Itoh N et al (2002) Fibroblast growth factor 18 influences proximal programming during lung morphogenesis. J Biol Chem 277 (25):22743–22749

134. Gonzales LW, Guttentag SH, Wade KC, Postle AD, Ballard PL (2002) Differentiation of human pulmonary type II cells in vitro by glucocorticoid plus camp. Am J Physiol Lung Cell Mol Physiol 283(5):L940–L951
135. Huang SX, Green MD, de Carvalho AT, Mumau M, Chen YW, D'Souza SL et al (2015) The in vitro generation of lung and airway progenitor cells from human pluripotent stem cells. Nat Protoc 10(3):413–425
136. Antonica F, Kasprzyk DF, Opitz R, Iacovino M, Liao XH, Dumitrescu AM et al (2012) Generation of functional thyroid from embryonic stem cells. Nature 491(7422):66–71
137. Ambasudhan R, Talantova M, Coleman R, Yuan X, Zhu S, Lipton SA et al (2011) Direct reprogramming of adult human fibroblasts to functional neurons under defined conditions. Cell Stem Cell 9(2):113–118
138. Wapinski OL, Vierbuchen T, Qu K, Lee QY, Chanda S, Fuentes DR et al (2013) Hierarchical mechanisms for direct reprogramming of fibroblasts to neurons. Cell 155(3):621–635
139. Lujan E, Chanda S, Ahlenius H, Sudhof TC, Wernig M (2012) Direct conversion of mouse fibroblasts to self-renewing, tripotent neural precursor cells. Proc Natl Acad Sci U S A 109 (7):2527–2532
140. Pang ZP, Yang N, Vierbuchen T, Ostermeier A, Fuentes DR, Yang TQ et al (2011) Induction of human neuronal cells by defined transcription factors. Nature 476(7359):220–223
141. Huang P, Zhang L, Gao Y, He Z, Yao D, Wu Z et al (2014) Direct reprogramming of human fibroblasts to functional and expandable hepatocytes. Cell Stem Cell 14(3):370–384
142. Yu B, He ZY, You P, Han QW, Xiang D, Chen F et al (2013) Reprogramming fibroblasts into bipotential hepatic stem cells by defined factors. Cell Stem Cell 13(3):328–340
143. Zhu S, Rezvani M, Harbell J, Mattis AN, Wolfe AR, Benet LZ et al (2014) Mouse liver repopulation with hepatocytes generated from human fibroblasts. Nature 508(7494):93–97
144. Qian L, Huang Y, Spencer CI, Foley A, Vedantham V, Liu L et al (2012) In vivo reprogramming of murine cardiac fibroblasts into induced cardiomyocytes. Nature 485 (7400):593–598
145. Ieda M, Fu JD, Delgado-Olguin P, Vedantham V, Hayashi Y, Bruneau BG et al (2010) Direct reprogramming of fibroblasts into functional cardiomyocytes by defined factors. Cell 142 (3):375–386
146. Pereira CF, Chang B, Qiu J, Niu X, Papatsenko D, Hendry CE et al (2013) Induction of a hemogenic program in mouse fibroblasts. Cell Stem Cell 13(2):205–218
147. Batta K, Florkowska M, Kouskoff V, Lacaud G (2014) Direct reprogramming of murine fibroblasts to hematopoietic progenitor cells. Cell Rep 9(5):1871–1884
148. Bredenkamp N, Ulyanchenko S, O'Neill KE, Manley NR, Vaidya HJ, Blackburn CC (2014) An organized and functional thymus generated from FOXN1-reprogrammed fibroblasts. Nat Cell Biol 16(9):902–908
149. Cahan P, Li H, Morris SA, Lummertz da Rocha E, Daley GQ, Collins JJ (2014) Cellnet: network biology applied to stem cell engineering. Cell 158(4):903–915
150. Srivastava D, Berry EC (2013) Cardiac reprogramming: from mouse toward man. Curr Opin Genet Dev 23(5):574–578
151. Wang D, Haviland DL, Burns AR, Zsigmond E, Wetsel RA (2007) A pure population of lung alveolar epithelial type II cells derived from human embryonic stem cells. Proc Natl Acad Sci U S A 104(11):4449–4454
152. Xia Y, Nivet E, Sancho-Martinez I, Gallegos T, Suzuki K, Okamura D et al (2013) Directed differentiation of human pluripotent cells to ureteric bud kidney progenitor-like cells. Nat Cell Biol 15(12):1507–1515
153. Takebe T, Sekine K, Enomura M, Koike H, Kimura M, Ogaeri T et al (2013) Vascularized and functional human liver from an iPSC-derived organ bud transplant. Nature 499 (7459):481–484
154. Taguchi A, Kaku Y, Ohmori T, Sharmin S, Ogawa M, Sasaki H et al (2014) Redefining the in vivo origin of metanephric nephron progenitors enables generation of complex kidney structures from pluripotent stem cells. Cell Stem Cell 14(1):53–67

155. Takasato M, Er PX, Becroft M, Vanslambrouck JM, Stanley EG, Elefanty AG et al (2014) Directing human embryonic stem cell differentiation towards a renal lineage generates a self-organizing kidney. Nat Cell Biol 16(1):118–126
156. Lancaster MA, Renner M, Martin CA, Wenzel D, Bicknell LS, Hurles ME et al (2013) Cerebral organoids model human brain development and microcephaly. Nature 501 (7467):373–379
157. Nakano T, Ando S, Takata N, Kawada M, Muguruma K, Sekiguchi K et al (2012) Self-formation of optic cups and storable stratified neural retina from human ESCs. Cell Stem Cell 10(6):771–785
158. Snoeck HW (2015) Modeling human lung development and disease using pluripotent stem cells. Development 142(1):13–16

Part V
Bioengineering the Lung

Chapter 17
Regenerating the Respiratory Tract

Mei Ling Lim, Alexandra B. Firsova, Neus Feliu, Elena V. Kuevda, Philipp Jungebluth, and Paolo Macchiarini

Introduction

Despite recent diagnostic and therapeutic advances, many diseases which affect the respiratory system are either difficult to treat or nontreatable. Chronic end-stage respiratory diseases (*e.g.*, severe cystic fibrosis, chronic obstructive pulmonary disease, pulmonary hypertension, and hereditary or congenital conditions) are typically life threatening, obstruct normal breathing, and are non-reversible. At present, the only available curative treatment is allogeneic transplantation. However, the shortage of donor organs, long waiting periods, size mismatching and the need for lifelong immunosuppressants remain major drawbacks. These circumstances constitute an appeal to both clinicians and researchers to urgently realize and investigate new therapeutic options that might be available to the millions of patients that suffer from respiratory dysfunction. Regenerative medicine, including tissue engineering (TE) and cell therapy, appears to be a potentially promising therapeutic option for many of these patients. This chapter will review these

M.L. Lim • A.B. Firsova • N. Feliu • P. Jungebluth
Advanced Center for Translational and Regenerative Medicine (ACTREM), Division of Ear, Nose, and Throat, Department for Clinical Science, Intervention and Technology, Karolinska Institute, Stockholm, Sweden

E.V. Kuevda
International Research, Clinical and Education Center of Regenerative Medicine, Kuban State Medical University, Krasnodar, Russian Federation

P. Macchiarini, M.D., Ph.D. (✉)
International Research, Clinical and Education Center of Regenerative Medicine, Kuban State Medical University, Krasnodar, Russian Federation

Advanced Center for Translational and Regenerative Medicine (ACTREM), European Airway Institute, Division of Ear, Nose, and Throat (CLINTEC), Department for Clinical Science, Intervention and Technology, Karolinska Institute, Stockholm, Sweden
e-mail: paolo.macchiarini@ki.se

I. Bertoncello (ed.), *Stem Cells in the Lung*, Stem Cell Biology and Regenerative Medicine, DOI 10.1007/978-3-319-21082-7_17

different therapeutic approaches to understand how a reconstituted complex whole tissue or organ can remodel, regenerate, and repair via specific mechanisms, and will provide an overview of the innovative applications that nanotechnology can offer to accelerate the field of TE and cell therapy.

Developmental Biology of the Respiratory Tract

Development of the respiratory tract can be divided into five stages: embryonic, pseudoglandular, canalicular, saccular, and alveolar (Fig. 17.1). The tracheal and lung primordium appears as early as the fourth week of human gestation, arising from the foregut endoderm at the embryonic stage. The formation of lung structure proceeds at the pseudoglandular stage; which includes the branching process of the bronchial tree (Fig. 17.1). The formation of ducts, an increase in the number of capillaries and the appearance of the first alveolar epithelium occurs at the canalicular stage. Subsequently, at the saccular stage, terminal sacs and ducts develop and undergo primary septation [1, 2]. This is followed by secondary septation, (*i.e.*, forming and lengthening of the secondary septa) which helps to increase the gas-exchange surface area by thinning the extracellular matrix (ECM) and alveolar epithelium to generate alveoli (alveolar stage) [3, 4]. During this period, flattening

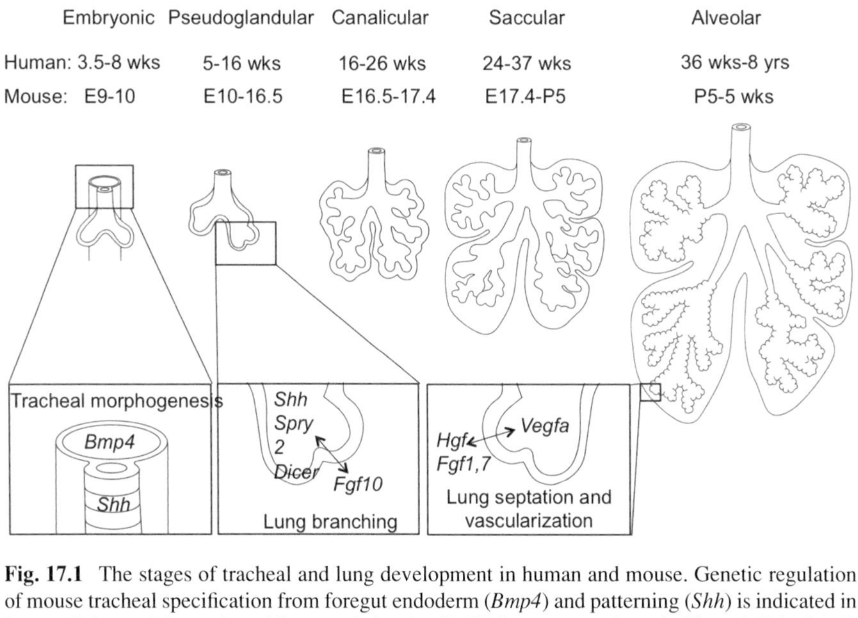

Fig. 17.1 The stages of tracheal and lung development in human and mouse. Genetic regulation of mouse tracheal specification from foregut endoderm (*Bmp4*) and patterning (*Shh*) is indicated in *bottom left* panel. Lung branching regulation by *Fgf10* from mesenchymal side and *Shh*, *Spry2*, *Tgfb*, and *Dicer* from epithelial side is specified in *bottom middle* panel. Lung vascularization by *Vegf-a* from epithelial cells and lung septation regulation by *Hgf*, *Fgf1,7* from mesenchymal and endothelial cells is indicated in *bottom right* panel

of type I alveolar epithelial cells (AEC-I), an increase in the number of type II alveolar epithelial cells (AEC-II), as well as vasculature development and capillary wall thinning occur [5, 6]. In humans, the alveolarization process can take around two postnatal years to complete (Fig. 17.1).

At different stages, respiratory tissue specification is predominantly regulated by the cellular compartment, relying on epithelial–mesenchymal crosstalk [7]. It involves interactions between different cell representatives of lung epithelium and mesenchyme by gene expression and paracrine effects, extensively studied in mouse models. The dorsal-ventral endodermal patterning during primordial bud separation is controlled by bone morphogenetic protein 4 (*Bmp4*) signaling in mouse [8] (Fig. 17.1). Differentiation of tracheal epithelial cells (ciliated, non-ciliated secretory, basal, and small mucus granule cells) occurs in a polarity-based cartilage-oriented manner [9]. The development and patterning of C-shaped hyaline cartilage rings in the trachea is regulated by sonic hedgehog (*Shh*) expression [10] (Fig. 17.1). The rings are connected by smooth muscle on the posterior side. The ECM defines the unique mechanistic properties of the trachea which are critical for bioengineering the organ.

In the lung, the branching of the bronchial tree was demonstrated to be controlled by fibroblast growth factor 10 (*Fgf10*) expressed in mesenchymal cells [11, 12], and Sonic hedgehog (*Shh*), Sprouty2 (*Spry2*) and *Dicer1* expressed in epithelial cells during pseudoglandular stage [13–16] (Fig. 17.1). At later stages, alveolar wall remodeling involves redistribution of tissue mass and stretching of septa initiated by the proliferation of interstitial fibroblasts regulated by PDGF-A [17]. Further proliferation of AEC-II is known to be triggered by the factors released by endothelial and mesenchymal cells: fibroblast growth factors 1 and 7 (FGF1, FGF7), hepatocyte growth factor (HGF), and retinoic acid (RA, the active metabolite of vitamin A) [18–20] (Fig. 17.1). Transforming growth factor beta (TGF-β) is also hypothesized to be involved in AEC differentiation [21]. Respiratory blood vessel formation is highly dependent on vascular endothelial growth factor A (VEGF-A), produced by AECII [20, 22] (Fig. 17.1). Inhibition of VEGF receptors causes both alveolar and endothelial cell apoptosis and enlargement of lung airspaces [23], which suggests that lung alveolar maintenance is highly dependent on the VEGF pathway. The knowledge gained from understanding the roles of different regulatory pathways and paracrine factors or chemical agents (e.g., dexamethasone, RA [24, 25]) that could affect lung development and structure formation, are valuable for the field of regenerative medicine studies.

Endogenous Progenitor Stem Cells

Progenitor and stem cells are involved in respiratory tissue regeneration and homeostasis in the respiratory system in cases of injury or disease. Various cell types can be affected by majority of lung diseases, and therefore different progenitor cell types would be involved in restoration and repopulation of the trachea and lung. Apart from epithelial cells, there are also interstitial fibroblasts, vascular, lymphatic and neuronal cells that might be damaged by respiratory diseases, with ECM also being affected. There are endogenous and exogenous sources of stem cells for the lung [26]. Among endogenous epithelial stem cells there are basal epithelial cells, variant club cells, bronchoalveolar stem cells, and AEC-II, that are hypothesized to be precursors of most lung epithelial cells [27]. Recent studies indicate that endogenous progenitors also include side population (SP) cells with mesenchymal (chondrocyte, osteocyte, and adipocyte), endothelial, and potentially epithelial differentiation capacity [28–30]. Isolation of endogenous cells from donors for therapeutic and bioengineering purposes is rather complicated since they are not easily accessible; therefore leading to the consideration of exogenous stem-cell sources as an alternative.

Exogenous Progenitor and Stem Cells

Prospective exogenous stem-cell treatments have to target regeneration of ECM and the multiple lung cell types (including epithelial, mesenchymal, and endothelial) that can be affected in respiratory diseases. There are several donor tissues e.g., bone marrow (BM), peripheral blood (PB), amniotic fluid (AF), or umbilical cord blood (UCB) that could serve as sources of exogenous stem and progenitor cells, including endothelial and mesenchymal progenitor cells [26]. Bone marrow, PB, AF, and UCB were reported to contain progenitor cells for blood, endothelium, and mesenchyme, which can contribute differently to the respiratory tract regeneration (Table 17.1) [31–36]. AF-derived stem cells were demonstrated to possess epithelial differentiation potential [37–39]. Cells from all these sources exhibit reparative properties in various lung disease models [37, 38, 40–44]. They can be involved in the regeneration process directly (proliferating in vivo, differentiating towards the targeted cell type at the injury site, or modulating ECM, see Table 17.1) or indirectly (contributing to lung homeostasis *via* epithelial–mesenchymal crosstalk) [45]. This involvement suggests that donor-derived stem cells can be a suitable tool for understanding the mechanism of injury repair [46–48], and offer possibilities in therapeutic application.

Tissue Engineering

The key elements to consider in TE are the production of a suitable biocompatible matrix that preserves spatial geometric architecture, recapitulates the biomechanical properties of the native tissue, and also retains bioactive cues. The matrix can be based on biological tissue or synthetic materials. In the case of biological matrices, donor tissue is treated and/or manipulated *via* various methodologies aiming for the removal of antigenic components. This so-called decellularization process is highly

Table 17.1 Examples of exogenous stem cells that can contribute to lung regeneration

Stem cells	Source	Cell fate	Isolation	Effects in lung	Reference
MSC	BM, PB, UCB, placenta, adipose tissue	Bone, cartilage, fat, fibroblasts	Biopsy or blood donation, high numbers in donated tissues, distinguished cell population	Contribute to fibrocytes, support ECM, involved in immunomodulation	[31, 35, 46, 47, 100–102]
EPC	BM, PB, UCB	Endothelium	Biopsy or blood donation, low numbers in donated tissues, no uniform markers for isolation	Contribute to vascularization	[32, 103–106]
HSC		T cells and B cells, monocytes, macrophages, neutrophils, eosinophils, mast cells, erythrocytes, megakaryocytes	Biopsy or blood donation, high numbers in donated tissues, distinguished cell population	Contribute to blood pool	[32, 34, 104, 107]
AFSC	Amniotic fluid	Multipotent (including MSC and EPC), epithelial	Amniocentesis (potential side effects to donor), multiple cell types	Contribute to bronchial and alveolar epithelial cells	[36–38]

critical since it determines the characteristics of the matrix and the fate of the seeded cells.

Various physical methods (perfusion, agitation, static, sonication), different flow rates, application temperature, different reagents (DNase, sodium deoxycholate, Triton-X, sodium dodecyl sulfate (SDS), (3)3-[(3-cholamidopropyl) dimethylammonio]-1-propanesulfonate (CHAPS)) and processing period are used to rebuild *de novo* tissues and organs that are non-immunogenic. This is clinically important for use in transplantation medicine. Although unlike biological matrices that retain various bioactive characteristics due to preservation of ECM, synthetic materials may also serve as basic scaffolds in TE.

Synthetic scaffolds can be made from biodegradable or non-degradable materials, and engineering procedures range from moulding to electrospinning, 3-D printing, and others. Because synthetic scaffolds are not derived from donor tissue, they do not pose a contamination risk during development, and can be rapidly produced and reproduced using relatively economical engineering processes. On the other hand, biological scaffolds hold significant bioactive information and cues that are important for the correct engraftment, proliferation, and differentiation of seeded cells. Because these cues are entirely lacking in synthetic scaffolds, this complicates their use in vivo or in clinical translation. New strategies and concepts are required to overcome these hurdles. The design of synthetic matrices using nanotechnology approaches may open a new direction in this so-far dead-end street of synthetic scaffolds.

Tracheal Bioengineering

Disorders affecting more than 50 % of the entire length of the adult trachea (or 1/3 in children) are inoperable due to the lack of reconstructive tissue. During the last decade, various surgical techniques and experimental methods have been investigated [49]. All of them have eventually failed due to time consuming protocols, heavy immunosuppression, or extremely difficult surgical procedures. Tissue engineering may provide a potential solution to this challenging situation. Based on experimental in vitro and in vivo animal data, the first-in-man transplantations of tissue-engineered tracheae have recently been performed, using both decellularized and synthetic scaffolds [50, 51]. The early and intermediate success, recently published in our 5-year-follow-up demonstrates the feasibility of this methodology [52]. Long-term data and clinical trials will provide significant data that may prove the real clinical meaning of this novel, currently rather conceptual strategy for tracheal replacement and reconstruction. Currently, a number of clinical as well as experimental research groups are investigating the use of both biological decellularized and synthetic tracheal scaffolds [50, 51, 53, 54]. A report from the recent expert group meeting [55] came to the conclusion that at present there is no consensus with regard to treatment methods for patients that suffer from complex tracheal disorders (Table 17.2). Even though the trachea appears to be a simple tube, its reconstruction or replacement remains challenging. Thus, research regarding novel methods of tracheal tissue regeneration, cell–surface interactions, and in vivo transfers are necessary to advance clinical frontiers in this challenging field.

Lung Bioengineering

Bioengineering of the lungs is complex, since this organ contains an intricate structure and a plethora of functions. There are over 60 different pulmonary cell types with multifaceted functions, an intricate alveolus-capillary interface to maintain gas exchange, and a fine network of microvasculature that must resist thrombosis and preserve elasticity for lung compliance to be adaptable for pressure–volume changes [56, 57]. One of the major challenges for bioengineering a three-dimensional lung structure is preserving or re-creating a delicate ECM with elasticity required for alveolar expansion, and thinness for efficient gas exchange.

The ECM microenvironment has a prominent role in identifying putative lung progenitor cells and driving cellular fate to maintain tissue integrity [58–60].

Table 17.2 The clinical experience of transplanting biological and synthetic tracheae

Method	Method	Pros & Cons	Application field
Allotransplantation	*Donor organ*		
All donor and the majority stent dependent	Fresh organ	*Pros*: processing time	Some few reports in human
		Cons: immunosuppression needed, surgically challenging	
	Cryopreserved, irradiated	*Pros*: no need for immunosuppression	Some few reports in human
		Cons: initial stenting	
	Chemical fixation	*Pros*	– 100 Adults
		– No need for immunosuppression	– 31 Children
		Cons	– Only one cancer patient
		– Not applicable for circumferential replacement	
		– Relies on local reconstructive tissue	
		– Long processing	
		– Permanent stenting	
	Fresh aortic allograft	*Pros*: processing time, no need for immunosuppression	Initial clinical application in six human cases
		Cons: requires permanent stenting	
Composites	In vivo allograft epithelialization	*Pros*	One clinical case
		– No ex vivo culture	
		Cons	
		– Relies on a donor trachea	
		– Processing time	
		– Initial immunosuppression	
	Polypropylene mesh covered with collagen sponge	*Pros*	Initial clinical application (patch)
		– No need for immunosuppression	
		Cons	
		– No long segments	
Tissue-engineered transplantation	*Biological* scaffold		
	Decellularized human trachea	*Pros*	Various patients with circumferential and partial replacement in both benign diseases and malignancies
		– No need for immunosuppression	
		– Biodegradable	
		– Maintained ECM	
		– Cell homing	
		– Biocompatibility	
		– Proangiogenicity	

(continued)

Table 17.2 (continued)

Method	Method	Pros & Cons	Application field
		Cons	
		– Donor dependent	
		– Ex vivo processing	
		– Stenting in the intermediate and long-term follow up	
	Decellularized porcine Jejunum	*Pros*	Three clinical cases (Patch), treating malignancies
		– Biocompatibility	
		Cons	
		– Mechanical strength	
		– Shortage of donor	
		– Processing time	
		– Immunogenicity?	
	Synthetic scaffold		
	POSS covalently bonded to PCU	*Pros*	One clinical case (circumferential entire trachea + bifurcation)
		– Customized	
		– Rapid and cheap production	
		Cons	
		– Lack of proangiogenic factors	
		– Mechanical properties	
		– Tissue integration	
	PET/PU (nanofiber)	*Pros*	Various clinical cases with different outcomes
		– Customized	
		– Mimicking ECM structure	
		– Allows for cell adhesion and migration	
		– Rapid and cheap production	
		Cons	
		– Lack of pro-angiogenic factors	
		– Mechanical properties	

In turn, the composition of ECM is also highly dependent on, and can be manipulated by, the surrounding cell types. The non-alveolar lung tissue consists of approximately 50 % ECM proteins [58]. The predominant ECM proteins, collagens, and elastin fibers provide structural and mechanical strength while laminin, fibronectin, and glycosaminoglycans (GAGs) are essential for cell attachment, proliferation, and differentiation [61–63]. However, the composition of ECM is highly dependent on the applied decellularization method [64–66]. Aside from preserving ECM composition, an ideal decellularization protocol should also preserve the gross native architecture of the lung (Fig. 17.2) that allows the scaffold to be

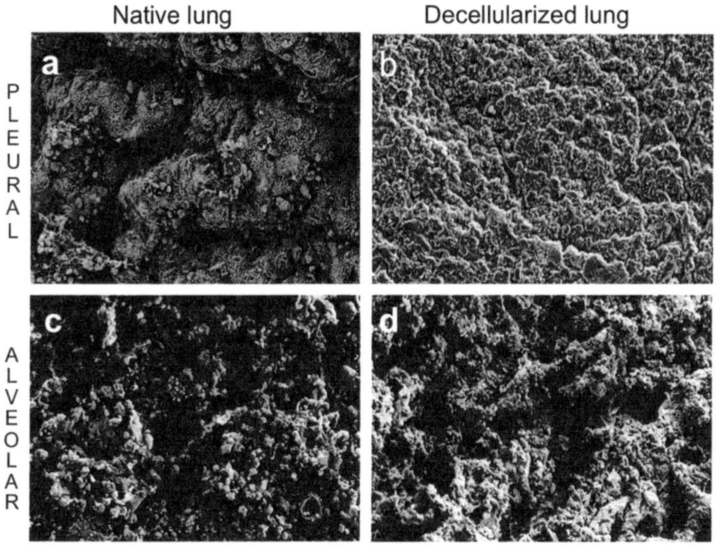

Fig. 17.2 An ideal decellularization protocol maintains gross native architecture. Scanning electron microscopy evaluation of pleural (**a**, **b**) and alveolar (**c**, **d**) lung surfaces before and after decellularization with 1 % deoxycholate and 0.1 % Triton-X-100. Scale bar 20 μm

physiologically ventilated and perfused [65]. Early experimental in vivo data do suggest that even complex structures such as the lung can be engineered. However, these experiments were performed in immunodeficient rodent models that cannot be immediately translated into a clinical setting because they do not address the most relevant clinical questions and concerns about graft rejection and the need for immunosuppression. Nonetheless, this type of experiment using decellularized lung tissues can be used to investigate and further understand various pulmonary cell–surface interactions, physiological changes, cell fate, and organ development. Besides, TE technologies can be applied to investigate lung disease models, such as elastase or bleomycin-induced lung injuries and the underlying pathohistological changes in the ECM [59, 67].

In Vitro Bioassay Systems

In vitro assays using decellularized lung scaffolds [68, 69], 2-D monolayer culture systems [70], or 3-D matrices such as matrigel [71], or porcine skin gelatin (Gelfoam) [72] have been used to assess the reparative and regenerative capacity of candidate endogenous lung stem cells and exogenous stem cells. Huh and colleagues [73] have developed a lung-on-a-chip, to create a combination of different cell types and a mechanical force interface to simulate living organs. Biomimetic microsystems of this type are useful to determine if implanted cells can release therapeutic and immunomodulatory substances at an adaptive dynamic rate, as judged by the cellular feedback mechanism. This can also be an accurate alternative to "replace, reduce, and refine" animal studies for predicting lung responses [73].

Ex Vivo Bioreactor System

Numerous research groups have successfully utilized three-dimensional models (trachea and *en bloc* lung) to characterize biocompatibility and biomechanics after tissue decellularization and recellularization. These models can potentially be used as tools for researching disease models, cell therapy, cell differentiation and fate, and drug application. The choice of stem cells for recellularizing tissues or organs has been a major focus of investigation in TE [74]. Cell types such as embryonic [75, 76] and induced pluripotent stem cells [68], amniotic fluid stem cells, mesenchymal stromal cells [77] or peripheral blood and bone-marrow-derived mononuclear cells are currently under investigation for their potential uses in different models using mouse, rat, porcine, non-human primate and human cadaveric donors. Furthermore, these ex vivo systems enable one to create simple tools to examine microenvironmental variables such as scaffold stiffness, mechanical ventilation, growth factors, pH, *etc.* and influence cell adhesion, proliferation, differentiation, or transdifferentiation [60, 78–80] (Fig. 17.3). These systems will also help to identify those factors that are important for providing the complex structural and biophysical cues required for tissue remodeling and regeneration.

Tumour Initiating Cells in the Respiratory Tract

In some circumstances, patients who opt for potential therapeutic options utilising cell therapies or tissue-engineered scaffolds (*e.g.*, tissue-engineered tracheal replacements) could suffer from recurring tumours. According to the cancer stem-cell paradigm it is believed that dysfunctional resident stem-like cells undergo oncogenic transformation giving rise to tumor-initiating cells which generate benign or malignant tumors [81]. Activated cancer-associated fibroblasts are also an integral component of the tumor microenvironment, influencing tumor growth and progression (e.g., by secreting growth factors, altering tissue metabolism and remodeling ECM) [82, 83]. Experimental models for the analysis of the interaction of airway tumor initiating cells with biomatrix scaffolds are essential in order to devise optimal therapeutic protocols employing tissue engineering technologies in lung cancer patients. Although various mouse models and assays have been used to identify adult lung stem and progenitor cells, and analyze their behavior in the normal and diseased lung [45, 84–86], the identity and properties of human airway stem/progenitor cells and tumor initiating cells are less clear.

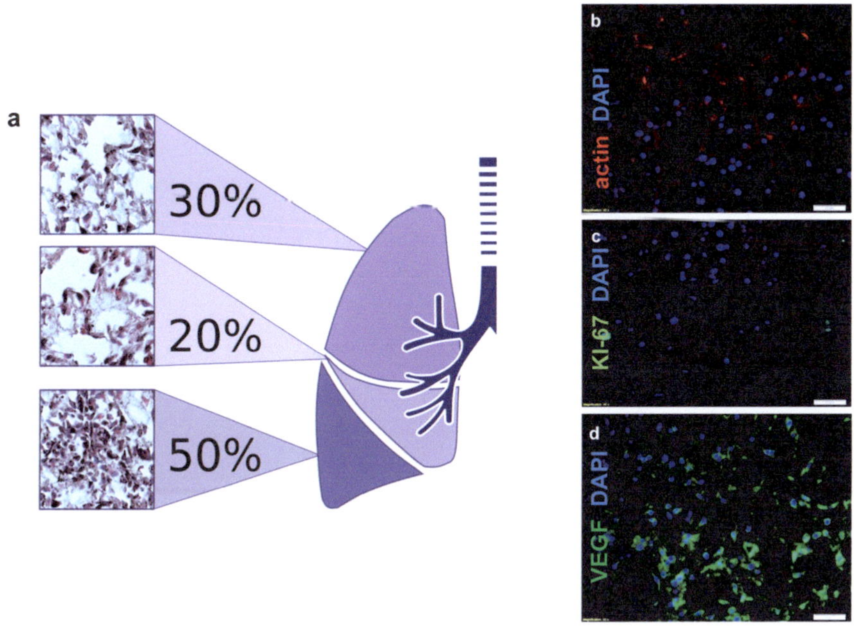

Fig. 17.3 Recellularizing ex vivo lung scaffold with plastic adherent bone-marrow-derived MSCs (BM-MSCs). Cells at passage 5 were delivered *via* pulmonary artery perfusion at 1 mL/min in a bioreactor for 7 days and were evaluated. (**a**) Cellular biodistribution on whole organ scaffold with bone-marrow-derived plate-adherent stem cells. Haematoxylin and eosin staining indicates the presence of reseeded cells on the scaffold. The lower lobe has the highest (50 %) number of detected cells. (**b**–**c**) Cell-to-matrix characterization of recellularized lung; (**b**) cytoskeleton is stained using actin antibody in *red*, (**c**) cell proliferation is indicated using Ki-67 antibody in *green* and (**d**) vasculogenesis marker VEGF presence is indicated in *green*. Scale bar 20 μm

We have recently isolated and characterized a subpopulation of mesenchymal stem-like cells from a human tracheal mucoepidermoid tumor (MEi cells) [87]. On serial passaging, these MEi tracheal cancer cells display a similar immunophenotypic signature profile and mesenchymal tri-lineage differentiation capacity as bone marrow-derived MSCs (BM-MSCs). They also express β-III tubulin (ectodermal marker), GATA6 (endodermal marker); but unlike BM-MSCs, MEi cells express the mucoepidermoid tumor-associated markers MUC1 and MAML2 (89 % and 95 %, respectively) (Fig 17.4). Notably, gene expression profiling and enrichment analysis using the Database for Annotation, Visualization and Integrated Discovery (DAVID) Gene Functional Classification Tool revealed that signaling pathways in cancer, purine metabolism and ECM receptor interactions were significantly enriched in MEi cells compared to BM-MSCs. This MEi human tracheal stem-cell-like population will likely be a useful reagent in studies aimed at elucidating mechanisms of initiation, development and/or progression of upper respiratory tract cancer; and in preclinical studies analyzing the interaction of cancer cells with tissue-engineered biomatrix scaffolds.

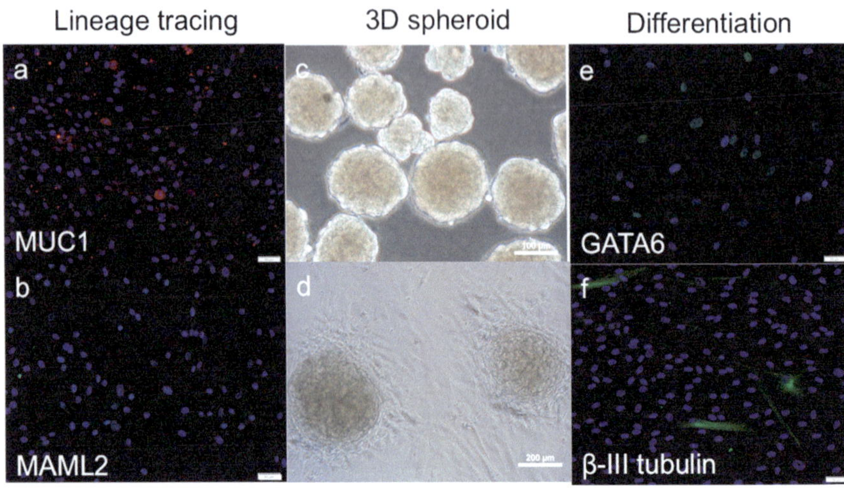

Fig. 17.4 In vitro assays to determine the cellular behavior of mesenchymal stem cell-like mucoepidermoid tumor (MEi cells). Lineage tracing assays (**a**) MUC1, (**b**) MAML2; 3D spheroid assay (**c**) spheroid forming ability, (**d**) spheroid outgrowths; differentiation assays (**e**) GATA6, (**f**) β-III tubulin

Nanomaterials in Regenerative Medicine

Nanotechnology is a rapidly growing discipline that offers a variety of applications with an important beneficence in medicine [88] and is expected to make a contribution to cancer prevention, detection, diagnosis, treatment, and imaging [89]. This includes the development of high-resolution in vivo imaging methodologies and the development of artificial organs, comprising biomimetic scaffolds and nanomaterials, among others (Fig. 17.5). The novel functional properties of nanomaterials make them attractive for both TE and cell therapy as it offers the possibility for tracking cells, for drug delivery, and for the design of long-term and better-performing scaffolds [90, 91]. For instance, nanomaterials could be designed to mimic certain features of the ECM to encapsulate and release growth factors and drugs in a controlled manner and promote tissue growth and regeneration at the site of interest. However, for their implementation in clinics, it is crucial to ensure the safety of these materials. Therefore, careful toxicity assessment is needed in order to provide knowledge for the safe design of these materials for such applications [92, 93].

Donor-derived stem cells have revolutionized the field of regenerative medicine. Their efficacy has been proven in a wide range of applications and emerged as a promising platform for a variety of clinical treatments [51, 52]. However, further preclinical studies are needed to determine the underlying mechanisms of in situ tissue regeneration. Indeed, the key problems include the lack of understanding of homing, therapeutic, and reparative mechanisms in vivo [94]. Understanding how stem cells are able to repair defects is of particular importance and studies to determine their fate and behavior in tissue repair in vivo are needed. Nanotechnology could provide new insight into these important and remaining questions.

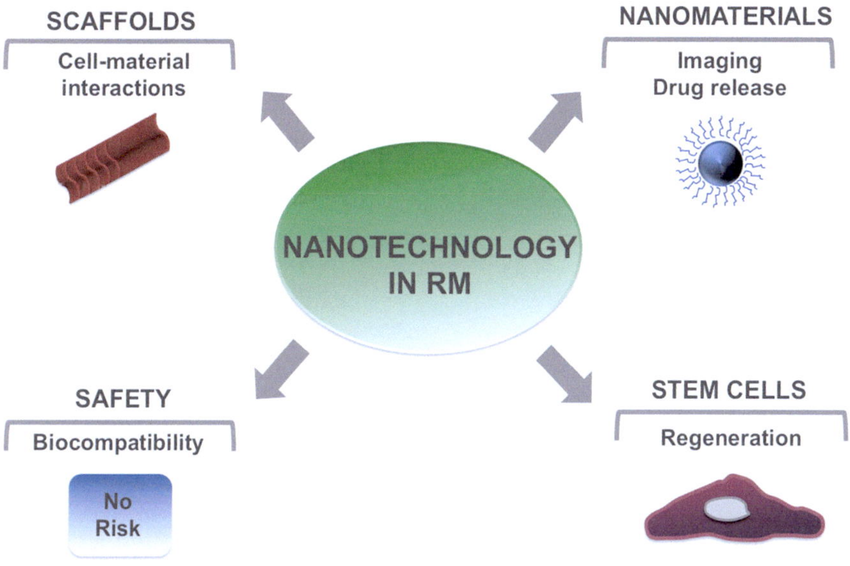

Fig. 17.5 Nanotechnology inputs in the field of regenerative medicine (RM)

In the recent years, efforts have been made to develop techniques based on nanoparticle-labeling of stem cells to improve the beneficial effects of stem-cell therapy. The imaging properties of nanomaterials will facilitate the identification of transplanted MSCs in vivo; and permit quantitative monitoring of their regenerative capabilities in a non-invasive manner and at high-resolution (Fig. 17.6). Several types of nanoparticles (NPs) have already been used in stem-cell studies for in vitro and in vivo imaging [95, 96]. Indeed, in stem-cell-based therapies, silica and magnetic NPs have been utilized to image and guide stem cells to their target [33]. By applying a magnetic field, labeled cells could be guided to accumulate at the site of interest and stimulate regeneration in a more efficient manner than traditional cell delivery approaches such as administering cells by intravenous or local injection. In this manner, higher cell retention could be achieved, enabling successful tissue regeneration by optimizing the engraftment of transplanted cells [97, 98]. Remarkably, in a recent study, Wu and co-workers successfully identified labeled cells in situ in the terminal bronchioles following transplantation and were able to quantify their engraftment and regeneration capabilities 1 week after intravenous administration [99]. Using this approach, the understanding of in vivo fate and homing capabilities of transplanted MSCs could be provided. Furthermore, understanding which physicochemical parameters of nanomaterials trigger specific biological responses under in vivo conditions remains a challenge. More efforts aimed at understanding the mechanisms of cell–material interactions at the nanoscale level are needed for the safe-smart-design of these materials for regenerative medicine applications. This knowledge could later be applied to treatment of other diseases in the field of regenerative medicine.

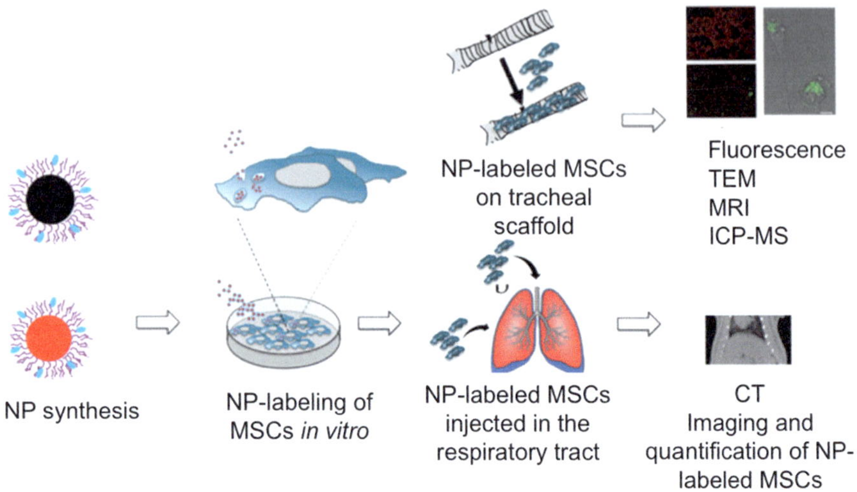

Fig. 17.6 Schematic representation of the nanoparticle (NP)-labeled MSCs and example of their potential application in tissue engineering and cell therapy. The nanoparticles exhibit novel physicochemical properties that following endocytosis by cells in vitro and applied to coat organ scaffolds or injected into a native organ, can thus be monitored with different imaging techniques (fluorescence, transmission electron microscopy, magnetic resonance imaging, inductively coupled plasma mass spectrometry or computerized tomography scanning) and provide specific treatment among others

Conclusion

The field of regenerative medicine holds great promise for the treatment of many respiratory diseases. We believe that TE- and cell-based therapies can safely and efficiently improve the lives of millions of people. We have reviewed several potential applications of nanotechnology in regenerative medicine, focusing on stem-cell labeling and tracking, emphasizing its potential to accelerate the safe and routine use of this technology in lung regenerative medicine. The knowledge obtained from these studies can provide valuable information on how to improve local or systemic effects, understand the biodistribution of transplanted cells, and evaluate the complexity of their migration, homing, and engraftment for restoration of lung function outcomes in patients.

References

1. McGowan SE (1992) Extracellular matrix and the regulation of lung development and repair. FASEB J 6:2895–2904
2. Burri PH (2006) Structural aspects of postnatal lung development—alveolar formation and growth. Biol Neonate 89:313–322

3. Burri PH (1997) Structural aspects of prenatal and postnatal development and growth of the lung. In: McDonald JA (ed) Lung growth and development. Marcel Decker, New York, pp 1–35
4. McGowan SE, Snyder JM (2004) Development of alveoli. In: Harding R, Pinkerton KE, Plopper CG (eds) The lung development, aging and environment. Elsevier Academic Press, London, pp 55–73
5. Burri PH (1974) The postnatal growth of the rat lung. 3. Morphology. Anat Rec 180:77–98
6. Kauffman SL, Burri PH, Weibel ER (1974) The postnatal growth of the rat lung. II. Autoradiography. Anat Rec 180:63–76
7. Firsova A, Hyakumura T, Cole TJ, Mollard R (2010) Epithelial-mesenchymal interactions during lung development and their potential relevance to lung repair. In: Polak J (ed) Cell therapy for lung disease. Imperial College Press, London, pp 91–124
8. Domyan ET, Ferretti E, Throckmorton K, Mishina Y, Nicolis SK, Sun X (2011) Signaling through BMP receptors promotes respiratory identity in the foregut via repression of Sox2. Development 138:971–981
9. Plopper CG, Alley JL, Weir AJ (1986) Differentiation of tracheal epithelium during fetal lung maturation in the rhesus monkey Macaca mulatta. Am J Anat 175:59–71
10. Miller LA, Wert SE, Clark JC, Xu Y, Perl AK, Whitsett JA (2004) Role of Sonic hedgehog in patterning of tracheal-bronchial cartilage and the peripheral lung. Dev Dyn 231:57–71
11. Bellusci S, Grindley J, Emoto H, Itoh N, Hogan BL (1997) Fibroblast growth factor 10 (FGF10) and branching morphogenesis in the embryonic mouse lung. Development 124:4867–4878
12. Min H, Danilenko DM, Scully SA, Bolon B, Ring BD, Tarpley JE, DeRose M, Simonet WS (1998) Fgf-10 is required for both limb and lung development and exhibits striking functional similarity to Drosophila branchless. Genes Dev 12:3156–3161
13. Metzger RJ, Klein OD, Martin GR, Krasnow MA (2008) The branching programme of mouse lung development. Nature 453:745–750
14. Mailleux AA, Tefft D, Ndiaye D, Itoh N, Thiery JP, Warburton D, Bellusci S (2001) Evidence that SPROUTY2 functions as an inhibitor of mouse embryonic lung growth and morphogenesis. Mech Dev 102:81–94
15. Pepicelli CV, Lewis PM, McMahon AP (1998) Sonic hedgehog regulates branching morphogenesis in the mammalian lung. Curr Biol 8:1083–1086
16. Harris KS, Zhang Z, McManus MT, Harfe BD, Sun X (2006) Dicer function is essential for lung epithelium morphogenesis. Proc Natl Acad Sci U S A 103:2208–2213
17. Bostrom H, Willetts K, Pekny M et al (1996) PDGF-A signaling is a critical event in lung alveolar myofibroblast development and alveogenesis. Cell 85:863–873
18. Massaro GD, Massaro D (1997) Retinoic acid treatment abrogates elastase-induced pulmonary emphysema in rats. Nat Med 3:675–677
19. Yen CC, Yang SH, Lin CY, Chen CM (2006) Stem cells in the lung parenchyma and prospects for lung injury therapy. Eur J Clin Invest 36:310–319
20. Yamamoto H, Yun EJ, Gerber HP, Ferrara N, Whitsett JA, Vu TH (2007) Epithelial-vascular cross talk mediated by VEGF-A and HGF signaling directs primary septae formation during distal lung morphogenesis. Dev Biol 308:44–53
21. Chen H, Zhuang F, Liu Y et al (2008) TGF-β receptor II in epithelia versus mesenchyme plays distinct roles in the developing lung. Eur Respir J 32:285–295
22. Ng YS, Rohan R, Sunday ME, Demello DE, D'Amore PA (2001) Differential expression of VEGF isoforms in mouse during development and in the adult. Dev Dyn 220:112–121
23. Kasahara Y, Tuder RM, Taraseviciene-Stewart L, Le Cras TD, Abman S, Hirth PK, Waltenberger J, Voelkel NF (2000) Inhibition of VEGF receptors causes lung cell apoptosis and emphysema. J Clin Invest 106:1311–1319
24. Massaro GD, Massaro D (1992) Formation of alveoli in rats: postnatal effect of prenatal dexamethasone. Am J Physiol 263:L37–L41

25. Massaro GD, Massaro D (1996) Postnatal treatment with retinoic acid increases the number of pulmonary alveoli in rats. Am J Physiol 270:L305–L310
26. Warburton D, Perin L, Defilippo R, Bellusci S, Shi W, Driscoll B (2008) Stem/progenitor cells in lung development, injury repair, and regeneration. Proc Am Thorac Soc 5:703–706
27. Olsson F, Denham M, Cole TJ, Hooper SB, Mollard R (2007) Deriving respiratory cell types from stem cells. Curr Stem Cell Res Ther 2:197–208
28. Irwin D, Helm K, Campbell N et al (2007) Neonatal lung side population cells demonstrate endothelial potential and are altered in response to hyperoxia-induced lung simplification. Am J Physiol Lung Cell Mol Physiol 293:L941–L951
29. Giangreco A, Shen H, Reynolds SD, Stripp BR (2004) Molecular phenotype of airway side population cells. Am J Physiol Lung Cell Mol Physiol 286:L624–L630
30. Majka SM, Beutz MA, Hagen M, Izzo AA, Voelkel N, Helm KM (2005) Identification of novel resident pulmonary stem cells: form and function of the lung side population. Stem Cells 23:1073–1081
31. Kuwana M, Okazaki Y, Kodama H, Izumi K, Yasuoka H, Ogawa Y, Kawakami Y, Ikeda Y (2003) Human circulating CD14+ monocytes as a source of progenitors that exhibit mesenchymal cell differentiation. J Leukoc Biol 74:833–845
32. Zhang M, Huang B (2012) The multi-differentiation potential of peripheral blood mononuclear cells. Stem Cell Res Ther 3:48
33. Lu CW, Hung Y, Hsiao JK et al (2007) Bifunctional magnetic silica nanoparticles for highly efficient human stem cell labeling. Nano Lett 7:149–154
34. Tse W, Laughlin MJ (2005) Umbilical cord blood transplantation: a new alternative option. Hematology Am Soc Hematol Educ Program 377–383
35. Kumar BM, Yoo JG, Ock SA et al (2007) In vitro differentiation of mesenchymal progenitor cells derived from porcine umbilical cord blood. Mol Cells 24:343–350
36. De Coppi P, Bartsch G Jr, Siddiqui MM et al (2007) Isolation of amniotic stem cell lines with potential for therapy. Nat Biotechnol 25:100–106
37. Carraro G, Perin L, Sedrakyan S et al (2008) Human amniotic fluid stem cells can integrate and differentiate into epithelial lung lineages. Stem Cells 26:2902–2911
38. Moodley Y, Ilancheran S, Samuel C, Vaghjiani V, Atienza D, Williams ED, Jenkin G, Wallace E, Trounson A, Manuelpillai U (2010) Human amnion epithelial cell transplantation abrogates lung fibrosis and augments repair. Am J Respir Crit Care Med 182:643–651
39. Sessarego N, Parodi A, Podesta M et al (2008) Multipotent mesenchymal stromal cells from amniotic fluid: solid perspectives for clinical application. Haematologica 93:339–346
40. Abe S, Boyer C, Liu X, Wen FQ, Kobayashi T, Fang Q, Wang X, Hashimoto M, Sharp JG, Rennard SI (2004) Cells derived from the circulation contribute to the repair of lung injury. Am J Respir Crit Care Med 170:1158–1163
41. Anjos-Afonso F, Siapati EK, Bonnet D (2004) In vivo contribution of murine mesenchymal stem cells into multiple cell-types under minimal damage conditions. J Cell Sci 117:5655–5664
42. Ishizawa K, Kubo H, Yamada M, Kobayashi S, Numasaki M, Ueda S, Suzuki T, Sasaki H (2004) Bone marrow-derived cells contribute to lung regeneration after elastase-induced pulmonary emphysema. FEBS Lett 556:249–252
43. Spees JL, Whitney MJ, Sullivan DE, Lasky JA, Laboy M, Ylostalo J, Prockop DJ (2008) Bone marrow progenitor cells contribute to repair and remodeling of the lung and heart in a rat model of progressive pulmonary hypertension. FASEB J 22:1226–1236
44. Yamada M, Kubo H, Kobayashi S, Ishizawa K, Numasaki M, Ueda S, Suzuki T, Sasaki H (2004) Bone marrow-derived progenitor cells are important for lung repair after lipopolysaccharide-induced lung injury. J Immunol 172:1266–1272
45. Kajstura J, Rota M, Hall SR et al (2011) Evidence for human lung stem cells. N Engl J Med 364:1795–1806
46. Spooner BS, Wessells NK (1970) Mammalian lung development: interactions in primordium formation and bronchial morphogenesis. J Exp Zool 175:445–454

47. Wessells NK (1970) Mammalian lung development: interactions in formation and morphogenesis of tracheal buds. J Exp Zool 175:455–466
48. Balasubramaniam V, Ryan SL, Seedorf GJ, Roth EV, Heumann TR, Yoder MC, Ingram DA, Hogan CJ, Markham NE, Abman SH (2010) Bone marrow-derived angiogenic cells restore lung alveolar and vascular structure after neonatal hyperoxia in infant mice. Am J Physiol Lung Cell Mol Physiol 298:L315–L323
49. Grillo HC (2002) Tracheal replacement: a critical review. Ann Thorac Surg 73:1995–2004
50. Macchiarini P, Jungebluth P, Go T et al (2008) Clinical transplantation of a tissue-engineered airway. Lancet 372:2023–2030
51. Jungebluth P, Alici E, Baiguera S et al (2011) Tracheobronchial transplantation with a stem-cell-seeded bioartificial nanocomposite: a proof-of-concept study. Lancet 378:1997–2004
52. Gonfiotti A, Jaus MO, Barale D et al (2014) The first tissue-engineered airway transplantation: 5-year follow-up results. Lancet 383:238–244
53. Berg M, Ejnell H, Kovács A et al (2013) Replacement of a tracheal stenosis with a tissue-engineered human trachea using autologous stem cells: a Case Report. Tissue Eng Part A 20:389–397
54. Elliott MJ, De Coppi P, Speggiorin S et al (2012) Stem-cell-based, tissue engineered tracheal replacement in a child: a 2-year follow-up study. Lancet 380:994–1000
55. Weiss DJ, Elliott M, Jang Q, Poole B, Birchall M (2014) Tracheal bioengineering: the next steps. Proceeds of an international society of cell therapy pulmonary cellular therapy signature series workshop, Paris, France, April 22, 2014. Cytotherapy 16:1601–1613
56. Calle EA, Ghaedi M, Sundaram S, Sivarapatna A, Tseng MK, Niklason LE (2014) Strategies for whole lung tissue engineering. IEEE Trans Biomed Eng 61:1482–1496
57. Nagao RJ, Ouyang Y, Keller R, Lee C, Suggs LJ, Schmidt CE (2013) Preservation of capillary-beds in rat lung tissue using optimized chemical decellularization. J Mater Chem B Mater Biol Med 1:4801–4808
58. Price AP, England KA, Matson AM, Blazar BR, Panoskaltsis-mortari A (2010) Development of a decellularized lung bioreactor system for bioengineering the lung: the matrix reloaded. Tissue Eng Part A 16:2581–2591
59. Parker MW, Rossi D, Peterson M, Smith K, Sikström K, White ES, Connett JE, Henke CA, Larsson O, Bitterman PB (2014) Fibrotic extracellular matrix activates a profibrotic positive feedback loop. J Clin Invest 124:1622–1635
60. Soucy PA, Werbin J, Heinz W, Hoh JH, Romer LH (2011) Microelastic properties of lung cell-derived extracellular matrix. Acta Biomater 7:96–105
61. Petersen TH, Calle EA, Colehour MB, Niklason LE (2012) Matrix composition and mechanics of decellularized lung scaffolds. Cells Tissues Organs 195:222–231
62. Hill RC, Calle EA, Dzieciatkowska M, Niklason LE, Hansen KC (2015) Quantification of extracellular matrix proteins from a rat lung scaffold to provide a molecular readout for tissue engineering. Mol Cell Proteomics 14(4):961–973
63. Ghaedi M, Calle EA, Mendez JJ, Gard AL, Balestrini J, Booth A, Bove PF, Gui L, White ES, Niklason LE (2013) Human iPS cell-derived alveolar epithelium repopulates lung extracellular matrix. J Clin Invest 123:4950–4962
64. Wallis JM, Borg ZD, Daly AB, Deng B, Ballif BA, Allen GB, Jaworski DM, Weiss DJ (2012) Comparative assessment of detergent-based protocols for mouse lung de-cellularization and re-cellularization. Tissue Eng Part C Methods 18:420–432
65. Daly KA, Liu S, Agrawal V, Brown BN, Johnson SA, Medberry CJ, Badylak SF (2012) Damage associated molecular patterns within xenogeneic biologic scaffolds and their effects on host remodeling. Biomaterials 33:91–101
66. Gilpin SE, Guyette JP, Gonzalez G, Ren X, Asara JM, Mathisen DJ, Vacanti JP, Ott HC (2014) Perfusion decellularization of human and porcine lungs: bringing the matrix to clinical scale. J Hear Lung Transplant 33:298–308
67. Wagner DE, Bonenfant NR, Parsons CS et al (2014) Comparative decellularization and recellularization of normal versus emphysematous human lungs. Biomaterials 35:3281–3297

68. Gilpin SE, Ren X, Okamoto T, Guyette JP, Mou H, Rajagopal J, Mathisen DJ, Vacanti JP, Ott HC (2014) Enhanced lung epithelial specification of human induced pluripotent stem cells on decellularized lung matrix. Ann Thorac Surg 98:1721–1729
69. Shojaie S, Ermini L, Ackerley C et al (2015) Acellular lung scaffolds direct differentiation of endoderm to functional airway epithelial cells: requirement of matrix-bound HS proteoglycans. Stem Cell Reports 4:419–430
70. Lim ML, Jungebluth P, Sjöqvist S, Nikdin H, Rós K (2013) Decellularized feeders: an optimized method for culturing pluripotent cells. Stem Cells Transl Med 2:975–982
71. Zhang WJ, Lin QX, Zhang Y et al (2011) The reconstruction of lung alveolus-like structure in collagen-matrigel/microcapsules scaffolds in vitro. J Cell Mol Med 15:1878–1886
72. Andrade CF, Wong AP, Waddell TK, Keshavjee S, Liu M (2007) Cell-based tissue engineering for lung regeneration. Am J Physiol Lung Cell Mol Physiol 292:L510–L518
73. Huh D, Matthews BD, Mammoto A, Montoya-Zavala M, Hsin HY, Ingber DE (2010) Reconstituting organ-level lung functions on a chip. Science 328:1662–1668
74. Calle EA, Petersen TH, Niklason LE (2011) Procedure for lung engineering. J Vis Exp (49). pii: 2651. doi: 10.3791/2651
75. Lecht S, Stabler CT, Rylander AL, Chiaverelli R, Schulman ES, Marcinkiewicz C, Lelkes PI (2014) Biomaterials Enhanced reseeding of decellularized rodent lungs with mouse embryonic stem cells. Biomaterials 35:3252–3262
76. Nichols JE, Niles J, Riddle M et al (2013) Production and assessment of decellularized pig and human lung scaffolds. Tissue Eng Part A 19:2045–2062
77. Bonvillain RW, Danchuk S, Sullivan DE et al (2012) A nonhuman primate model of lung regeneration: detergent-mediated decellularization and initial in vitro recellularization with mesenchymal stem cells. Tissue Eng Part A 18:2437–2452
78. Nichols JE, Niles JA, Cortiella J (2009) Design and development of tissue engineered lung. Organogenesis 5:57–61
79. Kubo H (2011) Molecular basis of lung tissue regeneration. Gen Thorac Cardiovasc Surg 59:231–244
80. Tsuchiya T, Balestrini JL, Mendez J, Calle EA, Zhao L, Niklason LE (2014) Influence of pH on extracellular matrix preservation during lung decellularization. Tissue Eng Part C Methods 20:1028–1036
81. Salcido CD, Larochelle A, Taylor BJ, Dunbar CE, Varticovski L (2010) Molecular characterisation of side population cells with cancer stem cell-like characteristics in small-cell lung cancer. Br J Cancer 102:1636–1644
82. Kalluri R, Zeisberg M (2006) Fibroblasts in cancer. Nat Rev Cancer 6:392–401
83. Cirri P, Chiarugi P (2012) Cancer-associated-fibroblasts and tumour cells: a diabolic liaison driving cancer progression. Cancer Metastasis Rev 31:195–208
84. Barkauskas C, Cronce M, Rackley C, Bowie E, Keene D, Stripp B, Randell S, Noble P, Hogan B (2013) Type 2 alveolar cells are stem cells in adult lung. J Clin Invest 123:3025–3036
85. McQualter JL, Yuen K, Williams B, Bertoncello I (2010) Evidence of an epithelial stem/progenitor cell hierarchy in the adult mouse lung. Proc Natl Acad Sci U S A 107:1414–1419
86. Lau AN, Goodwin M, Kim CF, Weiss DJ (2012) Stem cells and regenerative medicine in lung biology and diseases. Mol Ther 20:1116–1130
87. Lim ML, Ooi BNS, Jungebluth P et al (2014) Characterization of stem-like cells in mucoepidermoid tracheal paediatric tumor. PLoS One 9, e107712
88. Rivera-Gil P, Jimenez de Aberasturi D, Wulf V et al (2013) The challenge to relate the physicochemical properties of colloidal nanoparticles to their cytotoxicity. Acc Chem Res 46:743–749
89. Riehemann K, Schneider SW, Luger TA, Godin B, Ferrari M, Fuchs H (2009) Diagnostics and drug delivery nanomedicine—challenge and perspectives. Angew Chem Int Ed Engl 48:872–897

90. Chaudhury K, Kumar V, Kandasamy J, RoyChoudhury S (2014) Regenerative nanomedicine: current perspectives and future directions. Int J Nanomedicine 9:4153–4167
91. Fadeel B, Feliu N, Vogt C, Abdelmonem AM, Parak WJ (2013) Bridge over troubled waters: understanding the synthetic and biological identities of engineered nanomaterials. Wiley Interdiscip Rev Nanomed Nanobiotechnol 5:111–129
92. Feliu N, Fadeel B (2010) Nanotoxicology: no small matter. Nanoscale 2:2514–2520
93. Soenen SJ, Manshian B, Doak SH, De Smedt SC, Braeckmans K (2013) Fluorescent non-porous silica nanoparticles for long-term cell monitoring: cytotoxicity and particle functionality. Acta Biomater 9:9183–9193
94. Jungebluth P, Macchiarini P (2014) Airway transplantation. Thorac Surg Clin 24:97–106
95. Lathrop MJ, Brooks EM, Bonenfant NR et al (2014) Mesenchymal stromal cells mediate Aspergillus hyphal extract-induced allergic airway inflammation by inhibition of the Th17 signaling pathway. Stem Cells Transl Med 3:194–205
96. Park KS, Tae J, Choi B et al (2010) Characterization, in vitro cytotoxicity assessment, and in vivo visualization of multimodal, RITC-labeled, silica-coated magnetic nanoparticles for labeling human cord blood-derived mesenchymal stem cells. Nanomedicine 6:263–276
97. Pickard MR, Barraud P, Chari DM (2011) The transfection of multipotent neural precursor/stem cell transplant populations with magnetic nanoparticles. Biomaterials 32:2274–2284
98. Gil S, Correia CR, Mano JF (2015) Magnetically Labeled cells with surface-modified Fe3 O4 spherical and rod-shaped magnetic nanoparticles for tissue engineering applications. Adv Healthc Mater 4(6):883–891. doi:10.1002/adhm.201400611
99. Wu TJ, Tzeng YK, Chang WW, Cheng CA, Kuo Y, Chien CH, Chang HC, Yu J (2013) Tracking the engraftment and regenerative capabilities of transplanted lung stem cells using fluorescent nanodiamonds. Nat Nanotechnol 8:682–689
100. Phinney DG, Kopen G, Isaacson RL, Prockop DJ (1999) Plastic adherent stromal cells from the bone marrow of commonly used strains of inbred mice: variations in yield, growth, and differentiation. J Cell Biochem 72:570–585
101. Aguilar S, Nye E, Chan J, Loebinger M, Spencer-Dene B, Fisk N, Stamp G, Bonnet D, Janes SM (2007) Murine but not human mesenchymal stem cells generate osteosarcoma-like lesions in the lung. Stem Cells 25:1586–1594
102. Chang YS, Oh W, Choi SJ, Sung DK, Kim SY, Choi EY, Kang S, Jin HJ, Yang YS, Park WS (2009) Human umbilical cord blood-derived mesenchymal stem cells attenuate hyperoxia-induced lung injury in neonatal rats. Cell Transplant 18:869–886
103. Wu X, Lensch MW, Wylie-Sears J, Daley GQ, Bischoff J (2007) Hemogenic endothelial progenitor cells isolated from human umbilical cord blood. Stem Cells 25:2770–2776
104. Yoder MC, Mead LE, Prater D, Krier TR, Mroueh KN, Li F, Krasich R, Temm CJ, Prchal JT, Ingram DA (2007) Redefining endothelial progenitor cells via clonal analysis and hematopoietic stem/progenitor cell principals. Blood 109:1801–1809
105. Asahara T, Masuda H, Takahashi T, Kalka C, Pastore C, Silver M, Kearne M, Magner M, Isner JM (1999) Bone marrow origin of endothelial progenitor cells responsible for postnatal vasculogenesis in physiological and pathological neovascularization. Circ Res 85:221–228
106. Balasubramaniam V, Mervis CF, Maxey AM, Markham NE, Abman SH (2007) Hyperoxia reduces bone marrow, circulating, and lung endothelial progenitor cells in the developing lung: implications for the pathogenesis of bronchopulmonary dysplasia. Am J Physiol Lung Cell Mol Physiol 292:L1073–L1084
107. Orkin SH (2000) Diversification of haematopoietic stem cells to specific lineages. Nat Rev Genet 1:57–64

Chapter 18
Acellular Lung Scaffolds in Lung Bioengineering

Darcy E. Wagner, Franziska E. Uhl, and Daniel J. Weiss

Introduction

Chronic lung diseases such as chronic obstructive pulmonary disease (COPD) and idiopathic pulmonary fibrosis (IPF) are significantly increasing in prevalence and are predicted to be an increased major worldwide healthcare burden [1, 2]. There are currently no cures for these diseases and end-stage disease is associated with high mortality. While newly approved pharmaceutical interventions, such as Pirfenidone for IPF, have generated some excitement, it remains to be seen how effective these will be on a larger scale, and for what proportion of patients these treatments are suitable. For most patients, the only available treatment option at end-stage disease remains lung transplantation. However, there are not enough donor lungs to meet current transplantation needs and there are further complications associated with lung transplantation. Transplantation recipients require life-long immunosuppression and the 5-year survival after lung transplantation remains approximately 50 % [1, 2]. Alternative options are therefore desperately needed for this patient population.

One active and promising area of research is the generation of pulmonary tissue using ex vivo methods. The basic concept is that a scaffold of either biologic or artificial origin could be seeded with an appropriate cell source to regenerate functional lung tissue for subsequent transplantation (Fig. 18.1). While both of these techniques are still in their relatively early stages, one of their purported benefits is that either biologic or synthetic scaffolds could be recellularized with autologous cells, thus minimizing the immunological complications which

D.E. Wagner
Comprehensive Pneumology Center, Munich, Germany

F.E. Uhl • D.J. Weiss (✉)
University of Vermont College of Medicine, Burlington, VT, USA
e-mail: dweiss@uvm.edu

I. Bertoncello (ed.), *Stem Cells in the Lung*, Stem Cell Biology and Regenerative Medicine, DOI 10.1007/978-3-319-21082-7_18

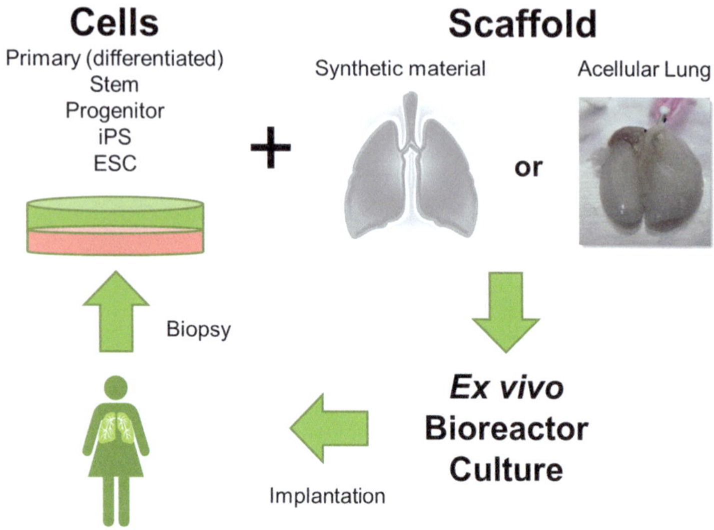

Fig. 18.1 Schematic of ex vivo organ engineering. Autologous cells are obtained by a biopsy from the eventual transplant recipient and expanded in ex vivo culture. A scaffold, either synthetic or an acellular lung, is manufactured and repopulated ex vivo by the usage of a bioreactor to create a functional tissue suitable for re-implantation

typically accompany lung transplantation. While this approach is not yet feasible in lungs, similar approaches have already been used clinically in simpler tissues including trachea, skin, and bone [3, 4]. Lung tissue, however, is a considerably more structurally complex organ and consists of a variety of cell types which must be functionally recapitulated in vivo. Due to these intrinsic differences in complexity between tissues, progress has significantly lagged behind the advances made in other organs. Synthetic scaffolds made from biocompatible or natural polymers are one potential option. A number of different materials and manufacturing technologies have already been evaluated for lung and will be discussed in more detail in this chapter.

An exciting new and active area of research involves the use of acellular lung scaffolds derived from cadaveric or failed transplant lungs. Acellular tissue is generated by removing cells from the native organ while preserving the 3D macroarchitecture and the majority of the extracellular matrix (ECM) proteins [5–17]. Whole organ decellularization as a platform for organ regeneration was first described in the heart in 2008 [15] and beginning in 2010, several groups described similar techniques in lung [18–23]. The use of acellular lungs has since expanded beyond their usage in regenerative medicine and has become an incredibly powerful in vitro tool for studying cell–ECM interactions or the impact of diseased matrix on cellular behavior [24–29].

This chapter discusses the status of current areas of research investigating ex vivo regeneration of lung tissue, and includes a discussion of concepts learned from the literature on ex vivo tissue culture and organ preservation.

Engineering a Scaffold

Designing and Manufacturing a Bioartificial Scaffold

Proposed bioartificial scaffolds for lung engineering have been manufactured by various techniques and from a variety of different materials. An overview of the current approaches is given in Table 18.1. In general two different methods of scaffold generation can be distinguished: additive (layer-by-layer or unit-by-unit generation) and subtractive methods (generation by removing material to form the final scaffold). Additive techniques benefit from the possibility to generate scaffolds with interconnecting pores. Depending on the resolution needed, however, these techniques may lead to long fabrication times. Examples for additive techniques are rapid prototyping and 3D bioprinting [30]. Subtractive methods such as porogen forming techniques and sphere-templating have also shown promising initial results [31, 32] but are more limited regarding scaffold design. There are various other methods to produce scaffolds for tissue engineering purposes like solvent casting, particulate leaching, melt molding, or freeze drying. Synthetic materials used thus far are polymers like polyglycolic acid (PGA), poly-lactic-co-glycolic acid (PLGA), poly-L-lactic-acid (PLLA), polyurethane (PU), and polyvinyl (PV) in order to match the mechanical properties of lung tissue. Hydrogels made of collagen I, gelatin, Matrigel, alginate, fibrinogen–fibronectin–vitronectin combinations or PGA combined with Pluronic F-127 have been used as scaffolds as well [33, 34]. Further, synthetic scaffolds can be loaded with growth factors, ECM components (e.g., collagen or whole lung extracts) or peptide sequences known to facilitate cell attachment (e.g., RGD) [35].

The lung has a highly complex structure with varying structural composition and mechanical properties which are still unable to be completely recapitulated using synthetic approaches. While scaffolds fabricated via foaming techniques are structurally similar to peripheral lung tissue (especially the alveoli), they lack a vascular system and innervations. It is also difficult to tune the various mechanical properties needed throughout the lung for proper breathing motions. Additionally, the challenge of scaffold recellularization to create a fully functional organ has not yet been achieved. Thus, the use of the current methods exclusively may not solve the issue of whole lung replacement, but there are many areas for improvement which can still be explored.

Acellular Scaffolds

Synthetic scaffolds could one day be accurately and precisely manufactured for the macro- and microarchitecture required for ex vivo lung bioengineering. However, the instructional cues which are needed on the scaffold for critical events such as initial cell attachment, potential cell-specific attachment cues, and differentiation

Table 18.1 Artificial scaffolds used for Tissue Engineering of the lung

Ref.	Scaffold	Study objective	In vitro/in vivo	Cell source seeding	Benefits/limitations
[83]	Polyglycolic acid and Pluronic F-127 hydrogel	Isolation and characterization of somatic lung progenitor cells and the promotion of alveolar tissue growth	In vitro, in vivo	Somatic lung progenitor cells	De novo generation of lung tissue, limited amount of cell source
[118]	PDLLA scaffolds (3D foam)	Suitability of scaffold for distal lung tissue engineering	In vitro	MLE-12	No toxic effects, supports cell growth
[119]	Matrigel, synthetic poly-lactic-co-glycolic acid (PLGA) and poly-L-lactic-acid (PLLA) porous foams and nanofibrous matrices	Engineering of 3D pulmonary tissue constructs	In vitro	Murine embryonic day 18 fetal pulmonary cells	Limited amount of cell source; use of mixed populations
[120]	Collagen I gel	Effects of FGF10, FGF7, and FGF2 on murine fetal pulmonary cells to control epithelial cell behavior	In vitro	Fetal murine pulmonary cells	Addition of growth factors to influence cell proliferation and differentiation possible
[84]	Gelfoam sponge (collagen I)	Application for cell-based lung regeneration	In vivo	Fetal lung cells (d19)	No severe local inflammatory response, sponge degraded after several months
[31]	Patterned porous hydrogel from poly(2-hydroxyethyl methacrylate) (poly(HEMA)), Collagen I treated	Suitability of scaffold fabrication method to design scaffolds with different macro- and microstructure, cell attachment	In vitro	Mouse skeletal myoblast cells	Tailoring of scaffold macro- and microstructure possible
[85]	Matrigel plug combined with FGF2-loaded polyvinyl sponges	Generation of vascularized pulmonary tissue constructs	In vivo, subcutaneous	Fetal pulmonary cells (d17.5)	limited amount of cell source, induction of a host inflammatory response
[35]	Fibrinogen–fibronectin–vitronectin hydrogel	Ability of hydrogel to promote cell engraftment	In vivo	Primary ovine lung mesenchymal cells	Mechanical properties similar to lung, safe cell delivery, no atelectasis and scarring

[121]	Cylindrical-shaped bronchioles, fibroblasts embedded into collagen I with ASM cells on the outer surface, epithelial cells on the inner surface (lumen)	Development of a tissue-engineered bronchiole	In vitro	Human lung primary cells (fibroblasts, airway smooth muscle and epithelial cells)	Cell–cell interactions and airway remodeling events can be studied
[74]	Cellularized collagen matrix (Gelfoam)	Treatment of postpneumonectomy space	In vivo	BMSC	Cells only survive in early time points inside the scaffold (until d7)
[34]	Nanofibers coated with lung extracts	Effect of lung extract and modulus on cell differentiation	In vitro	Bone marrow cells	High throughput possible
[33]	Collagen-elastin hydrogels	Effect of elastin and cell seeding on hydrogel stiffness, Suitability as building blocks for lung engineering	In vitro	Lung fibroblasts	Adaption of mechanical properties to the lung tissue (alveolus)
[86]	Matrigel, decellularized rat lung slices	Potential of AF-MSCs to generate lung precursor cells	In vitro	AF-MSC	AF-MSC are a possible cell source for cell therapy, differentiation on lung matrix more effective, limited amount of cell source, lack of cell proliferation
[32]	Gelatin/microbubble-scaffolds	Vascularization of artificial scaffold for lung tissue engineering	In vitro, in vivo, in ovo, subcutaneous implantation in SCID mice	mPSC	Scaffolds promote angiogenesis and differentiation to alveolar pneumocytes

Abbreviations: *AF-MSC* amniotic fluid mesenchymal stem cells, *ASM* airway smooth muscle cells, *BMSC* bone marrow stem cells, *FGF* fibroblast growth factor, *HEMA* 2-hydroxyethyl methacrylate, *MLE-12* mouse epithelial cell line, *mPSC* mouse pulmonary stem/progenitor cells, *PDLLA* Poly-DL-lactic-acid, *PLGA* poly-lactic-co-glycolic acid, *PLLA* poly-L-lactic-acid, *SCID* severe combined immunodeficient cues are not known. Furthermore, if these criteria were known, the lung scaffold would likely also need to be engineered with a material and manufacturing process selection which matched the mechanical and gas diffusion properties of native lung. This makes the engineering of a completely synthetic scaffold daunting. While synthetic materials could be engineered to include specific integrin binding sites to enhance cell adhesion (e.g., Arg-Gly-Asp (RGD) binding sites), it remains unknown what specific integrin binding sites need to be included and in what spatial arrangement they need to be. On the other hand, acellular scaffolds retain many of the native integrin binding sites in their correct spatial arrangement, and decellularization processes preserve the general organ architecture and ECM composition. Lung ECM has also long been known to provide instructional cues during prenatal development, postnatal tissue regeneration, remodeling responses following injury, and general tissue homeostasis [36–40]. Similarly, acellular scaffolds have been shown to have biologically inductive clues [21, 22, 27, 41, 42]. While hybrid materials, consisting of synthetic and acellular matrix components, are also an attractive possibility, these concepts are in their infancy. Hybrid materials could be utilized to enhance cell adhesion and biological activity while taking advantage of the ability to more precisely manufacture scaffolds or scaffold components with synthetic materials [21, 34, 35, 43, 44]. Differences between acellular and synthetic scaffold approaches are summarized in Table 18.2. Owing to the current advantages of acellular scaffolds, we will focus our discussion in the remainder of this chapter on their manufacture, assessment, and usage.

Decellularization

Methods of Decellularization

The derivation of a cell-free ECM is not a new concept. Lwebuga-Mukasa and colleagues first described the generation of acellular lung scaffolds in 1986 for the study of rat type II alveolar epithelial (AEII) cell behavior on a native basement membrane [45]. This technique was heavily explored in simple tissues in the 1990s and early 2000s [46–49] and has made strides into the clinic. Acellular biologic scaffolds have been created from a variety of different simple tissues, including skin, esophagus, and trachea [4]. Decellularization was first applied to complex tissues using whole organ perfusion decellularization in heart in 2008 [15]. Beginning in 2010, several groups described similar techniques in lung [18–23] and since this time, the field has grown rapidly.

The basic goal of any decellularization technique is to remove the endogenous cell population while retaining the macroarchitecture of the organ or the tissue, along with the ECM composition (Fig. 18.2). Maintenance of mechanical tissue properties is also thought to be critical in evaluating decellularization protocols. A variety of methods have been described to decellularize tissue. Most commonly, a

Table 18.2 Comparison of biologic vs. synthetic scaffold approach for ex vivo bioengineering

	Biologic (acellular) scaffold		Synthetic scaffold		Potential hybrid design	
Differentiation and engraftment cues	+	Retains native integrin binding sites	–	Lacks specific integrin binding sites (must be engineered into scaffolds)	+	Could be engineered with specific ECM components or engineered integrin sites
Immunogenicity	+	Antigen removal during decellularization	+/–	Unknown/variable depending on material	+/–	Unknown/variable depending on the artificial matrix material chosen
Manufacturability	+	Native architecture largely retained	–	Complex architecture possible	–	Complex architecture possible
	–	Large variability between donor scaffolds	+	Precise control possible (i.e., repeatability)	+	More ability than acellular to be controlled, but ECM incorporation introduces a degree of variability
Long-term storage	–	Degradation with long-term storage	+	Improved storage stability	+/–	Improved storage stability, but would likely loose biologic activity under long-term storage

series or combination of detergents, solvents, acids/bases, and hypotonic or hypertonic solutions are used to remove the majority of cellular components. Alternative methods include physical methods such as freeze/thaw cycles and/or biological agents such as enzymatic treatment [48]. Methods of decellularization are comprehensively reviewed elsewhere [50]. In general, most protocols last from 1 to 7 days.

There are a variety of published reports on techniques for decellularizing mouse, rat, porcine nonhuman primate and human lungs (Tables 18.3 and 18.4). In the lung, maintenance of both large and small airways and vessels is critical, in addition to the more delicate structures such as alveolar, capillary, and lymphatic systems which can be damaged through the use of excessive pressure during decellularization [51]. Perfusion decellularization has been most commonly utilized for whole lung decellularization, but there have also been reports of excising segments or slices from native lung and decellularizing these smaller segments [28, 29, 52, 53]. Detergents are the most commonly utilized decellularization agents used in perfusion based lung decellularization. There are several studies which have directly compared differences between these methods, and endpoint comparisons included assessment of proteomic composition, the mechanical properties of the final acellular scaffold, and recellularization efficacy [23, 41, 52]. The most commonly utilized detergents for lung are either the ionic detergents sodium deoxycholate (SDC) and sodium dodecyl sulfate (SDS), which are often used in

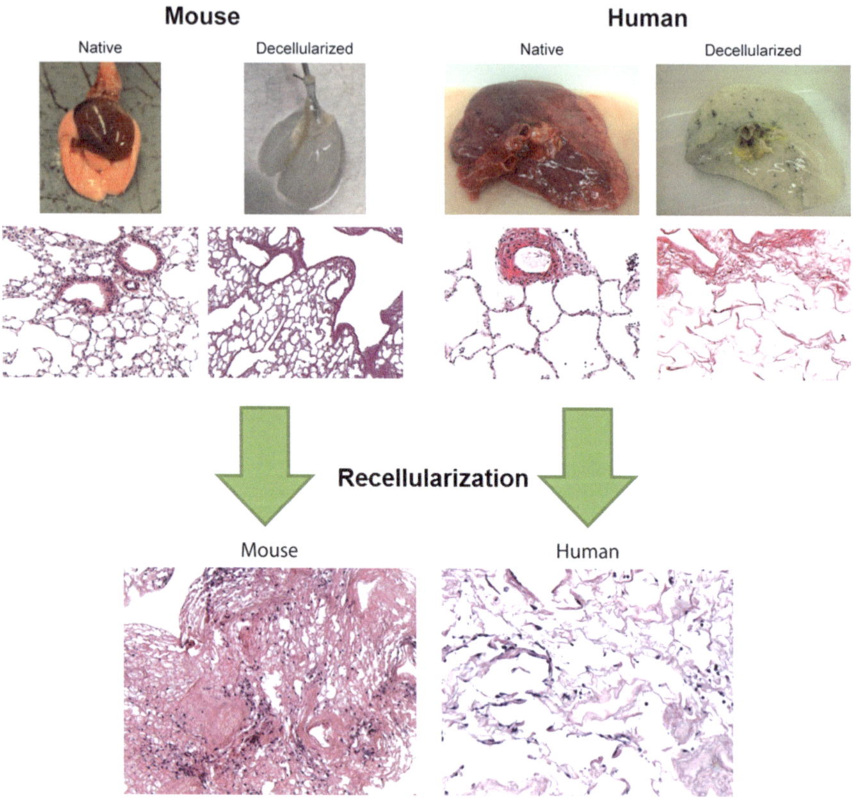

Fig. 18.2 Overview of the decellularization and recellularization process. Representative images of native and decellularized lungs from mice and humans (*upper panel*) demonstrating loss of pigmentation following decellularization, whereby the lungs become translucent white in color. H&E staining reveals complete cellular removal and gross maintenance of histological architecture. Histological analysis following recellularization with murine alveolar epithelial cells (C10) (*left*) and human bronchial epithelial cells (HBE) (*right*) into acellular mouse and human lung slices. Cells can be seen to have attached to the acellular lungs after 1 day of slice culture

combination with the nonionic detergent Triton X-100 [18, 22–27, 29, 41, 54–58]. Zwitterionic detergents such as 3-[(3-cholamidopropyl)dimethylammonio]-1-propanesulfonate (CHAPS) have also been used [19, 52, 59, 60], but some reports demonstrate that these may be more damaging than ionic or nonionic detergents due to their efficiency in denaturing proteins [48]. Many protocols also incorporate additional rinses and incubations for the purpose of removing organic components which are difficult to remove with the other detergents. The most commonly utilized additional steps are the use of hypertonic solution for lysis of cells (e.g., 1 M NaCl), or DNase/RNase to clear residual DNA and RNA. While both vascular-only perfusion and a combination of vascular and airway perfusion have produced acellular scaffolds capable of supporting recellularization, there is no consensus on the best route of administration and removal of decellularization agents.

Table 18.3 Compiled studies of ex vivo lung bioengineering using rodent and small primate decellularized whole lung scaffolds

Ref.	Scaffold	Study objective	Method of decellularization	Length of decellularization process	Endpoint assessments
[122]	Alveolar basement membrane (calf, dog, rabbit, adult/newborn rat)	Basement membrane	Filtered distal lung homogenate, saline, 4 % Triton X-100 with protease inhibitors, $NaHCO_3$ rinse, distilled H_2O rinse	26–52 h depending on homogenate volume	Histology, IF, SEM, amino acid analysis, carbohydrate analysis
[45]	Acellular alveolar vs. amniotic basement membranes	Differentiation on different basement membranes	Distilled H_2O, 0.1 % Triton X-100, 2 % SDC, NaCl, pancreatic DNase Type 1S	>2 days	Cell attachment and morphology
[20]	Mouse (female C57/BL6) acellular lungs	Effect of matrix on spatial engraftment of E17 fetal lung homogenate	Airway and vascular perfusion: distilled H_2O, 0.1 % Triton X-100, 2 % SDC, 1 M NaCl, porcine pancreatic DNase	3 days (approximately 63 h)	Histology, quantification of ECM proteins, IF, SEM, function with Flexivent, bioreactor with fetal type II cells
[19]	Rat acellular lungs (male Fischer 344)	Development of bioartificial lung for orthotopic transplantation	Vascular perfusion only (1–5 mL/min with less than 20 mmHg arterial pressure) CHAPS, NaCl, EDTA, PBS	4 h	Histology, IF, DNA quantification assay, collagen assay, GAG assay, western blots, SEM, TEM, micro-CT imaging
[21]	Rat acellular lung (Sprague Dawley)	Comparison of matrices including decellularized rat lung in ability to support mESCs	Fast freeze/thaw cycles, 1 % SDS, DNase, RNase, PBS, Penicillin/Streptomycin, Amphotericin, DMEM	>6 weeks	Quantification of DNA, IH, confocal microscopy, flow cytometry, 2 photon microscopy, presence of SP-A
[18]	Rat acellular lung (Sprague Dawley)	Development of bioartificial lung for orthotopic transplantation	Vascular perfusion only: pulmonary artery pressure kept constant at 80 cmH_2O, heparinized PBS with 0.1 % SDS, deionized water, Triton X-100, and PBS with Penicillin, Streptomycin, Amphotericin B	3 days (approximately 75 h) including incubation with antibiotics	Histology, morphology, mechanical function, fluoroscopy, gas exchange, transplantation, protein analysis

(continued)

Table 18.3 (continued)

Ref.	Scaffold	Study objective	Method of decellularization	Length of decellularization process	Endpoint assessments
[58]	Rat acellular lung (Sprague Dawley)	Orthotopic transplantation	Vascular perfusion only: pulmonary artery pressure kept constant at 80 cmH_2O, heparinized PBS with 0.1 % SDS, deionized water, Triton X-100, and PBS with Penicillin, Streptomycin, Amphotericin B	3 days (approximately 75 h) including incubation with antibiotics	Histology, IH, morphology, fluoroscopy, functional analysis, transplantation of seeded lungs with fetal pulmonary cells and Pulmonary artery and vein with endothelial cells
[123]	Rat acellular liver and lung (Lewis)	Cellular differentiation on 3D in vitro scaffold	Lung lobes cut into 300 μm thick, 0.5 % Triton X-100, 10 mM ammonia, mechanical disruption, PBS, distilled water	N/A	Histology, TEM, environmental scanning, PCR, IH, liquid chromatography with tandem mass spectrometry
[22]	Mouse acellular lung (C57BL/6; BALB/C)	Initial binding and recellularization of MSCs in acellular scaffold; directed seeding with integrin blocking	Same as [20]	3 days (approximately 72 h)	Histology, IF, EM, dye perfusion to assess vascular continuity, mass spectrometry, western blot, lung mechanics with Flexivent, inoculation of bone-marrow-derived MSCs
[23]	Mouse acellular lung and lung slices (BALB/C)	Comparison of detergent-based decellularization protocols	Airway and vascular perfusion. 3 different protocols assessed: (1) [20]; (2) [18]; (3) [19]	3 days (approximately 72 h)	IH, mass spectrometry, western blot, mechanics with Flexivent, gelatinase assay, DNase, RNase, comparative recellularization with MSCs and C10s
[55]	Normal rhesus macaque acellular lung	Initial binding and recellularization of MSCs in acellular scaffold	Airway and vascular perfusion: PBS, EDTA, Penicillin/Streptomycin at initial harvest: pulmonary artery: PBS + Heparin + Sodium Nitroprusside with pressures 25–30 mmHg; then trachea and vasculature: deionized H_2O [20]	2–3 days (approximately 48–72 h)	Histology, morphology, IH, western blot, genomic DNA, proteomics, seeding with bone marrow and adipose derived rhesus MSCs

[57]	Mouse acellular lung and lung slices (C57/BL6)	Seeding with and differentiation of mESCs-derived endodermal lung precursors	Same as [20]	3 days (approximately 72 h)	Phenotypic assessment of mESCs and their capacity for differentiation
[56]	Mouse acellular lung (C57BL/6)	Comparison of timing of decellularization, pre-coating, and support of mESCs and their differentiation capacity into alveolar epithelial cells	Same as [20]	1 vs. 3 days (approximately 24 h vs. 50 h)	Histology, morphology, EM, western blot, gelatinase assay, IF, mechanical properties with Flexivent, viability of differentiated mESCs seeded into the scaffold, subcutaneous implantation of scaffold
[60]	Rat acellular lung	Comparison of different detergent-based protocols	Two approaches: (1) [19]; (2) NaCl, EDTA, SDS	4 h	Histology, collagen assay, elastin assay, GAG assay, DNA assay, mechanical testing with linear strips
[99]	Rat acellular lung	Creation of perfusable human lung cancer nodules	Same as [18]	3 days (approximately 75 h) including incubation with antibiotics	Recellularization with human A549, H460, or H1299 and cultured with perfused, oxygenated media for 7–14 days
[54]	Mouse acellular lung and lung slices (C57BL/6)	Effect of time to necropsy, length of storage, and two different methods of sterilization	Same as [20]	3 days (approximately 72 h)	Histology, IH, morphology, mass spectrometry, seeded lungs with MSCs and C10 epithelial cell line
[24]	Mouse acellular lung and lung slices (C57BL/6)	Effect of recipient age and injury on de- and recellularization	Same as [20]	3 days (approximately 72 h)	Histology, IH, mass spectrometry, inoculation with MSCs and C10 epithelial cell line
[69]	Rat and mouse acellular lung slices	Engraftment and survival of fibroblasts through a β1-integrin and FAK-dependent pathway through ERK	Same as [19]	1 Day	Histology, SEM, immunohistochemistry, DNA assay, recellularization with mouse A9 cells

(continued)

Table 18.3 (continued)

Ref.	Scaffold	Study objective	Method of decellularization	Length of decellularization process	Endpoint assessments
[41]	Rat, porcine, human acellular lungs	Evaluation of the capacity of acellular lung scaffolds to support recellularization with lung progenitors derived from human induced pluripotent stem cells (iPSCs)	Same as [18]	1 Day	Histology, IF, quantitative PCR, implantation of rat acellular scaffold recellulalarized with iPSC, blood gas analysis in orthotopic transplantation
[124]	Rat acellular lung	Comparison of different perfusions (manual, trachea + vasculature constant flow, vasculature only constant pressure)	Airway and vascular perfusion: PBS, 0.1 % Triton X-100, 2 % SDC, NaCl, Pancreatic DNase, $MgSO_4$, $CaCl_2$, PBS with Penicillin/Streptomycin and Amphotericin B	Up to 3 days	Histology, nucleic acid detection, IF, western blot, inoculation with C10 epithelial cell line
[86]	Rat acellular lung	potential of AF-MSCs to generate lung precursor cells in the decellularized lung	Same as [124]	Approximately 1 day	Viability assay (TUNEL staining), IF, quantitative PCR
[62]	Mouse acellular lung	Measurement of local stiffness by AFM	Same as [21]	>6 weeks	IH, IF, SEM, 2 photon microscopy AFM local stiffness measurement
[125]	Rat acellular lung	Influence of pH during decellularization on ECM	Same as [19]	4 + 48 h after treatment	Histology, IH, DNA quantification assay, collagen assay, GAG assay, western blot, TEM, subcutaneous implantation

Abbreviations: *A9* transformed subcutaneous murine fibroblasts, *AFM* atomic force microscopy, *C10* immortalized murine alveolar epithelial cells, $CaCl_2$ calcium chloride, *CHAPS* 3-[(3-cholamidopropyl)dimethylammonio]-1-propanesulfonate hydrate, cmH_2O centimeters of water (pressure), *CT* computed tomography, *DMEM* Dulbecco's modified Eagle's medium, *DNase* deoxyribonuclease, *DNA* deoxyribonucleic acid, *E17* embryonic day 17, *EDTA* ethylenediaminetetraacetic acid, *FBS* fetal bovine serum, *GAG* glycosaminoglycan, H_2O water, *IF* immunofluorescence, *IH* immunohistochemistry, *mESCs* murine embryonic stem cells, $MgSO_4$ magnesium sulfate, *mM* millimolar, *mmHg* millimeters of mercury, *MSC* mesenchymal stem cells, *NaCl* sodium chloride, $NaHCO_3$ sodium biocarbonate, *PBS* phosphate buffered saline, *PCR* polymerase chain reaction, *RNase* ribonuclease, *SEM* scanning electron microscopy, *SP-A* surfactant protein A, *TEM* transmission electron microscopy

Table 18.4 Summary of decellularization methods for human and porcine lungs

Ref.	Species	Decellularization agents	Perfusion parameters	Instillation route	Days
[19]	Human	CHAPS, NaCl, and EDTA	Constant pressure (25 mmHg)	Airway and vasculature	1
[27]	Human/ IPF	Triton X-100, SDC, NaCl, DNase	Unspecified	Airway and vasculature	3
[126]	Human/ porcine	SDS, Triton X-100	Constant pressure (30 cmH_2O)	Vascular	4–7
[52]	Human/ porcine	(a) SDS; (b) CHAPS; (c) Tween-20, SDC, peracetic acid	None—lung segments and agitation	N/A	1
[44]	Human/ porcine	Freeze/thaw; graded SDS perfusion	Varying flow rates (100–500 mL/h)	Airway and vasculature	7
[61]	Porcine	Triton X-100, SDC, NaCl	12–25 mL/min (15 mmHg)	Airway and vasculature	1
[29]	Human/ IPF	SDS, Triton X-100, NaCl	None—thin lung slices	N/A	2
[26]	Human/ porcine	Triton X-100, SDC, NaCl, DNase Peracetic Acid	Constant flow rates (1 L, 2 L, 3 L/min)	Airway and vasculature	3
[25]	Human/ COPD	Triton X-100, SDC, NaCl, DNase Peracetic Acid	Constant flow rate 2 L/min	Airway and vasculature	3
[73]	Human/ rat	SDS, Triton X-100	Constant pressure (50 cmH_2O)	Vascular	4–7
[41]	Human/ porcine/ rat	(a) SDS; (b) SDC; (c) CHAPS	Constant pressure (30 cmH_2O)	Vascular	4–7

Abbreviations: *CHAPS* 3-[(3-cholamidopropyl)dimethylammonio]-1-propanesulfonate, *COPD* chronic obstructive pulmonary disease, *IPF* idiopathic pulmonary fibrosis, *NaCl* sodium chloride, *SDC* sodium deoxycholate, *SDS* sodium dodecyl sulfate

How differences in protocols and routes of administration for decellularization reagents might affect recellularization protocols or potential immunogenicity of implanted scaffolds is not yet known. There is currently no set of standards for demonstrating that a protocol has generated an optimal acellular scaffold. However, Crapo et al. proposed three minimal criteria: (1) <50 ng dsDNA per 1 mg ECM dry weight; (2) <200 bp DNA fragment length; (3) absence of visible nuclear content in histological sections by 4′,6-diamidino-2-phenylindole (DAPI) or hematoxylin-eosin (H&E) staining [50]. However, these are generic criteria for all acellular scaffolds and there are tissue and organ-specific requirements, such as preservation of mechanical properties that are likely important for lung. Furthermore, differences in retention of ECM components and mechanics have also been observed [23, 52] and these may be critical criteria in establishing lung-specific guidelines which must be met with the various protocols utilized in different laboratories.

Scaling Up Decellularization Protocols for the Clinic

Scaling up decellularization protocols from rodent lungs to potential clinical sources (e.g., large animal xenogeneic sources: e.g., porcine or human scaffolds) presents a new set of further challenges. In addition to anatomical differences, there are practical differences in handling organs of this size and it is not a simple matter of scaling up volumes. While rodent and macaque lungs have been decellularized by hand, higher pressures and volumes must be utilized for sufficient inflation of perfusion pathways (e.g., vasculature, airways, etc.) in larger organs. This ensures that perfused solutions reach distal airspaces and capillary beds and that the ensuing cellular debris is cleared from the lungs. All of the published protocols to date for decellularizing whole large animal or human lungs utilize perfusion pumps to generate acellular scaffolds which can support recellularization [19, 20, 25–27, 41, 44], and a recent report demonstrates a potential automated scheme which minimizes many of the practical issues [61]. While not a model for clinical translation, human and porcine lung segments have also been decellularized using small segments in order to improve high throughput study [28, 29, 52, 53] (Table 18.4). There are a variety of techniques which have been reported for assessing the efficacy of the decellularization protocol as well as for characterizing the remaining scaffold. Most reports characterize scaffolds using histologic, immunofluorescent staining, and DNA detection/quantification (Fig. 18.2). We will next discuss these endpoint assessments.

Residual Extracellular Matrix and Other Proteins

Owing to the importance of ECM components, retention of key ECM components is a critical parameter to assess as an endpoint when evaluating potential decellularization protocols. The precise combination of ECM proteins that must be retained to preserve the ability of the acellular scaffold to give organotypic cues for cellular differentiation and functional tissue level assembly remains unknown. The major structural and functional molecules in the ECM include proteins such as collagens, elastin, fibronectin, and laminins as well as a variety of glycoproteins including glycosaminoglycans (GAGs). Collagens are the chief structural components of the lung and are responsible for overall mechanical strength while elastin gives the lung its elastic properties of reversible distension and intrinsic recoil. GAGs help control macromolecular and cellular movement across the basal lamina and may also play a role in the mechanical integrity of the lung, although less is known about their exact role, matrix molecules are generally highly conserved proteins in eukaryotic organisms and therefore it is generally thought that these scaffolds will have minimal to no immune response if used in a xenogeneic context. This may theoretically explain the lack of an adverse immune response seen in xenotransplantation of other decellularized organs such as skin, trachea, and esophagus [4, 7, 11, 12].

There are a variety of techniques which have been used to evaluate ECM components, including histology, immunohistochemistry, western blotting, mass spectrometry-based proteomics, and component-specific assays such as Sircol Collagen Assay, Fastin elastase, etc. (Table 18.3). The majority of lung decellularization techniques result in significant loss of elastin and sulfated GAGs in all species studied thus far [18–20, 22–24, 54–56]. In head-to-head comparison studies of lung decellularization protocols, SDS and SDC have been found to retain more elastin as compared to CHAPS-based protocols [23, 60]. However, despite the differences in retention of ECM components, inoculated cells appear to behave similarly in the recellularization/repopulation assays currently used (including histological and immunofluorescence evaluation). Therefore, it remains unknown if there is an optimal decellularization protocol, and if so, which is best suited for translation to the clinic.

A recently emerging trend is the use of mass spectrometry proteomic analysis to help delineate differences between protein loss and retention in protocols or in scaffold source [22–26, 41]. This assessment has also been used to aid in the selection of optimal protocol parameters such as flow rates or pressures [26]. For example, proteomic analysis can help delineate the impact of changes in protocols during different steps, decellularization agents, or in decellularization parameters (e.g., flow rate, pressure, rinse volumes, etc.) by quantifying or semiquantitatively assessing which choices preferentially retain certain ECM components or minimize/maximize retention of cellular-associated proteins [25–27]. In addition to detecting ECM composition and residual proteins in acellular scaffolds, it has been used for distinguishing differences between decellularization methods or lung origin, including disease states or donor age [23–27, 41, 53, 55]. These assessments also yield critical information for those studying cell–ECM interactions as it can help delineate differences in the underlying matrix.

One particularly striking and consistent result amongst the various groups utilizing this analytical approach is the amount and breadth of non-ECM proteins detected in the scaffold following decellularization. In particular, cytoskeletal elements and cell-associated proteins appear to be retained in the scaffolds, while in general, lesser secreted proteins are detected. This suggests that transmembrane proteins and their associated cytoskeletal elements may remain anchored to the ECM with currently used decellularization protocols. The impact of these residual proteins on recellularization, including potential immunogenicity remains unknown. Furthermore, in the current reports, proteomic assessment has been limited in scope and generally only the most abundantly expressed proteins are reported.

Mechanical Assessments of Decellularized Scaffolds

A variety of in vitro assessments have been utilized to assess the mechanical properties of acellular scaffolds. Investigators have explored both micro-[62, 63]

and macroscale [22, 23, 44, 64, 65] mechanical measurements of acellular lungs as well as force tension relationships in linear strips of decellularized lungs [52, 60]. While techniques such as atomic force microscopy (AFM) are useful in obtaining topographical information and initially assessing mechanical properties of the scaffolds [27, 62, 63, 65], these results have yet to be correlated to recellularization or functional performance. Traditional lung mechanics testing of acellular scaffolds has shown that in the absence of cells and surfactants, acellular scaffolds are stiffer than their naïve counterparts [22]. Introduction of exogenous surfactant into the acellular scaffolds can partially restore lung compliance [22]. This is an important finding and indicates that during recellularization strategies, serial measurements of lung mechanics could be used as a noninvasive and nondestructive means to assess functionality of the regenerating scaffold. For example, decreases in elastance could be used as a measurement of *de novo* surfactant production. However, as acellular lungs are often leaky following decellularization, interpreting results in this context can be challenging [44]. The importance and challenges of measuring mechanical properties in ex vivo bioengineering is discussed in more detail in the review by Suki [66].

Recellularization

Recellularization of Acellular Scaffolds for Bioengineering New Lung

The lung is a complex organ with a variety of different functions. These include gas exchange, immune system surveillance, and ciliary clearance of inhaled foreign objects. In order to accomplish all of these diverse functions, lung tissue utilizes a variety of different cell types, all of which uniquely contribute to some critical aspect of lung function [67]. Following a variety of acute injuries, such as infection or chemical insult, the lung has the capability to repair itself through activation of endogenous regeneration. The heterogeneous cell population of the lung is replenished by resident stem or progenitor cells, which differentiate into the various adult cell types [68]. Once implanted, it is thought that any ex vivo regeneration strategy must recapitulate these functions, whether it is through a completely biological strategy (i.e., functioning tissue) or some combined artificial and biological solution. It is therefore likely that lung tissue grown ex vivo require some minimal restoration of these subtypes so that it will function once transplanted.

While a variety of cell sources are being investigated for recellularizing acellular and artificial scaffolds, obtaining sufficient cell numbers with any source remains a significant open question. The ideal solution is thought to be the usage of an autologously derived source of cells to minimize post-transplantation immune complications which are a significant cause of morbidity in transplanted patients. One potential source is the use of fully differentiated primary adult cells. However, these cells may not have sufficient replicative capacity to fully recellularize the organ, plus, normal repair and regeneration following normal lung injury (e.g., illness) may not be possible. Nevertheless, these sorts of repopulation studies may shed light on recellularization strategies using other cell types. It remains unknown if multiple cell types could be isolated from the eventual transplant recipient, grown to sufficient numbers ex vivo and then used in a recellularization approach to restore functionality. While it has been shown that a strategy such as integrin blocking can be used to direct initial cell engraftment of a single cell population [22, 69], scaling this clinically and further adding the complex challenge of uniquely directing the right cell population to a specific architectural location would be challenging. Alternatively, autologous endogenous lung progenitor cells from the various compartments could be utilized (e.g., distal and proximal epithelial progenitor cells, endothelial progenitor cells, etc.) along with stromal cells to recellularize acellular scaffolds. However, the same challenges of obtaining sufficient cell numbers for an initial seeding strategy and directing cells to their correct compartment remain. In both instances, it remains unknown if normal cells could be obtained from a patient with a preexisting lung disease or if isolated diseased cells could be gene-corrected prior to subsequent recellularization. Recent work indicates that the scaffold may more significantly contribute to phenotype than cell-origin. Fibrotic scaffolds were found to induce a pro-fibrotic profile, independent of whether normal or IPF-derived human fibroblasts were used in repopulation assays, whereas the normal lung scaffold did not induce a pro-fibrotic profile if either cell type was used [29]. An allogeneic cell source could also be used, but this re-introduces the potential for immune complications following transplantation. Furthermore, the identification of *bona fide* distal airway lung progenitor cells in the adult human lung remains controversial.

A potentially more appealing autologous approach is the use of induced pluripotent stem cells (iPS) which are derived from reprogramming somatic cells to a stem-cell-like state. While iPS cells avoid the ethical controversies surrounding the use of embryonic stem cells (ESCs)—stem cells derived from the inner blastocyst of in vitro fertilized embryos—iPS cells have been shown to retain epigenetic memory of their tissue origin and have been shown to form teratomas [70]. iPS cells are typically derived from dermal fibroblasts and thus, differentiating them into the various lung cell types has been challenging. However, despite this limitation, recent work has demonstrated that human iPS cells can be differentiated into cells expressing a distal pulmonary epithelial cell immunophenotype and seeded into acellular human lung scaffolds [71–73]. These results further encourage the use of this approach in moving towards the clinic.

Other potential approaches include the use of fetal homogenates or ESCs. As previously mentioned, ethical concerns remain for either of these approaches, as well as the potential for teratoma formation with ESCs. While initial studies have shown that ESCs can engraft in acellular murine lungs [21, 57], seeding into acellular lungs was not sufficient to induce differentiation. Optimized in vitro differentiation protocols must be used in conjunction with seeding and repopulation strategies. Significantly, ESC-derived murine Nkx2-1GFP+ progenitor cells were able to recellularize acellular murine lungs and form alveolar structures, while in contrast, seeding with undifferentiated ESCs resulted in nonspecific cell masses in distal regions of acellular lungs. Fetal homogenates have the distinct advantage of containing all the necessary cell populations, and have been shown to have some capacity for self-assembly. These cells have been successfully used in the current rodent models of ex vivo regeneration and transplantation. However, in both instances, ethical concerns remain in obtaining these cells and the need for immunosuppressive drug treatment post-transplantation remains unknown. Tables 18.5 and 18.6 summarize recellularization approaches in animal and human models and the phenotype adopted by seeded cells.

Implantation of Recellularized Scaffolds

Important proof of concept studies have shown that recellularized scaffolds can be implanted and participate in gas exchange for short time periods. Decellularized rat lungs re-endothelialized with human umbilical vein endothelial cells (HUVEC) and recellularized with fetal rat lung homogenates and A549 epithelial cells were transplanted into rats that had undergone previous pneumonectomy [18, 19]. While the ex vivo regenerated lungs were shown to contribute to gas exchange following transplantation, the transplants developed significant pulmonary edema and/or hemorrhage resulting in respiratory failure after several hours. In a subsequent study, survival for 14 days was achieved after implantation but lung function progressively declined and the histologic appearance of the graft at necropsy demonstrated significant atelectasis and indications of fibrotic-like alterations [58]. A third study also confirmed the feasibility of short-term survival (60 min) following orthotopic transplantation of a rat lung recellularized with iPS cells [73]. Transplanted grafts were perfused and partial pressure of carbon dioxide in the blood was maintained within normal limits over the observation period. However, blood gas measurements were taken from the left pulmonary vein and represent a mix of blood which had perfused both the left (bioartificial lung) and the naïve lung, and thus is likely not representative of active gas exchange in the transplanted lobe. Additionally, occasional alveolar hemorrhage was observed. Despite these limitations, these studies, nonetheless, provide proof of concept that acellular lungs can be recellularized, surgically implanted, and might minimally participate in gas exchange. However, they also demonstrate the significant challenges that remain in translating towards the clinic. A recellularized acellular lung needs to meet a number of functional requirements in order to be clinically transplantable: adequate gas exchange, waste transport, unidirectional mucociliary clearance, and the ability to maintain physiologic airway pressures and volumes. Thus far, there has been a compartmentalized approach to the respiratory system, separating regeneration of the trachea, vasculature, proximal airways, and distal lung. An animal model which accomplishes restoration of all of these functions has

Table 18.5 Distribution and phenotype of cells seeded onto animal models of acellular scaffolds

Ref.	Cells used for seeding	Scaffold	Route	Duration	Distribution	Final phenotype
[45]	ATII	Acellular alveolar vs. amniotic basement membranes	Direct seeding	8 days	N/A	Alveolar matrices: ATI; amnionic membranes ATII
[21]	mESC	Rat (Sprague Dawley) acellular lung	Trachea	21 days	Proximal-distal regiospecific CC10, proSP-C expression	Tracheobronchial: CC10, Ck18; distal lung: proSP-C, CD31, PDGFRα
[18]	HUVEC (DsRed)	Rat acellular lung	Pulmonary artery	9 days	All vessels	Endothelial cells
	A549	Rat acellular lung	Trachea	9 days	Airways/alveoli	Airway/alveolar epithelium
	HUVEC (DsRed)	Rat acellular lung	Pulmonary artery	9 days	Entire vasculature	Endothelial cells
	Rat fetal lung cells (GD19-20)	Rat acellular lung	Trachea	9 days	Airways/alveoli	proSP-A, proSP-C, TTF-1/Nkx2.1 (ATII); T1α (ATI); Vimentin (fibroblast)
[19]	Neonatal (7d) lung epithelial cells (rat)	Rat acellular lungs (Fischer 344)	Trachea	8 days	Alveolar, small airways	CCSP (Clara cell), proSP-C (ATII), Aqp5 (ATI), Ck14 (basal cell)
	Lung vascular endothelium (rat)	Rat acellular lungs (Fischer 344)	Pulmonary artery	7 days	Microvascular	CD31
[20]	Fetal lung (E17)	Mouse acellular lungs	Tracheal	7 days	Alveolar	$Ck18^{+}$/proSP-C^{+} (ATII); CD11b, Aqp5, CCSP, CD31, and vimentin
[22]	mBM-MSCs	Mouse acellular lung	Trachea	28 days	Parenchymal > airway (squamous)	MSCs: no evidence for transdifferentiation
	C10—hATII (non-tumorigenic)	Mouse acellular lung	Trachea	28 days	Parenchymal	N/A

(continued)

Table 18.5 (continued)

Ref.	Cells used for seeding	Scaffold	Route	Duration	Distribution	Final phenotype
[58]	Rat fetal (GD17-20) pneumocytes	Rat acellular lung	Trachea	14 days	Alveolar/distal bronchioles > trachea/bronchi	CCSP (airways); TTF-1, proSP-C (alveolar)
	HUVEC	Athymic nude rat	Pulmonary artery	14 days	Proximal to distal vasculature	$CD31^+$
[123]	Mouse ATII (primary or P2)	Acellular lung microscaffold	Direct seeding	22 days	Alveolar	proSP-C/SP-C (ATII-like, from primary or cultured ATII); Aqp5, PDPN (ATI); CCSP (primary)
[23]	mBM-MSCs	Mouse acellular lung	Trachea	14 days	Alveolar	MSCs
	C10-hAECII (non-tumorigenic)	Mouse acellular lung	Trachea	14 days	Large and small Airways	Squamous morphology
[55]	Rhesus BM-MSCs	Rhesus macaque	Secondary bronchus	7 days	Alveolar septae, terminal bronchioles, respiratory bronchioles	MSCs phenotype
	Rhesus AD-MSCs	Rhesus macaque	Secondary bronchus	7 days	Alveolar septae, terminal and respiratory bronchioles	MSCs phenotype
[56]	mESCs diff. to $TTF\text{-}1^{pos}$/$proSP\text{-}C^{pos}$	Mouse acellular lung	Immersion	14 Day	Alveolar	TTF-1/Nkx2.1, proSP-C (alveolar); PDFGRα (mesenchymal)
[57]	mESCs	Mouse acellular lung	Trachea	10 days	Hypercellular sheets (alveolar)	Ciliated cells (airways); $T1\alpha^{neg}$ (alveoli)
	$Nkx2.1^{GFP}$	Mouse acellular lung	Trachea	10 days	Alveolar	Nkx2.1/T1α (alveoli)
[54]	Mouse BM-MSCs	Mouse acellular lung	Trachea	28 days	Alveolar	N/A
	C10-mATII (non-tumorigenic)	Mouse acellular lung	Trachea	28 days	Alveolar	N/A

[24]	mBM-MSCs	Mouse acellular lung	Trachea	28 days	Alveolar	N/A
	C10-mATII (non-tumorigenic)	Mouse acellular lung	Trachea	28 days	Alveolar	N/A
[69]	A9—murine immortalized fibroblasts	Rat and mouse acellular lung	Slices	14 days	Alveolar	ProCol1α1
[73]	Human iPS-derived epithelial progenitors and iPS-derived endothelium	Rat whole lung	Tracheal and pulmonary artery	10 days of iPS/derived endothelial cell culture under perfusion, followed by iPS-derived epithelial progenitor cell installation	Alveolar and vascular	Epithelial: Nkx2.1, T1-α, CCSP
						Endothelial: CD31

Abbreviations: *A549* adenocarcinomic human alveolar epithelial cell line, *ATI/II* alveolar epithelial cell type I/II, *Aqp* aquaporin, *C10* human colorectal adenocarcinoma epithelial cell line, *CC10* Clara cell-specific protein, *CCSP* Clara cell secretory protein, *CD* cluster of differentiation, *Ck* cytokeratin, *E17* embryonic day 17, *h* human, *HUVEC* human umbilical vein endothelial cell, *iPS* induced pluripotent stem cell, *m* murine, *mBM-MSC* murine bone-marrow-derived mesenchymal stem cell, *mESC* murine embryonic stem cell, *MSC* mesenchymal stem cell, *Nkx2.1* NK2 homeobox 1, *P2* passage number 2, *PDGFR* platelet-derived growth factor receptor, *PDPN* podoplanin, *Col1α1* collagen type 1 α1, *SP-A/C* surfactant protein A/C, *TTF-1* thyroid transcription factor 1

Table 18.6 Recellularization studies in porcine and human acellular scaffolds

Ref.	Species	Lung origin	Recellularization technique	Cell types	Summary
[19]	Human	Normal	Segment incubated with cells	A549; endothelial progenitor cells	Human scaffolds support recellularization
[27]	Human	Normal/IPF	Thin slices	hLF	hLFs on IPF scaffolds increase α-SMA expression
[71]	Human	Normal	Thin slices	hiPS-ATII, ATII	hiPS-ATII cells adhered to acellular lung matrix and a subpopulation differentiated to ATI phenotype
[126]	Human/porcine	Normal	Thin slices; single lobe recellularization SAEC and pressure-controlled perfusion	HUVEC, SAEC, PAEC	Cell survival for 3 days in an acellular perfused upper right lobe recellularized with PAECs
[52]	Human/porcine	Normal	Thin slices	hMRC-5, hSAEC	Acellular and human porcine lungs supported cell viability
[42]	Human	Normal	Thin slices and rotating bioreactor	hiPS-ATII, ATII	Rotating bioreactor culture at ALI enhanced ATII to ATI differentiation
[127]	Human	Normal	Thin slices	hBM-MSCs, hAT-MSCs	Both MSC populations attached to the acellular lung matrix; culture on acellular lung matrices enhanced SPC, AQP5, and Caveolin-1
[44]	Human/porcine	Normal	Excised segments (~0.5 cm^3) injected with cells	mESC, hFLC, BMMSCs, hAEC	Porcine/human lungs supported cellular attachment; 1 % SDS protocol minimized T-cell activation
[29]	Human	Normal/IPF	Thin slices	Normal and IPF-derived hLF	ECM contributed more significantly to IPF phenotype rather than cell-origin
[97]	Human/porcine	Normal	Physiologic instillation and thin slices	hMSC, hBE, CBF12, hLF	Normal acellular porcine and human lungs are nontoxic
[25]	Human	Normal/COPD	Physiologic instillation and thin slices	hMSC, hBE, CBF12, hLF	COPD lungs do not support viability comparable to normal lungs

Abbreviations: *α-SMA* alpha smooth muscle actin, *ALI* air–liquid-interface, *BMMSCs* porcine bone-marrow-derived mesenchymal stem cells, *CBF12* human endothelial progenitor cells, *COPD* chronic obstructive pulmonary fibrosis, *hAEC* human alveolar epithelial cells, *hAT-MSC* human adipose derived mesenchymal stem cells, *hBE* human bronchial epithelial cells, *hBM-MSC* human bone-marrow-derived mesenchymal stem cells, *hFLC* human fetal lung cells, *hiPS* human induced pluripotent cells, *hLF* human lung fibroblasts, *hMSC* human bone-marrow-derived mesenchymal stem cells, *HUVEC* human umbilical vein endothelial cells, *IPF* idiopathic pulmonary fibrosis, *mESC* murine embryonic stem cells, *PAEC* pulmonary alveolar epithelial cells, *SAEC* human small airway epithelial cells not been achieved and it will likely be several years before this can be accomplished.

Most recellularized artificial scaffolds have been explored in a limited context and primarily in vitro (Table 18.1). The main research focus up until now has been on the structural development of candidate scaffold designs and materials for lung tissue engineering. There have also been studies investigating cell differentiation of progenitor cells to generate sufficient amounts of cells to repopulate the artificial/decellularized organ. Implantation of artificial scaffolds has only been performed in a few cases and was performed either subcutaneously [32] or into the pleural cavity following pneumonectomy [74]. The distal lung has been the predominant research focus to date. However, no candidate scaffolds have yet to include considerations for vascularization and therefore it remains unclear if these scaffolds could be viable once transplanted. Thus far, there have been no reports of an attempt to transplant a whole bioartificial lung.

Immunogenicity of Implanted Scaffolds

The a priori assumption for clinical use of decellularized lung scaffolds is that acellular scaffolds will be nonimmunogenic because the cellular material has been removed, including cell-associated immunogens, such as Toll-like receptors (TLR) and enzymes associated with xenogeneic immunogenicity, such as (alpha1,3) galactose. However, some ECM and other proteins identified in the remaining decellularized scaffolds are known to be immunogenic [46, 75–77]. This issue has not yet been adequately studied or resolved. Further, cells inoculated into decellularized scaffolds secrete ECM and other proteins [22]. Thus, inoculated cells may considerably remodel the scaffold and generate their own basement membrane, shielding the denuded basement membrane, which can be immunogenic [78].

Some of these remaining proteins may also be beneficial with regard to their ability to induce an immune response. A growing body of literature suggests that decellularized scaffolds can polarize macrophages to the anti-inflammatory M2 phenotype, which is viewed as a more permissive, regenerative phenotype [4, 7, 46, 79–81]. Further, recent work in lung repair and regeneration has demonstrated the critical role that the immune system has in orchestrating normal repair and regeneration in adult lungs [82]. To date, with the exception of the use of fetal homogenates, no recellularization studies have included immune cells. Thus, it is unknown whether retention of these immunogenic components may actually be beneficial in a regeneration strategy. However, one study of interest demonstrated that co-culture of a recellularized acellular lung slice with CD206+ macrophages was found to increase fibroblast proliferation and prolong survival [28]. While this study was a model of disease, it demonstrates the critical role that one immune cell population can have on recellularization in the scaffolds through orchestrating paracrine signaling.

Some groups have implanted recellularized artificial scaffolds. Cortiella and colleagues used PGA and Pluronic F-127 hydrogels and showed that the foreign body response was reduced by usage of the latter [83]. A Gelfoam sponge recellularized with fetal lung cells induced no severe local immune response [84] while a Matrigel plug combined with FGF2-loaded polyvinyl sponge did [85]. This provides evidence that the scaffold material is in part responsible for triggering the immune reaction of the recipient, and that usage of natural matrices like collagen I, fibrinogen–fibronectin–vitronectin, and gelatin seem to be less inflammatory than PGA or polyvinyl [35, 74, 86].

Environmental Factors in Ex Vivo Lung Regeneration

The majority of published work focuses on decellularization methods, lung origin (i.e., disease state or age), and cell sources. There have been limited investigations into the addition or supplementation of exogenous growth factors to scaffolds, and especially a lack of studies examining the role of environmental cues, such as mechanical stretch or oxygen control, in generating functional lung tissue. Despite the presumed importance that factors such as mechanical stimuli and oxygen tension will have in regeneration schemes, they have remained largely unexplored in acellular scaffolds. These critical factors are known to play roles in both embryonic development and post natal repair and regeneration [87–89].

Traditional in vitro cell culture is performed at 20 % oxygen, however, physiologic oxygen levels in individual cells vary depending on the tissue type, tissue density, and cell/tissue proximity to blood vessels [90]. It has long been known that hypoxia can mediate angiogenesis and that vascular endothelial growth factor (VEGF) expression is upregulated in hypoxia [91]. During embryonic development, the lung environment is hypoxic (1–5 % oxygen) [90] and lower oxygen tension levels have been shown to positively influence in vitro differentiation. Lowering oxygen tension to levels typically encountered by cells in the developing embryo has been shown to enhance in vitro differentiation of ESC and iPS cells to Nkx2-1^{+} lung/thyroid progenitor cells [92]. Further studies of cellular differentiation in acellular scaffolds are needed to clarify the potential role of oxygen tension in an ex vivo regeneration strategy.

There is also a large and growing body of literature that delineates the importance of mechanical stimuli on embryonic lung development as well as in normal and diseased tissue repair and regeneration in vivo and ex vivo [88, 89]. Mechanical stretch is known to induce upregulation of surfactant protein C (SP-C) mRNA and protein expression in ATII cells, while shear stress on endothelial cells is critical for VEGF expression [93]. Several studies have examined the effect of mechanotransduction on fetal or adult lung cells in vitro; [93–96] but there is no available information on effects of stretch on development of lung epithelial tissue from embryonic or adult stem cells or from endogenous lung progenitor cells. We have observed upregulation of lung epithelial genes in murine bone-marrow-derived mesenchymal stem cells seeded into acellular mouse lungs and ventilated (Wagner et al. unpublished data). In particular, we found that SP-C mRNA was significantly upregulated at physiologic tidal volumes; a result we also observed in human ATII cells ventilated in small segments of acellular human lung (Wagner et al. unpublished data) using an artificial pleural coating on excised acellular segments, permitting ventilation [97]. While perfusion parameters have not yet been studied in detail, cultivation of a recellularized human lobe was done under perfusion conditions [41] and a rotating bioreactor culture was found to have positive effects on iPS cells differentiating into distal lung epithelial cells [42].

In addition to utilizing a scaffold from a suitable source and using an optimized decellularization protocol, precise control of the mechanical and gaseous environment with bioreactor technologies (e.g., stimuli mimicking stretch from breathing and shear stress induced by blood flow or breathing) will be necessary for a successful regeneration scheme.

Lessons Learned from Ex Vivo Organ and tissue Culture

Despite rigorous research efforts, it remains challenging to keep normal, healthy tissue slices and organ explants viable. Most in vitro studies (i.e., lung slices) are not kept longer than a few days while the difficulties in maintaining adequate tissue viability for more than a few hours for candidate donor lungs for transplantation is known to be extremely difficult and is a major limiting factor in maximizing the number of organs available for transplantation. In both of these research areas, tissue slices and organs are generally derived from healthy tissue sources, which are the end goals of an ex vivo strategy. Thus, the challenge of generating functional lung tissue ex vivo is even more daunting considering the challenges experienced in these fields.

Despite the fact that it is widely regarded that sophisticated bioreactor technologies will be needed for ex vivo lung tissue regeneration, there have been limited reports to date examining the effect of the various parameters which could be controlled in bioreactors (e.g., oxygen tension, mechanical ventilation, and vascular perfusion). Additional factors, such as optimal media formulation, have also been minimally explored in the current literature. However, several studies have strongly established the groundwork and the necessity of incorporating bioreactor technologies with ex vivo schemes to maintain or enhance phenotypes. Culture of hATII cells and hiPS-ATII cells in a rotating bioreactor at air–liquid-interface (ALI) was found to be beneficial in maintaining the phenotypic expression of distal epithelial lung cells [42]. In whole lungs or lobes, limited data is available on the viability of cells following recellularization. A single study demonstrated that human small airway epithelial cells (SAECs) instilled into a whole acellular human lung lobe could be maintained for 3 days with constant media perfusion [41]. However, a major limiting factor in both of these studies is that only short time points were analyzed and longer ex vivo schemes will likely be necessary for generating functional lung tissue [30]. One resource which may be beneficial in guiding the development of optimal lung bioengineering strategies is the ex vivo organ and tissue culture literature describing practices and strategies utilized in those fields for optimizing and maintaining the viability of tissue and organs.

Acellular Lungs as Ex Vivo Models of Disease

In addition to their potential use as scaffolds for tissue engineering, there has been rapid growth in the use of acellular lungs as ex vivo models which more closely recapitulate diseased in vivo environments. These experiments provide a new opportunity for insight into cell–ECM interactions capable of driving disease phenotypes. Human fibroblasts from normal human lungs seeded onto acellular scaffolds derived from fibrotic lungs were found to increase their alpha-SMA expression [27], and the ECM was found to contribute more significantly to IPF correlated gene expression changes in fibroblasts rather than cell-origin (i.e., from IPF or normal lungs) [29]. However, many cell-associated proteins, characteristic of pulmonary fibrosis (e.g., TGF-beta, Ctnnb1, etc.) are retained in decellularized mouse lungs following bleomycin injury [24]. In addition, ECM-associated proteins and matrikines (ECM derived peptides which are liberated by partial proteolysis of ECM macromolecules) are detectable by proteomic approaches following decellularization [22, 23, 25–27, 53, 54]. These proteins, in addition to the detected ECM components, may significantly contribute to the phenotypic changes observed by several groups in recellularization assays. In particular, observation of acquisition of a more fibrotic phenotype by normal fibroblasts in acellular human IPF lungs, may be attributed to these residual proteins, rather than the ECM components alone [27, 29].

Similarly, in acellular lungs derived from murine models of emphysema and from human patients with COPD, cells were unable to remain comparably viable as the same cells seeded into healthy acellular scaffolds [24, 25]. This suggests that either the matrix is impaired in COPD or that the residual protein composition is significantly altered as compared to normal acellular lungs. These studies generate exciting insight into the potential role of the matrix and matrix-associated proteins in driving disease phenotypes and provide proof of concept for use of acellular lungs as a novel platform for studying cell–matrix interactions.

A further novel use of acellular scaffolds in disease models has been utilized to study the role of macrophages in IPF using a Transwell culture setup of thin acellular lung slices recellularized in the Transwell insert, with macrophage co-culture [28]. Decellularized mouse lung slices seeded with murine fibroblasts were co-cultured with $CD206^+$ or $CD206^-$ macrophages from day 14 of murine lungs following bleomycin-induced lung injury (or in the absence of macrophages). $CD206^+$ macrophages were found to increase fibroblast proliferation and survival in the lung slices. However, there was no induction of α-SMA expression. Nonetheless, this study takes advantage of the ability to selectively study cells and cell combinations in isolation using acellular lungs. Similarly, the human fibroblast cell line MRC5 was seeded onto slices of normal human decellularized lung slices and stimulated with rhCHI3L1, a prototypic-chitinase-like protein recently shown to be elevated in human IPF. The addition of rhCHI3L1 induced α-SMA expression in the MRC5 cells and they adopted a contractile phenotype, as assessed by histology [28].

In addition to repopulation assays, it has also been suggested that recellularized acellular scaffolds could also be used for studying infectious diseases [98] and used as models for cancer development [99]. Thus, studies to date have likely only begun to demonstrate the utility of acellular tissue as ex vivo models of disease which more closely recapitulate in vivo microenvironments than traditional in vitro setups.

Precision Cut Tissue Slices

"Precision cut tissue slices" for ex vivo analysis have been used since the mid-nineteen eighties, when Smith et al. first reported on liver tissue that was sliced into 250 μm thin sections with low variation in thickness (<5 %) [100]. Highly delicate slices (thickness in general 25–300 μm) were fabricated with a device called a Krumdieck tissue slicer. This device overcame the variations in thickness previously seen due to manual cutting of tissue with a razorblade. This thickness also reduced the risk of malnutrition and lack of oxygenation for cells inside the tissue slice [101, 102]. Tissue slicers (Krumdieck or devices from Alabama Research and Development or Leica) use a core, drilled from the tissue that is to be sliced, and generate slices by cutting this core with a knife rotating perpendicular to the core axis. Another possibility for slice generation is the use of vibratomes (e.g., Leica, Zeiss), using a vibrating knife, thereby reducing mechanical impact to the tissue [103]. Several organs have been used to produce tissue slices including brain, heart, liver, kidney, and lung [102–108].

In general, the stiffness of most organ tissue is itself sufficient for slice generation. However, as lung tissue requires high elasticity for breathing movements and high surface area to volume for gas exchange, its density is low compared to other organs. Therefore, it needs to be filled with a supporting material in order to be sliced. One commonly used material to infiltrate the lungs is low melting agarose (used at 37 °C between 1 and 3 % w/v). After allowing the filled lungs to cool and the agarose to gel, the lung can be sliced. Slices have been used in diverse studies, some of which are listed in Table 18.7. Typical experimental durations have been reported in the range of 24–72 h [101, 102]. Using the current techniques, slice cultures seem to decrease in viability after 72 h and thus there is currently no possibility of long-term cultivation. Prolongation of this cultivation period would greatly expand the repertoire of studies which can currently be conducted using this technique. Some possible avenues of exploration of major interest would be the study of disease development or tissue regeneration.

Table 18.7 Studies applying slice cultivations

Study type	Ref.
Pulmonary physiology	[108, 128–134]
Pharmacology	[128, 135–139]
Pathogenesis	[140, 141]
Toxicity	[107, 142]
Cellular effects of mechanical stretch	[108, 129, 134, 143]
Cytokine release	[142]
Viral infection and gene transfer	[144, 145]
Viral exacerbations	[146, 147]

Ex Vivo Maintenance of Explanted Organs for Transplantation

The shortage of donor organs is a major limiting factor in the treatment options for end-stage lung disease patients. It is further currently impossible to fully mimic all the diverse lung functions in a sustainable and practical manner (i.e., portable) with manmade technical devices. In addition to traditional allogeneic transplantation, one alternative approach could be to use intact xenogeneic organs to restore the function of complex organs as has been done with liver and kidney (Butler and McAnulty refs).

In the lung, progress has been slower. Cypel and colleagues investigated whether ventilation of explanted lungs for up to 4 h with subsequent inflation to full capacity and storage in 4 °C Perfadex solution could improve transplantation outcomes [109]. Although not reaching statistical significance, the incidence of primary graft dysfunction 72 h after transplantation was lower in the ex vivo perfusion group (15 %) compared to the control group (30 %, $p = 0.11$). This ex vivo perfusion at 37 °C (normothermic) and storage at 4 °C has been found to have no drawbacks on transplantation outcome compared to normal donor lungs and therefore seems to be very promising. Still there is no consensus about which solution is the best to use.

The US Food and Drug Administration (FDA) recently approved the "XVIVO Perfusion System with STEEN Solution" (XVIVO Perfusion Inc. Englewood, Colorado, USA) which has been shown to increase the time for evaluation of the functional suitability of a donor organ for transplantation. Donor lungs are kept at body temperature while flushing the vasculature up to 4 h with a sterile solution (STEEN Solution). STEEN solution is a normal oncotic pressure solution containing human serum albumin, dextran, and a low K^+ concentration. This solution is designed to prevent edema formation, thrombogenesis, and vascular spasm under normothermic conditions. Lungs remain ventilated during the evaluation period and cells are thus maintained in more physiologic oxygen levels. Waste products are removed by flushing of the vasculature. With this technique, a proportion of organs once regarded as nonideal can become suitable for transplantation with similar rejection and 12-month survival rates compared to optimal donor organs.

For basic and translational research efforts, these studies are of major importance because they elucidate critical parameters, including perfusate content, temperature, and perfusion rates and pressures that need to be controlled and optimized for long-term cultivation of organs. It is likely that many of these parameters will also be critical to control in ex vivo organ culture techniques, regardless of whether it is simply for longer ex vivo culture for basic science or for preservation/maintenance for candidate transplant organs or tissue engineering schemes.

Discussion and Outlook

While the prospect of utilizing acellular lung scaffolds clinically may still seem like science fiction to many, the progress made in the last few years has rapidly indicated that this may be a viable option in the not so distant future. The difficulties encountered in maintaining ex vivo viability of freshly explanted healthy organs highlight many of the challenges which the ex vivo regeneration field faces, in addition to those unique to the field. Ex vivo whole organ cultures experience decreases in viability, selective survival of specific cell types, and loss of phenotypic expression over time with current techniques. Many of these same problems may plague ex vivo bioengineering strategies. Even in very thin tissue slice models of naïve tissue, where lack of nutrition and oxygenation is theoretically not of major concern, cells can only maintain their functionality, proliferative capacity, and viability for short periods of time (up to 72 h). This is exacerbated in cultivating whole organs, such as lung, where the need for proper control of medium oxygenation, osmolarity, pH, ventilation, and tissue perfusion in three dimensions is required to keep the tissue viable for long-term cultivation. This likely reflects the combination of a number of factors at play, whose importance we may not currently be fully aware of.

Chiefly among these may be media formulation. Currently, media formulations which have been optimized for two-dimensional (2D) cultivation of homogenous cell populations are utilized in cultivation of both precision cut tissue slices and recellularized acellular tissue slices. However, the media composition needed for whole organ cultivation needs to be optimized for multiple cell types and it remains unclear if the media formulations which are viewed as optimal in 2D are even optimal in that setup. Furthermore, stem and progenitor cells should sustain their capability to differentiate and replenish damaged or absent cell compartments and using a media which pushes these populations into a differentiated state may not be desirable. Therefore, the media formulation used should somehow be able to serve multiple roles simultaneously. To achieve this, different cell types and stem cells initially seeded into acellular scaffolds in an undifferentiated state may require the timed sequential addition of different growth factors, nutrients, and amino acids to regulate signaling pathways involved in cellular proliferation and differentiation.

Means of surveying and controlling the cultivation conditions and media formulation are needed. The knowledge from bioprocess engineering may help to fill the knowledge gap in the needs of whole organ cultivation and ex vivo bioengineering of lung. To date, no study has been conducted addressing the composition of organ-specific cultivation medium supporting long-term cultivation and cellular maintenance in recellularizing lung scaffolds.

Additionally, currently used cultivation conditions for either ex vivo naïve tissue or recellularized acellular scaffolds do not even remotely resemble the in vivo environment. These environments lack proper mechanical (stretch) and environmental stimuli (contact to certain media/air). For example in the lung, it has been shown that isolated ATII cells in tissue culture lose SP-C expression over time and transdifferentiate into alveolar epithelial type I (ATI) cells [110, 111]. A similar decrease in SP-C expression was observed when we cultured naïve murine and human lung tissue slices for 7 days in submerged culture (Uhl et al. unpublished data). As it is known that mechanical stimulation induces SP-C expression in ATII cells [112–114], this suggests, that ventilation of whole organ cultures or stretching of lung slices may be necessary to retain ATII cells in their progenitor state. On the other hand, nonphysiologic ventilation may cause alveolar epithelial cell damage. In a healthy organ the tolerance of cells to mechanical stimuli may be different to that in disease. Further, we know that the mechanics of the acellular lung are dramatically different than naïve lung [22], even despite administration of exogenous surfactant. This indicates the importance of maintaining precise control of the environmental parameters during the whole regenerative scheme.

Reseeding of decellularized matrices has currently been limited to only a few different cell types and often times in monoculture. Each additional cell type adds complexity, making interpretation of results utilizing homogenates or multiple cell types challenging. Usage of stem and progenitor cells (e.g., embryonic stem cells, mesenchymal stem cells, or iPS cells) is appealing for recellularization strategies as these cells can potentially differentiate into the multitude of cell types needed in a specific area of the scaffold. The potential for this approach was demonstrated with the use of ESC-derived murine Nkx2-1^{GFP+} in acellular lungs. These cells repopulated distal airspaces and a subpopulation differentiated into Nkx2-1^{GFP-} and acquired a morphology characteristic of ATI cells and expressed the phenotypic ATI marker podoplanin (T1α) [57]. While encouraging, the necessity of regenerating the multitude of cell types in the lung remains a challenge.

A clinical translation scheme of recellularization of decellularized organs will require precise process control. Metrics for assessing successful decellularization need to be established and a consistent decellularization scheme should be utilized. During recellularization phases, the initial seeding may be accomplished by attachment followed by migration and/or proliferation. We have observed that during this initial seeding phase, physiologic ventilation and perfusion are not feasible until cells have adequately adhered to the scaffold, and in fact, inclusion of these stimuli may even be detrimental. The initial properties and composition of the organ are not comparable to the in vivo situation. For example, in decellularized lungs, there is a lack of surfactant in the alveoli prior to recellularization, and this dramatically effects mechanical properties [77]. It has also been shown that there is a loss of ECM components, such as elastin, following perfusion decellularization using most protocols [115]. The effect of the loss of these ECM components on initial engraftment and subsequent recellularization and regeneration remain unknown, but may be critical to the success of a regenerative scheme. The importance of preserving the native integrin binding sites in recellularization schemes has already been demonstrated and cells can be directed to certain ECM binding sites through integrin blocking [22]. It has been shown that fibroblasts seeded into acellular mouse lungs utilize a β_1-integrin-dependent pathway and thus preservation of these integrin epitopes seems to be critical. As an accessory technology, a collagen I and Matrigel solution has been used as a pretreatment to coat the decellularized lungs via the trachea before cell seeding to enhance engraftment [56]. Alternatively, cells have also been injected in a hydrogel (Pluronic-F127) for recellularization [21, 44]. Addressing the question of how the matrix should be prepared before inoculation might be an extremely important aspect not yet explored in detail.

There are still major hurdles to overcome for ex vivo engineering. Using state-of-the-art ex vivo preservation techniques, freshly explanted organs, such as kidney and liver, can only maintain viability and function for 5 days [116, 117]. Perfusion at physiologic flow rates is needed in conjunction with the appropriate perfusates tailored in their chemical composition for lungs. This will also be essential for ex vivo recellularization strategies using acellular or synthetic scaffolds and unfortunately, these approaches are not yet mature enough for use. As cells first need to be distributed by migration and likely undergo differentiation inside the matrix, optimal media composition and environmental stimuli will be crucial for ex vivo bioengineering strategies. In order to control for and adapt these stimuli to the regenerating organ during the cultivation period, a range of ancillary technologies need to be integrated and developed into existing bioreactor technologies (e.g., sensors, pumps, and analytic and process control systems). While the road to translating acellular scaffolds into the clinic is long, steady progress has been made in this relatively young field and it has a promising future.

References

1. Lopez AD, Shibuya K, Rao C, Mathers CD, Hansell AL, Held LS et al (2006) Chronic obstructive pulmonary disease: current burden and future projections. Eur Respir J 27 (2):397–412
2. Eisner MD, Anthonisen N, Coultas D, Kuenzli N, Perez-Padilla R, Postma D et al (2010) An official American Thoracic Society public policy statement: novel risk factors and the global burden of chronic obstructive pulmonary disease. Am J Respir Crit Care Med 182 (5):693–718
3. Macchiarini P, Jungebluth P, Go T, Asnaghi MA, Rees LE, Cogan TA et al (2008) Clinical transplantation of a tissue-engineered airway. Lancet 372(9655):2023–2030
4. Keane TJ, Badylak SF (2014) Biomaterials for tissue engineering applications. Semin Pediatr Surg 23(3):112–118

5. Badylak SF, Weiss DJ, Caplan A, Macchiarini P (2012) Engineered whole organs and complex tissues. Lancet 379(9819):943–952
6. Baiguera S, Del Gaudio C, Jaus MO, Polizzi L, Gonfiotti A, Comin CE et al (2012) Long-term changes to in vitro preserved bioengineered human trachea and their implications for decellularized tissues. Biomaterials 33(14):3662–3672
7. Fishman JM, De Coppi P, Elliott MJ, Atala A, Birchall MA, Macchiarini P (2011) Airway tissue engineering. Expert Opin Biol Ther 11(12):1623–1635
8. Haag J, Baiguera S, Jungebluth P, Barale D, Del Gaudio C, Castiglione F et al (2012) Biomechanical and angiogenic properties of tissue-engineered rat trachea using genipin cross-linked decellularized tissue. Biomaterials 33(3):780–789
9. Haykal S, Soleas JP, Salna M, Hofer SO, Waddell TK (2012) Evaluation of the structural integrity and extracellular matrix components of tracheal allografts following cyclical decellularization techniques: comparison of three protocols. Tissue Eng Part C Methods 18 (8):614–623
10. Hinderer S, Schesny M, Bayrak A, Ibold B, Hampel M, Walles T et al (2012) Engineering of fibrillar decorin matrices for a tissue-engineered trachea. Biomaterials 33(21):5259–5266
11. Jungebluth P, Bader A, Baiguera S, Moller S, Jaus M, Lim ML et al (2012) The concept of in vivo airway tissue engineering. Biomaterials 33(17):4319–4326
12. Jungebluth P, Moll G, Baiguera S, Macchiarini P (2012) Tissue-engineered airway: a regenerative solution. Clin Pharmacol Ther 91(1):81–93
13. Krawiec JT, Vorp DA (2012) Adult stem cell-based tissue engineered blood vessels: a review. Biomaterials 33(12):3388–3400
14. Orlando G, Baptista P, Birchall M, De Coppi P, Farney A, Guimaraes-Souza NK et al (2011) Regenerative medicine as applied to solid organ transplantation: current status and future challenges. Transpl Int 24(3):223–232
15. Ott HC, Matthiesen TS, Goh SK, Black LD, Kren SM, Netoff TI et al (2008) Perfusion-decellularized matrix: using nature's platform to engineer a bioartificial heart. Nat Med 14 (2):213–221
16. Totonelli G, Maghsoudlou P, Garriboli M, Riegler J, Orlando G, Burns AJ et al (2012) A rat decellularized small bowel scaffold that preserves villus-crypt architecture for intestinal regeneration. Biomaterials 33(12):3401–3410
17. Wertheim JA, Baptista PM, Soto-Gutierrez A (2012) Cellular therapy and bioartificial approaches to liver replacement. Curr Opin Organ Transplant 17(3):235–240
18. Ott HC, Clippinger B, Conrad C, Schuetz C, Pomerantseva I, Ikonomou L et al (2010) Regeneration and orthotopic transplantation of a bioartificial lung. Nat Med 16(8):927–933
19. Petersen TH, Calle EA, Zhao L, Lee EJ, Gui L, Raredon MB et al (2010) Tissue-engineered lungs for in vivo implantation. Science 329(5991):538–541
20. Price AP, England KA, Matson AM, Blazar BR, Panoskaltsis-Mortari A (2010) Development of a decellularized lung bioreactor system for bioengineering the lung: the matrix reloaded. Tissue Eng Part A 16(8):2581–2591
21. Cortiella J, Niles J, Cantu A, Brettler A, Pham A, Vargas G et al (2010) Influence of acellular natural lung matrix on murine embryonic stem cell differentiation and tissue formation. Tissue Eng Part A 16(8):2565–2580
22. Daly AB, Wallis JM, Borg ZD, Bonvillain RW, Deng B, Ballif BA et al (2012) Initial binding and recellularization of decellularized mouse lung scaffolds with bone marrow-derived mesenchymal stromal cells. Tissue Eng Part A 18(1–2):1–16
23. Wallis JM, Borg ZD, Daly AB, Deng B, Ballif BA, Allen GB et al (2012) Comparative assessment of detergent-based protocols for mouse lung de-cellularization and re-cellularization. Tissue Eng Part C Methods 18(6):420–432
24. Sokocevic D, Bonenfant NR, Wagner DE, Borg ZD, Lathrop MJ, Lam YW et al (2013) The effect of age and emphysematous and fibrotic injury on the re-cellularization of de-cellularized lungs. Biomaterials 34(13):3256–3269

25. Wagner DE, Bonenfant NR, Parsons CS, Sokocevic D, Brooks EM, Borg ZD et al (2014) Comparative decellularization and recellularization of normal versus emphysematous human lungs. Biomaterials 35(10):3281–3297
26. Wagner DE, Bonenfant NR, Sokocevic D, DeSarno MJ, Borg ZD, Parsons CS et al (2014) Three-dimensional scaffolds of acellular human and porcine lungs for high throughput studies of lung disease and regeneration. Biomaterials 35(9):2664–2679
27. Booth AJ, Hadley R, Cornett AM, Dreffs AA, Matthes SA, Tsui JL et al (2012) Acellular normal and fibrotic human lung matrices as a culture system for in vitro investigation. Am J Respir Crit Care Med 186(9):866–876
28. Zhou Y, Peng H, Sun H, Peng X, Tang C, Gan Y et al (2014) Chitinase 3-like 1 suppresses injury and promotes fibroproliferative responses in Mammalian lung fibrosis. Sci Transl Med 6(240):240ra76
29. Parker MW, Rossi D, Peterson M, Smith K, Sikstrom K, White ES et al (2014) Fibrotic extracellular matrix activates a profibrotic positive feedback loop. J Clin Invest 124 (4):1622–1635
30. Murphy SV, Atala A (2014) 3D bioprinting of tissues and organs. Nat Biotechnol 32 (8):773–785
31. Bryant SJ, Cuy JL, Hauch KD, Ratner BD (2007) Photo-patterning of porous hydrogels for tissue engineering. Biomaterials 28(19):2978–2986
32. Ling TY, Liu YL, Huang YK, Gu SY, Chen HK, Ho CC et al (2014) Differentiation of lung stem/progenitor cells into alveolar pneumocytes and induction of angiogenesis within a 3D gelatin—microbubble scaffold. Biomaterials 35(22):5660–5669
33. Dunphy SE, Bratt JA, Akram KM, Forsyth NR, El Haj AJ (2014) Hydrogels for lung tissue engineering: biomechanical properties of thin collagen-elastin constructs. J Mech Behav Biomed Mater 38:251–259
34. Fischer SN, Johnson JK, Baran CP, Newland CA, Marsh CB, Lannutti JJ (2011) Organ-derived coatings on electrospun nanofibers as ex vivo microenvironments. Biomaterials 32 (2):538–546
35. Ingenito EP, Sen E, Tsai LW, Murthy S, Hoffman A (2010) Design and testing of biological scaffolds for delivering reparative cells to target sites in the lung. J Tissue Eng Regen Med 4 (4):259–272
36. Fernandes H, Mentink A, Bank R, Stoop R, van Blitterswijk C, de Boer J (2010) Endogenous collagen influences differentiation of human multipotent mesenchymal stromal cells. Tissue Eng Part A 16(5):1693–1702
37. Lin YM, Zhang A, Rippon HJ, Bismarck A, Bishop AE (2010) Tissue engineering of lung: the effect of extracellular matrix on the differentiation of embryonic stem cells to pneumocytes. Tissue Eng Part A 16(5):1515–1526
38. Mariani TJ, Sandefur S, Pierce RA (1997) Elastin in lung development. Exp Lung Res 23 (2):131–145
39. Nguyen NM, Senior RM (2006) Laminin isoforms and lung development: all isoforms are not equal. Dev Biol 294(2):271–279
40. Rippon HJ, Polak JM, Qin M, Bishop AE (2006) Derivation of distal lung epithelial progenitors from murine embryonic stem cells using a novel three-step differentiation protocol. Stem Cells 24(5):1389–1398
41. Gilpin SE, Guyette JP, Gonzalez G, Ren X, Asara JM, Mathisen DJ et al (2014) Perfusion decellularization of human and porcine lungs: bringing the matrix to clinical scale. J Heart Lung Transplant 33(3):298–308
42. Ghaedi M, Mendez JJ, Bove PF, Sivarapatna A, Raredon MS, Niklason LE (2014) Alveolar epithelial differentiation of human induced pluripotent stem cells in a rotating bioreactor. Biomaterials 35(2):699–710
43. Ingenito EP, Tsai L, Murthy S, Tyagi S, Mazan M, Hoffman A (2012) Autologous lung-derived mesenchymal stem cell transplantation in experimental emphysema. Cell Transplant 21(1):175–189

44. Nichols JE, Niles J, Riddle M, Vargas G, Schilagard T, Ma L et al (2013) Production and assessment of decellularized pig and human lung scaffolds. Tissue Eng Part A 19 (17–18):2045–2062
45. Lwebuga-Mukasa JS, Ingbar DH, Madri JA (1986) Repopulation of a human alveolar matrix by adult rat type II pneumocytes in vitro. A novel system for type II pneumocyte culture. Exp Cell Res 162(2):423–435
46. Badylak SF, Gilbert TW (2008) Immune response to biologic scaffold materials. Semin Immunol 20(2):109–116
47. Badylak SF, Taylor D, Uygun K (2011) Whole-organ tissue engineering: decellularization and recellularization of three-dimensional matrix scaffolds. Annu Rev Biomed Eng 13:27–53
48. Gilbert TW, Sellaro TL, Badylak SF (2006) Decellularization of tissues and organs. Biomaterials 27(19):3675–3683
49. Wainwright DJ (1995) Use of an acellular allograft dermal matrix (AlloDerm) in the management of full-thickness burns. Burns 21(4):243–248
50. Crapo PM, Gilbert TW, Badylak SF (2011) An overview of tissue and whole organ decellularization processes. Biomaterials 32(12):3233–3243
51. Guyette JP, Gilpin SE, Charest JM, Tapias LF, Ren X, Ott HC (2014) Perfusion decellularization of whole organs. Nat Protoc 9(6):1451–1468
52. O'Neill JD, Anfang R, Anandappa A, Costa J, Javidfar J, Wobma HM et al (2013) Decellularization of human and porcine lung tissues for pulmonary tissue engineering. Ann Thorac Surg 96(3):1046–1055, discussion 1055-6
53. Nakayama KH, Lee CC, Batchelder CA, Tarantal AF (2013) Tissue specificity of decellularized rhesus monkey kidney and lung scaffolds. PLoS One 8(5):e64134
54. Bonenfant NR, Sokocevic D, Wagner DE, Borg ZD, Lathrop MJ, Lam YW et al (2013) The effects of storage and sterilization on de-cellularized and re-cellularized whole lung. Biomaterials 34(13):3231–3245
55. Bonvillain RW, Danchuk S, Sullivan DE, Betancourt AM, Semon JA, Eagle ME et al (2012) A nonhuman primate model of lung regeneration: detergent-mediated decellularization and initial in vitro recellularization with mesenchymal stem cells. Tissue Eng Part A 18 (23–24):2437–2452
56. Jensen T, Roszell B, Zang F, Girard E, Matson A, Thrall R et al (2012) A rapid lung de-cellularization protocol supports embryonic stem cell differentiation in vitro and following implantation. Tissue Eng Part C Methods 18(8):632–646
57. Longmire TA, Ikonomou L, Hawkins F, Christodoulou C, Cao Y, Jean JC et al (2012) Efficient derivation of purified lung and thyroid progenitors from embryonic stem cells. Cell Stem Cell 10(4):398–411
58. Song JJ, Kim SS, Liu Z, Madsen JC, Mathisen DJ, Vacanti JP et al (2011) Enhanced in vivo function of bioartificial lungs in rats. Ann Thorac Surg 92(3):998–1005, discussion 1005-6
59. Petersen TH, Calle EA, Colehour MB, Niklason LE (2011) Matrix composition and mechanics of decellularized lung scaffolds. Cells Tissues Organs 195(3):222–231
60. Petersen TH, Calle EA, Colehour MB, Niklason LE (2012) Matrix composition and mechanics of decellularized lung scaffolds. Cells Tissues Organs 195(3):222–231
61. Price AP, Godin LM, Domek A, Cotter T, D'Cunha J, Taylor DA et al (2015) Automated decellularization of Intact, human-sized lungs for tissue engineering. Tissue Eng Part C Methods 21(1):94–103
62. Melo E, Garreta E, Luque T, Cortiella J, Nichols J, Navajas D et al (2014) Effects of the decellularization method on the local stiffness of acellular lungs. Tissue Eng Part C Methods 20(5):412–422
63. Melo E, Cardenes N, Garreta E, Luque T, Rojas M, Navajas D et al (2014) Inhomogeneity of local stiffness in the extracellular matrix scaffold of fibrotic mouse lungs. J Mech Behav Biomed Mater 37:186–195

64. Uriarte JJ, Nonaka PN, Campillo N, Palma RK, Melo E, de Oliveira LV et al (2014) Mechanical properties of acellular mouse lungs after sterilization by gamma irradiation. J Mech Behav Biomed Mater 40:168–177
65. Nonaka PN, Uriarte JJ, Campillo N, Melo E, Navajas D, Farre R et al (2014) Mechanical properties of mouse lungs along organ decellularization by sodium dodecyl sulfate. Respir Physiol Neurobiol 200:1–5
66. Suki B (2014) Assessing the functional mechanical properties of bioengineered organs with emphasis on the lung. J Cell Physiol 229(9):1134–1140
67. Morrisey EE, Hogan BL (2010) Preparing for the first breath: genetic and cellular mechanisms in lung development. Dev Cell 18(1):8–23
68. Kotton DN, Morrisey EE (2014) Lung regeneration: mechanisms, applications and emerging stem cell populations. Nat Med 20(8):822–832
69. Sun H, Calle E, Chen X, Mathur A, Zhu Y, Mendez J et al (2014) Fibroblast engraftment in the decellularized mouse lung occurs via a beta1-integrin-dependent, FAK-dependent pathway that is mediated by ERK and opposed by AKT. Am J Physiol Lung Cell Mol Physiol 306 (6):L463–L475
70. Miura K, Okada Y, Aoi T, Okada A, Takahashi K, Okita K et al (2009) Variation in the safety of induced pluripotent stem cell lines. Nat Biotechnol 27(8):743–745
71. Ghaedi M, Calle EA, Mendez JJ, Gard AL, Balestrini J, Booth A et al (2013) Human iPS cell-derived alveolar epithelium repopulates lung extracellular matrix. J Clin Invest 123 (11):4950–4962
72. Huang SX, Islam MN, O'Neill J, Hu Z, Yang YG, Chen YW et al (2014) Efficient generation of lung and airway epithelial cells from human pluripotent stem cells. Nat Biotechnol 32 (1):84–91
73. Gilpin SE, Ren X, Okamoto T, Guyette JP, Mou H, Rajagopal J et al (2014) Enhanced lung epithelial specification of human induced pluripotent stem cells on decellularized lung matrix. Ann Thorac Surg 98(5):1721–1729
74. Tsunooka N, Hirayama S, Medin JA, Liles WC, Keshavjee S, Waddell TK (2011) A novel tissue-engineered approach to problems of the postpneumonectomy space. Ann Thorac Surg 91(3):880–886
75. Franz S, Rammelt S, Scharnweber D, Simon JC (2011) Immune responses to implants—a review of the implications for the design of immunomodulatory biomaterials. Biomaterials 32 (28):6692–6709
76. Shaw AS, Filbert EL (2009) Scaffold proteins and immune-cell signalling. Nat Rev Immunol 9(1):47–56
77. Daly KA, Liu S, Agrawal V, Brown BN, Johnson SA, Medberry CJ et al (2012) Damage associated molecular patterns within xenogeneic biologic scaffolds and their effects on host remodeling. Biomaterials 33(1):91–101
78. Iwata T, Philipovskiy A, Fisher AJ, Presson RG Jr, Chiyo M, Lee J et al (2008) Anti-type V collagen humoral immunity in lung transplant primary graft dysfunction. J Immunol 181 (8):5738–5747
79. Brown BN, Londono R, Tottey S, Zhang L, Kukla KA, Wolf MT et al (2012) Macrophage phenotype as a predictor of constructive remodeling following the implantation of biologically derived surgical mesh materials. Acta Biomater 8(3):978–987
80. Brown BN, Ratner BD, Goodman SB, Amar S, Badylak SF (2012) Macrophage polarization: an opportunity for improved outcomes in biomaterials and regenerative medicine. Biomaterials 33(15):3792–3802
81. Brown BN, Valentin JE, Stewart-Akers AM, McCabe GP, Badylak SF (2009) Macrophage phenotype and remodeling outcomes in response to biologic scaffolds with and without a cellular component. Biomaterials 30(8):1482–1491
82. Hogan BL, Barkauskas CE, Chapman HA, Epstein JA, Jain R, Hsia CC et al (2014) Repair and regeneration of the respiratory system: complexity, plasticity, and mechanisms of lung stem cell function. Cell Stem Cell 15(2):123–138

83. Cortiella J, Nichols JE, Kojima K, Bonassar LJ, Dargon P, Roy AK et al (2006) Tissue-engineered lung: an in vivo and in vitro comparison of polyglycolic acid and pluronic F-127 hydrogel/somatic lung progenitor cell constructs to support tissue growth. Tissue Eng 12 (5):1213–1225
84. Andrade CF, Wong AP, Waddell TK, Keshavjee S, Liu M (2007) Cell-based tissue engineering for lung regeneration. Am J Physiol Lung Cell Mol Physiol 292(2):L510–L518
85. Mondrinos MJ, Koutzaki SH, Poblete HM, Crisanti MC, Lelkes PI, Finck CM (2008) In vivo pulmonary tissue engineering: contribution of donor-derived endothelial cells to construct vascularization. Tissue Eng Part A 14(3):361–368
86. Vadasz S, Jensen T, Moncada C, Girard E, Zhang F, Blanchette A et al (2014) Second and third trimester amniotic fluid mesenchymal stem cells can repopulate a de-cellularized lung scaffold and express lung markers. J Pediatr Surg 49(11):1554–1563
87. Colom A, Galgoczy R, Almendros I, Xaubet A, Farre R, Alcaraz J (2014) Oxygen diffusion and consumption in extracellular matrix gels: implications for designing three-dimensional cultures. J Biomed Mater Res A 102(8):2776–2784
88. Ingber DE (2006) Mechanical control of tissue morphogenesis during embryological development. Int J Dev Biol 50(2–3):255–266
89. Ingber DE (2006) Cellular mechanotransduction: putting all the pieces together again. FASEB J 20(7):811–827
90. Simon MC, Keith B (2008) The role of oxygen availability in embryonic development and stem cell function. Nat Rev Mol Cell Biol 9(4):285–296
91. Shweiki D, Itin A, Soffer D, Keshet E (1992) Vascular endothelial growth factor induced by hypoxia may mediate hypoxia-initiated angiogenesis. Nature 359(6398):843–845
92. Garreta E, Melo E, Navajas D, Farre R (2014) Low oxygen tension enhances the generation of lung progenitor cells from mouse embryonic and induced pluripotent stem cells. Physiol Rep 2(7)
93. Sanchez-Esteban J, Tsai SW, Sang J, Qin J, Torday JS, Rubin LP (1998) Effects of mechanical forces on lung-specific gene expression. Am J Med Sci 316(3):200–204
94. Boudreault F, Tschumperlin DJ (2010) Stretch-induced mitogen-activated protein kinase activation in lung fibroblasts is independent of receptor tyrosine kinases. Am J Respir Cell Mol Biol 43(1):64–73
95. Huang Z, Wang Y, Nayak PS, Dammann CE, Sanchez-Esteban J (2012) Stretch-induced fetal type II cell differentiation is mediated via ErbB1-ErbB4 interactions. J Biol Chem 287 (22):18091–18102
96. Weibel ER (2013) It takes more than cells to make a good lung. Am J Respir Crit Care Med 187(4):342–346
97. Wagner D, Fenn S, Bonenfant N, Marks E, Borg Z, Saunders P et al (2014) Design and synthesis of an artificial pulmonary pleura for high throughput studies in acellular human lungs. Cell Mol Bioeng 7(2):184–195
98. Crabbe A, Ledesma MA, Nickerson CA (2014) Mimicking the host and its microenvironment in vitro for studying mucosal infections by Pseudomonas aeruginosa. Pathog Dis 71(1):1–19
99. Mishra DK, Thrall MJ, Baird BN, Ott HC, Blackmon SH, Kurie JM et al (2012) Human lung cancer cells grown on acellular rat lung matrix create perfusable tumor nodules. Ann Thorac Surg 93(4):1075–1081
100. Smith PF, Gandolfi AJ, Krumdieck CL, Putnam CW, Zukoski CF 3rd, Davis WM et al (1985) Dynamic organ culture of precision liver slices for in vitro toxicology. Life Sci 36 (14):1367–1375
101. Sanderson MJ (2011) Exploring lung physiology in health and disease with lung slices. Pulm Pharmacol Ther 24(5):452–465
102. Liberati TA, Randle MR, Toth LA (2010) In vitro lung slices: a powerful approach for assessment of lung pathophysiology. Expert Rev Mol Diagn 10(4):501–508
103. Rice WL, Van Hoek AN, Paunescu TG, Huynh C, Goetze B, Singh B et al (2013) High resolution helium ion scanning microscopy of the rat kidney. PLoS One 8(3), e57051

104. Pilaz LJ, Silver DL (2014) Live imaging of mitosis in the developing mouse embryonic cortex. J Vis Exp (88)
105. De Kanter R, Olinga P, De Jager MH, Merema MT, Meijer DK, Groothius GM (1999) Organ slices as an in vitro test system for drug metabolism in human liver, lung and kidney. Toxicol In Vitro 13(4–5):737–744
106. Martin C, Uhlig S, Ullrich V (1996) Videomicroscopy of methacholine-induced contraction of individual airways in precision-cut lung slices. Eur Respir J 9(12):2479–2487
107. Parrish AR, Gandolfi AJ, Brendel K (1995) Precision-cut tissue slices: applications in pharmacology and toxicology. Life Sci 57(21):1887–1901
108. Davidovich N, Chhour P, Margulies SS (2013) Uses of remnant human lung tissue for mechanical stretch studies. Cell Mol Bioeng 6(2):175–182
109. Cypel M, Yeung JC, Liu M, Anraku M, Chen F, Karolak W et al (2011) Normothermic ex vivo lung perfusion in clinical lung transplantation. N Engl J Med 364(15):1431–1440
110. Ezzie ME, Crawford M, Cho JH, Orellana R, Zhang S, Gelinas R et al (2012) Gene expression networks in COPD: microRNA and mRNA regulation. Thorax 67(2):122–131
111. Flozak AS, Lam AP, Russell S, Jain M, Peled ON, Sheppard KA et al (2010) Beta-catenin/T-cell factor signaling is activated during lung injury and promotes the survival and migration of alveolar epithelial cells. J Biol Chem 285(5):3157–3167
112. Majumdar A, Arold SP, Bartolak-Suki E, Parameswaran H, Suki B (2012) Jamming dynamics of stretch-induced surfactant release by alveolar type II cells. J Appl Physiol (1985) 112 (5):824–831
113. Dietl P, Liss B, Felder E, Miklavc P, Wirtz H (2010) Lamellar body exocytosis by cell stretch or purinergic stimulation: possible physiological roles, messengers and mechanisms. Cell Physiol Biochem 25(1):1–12
114. Douville NJ, Zamankhan P, Tung YC, Li R, Vaughan BL, Tai CF et al (2011) Combination of fluid and solid mechanical stresses contribute to cell death and detachment in a microfluidic alveolar model. Lab Chip 11(4):609–619
115. Wagner DE, Bonvillain RW, Jensen T, Girard ED, Bunnell BA, Finck CM et al (2013) Can stem cells be used to generate new lungs? Ex vivo lung bioengineering with decellularized whole lung scaffolds. Respirology 18(6):895–911
116. Butler AJ, Rees MA, Wight DG, Casey ND, Alexander G, White DJ et al (2002) Successful extracorporeal porcine liver perfusion for 72 hr. Transplantation 73(8):1212–1218
117. McAnulty JF, Ploeg RJ, Southard JH, Belzer FO (1989) Successful five-day perfusion preservation of the canine kidney. Transplantation 47(1):37–41
118. Lin YM, Boccaccini AR, Polak JM, Bishop AE, Maquet V (2006) Biocompatibility of poly-DL-lactic acid (PDLLA) for lung tissue engineering. J Biomater Appl 21(2):109–118
119. Mondrinos MJ, Koutzaki S, Jiwanmall E, Li M, Dechadarevian JP, Lelkes PI et al (2006) Engineering three-dimensional pulmonary tissue constructs. Tissue Eng 12(4):717–728
120. Mondrinos MJ, Koutzaki S, Lelkes PI, Finck CM (2007) A tissue-engineered model of fetal distal lung tissue. Am J Physiol Lung Cell Mol Physiol 293(3):L639–L650
121. Miller C, George S, Niklason L (2010) Developing a tissue-engineered model of the human bronchiole. J Tissue Eng Regen Med 4(8):619–627
122. Kuttan R, Spall RD, Duhamel RC, Sipes IG, Meezan E, Brendel K (1981) Preparation and composition of alveolar extracellular matrix and incorporated basement membrane. Lung 159 (6):333–345
123. Shamis Y, Hasson E, Soroker A, Bassat E, Shimoni Y, Ziv T et al (2011) Organ-specific scaffolds for in vitro expansion, differentiation, and organization of primary lung cells. Tissue Eng Part C Methods 17(8):861–870
124. Girard ED, Jensen TJ, Vadasz SD, Blanchette AE, Zhang F, Moncada C et al (2013) Automated procedure for biomimetic de-cellularized lung scaffold supporting alveolar epithelial transdifferentiation. Biomaterials 34(38):10043–10055

125. Tsuchiya T, Balestrini JL, Mendez J, Calle EA, Zhao L, Niklason LE (2014) Influence of pH on extracellular matrix preservation during lung decellularization. Tissue Eng Part C Methods 20(12):1028–1036
126. Gilpin SE, Lung K, de Couto GT, Cypel M, Sato M, Singer LG et al (2013) Bone marrow-derived progenitor cells in end-stage lung disease patients. BMC Pulm Med 13:48
127. Mendez JJ, Ghaedi M, Steinbacher D, Niklason L (2014) Epithelial cell differentiation of human mesenchymal stromal cells in decellularized lung scaffolds. Tissue Eng Part A 20 (11–12):1735–1746
128. Morin JP, Baste JM, Gay A, Crochemore C, Corbiere C, Monteil C (2013) Precision cut lung slices as an efficient tool for in vitro lung physio-pharmacotoxicology studies. Xenobiotica 43 (1):63–72
129. Dassow C, Wiechert L, Martin C, Schumann S, Muller-Newen G, Pack O et al (2010) Biaxial distension of precision-cut lung slices. J Appl Physiol 108(3):713–721
130. Wright JL, Churg A (2008) Short-term exposure to cigarette smoke induces endothelial dysfunction in small intrapulmonary arteries: analysis using guinea pig precision cut lung slices. J Appl Physiol 104(5):1462–1469
131. Schleputz M, Uhlig S, Martin C (2010) Electric field stimulation of precision-cut lung slices. J Appl Physiol 110(2):545–554
132. Wyatt TA, Sisson JH, Allen-Gipson DS, McCaskill ML, Boten JA, DeVasure JM et al (2012) Co-exposure to cigarette smoke and alcohol decreases airway epithelial cell cilia beating in a protein kinase Cepsilon-dependent manner. Am J Pathol 181(2):431–440
133. Schnorbusch K, Lembrechts R, Brouns I, Pintelon I, Timmermans JP, Adriaensen D (2012) Precision-cut vibratome slices allow functional live cell imaging of the pulmonary neuroepithelial body microenvironment in fetal mice. Adv Exp Med Biol 758:157–166
134. Davidovich N, Huang J, Margulies SS (2012) Reproducible uniform equibiaxial stretch of precision-cut lung slices. Am J Physiol Lung Cell Mol Physiol 304(4):L210–L220
135. Moreno L, Perez-Vizcaino F, Harrington L, Faro R, Sturton G, Barnes PJ et al (2006) Pharmacology of airways and vessels in lung slices in situ: role of endogenous dilator hormones. Respir Res 7:111
136. Seehase S, Schleputz M, Switalla S, Matz-Rensing K, Kaup FJ, Zoller M et al (2011) Bronchoconstriction in non-human primates: a species comparison. J Appl Physiol 111 (3):791–798
137. Wohlsen A, Martin C, Vollmer E, Branscheid D, Magnussen H, Becker WM et al (2003) The early allergic response in small airways of human precision-cut lung slices. Eur Respir J 21 (6):1024–1032
138. Held HD, Martin C, Uhlig S (1999) Characterization of airway and vascular responses in murine lungs. Br J Pharmacol 126(5):1191–1199
139. Bussek A, Wettwer E, Christ T, Lohmann H, Camelliti P, Ravens U (2009) Tissue slices from adult mammalian hearts as a model for pharmacological drug testing. Cell Physiol Biochem 24(5–6):527–536
140. Zhou J, Alvarez-Elizondo MB, Botvinick E, George SC (2012) Local small airway epithelial injury induces global smooth muscle contraction and airway constriction. J Appl Physiol 112 (4):627–637
141. Vaira V, Fedele G, Pyne S, Fasoli E, Zadra G, Bailey D et al (2010) Preclinical model of organotypic culture for pharmacodynamic profiling of human tumors. Proc Natl Acad Sci U S A 107(18):8352–8356
142. Henjakovic M, Sewald K, Switalla S, Kaiser D, Muller M, Veres TZ et al (2008) Ex vivo testing of immune responses in precision-cut lung slices. Toxicol Appl Pharmacol 231 (1):68–76
143. Rausch SM, Haberthur D, Stampanoni M, Schittny JC, Wall WA (2011) Local strain distribution in real three-dimensional alveolar geometries. Ann Biomed Eng 39 (11):2835–2843

144. McBride S, Rannie D, Harrison DJ (2000) Gene transfer to adult human lung tissue ex vivo. Gene Ther 7(8):675–678
145. Nguyen DT, de Vries RD, Ludlow M, van den Hoogen BG, Lemon K, van Amerongen G et al (2013) Paramyxovirus infections in ex vivo lung slice cultures of different host species. J Virol Methods 193(1):159–165
146. Bauer CM, Zavitz CC, Botelho FM, Lambert KN, Brown EG, Mossman KL et al (2010) Treating viral exacerbations of chronic obstructive pulmonary disease: insights from a mouse model of cigarette smoke and H1N1 influenza infection. PLoS One 5(10), e13251
147. Marquardt A, Halle S, Seckert CK, Lemmermann NA, Veres TZ, Braun A et al (2011) Single cell detection of latent cytomegalovirus reactivation in host tissue. J Gen Virol 92 (Pt6):1279–1291

Part VI
Afterword

Chapter 19
Afterword: Prospects and Challenges in Lung Regenerative Medicine

Ivan Bertoncello

Introduction

In 2005 a workshop on 'Adult Lung Stem Cells, Lung Biology and Lung Disease' was convened at the University of Vermont to review current research in the field, and discuss how endogenous lung stem and progenitor cells and exogenously-derived reparative cells might be used to deliver clinical cell-based therapies for intractable lung diseases [1]. That workshop highlighted the lack of specific lung stem cell biomarkers and predictive in vitro stem cell assays, and clinically relevant animal models of human lung disease as major impediments to realising that goal.

The preceding chapters in this monograph summarise the remarkable progress since then in understanding how lung stem and progenitor cell compartments are organised and regulated during development and in the normal adult lung, and in animal models of lung disease and injury. The properties of exogenously derived reparative cells are also now better understood, and clinical cell therapy trials using mesenchymal stromal cells (MSC) and endothelial progenitor cells (EPC) to treat lung diseases are underway [2–4]. Embryonic stem cells (ESC) and patient-derived induced pluripotent cells (iPSC) are being used to model lung development and lung disease, in order to identify relevant therapeutic targets, and to develop gene correction protocols [5–7]. Lung bioengineering utilising acellular biomatrix scaffolds is also emerging as an exciting and fertile research focus.

While the progress made during the last decade provides a firm scientific foundation for the development of safe and effective stem cell-based therapies for lung disease, we are still faced with many of the practical problems and challenges identified in an earlier perspective [8].

I. Bertoncello (✉)
Lung Health Research Centre, Department of Pharmacology and Therapeutics, University of Melbourne, Melbourne, VIC, Australia
e-mail: IvanB@unimelb.edu.au

I. Bertoncello (ed.), *Stem Cells in the Lung*, Stem Cell Biology and Regenerative Medicine, DOI 10.1007/978-3-319-21082-7_19

Devising Disease-Specific Cell Therapies

The lung comprises as many as 60 cell types which are differentially affected in different lung diseases. Consequently, different cells types (likely more than one) will need to be targeted, in order to deliver cell therapies for different lung diseases of varying complexity.

For example, disease progression in COPD is characterised by airway remodelling, loss of elasticity, endothelial dysfunction, epithelial and goblet cell hyperplasia, and destruction of alveoli [9, 10]. Interstitial lung diseases are characterised by myofibroblast expansion, aberrant epithelial–mesenchymal differentiation, excessive matrix deposition, and failure of alveolar re-epithelialisation [11]. Asthma is primarily an inflammatory disorder, and its pathogenesis is complicated by the involvement of many environmental and genetic factors which act on the susceptible functionally deficient airway epithelium [12]. In cystic fibrosis (CF), different mutations of the CF transmembrane conductance regulator (*Cftr*) gene result in excessive mucus production by anion secreting airway epithelium, disrupting mucociliary clearance of inhaled pathogens and subsequent inflammation and deterioration of lung function [13, 14].

Endogenous Lung Stem and Progenitor Cells

The diverse proximal and distal airway, and alveolar epithelial cell types in the lung are maintained by both region specific and multipotent stem and progenitor cell sub-populations distributed along the proximal-distal axis of the airway tree. There are significant differences in the composition of these lung stem cell pools which are both developmentally determined and externally regulated; and different stem and progenitor cell types are recruited to repair the lung in response to injury or perturbation of differing severity. This has also been found to hold true for alveolar re-epithelialisation, where Type II alveolar epithelial cells (AECII) have long been regarded as the archetypal alveolar epithelial stem cell [15]. Here, recent studies have described the recruitment of novel rare multipotent lung stem cells termed distal airway stem cells (DASC) [16] or lineage negative epithelial stem/progenitor cells (LNEP) [17] which are inactive in the steady state but are mobilised, and migrate to repair severely injured airway and alveolar epithelia.

Researchers developing strategies for harnessing endogenous lung stem cells to repair the lung in different contexts will need to determine which regenerative cell types are 'fit for purpose' not only in different lung diseases or following different injuries, but also in different phases of disease, or recovery from injury.

The delivery of cell therapies by transplantation of autologous endogenous stem cells will be contingent on the development of culture systems for ex vivo stem cell expansion that will not compromise their regenerative potential. And, successful transplantation will be critically dependent on variables including cell dose, route of delivery, timing of stem cell transfer, and the regimen used to precondition recipients to accept engrafting cells. While it is feasible to control these variables in mouse disease and injury models, successful clinical transplantation of autologous endogenous lung stem cells remains an aspirational goal.

Little progress has been made in identifying suitable selectable markers, and in developing robust validated functional assays, for the prospective isolation and characterisation of human homologues of mouse endogenous lung stem cells. This objective is also confounded by factors including the individual biological variation in human subjects, and the limited availability of normal viable human lung tissue of high quality. Were these problems resolved, then scaling ex vivo expansion of defined endogenous human lung stem and progenitor cell types to generate sufficient autologous stem cells for clinical transplantation in a time-sensitive manner will be a major hurdle. And, allogeneic transplantation of endogenous lung stem cells will encounter the added problems of rejection and/or graft-versus-host responses.

The Stem Cell Microenvironment

The regenerative capacity of stem cells is determined by their dynamic bidirectional interaction with the niche microenvironments in which they reside [18]. Stem cell niche microenvironments also differ in composition during development and ageing and change in response to injury and disease progression, to recruit regenerative cells to repair the damaged lung or alternately, to exacerbate injury by impairing their reparative ability [19]. In specific diseases, or lung injuries it will be important to determine whether therapies are best delivered by manipulating regenerative cells or by repairing their microenvironment. This will require a precise understanding of temporal changes in stem cell microenvironments during disease progression, or following injury; and the factors and pathways mediating the crosstalk between mesenchymal, inflammatory cells and epithelial reparative cells in different phases of disease and injury. The inability to precisely map region-specific mesenchymal cell domains in the lung and to dissect the heterogeneous mesenchymal stromal cell compartment (other than broadly resolving myofibroblastic and lipofibroblastic cell types) have compromised our ability to define the role of specific mesenchymal cell types in regulating lung epithelial regeneration and repair. Recent studies utilising clonal cell lineage labelling and cell tracing to analyse spatial and temporal mesenchymal domains during lung development in mice [20, 21] will likely inform the design of strategies to do so in disease and injury models.

Mouse Models

A number of authors in this volume have pointed out major differences in the spatial distribution and organisation of regenerative cells, epithelial cell lineages and regional microenvironments in the human and mouse lung. It is also important to note that the common inbred strains of experimental mice used in respiratory research were initially developed to establish the existence and influence of genetic factors on the incidence of cancer [22]. Coincidentally, these genetic factors also influence traits including stem cell pool size and cycling characteristics [23, 24], differential leukocyte counts [25], immune surveillance and inflammatory responses, sensitivity to cytokines [26], susceptibility to cytotoxic drugs [27] and irradiation [28], and lung fibrosis [29]; and condition responses of mice to genetic manipulation [30]. Mouse cells also have telomeres which are five- to tenfold longer than human cells, endowing stem and progenitor cell in mice with a replicative capacity unmatched by their human counterparts [31]. These traits compromise the ability of preclinical mouse models to accurately model human lung diseases [29], as well as their ability to accurately assess the feasibility and predict the efficacy of clinical cellular therapies.

Mesenchymal Stromal Cell Therapies

Although clinical trials utilising MSC for the treatment of various diseases of the lung, as well as other organs, there is an ongoing debate as to whether clinical trials utilising these MSC-based therapies are too far ahead of the science [4]. The reviews of MSC-based therapies in this monograph, and by others [32] conclude that it is not always clear how the clinical benefit of MSC is mediated in individual settings; and that in most instances the efficacy of MSC-based therapies is likely explained by biological mechanisms other than the stem cell potential of engrafted MSC.

Some argue that such trials are premature, until better reference standards, definitive markers, and functional assays for the validation of the potency of expanded MSC become available. However, it is instructive to note that early clinical trials in bone marrow transplantation were not contingent on the availability of such reference standards, or on the precise delineation of the hematopoietic stem cell hierarchy [4]. On the contrary, progress relied on empirical design of clinical protocols, and on lessons learnt from exposition of problems in transplant patients not anticipated by experiments done mostly in mice that did not accurately model the behaviour of transplanted cells in patients with unique clinical problems, or exacerbated disease [33]. Therefore, history suggests that successful MSC-based cell therapies for lung disease will be best developed by continuing to carry out carefully designed clinical trials while maintaining a constant dialogue between basic scientists and clinicians [4].

Lung Bioengineering

A major challenge in employing cellular therapies to repair the lung, and in lung bioengineering, is how to avoid the mismatched growth of airway or alveolar epithelial cells, interstitial stromal cells, and vasculature leading to the formation of inefficient respiratory units with impaired gas exchange and persisting abnormalities. This will require expert knowledge of the cellular and molecular pathways that orchestrate tissue morphogenesis. A recent report of the formation of complex vascularised organ buds in mice generated by transplantation of self-assembled organoids formed in vitro by mesenchymal cell-driven condensation of heterotypic cell mixtures on a matrix scaffold [34] is a promising and exciting step forward in devising tissue bioengineering approaches relevant to lung regenerative medicine.

Lineage Commitment and Differentiation

Precise genetic tools continue to reveal the role of critical genes and signalling pathways in cell lineage specification of different lung epithelial stem and progenitor cells. For example, recent studies show that deletion of the histone methyltransferase, Ezh2, induces basal cell proliferation and reduces secretory epithelial cell differentiation during lung development [35, 36]. Another shows that repression of the Hippo/Yap signalling pathway enhances airway epithelial proliferation and inhibits sacculation in the foetal lung while causing airway hyperplasia in the adult lung [37]. Yet another, that differential expression of c-myb and Notch signalling via the Notch2 intracellular domain directs the differentiation of tracheal basal cells towards a ciliated or secretory cell phenotype respectively [38, 39].

These studies lead one to speculate whether gene editing technologies [40], or pharmacologic manipulation of specific regulator genes in regenerative cells could be used to arrest disease progression by altering the fate choices available to lung epithelial stem cells at branch-points of cell lineage commitment: for example, to attenuate secretory cell hyperplasia. For genetic diseases such as CF, the targeted correction of inherited mutations of the *Cftr* gene in CF patient-derived cell lines provides proof-of-principle for the potential efficacy of gene therapy [6]. However, clinical trials delivering corrected gene vectors to the CF lung by inhalation have not lived up to the original expectations [14]. It remains to be seen whether manipulating airway stem cells using gene editing tools [40] including transcription activator-like effector nucleases (TALENS) and the Crispr/cas9 system will be clinically effective. Implementing cell therapies using these technologies will require researchers to devise protocols that efficiently and effectively target a specific gene in a disease-relevant stem/progenitor cell subset to modify its behaviour long-term, while avoiding potentially deleterious off-target effects and interactions. This is an ambitious undertaking in experimental models, let alone in a clinical setting where lung stem and progenitor cell compartments are still poorly delineated.

Conclusion

In summary, although mouse models of lung disease are not perfect, experimentation in these models provides a sound scientific rationale for translational studies aimed at identifying endogenous lung stem and progenitor cells, and exogenous cells, which can be utilised to develop cell therapies for lung disease; and identify variables that must be considered in developing and optimising cell therapies for lung disease and lung injury.

References

1. Weiss DJ, Berberich MA, Borok Z, Gail DB, Kolls JK et al (2006) Adult stem cells, lung biology, and lung disease. Proc Am Thorac Soc 3:193–207
2. Antunes MA, Laffey JG, Pelosi P, Rocco PR (2014) Mesenchymal stem cell trials for pulmonary diseases. J Cell Biochem 115:1023–1032
3. Weiss DJ (2014) Concise review: current status of stem cells and regenerative medicine in lung biology and diseases. Stem Cells 32:16–25
4. Prockop DJ, Prockop SE, Bertoncello I (2014) Are clinical trials with mesenchymal stem/progenitor cells too far ahead of the science? Lessons from experimental hematology. Stem Cells 32:3055–3061
5. Hawkins F, Kotton DN (2015) Embryonic and induced pluripotent stem cells for lung regeneration. Ann Am Thorac Soc 12(Suppl 1):S50–S53
6. Crane AM, Kramer P, Bui JH, Chung WJ, Li XS et al (2015) Targeted correction and restored function of the CFTR gene in cystic fibrosis induced pluripotent stem cells. Stem Cell Reports 4:569–577
7. Wilson AA, Ying L, Liesa M, Segeritz CP, Mills JA et al (2015) Emergence of a stage-dependent human liver disease signature with directed differentiation of alpha-1 antitrypsin-deficient iPS Cells. Stem Cell Reports 4(5):873–885
8. Bertoncello I, McQualter JL (2010) Endogenous lung stem cells: what is their potential for use in regenerative medicine? Expert Rev Respir Med 4:349–362
9. Henson PM, Vandivier RW, Douglas IS (2006) Cell death, remodeling, and repair in chronic obstructive pulmonary disease? Proc Am Thorac Soc 3:713–717
10. Szilasi M, Dolinay T, Nemes Z, Strausz J (2006) Pathology of chronic obstructive pulmonary disease. Pathol Oncol Res 12:52–60
11. Horowitz JC, Martinez FJ, Thannickal VJ (2009) Mesenchymal cell fate and phenotypes in the pathogenesis of emphysema. COPD 6:201–210
12. Holgate ST, Davies DE (2009) Rethinking the pathogenesis of asthma. Immunity 31:362–367
13. George AM, Jones PM, Middleton PG (2009) Cystic fibrosis infections: treatment strategies and prospects. FEMS Microbiol Lett 300:153–164
14. Corvol H, Thompson KE, Tabary O, le Rouzic P, Guillot L (2015) Translating the genetics of cystic fibrosis to personalized medicine. Transl Res. pii: S1931–5244(15)00131-0. doi: 10.1016/j.trsl.2015.04.008. [Epub ahead of print]

15. Guillot L, Nathan N, Tabary O, Thouvenin G, Le Rouzic P et al (2013) Alveolar epithelial cells: master regulators of lung homeostasis. Int J Biochem Cell Biol 45:2568–2573
16. Zuo W, Zhang T, Wu DZ, Guan SP, Liew AA et al (2015) p63(+)Krt5(+) distal airway stem cells are essential for lung regeneration. Nature 517:616–620
17. Vaughan AE, Brumwell AN, Xi Y, Gotts JE, Brownfield DG et al (2015) Lineage-negative progenitors mobilize to regenerate lung epithelium after major injury. Nature 517:621–625
18. Voog J, Jones DL (2010) Stem cells and the niche: a dynamic duo. Cell Stem Cell 6:103–115
19. Volckaert T, De Langhe S (2014) Lung epithelial stem cells and their niches: Fgf10 takes center stage. Fibrogenesis Tissue Repair 7:8
20. Kumar ME, Bogard PE, Espinoza FH, Menke DB, Kingsley DM, Krasnow MA (2014) Mesenchymal cells. Defining a mesenchymal progenitor niche at single-cell resolution. Science 346:1258810
21. Lee JH, Kim CF (2014) Developmental biology. Mesenchymal progenitor panoply. Science 346:810–811
22. Staats J (1966) The laboratory mouse. In: Green EL (ed) Biology of the laboratory mouse. Dover, New York, pp 1–9
23. de Haan G, Van Zant G (1997) Intrinsic and extrinsic control of hematopoietic stem cell numbers: mapping of a stem cell gene. J Exp Med 186:529–536
24. Müller-Sieburg CE, Cho RH, Sieburg HB, Kupriyanov S, Riblet R (2000) Genetic control of hematopoietic stem cell frequency in mice is mostly cell autonomous. Blood 95:2446–2448
25. Russell ES, Bernstein SE (1966) Blood and blood formation. In: Green EL (ed) Biology of the laboratory mouse. Dover, New York, pp 608–622
26. Roberts AW, Foote S, Alexander WS, Scott C, Robb L, Metcalf D (1997) Genetic influences determining progenitor cell mobilization and leukocytosis induced by granulocyte colony-stimulating factor. Blood 89:2736–2744
27. Marsh JC (1976) The effects of cancer chemotherapeutic agents on normal hematopoietic precursor cells: a review. Cancer Res 36:1853–1882
28. Yuhas JM, Storer JB (1969) On mouse strain differences in radiation resistance: hematopoietic death and the endogenous colony-forming unit. Radiat Res 39:608–622
29. Walkin L, Herrick SE, Summers A, Brenchley PE, Hoff CM et al (2013) The role of mouse strain differences in the susceptibility to fibrosis: a systematic review. Fibrogenesis Tissue Repair 6:18
30. Rawlins EL, Perl AK (2012) The a"MAZE"ing world of lung-specific transgenic mice. Am J Respir Cell Mol Biol 46:269–282
31. Calado RT, Dumitriu B (2013) Telomere dynamics in mice and humans. Semin Hematol 50:165–174
32. Prockop DJ (2007) "Stemness" does not explain the repair of many tissues by mesenchymal stem/multipotent stromal cells (MSCs). Clin Pharmacol Ther 82:241–243
33. Thomas ED (1994) The evolution of the scientific foundation of marrow transplantation based on human studies. In: Forman SJ, Blume KG, Thomas ED (eds) Bone marrow transplantation. Blackwell Scientific Publications, Boston, pp 12–15
34. Takebe T, Enomura M, Yoshizawa E, Kimura M, Koike H et al (2015) Vascularized and complex organ buds from diverse tissues via mesenchymal cell-driven condensation. Cell Stem Cell 16:556–565
35. Snitow ME, Li S, Morley MP, Rathi K, Lu MM et al (2015) Ezh2 represses the basal cell lineage during lung endoderm development. Development 142:108–117
36. Galvis LA, Holik AZ, Short KM, Pasquet J, Lun AT et al (2015) Repression of Igf1 expression by Ezh2 prevents basal cell differentiation in the developing lung. Development 142:1458–1469
37. Lange AW, Sridharan A, Xu Y, Stripp BR, Perl AK, Whitsett JA (2015) Hippo/Yap signaling controls epithelial progenitor cell proliferation and differentiation in the embryonic and adult lung. J Mol Cell Biol 7:35–47

38. Pardo-Saganta A, Law BM, Tata PR, Villoria J, Saez B et al (2015) Injury induces direct lineage segregation of functionally distinct airway basal stem/progenitor cell subpopulations. Cell Stem Cell 16:184–197
39. Carraro G, Stripp BR (2015) A new notch for lung stem cells. Cell Stem Cell 16:107–109
40. Kmiec EB (2015) Is the age of genetic surgery finally upon us? Surg Oncol 24:95–99

Index

A
Accessory lobe, 128
Acellular scaffolds, 311–315
  animal models, 327–329
  human models, 330
Acute respiratory distress syndrome (ARDS), 213
Air–liquid interface (ALI), 104
Airway basal cells
  accumulating evidence, 101–102
  different types, 102
  K14, 99
  lineage tracing, 101
  maintain and repair, 100
  proliferation and differentiation, 102–104
Airway innervation
  ASM dysfunction, 51–52
  cellular targets, 51–53
  extrinsic and intrinsic, 56–57
  lungs developmental defects
    CCHS, 53–54
    DiGeorge syndrome, 55–56
    Phox2b, 55
    Sox10, 54–55
    Tbx1, 55–56
    WS-IV, 54–55
Airway smooth muscle (ASM), 173
  differentiation, 45
  innervation, 51
  peristaltic movement of, 52
  potential role, 51
Alveolar development
  angiogenic growth factors, 34–35
  HIF, 29–30
  nitric oxide perturbation, 29
  PECAM-1, 30
  VEGF, 28–29
Alveolar epithelial cells, 214. *See also* Type II alveolar epithelial cells (ATII)
Alveolar epithelial type I (ATI) cells, 338
Alveolar epithelium, 290
Alveolar macrophages (AMs)
  activated, 163
  in COPD, 163
  defence against inhaled pathogens, 161
  function, 161–162
  maintenance, 161
  proliferation, 161
  viral stimulation, 162–163
Amniotic fluid (AF), 292
Angiogenesis, 25, 130–131, 250
  ageing lung development, 130
  alveolar development, 28–30
  arrested alveolar growth/loss of alveoli, 30–31
  BPD, disrupted, 31–32
  lung development, 26–27
Anterior foregut endoderm (AFE), 272–274
Asthma, ECM alterations, 173
Atherosclerosis, 246
ATII. *See* Type II alveolar epithelial (ATII) cells
Axonal guidance cues (AGC), 33

B
Basal cells, 99–102
BASC. *See* Bronchoalveolar stem cells (BASC)

I. Bertoncello (ed.), *Stem Cells in the Lung*, Stem Cell Biology and Regenerative Medicine, DOI 10.1007/978-3-319-21082-7

Bioartificial scaffold, 311–313
Blood outgrowth endothelial cells (BOECs), 256
Bmp4, 10
Bone marrow (BM), 214, 292
Bone-marrow-derived progenitor cells (BMPC), 255
Bone marrow MSCs and pulmonary fibrosis, 218
BPD. *See* Bronchopulmonary dysplasia (BPD)
Branching morphogenesis
  cell biology
    extracellular matrix, 9
    movement, 8–9
    polarity, 8
  vs. epithelial progenitors
    Sox9, 4
    spatial relationships, 5
    temporal relationships, 4
  genetic control
    gene regulatory network, 12
    multiple signalings, 9–10
    phenotype nomenclature, 10–11
  morphological features, 5–6
    dichotomous and monopodial branching, 7
    lobation, 6–7
    organs, 7–8
    subroutines, 7
  termination, 12–13
Bronchiolitis obliterans (BO), 213
Bronchoalveolar stem cells (BASC), 250, 252
Bronchopulmonary dysplasia (BPD), 72–73, 256
  AGC, 33
  alveolarization and angiogenesis, 30–31
  antiangiogenic factors, 33–34
  disrupted angiogenesis, 31–32
  HIF signaling, 31

C
Central congenital hypoventilation syndrome (CCHS), 53–54
CF. *See* Cystic fibrosis (CF)
Chronic lung disease, implications for, 204–206
Chronic obstructive pulmonary disease (COPD), 162, 213, 256, 267, 309
  ECM alterations, 173–175
Cigarette smoke extract (CSE), 137
Circulating endothelial cells (CECs), 247, 248
Circulating progenitor cells (CPC)/ nonangiogenic-CPC ratio, 256
Club cells, 98, 113, 115
Club cell specific factor (CCSP), 214
Compensatory lung growth and remodeling
  innate potential for, 193–194
  pneumonectomy model of, 194–195
  structural basis of, 199–202
COPD. *See* Chronic obstructive pulmonary disease (COPD)
Cystic fibrosis (CF), 266

D
DASC. *See* Distal airway stem cells (DASC)
Database for Annotation Visualization and Integrated Discovery (DAVID) Gene Functional Classification Tool, 299
Decellularization and recellularization process, 316. *See also* Tissue decellularisation
Definitive endoderm (DE), 272
Delivering cell therapy, 228–230
DiGeorge syndrome, 55–56
Distal airway stem cells (DASC), 352
Distal bronchiolar airways
  ex vivo analyses, 117–118
  in vivo homeostasis and repair
    Club cell ablation, 115
    mice and rats, 114–115
    mouse models, 115
    murine airways, 116
    stem cells of, 116
    Wnt-responsive genes, 117
  stem cell niche
    components, 118, 119
    functions, 118
    in lung cancer, 121
    in murine models, 120
    notch signalling, 119–120
    regulation of, 118
    signalling pathways and factors, 120
Downstream hematopoietic progenitor cells, 243

E
ECFC. *See* Endothelial colony forming cells (ECFC)
ECM. *See* Extracellular matrix (ECM)
Embryonic stem cells (ESC), 265, 351
Endogenous lung-resident mesenchymal progenitors, 217–218
Endogenous lung stem and progenitor cells, 352–353

Endogenous progenitor stem cells, 291–292
Endoglin (CD105), 33–34
Endothelial and mesenchymal cells, 291
Endothelial cells
in epithelial stem cell niche, 154
proliferation, 244
Endothelial colony forming cells (ECFC), 248, 249, 256
Endothelial monocyte activating polypeptide II (EMAP-II), 33
Endothelial progenitor cells (EPCs), 82, 351
Epigenome-wide association scans (EWAS), 138
Epithelial cell injury, 214
Epithelial–mesenchymal crosstalk
branch timing clock, 18
endothelial-nerve-matrix, 21
growth factors, 17–18
Epithelial progenitors
vs. branching morphogenesis
Sox9, 4
spatial relationships, 5
temporal relationships, 4
fibrosis, 214
Epithelial stem cell niche
3D matrigel cultures, 152
endothelial cells, 154
environmental insults, 151–152
in vivo, regeneration, 152
lipofibroblasts, 152–153
mesenchymal progenitor cells, 153–154
tumour progression, 154–155
ESCs. *See* Embryonic stem cells (ESCs)
Exogenous progenitor and stem cells, 292, 293
Extracellular matrix (ECM), 290, 310
alterations, in lung disease, 172–174
healthy ageing lung, 135–136
influence on cellular homeostasis, 171
lung development, 9
organisation of, 169, 170
role
fetal lung development, 170–171
stem cell niche, 171–172
tissue decellularisation
advantages and disadvantages, 178–180
in vitro culture systems, 175, 176
lung development and functionality of, 174, 175
organ transplantation programmes, 177
scaffolds, 176–177
successfully research, 180
translational medical research milestones, 176
Extrinsic lung innervation development
ASM, 52
parasympathetic, 50
proportion, 51
Ex vivo bioengineering
acellular lungs, 334–335
biologic *vs.* synthetic scaffold approach, 315
decellularizing, 315–321
ex vivo lung regeneration, environmental factors in, 332–333
organ engineering, 309, 310
Ex vivo bioreactor system, 298, 299
Ex vivo lungs, 204–206

F
Fetal lung developments
amniotic fluid volume, 67
CDH, 68
ECM role, 170–171
evidence programming, 63–64
exposures
alcohol consumption, 69–70
corticosteroids, 70–71
intrauterine inflammation, 71
tobacco smoking and nicotine, 69
nutrition and growth, 64
during adult life, 65
animal studies, 65–66
conducting airways, 66
during infancy, 65
placental insufficiency, 66–67
Fibroblast growth activation factor receptor 1 (FGFR1) activation, 251, 252
Fibroblast growth factor 1 (FGF1), 291
Fibroblast growth factor-10 (*Fgf10*), *191*, 291
Fibroblast growth factors 7 (FGF1), 291
Fibrocytes and pulmonary fibrosis, 218
Fibrosis
and epithelial progenitors, 214
and extrapulmonary mesenchymal progenitors, 217–218
and mesenchymal progenitors, 214–217
Fibrotic lung diseases, 213. *See also* Fibrosis
Forkhead Box (Fox), 32
FoxM1, 255
Functional compensation, 194

H
Healthy ageing lung
during development, 127

Healthy ageing lung (*cont.*)
disease-induced premature phenotype, 137–138
effects, on competence, 133–135
epigenetic dysregulation, 138
extracellular matrix control, 135–136
forced vital capacity, 129
in humans, 129
intrinsic and extrinsic mechanisms, 129
multiple studies, 128
PNX, 128
regeneration vulnerable
AECII, 131–132
alveolarization, 130
angiogenesis, 130–131
interstitial fibroblasts, 130
lipofibroblast differentiation, 130
regenerative responses, 128
retinoic acid, 128
structural and functional decline, 129
telomere shortening, 136
transition, post-natal and post-PNX, 132–133
Hematopoietic stem and progenitor cells (HSPC), 245
Hepatocyte growth factor (HGF), 291
Heterogeneity, macrophage, 160–161
HIFs. *See* Hypoxia inducible factors (HIFs)
Homeostasis, tissue regeneration, 163–164
Homeostatic signaling pathways, 192
hPSCs. *See* Human pluripotent stem cells (hPSCs)
HSPC. *See* Hematopoietic stem and progenitor cells (HSPC)
Human pluripotent stem cells (hPSCs), 265
Human tracheal mucoepidermoid tumor (MEi cells), 299
Hypoxia, 31
Hypoxia inducible factors (HIFs)
alveolar development, 29–30
signaling, 31

I

Idiopathic pulmonary fibrosis (IPF), 174, 213, 225, 233–235, 268, 309
Immune regulation, 163–164
Induced pluripotent state (iPS), 265
Induced pluripotent stem cells (iPSCs), 265
Inflammation and tissue regeneration, 162–163
Innate lymphoid cells (ILCs), 160, 163–164
International Society for Cellular Therapy (ISCT), 226
Intrinsic lung innervation development
chick lung, 46
vs. extrinsic innervation, 50
NCC, 46
from neural crest cells, 47
neural crest-derived, 49
spatiotemporal patterns, 49
In vitro bioassay systems, 297–298
iPS. *See* Induced pluripotent state (iPS)
iPSCs. *See* Induced pluripotent stem cells (iPSCs)

L

Lineage negative epithelial stem/progenitor cells (LNEP), 352
Lipofibroblasts, 130, 152–153
Lipopolysaccharide (LPS), 255
Lobation, lung branching morphogenesis, 6–7
Lung and airway epithelial cells
alternative methods, 276–277
differentiation of PSCs
anterior foregut endoderm (AFE), 272–274
definitive endoderm (DE), 272
FGF2 and FGF10, 276
mature cell types, 275
proximal airway epithelium, 275
ventral anterior foregut, 273–274
ventralization strategy, 274
lung development
anterior foregut endoderm (AFE), 270, 271
definitive endoderm (DE), 269
organ domain specification, 270
proximodistal specification, 270
from pluripotent stem cells (PSCs)
CF, 266
COPD, 267
directed differentiation, 272–276
disease modeling, 266, 267
hPSC-derived cells, 266, 268, 269
Idiopathic pulmonary fibrosis, 268
lung regeneration, 265
respiratory system, developmental dynamics and architecture of, 269
type II alveolar epithelial (ATII) cells, 267, 268
vascular leakage, 266
Lung bioengineering, 296–297, 355
decellularized scaffolds
extracellular matrix, 322–323
mechanical assessments, 323–324

methods of, 314–321
scaling up decellularization protocols, 322
ex vivo organ, acellular lungs, 334–335
recellularized scaffolds
acellular scaffolds, 324–329
ex vivo lung regeneration, environmental factors in, 332–333
implantation of, 326, 331
implanted scaffolds, immunogenicity of, 331–332
scaffold
acellular scaffolds, 311–315
bioartificial scaffold, designing and manufacturing, 311–313
tissue culture, precision cut tissue slices, 335–336
Lung cancer, distal stem cells in, 121
Lung development
angiogenesis, 26–28
classical stages, 26
mesenchyme crosstalk
to epithelium, 17
growth factors in, 17–18
neural regulation
airway branching process, 45
airway innervation, 51–56
extrinsic innervation, 50–51
intrinsic innervation, 46–50
laryngotracheal groove forms, 44
Sonic hedgehog, 44
transcription factor expression, 43
Lung disease. *See also* Fetal lung developments; Neonatal lung developments
asthma, 173
COPD, 173–174
IPF, 174
Lung fibrosis. *See also* Fibrosis
endogenous lung-resident mesenchymal progenitors, 217–218
extrapulmonary mesenchymal progenitors, 217–218
mesenchymal epithelial unit, 215–216
Lung growth
and compensation, species differences in, 202–203
induction, translational challenges, 203–205
Lung parenchyma, 191
Lung regeneration and repair
chronic lung disease, 204–206
compensatory lung growth
innate potential for, 193–194
pneumonectomy model of, 194–195
structural basis of, 199–202
epithelial stem cell niche (*see* Epithelial stem cell niche)
ex vivo lungs, 204–206
ILC role, 163–164
lung growth
and compensation, species differences in, 202–203
translational challenges, 203–205
macrophage
and defence against inhaled pathogens, 161–162
heterogeneity, 160–161
resolution, 162–163
mechanical signal–response relationships, 196–199
mechanical stress modulates lung development, 192–193
quantifying regional mechanical stimuli, 199
reciprocal signalling, 149
resident lung MSC, 150–151
Lung regenerative medicine
devising disease-specific cell therapies, 352
endogenous lung stem and progenitor cells, 352–353
lineage commitment and differentiation, 355–356
lung bioengineering, 355
mesenchymal stromal cell therapies, 354
mouse models, 354
stem cell microenvironment, 353
Lung-resident mesenchymal progenitors, 214
Lung vascular growth, 29
Lung vascular regeneration and repair
in lung injury
BMPC, 255
BOECs, 256
BPD, 256
COPD, 256
circulating progenitor cells (CPC)/ nonangiogenic-CPC ratio, 256
ECFC, 256
FoxM1, 255
LPS, 255
mechanisms
atherosclerosis, 246
CECs, 247, 248
downstream hematopoietic progenitor cells, 243

Lung vascular regeneration and repair (*cont.*)
ECFC, 248, 249
endothelial cell proliferation, 244
hematopoietic stem cells, 243
HSPC, 245
human circulating blood cells, 245
neoangiogenesis, 246
PAEC, 249
plaque development, 246–247
proliferative potential, 244
re-endothelialization, 246
tissue homeostasis, 244
Ulex Europaeus Agglutinin 1 (UEA1), 245
in post-pneumonectomy lung growth
AECIIs, 251, 252
alveolar neovascularization, 252, 253
angiogenesis, 250
BASC, 250, 252
blood-borne $CD34^+$ endothelial cells, 252, 254, 255
cellular and molecular mechanisms, 250
Clara cell secreted protein (CCSP), 250
fibroblast growth activation factor receptor 1 (FGFR1) activation, 251, 252
green fluorescence protein (GFP), 253
MMP14, 251
PCECs, 251, 252
physiologic factors, 250
VEGFR2, 251, 252

M
Macrophage, 159
and defence against inhaled pathogens, 161–162
heterogeneity, 160–161
resolution, 162–163
Matrix metalloproteinase 14 (MMP14), 251
Mechanical signal–response relationships, 196–199
Mechanical stimuli, 199, 204
Mechanical stress modulates lung development, 192–193
Mechano-transduction pathways, 204
Mesenchymal cell infiltration, 213
Mesenchymal epithelial unit, 215–216
Mesenchymal stem cell (MSC), 214, 215
Mesenchymal stromal cell (MSC), 82, 351, 354
delivering cell therapy
mitochondria, 230
regenerative therapy approach, 229
source, 229–230
whole cells, microvesicles or conditioned medium, 230
endothelial cells, 154
epithelial stem cell niche (*see* Epithelial stem cell niche)
human studies
acute lung injury (ALI), 236–237
chronic obstructive pulmonary disease, 236
IPF, 236
obliterative bronchiolitis, 236
resident lung
comprises, 150
flow cytometry fractionation, 151
isolation, 150
tracing studies, 150, 151
therapeutic potential
immunosuppression and tolerance, 227
lung homeostasis and epithelial repair, 227–228
relative immune privilege, 226–227
translational research, 230–235
Mesenchyme
crosstalk (*see* Epithelial–mesenchymal crosstalk)
FGF10 signaling, 21
pericytes, 19
smooth muscle cells development, 19
TGF beta signaling, 20
Mouse tracheal epithelial cells (MTEC), 104
Murine lung stem cells, 117
Myoepithelial cells (MECs), 108

N
Nanoparticles (NPs), 301
Neoangiogenesis, 246
Neonatal lung developments
cell-based therapies, 82–83
environmental exposures, 81–82
nutrition, 78–79
postnatal environments, 71
preterm birth
BPD, 72
corticosteroids, 77–78
definition, 72
extremely preterm infants, 72
histological representation, 73
hyperoxia, 74–75
mechanical ventilation, 75–77
RDS, 73
very preterm infants, 72

respiratory infections
bacterial colonization, 80–81
viral, 79–80
Neuroepithelial bodies (NEB), 52
Neuroepithelial endocrine cells. *See* Pulmonary neuroendocrine cells (PNECs)
Nitric oxide (NO) perturbation, alveolar development, 29
Non-small cell lung cancer (NSCLC), 155

P
Partial pneumonectomy (PNX), 127–128
Patient-derived induced pluripotent cells (iPSC), 351
Patterning, 10
PECAM-1, 30
Pericytes, 19
Peripheral blood (PB), 292
Phox2b, 55
Plaque development, 246–247
Pluripotent stem cells (PSCs)
differentiation of
anterior foregut endoderm (AFE), 272
definitive endoderm (DE), 272
FGF2 and FGF10, 276
mature cell types, 275
proximal airway epithelium, 275
ventral anterior foregut, 273–274
ventralization strategy, 274
lung and airway epithelial cells
ATII cells, 267, 268
CF, 266
COPD, 267
directed differentiation, 272–276
disease modeling, 266, 267
hPSC-derived cells, 266, 268, 269
Idiopathic pulmonary fibrosis, 268
lung regeneration, 265
respiratory system, developmental dynamics and architecture of, 269
vascular leakage, 266
Pneumonectomy, compensatory lung growth, 194–195
Precision cut tissue slices, 335–336
Preterm birth, neonatal lung developments
BPD, 72
corticosteroids, 77–78
definition, 72
extremely preterm infants, 72
histological representation, 73
hyperoxia, 74–75
mechanical ventilation, 75–77
airway reactivity, 77
altering alveolar structure, 76–77
immaturity, 75
lung injury, 76
noninvasive strategies, 75–76
potential pathways, 77
RDS, 73
very preterm infants, 72
Proliferative potential, 244
Proximal airways, 97–98
cartilaginous trachea and bronchi, 98
intermediate cells, 105
SMG myoepithelial cells, 108
stem cells
dedicated and facultative, 98–99
niche, 104
signaling pathways, 102–104
SMG, 106–107
subtypes, 99–102
submucosal glands, 105–106
Proximal–distal patterning, 11
PSCs. *See* Pluripotent stem cells (PSCs)
Pulmonary artery endothelial cells (PAEC), 249
Pulmonary capillary endothelial cells (PCECs), 251, 252
Pulmonary fibrosis, 218, 235
Pulmonary neuroendocrine cells (PNECs), 52

R
Recombination activating gene (RAG), 163
Regenerative medicine, nanomaterials in, 300–302
Respiratory mechanical stresses, 191
Respiratory syncytial virus (RSV), 79–80
Respiratory tract regeneration
developmental biology, 290–291
endogenous progenitor stem cells, 291–292
exogenous progenitor and stem cells, 292, 293
ex vivo bioreactor system, 298, 299
in vitro bioassay systems, 297–298
lung bioengineering, 296–297
nanomaterials in regenerative medicine, 300–302
tissue engineering, 293–294
tracheal bioengineering, 294–296
tumour initiating cells, 298–300
Retinoic acid (RA), 203, 291

S
Scaffold remodeling, 204
Stem cell microenvironment, 353
Stem cell niche, ECM role, 171–172
Streptococcus pneumonia, 161
Submucosal glands (SMGs)
  hypertrophic, 105–106
  myoepithelial cells, 108
  potential participation, 106
  tubules, 105
Sudden infant death syndrome (SIDS), 53

T
TE. *See* Tissue engineering (TE)
Tgf signalings, 10
Thorax growth, 192–193
Tissue decellularisation. *See also* Extracellular matrix (ECM)
  advantages and disadvantages, 178–180
  lung development and functionality of, 174, 175
  organ transplantation programmes, 177
  scaffolds, 176–177
  successfully research, 180
  translational medical research milestones, 176
  in vitro culture systems, 175, 176
Tissue engineering (TE), 293–294
Tissue Engineering of lung, artificial scaffolds, 311–314
Tissue regeneration
  homeostasis, 163–164
  inflammation and, 162–163
Tracheal bioengineering, 294–296
Tracheoesophageal fistula/esophageal atresia (TEF/EA), 268
Transforming growth factor beta (TGF-β), 291
Translational leap, 230–231
Translational research, 230–235
Tumour initiating cells, 298–300
Type I alveolar epithelial cells (AEC-I), 291
Type II alveolar epithelial cells (ATII), 131, 267, 268, 291, 352

U
Ulex Europaeus Agglutinin 1 (UEA1), 245
Umbilical cord blood (UCB), 292

V
Vascular endothelial growth factor (VEGF)
  alveolar development, 28–29
  HIF signaling, in BPD, 31
  hypoxia, 31
Vascular endothelial growth factor A (VEGF-A), 291
Vascular endothelial growth factor receptor 2 (VEGFR2), 245
Vasculogenesis, 25
Velocardiofacial syndrome. *See* DiGeorge syndrome

W
Waardenburg syndrome type IV (WS-IV), 54–55
Wnt signaling, 10

MIX
Papier aus verantwortungsvollen Quellen
Paper from responsible sources
FSC® C105338

If you have any concerns about our products,
you can contact us on
ProductSafety@springernature.com

In case Publisher is established outside the EU,
the EU authorized representative is:
Springer Nature Customer Service Center GmbH
Europaplatz 3, 69115 Heidelberg, Germany

Printed by Libri Plureos GmbH
in Hamburg, Germany